Intermediate Algebra

MAT 1033

Charles P. McKeague

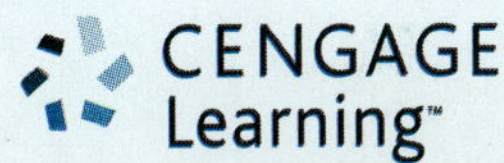

Australia • Brazil • Japan • Korea • Mexico • Singapore • Spain • United Kingdom • United States

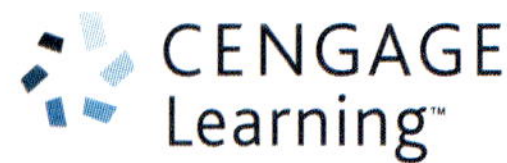

Intermediate Algebra: MAT 1033

Charles P. McKeague

Executive Editors:
Michele Baird
Maureen Staudt
Michael Stranz

Project Development Manager:
Linda deStefano

Senior Marketing Coordinators:
Sara Mercurio
Lindsay Shapiro

Production/Manufacturing Manager:
Donna M. Brown

PreMedia Services Supervisor:
Rebecca A. Walker

Rights & Permissions Specialist:
Kalina Hintz

Cover Image:
Getty Images*

* Unless otherwise noted, all cover images used by Custom Solutions, a part of Cengage Learning, have been supplied courtesy of Getty Images with the exception of the Earthview cover image, which has been supplied by the National Aeronautics and Space Administration (NASA).

ISBN-13: 978-0-495-83919-4

ISBN-10: 0-495-83919-1

Cengage Learning
5191 Natorp Boulevard
Mason, Ohio 45040
USA

Cengage Learning is a leading provider of customized learning solutions with office locations around the globe, including Singapore, the United Kingdom, Australia, Mexico, Brazil, and Japan. Locate your local office at:
international.cengage.com/region

Cengage Learning products are represented in Canada by Nelson Education, Ltd.

For your lifelong learning solutions, visit **custom.cengage.com**

Visit our corporate website at **cengage.com**

Printed in the United States of America

For additional help with math please log into

www.mathtv.com

Then click on your school

Brief Contents

Contents

Chapter 4 Rational Exponents and Roots 215

Chapter 5 Quadratic Equations 263

Preface to the Instructor

This book has been created specifically for the Math 1033 course at Miami Dade College. This custom textbook is intended for a college level course in intermediate algebra in order to prepare students for college algebra or precalculus. Any and all comments or suggestions for this text can be emailed directly to us at MDC@xyztextbooks.com.

Features of the Book

Practice Problems The Practice Problems, with their answers and solutions, are the key to moving students through the material in this book. We call it the EPAS system for Example, Practice Problem, Answer, Solution. Here is how it works: In the margin, next to each example, is a Practice Problem. Students begin by reading through the text, stopping after they have read through an example. Then they work the Practice Problem in the margin next to the example. When they are finished, they check the answers to the Practice Problem at the bottom of the page. If they have made a mistake, they try the problem again. If they still cannot get it right, they look at the solution to the Practice Problem in the back of the book. After working their way through the section in this manner, they are ready to start on the problems in the Problem Set.

Early Coverage of Graphing The material on graphing equations in two variables begins in Chapter 3.

Complete Coverage of Functions Functions are introduced in Section 2.5, early in the course. Function notation is covered in Section 2.6. The concept of a function is then carried through to the remaining chapters in the book.

Using Technology Scattered throughout the book is material that shows how graphing calculators, spreadsheet programs, and computer graphing programs can be used to enhance the topics being covered. This material is easy to find because it appears under the heading "Using Technology."

Blueprint for Problem Solving The Blueprint for Problem Solving is a detailed outline of the steps needed to successfully attempt application problems. Intended as a guide to problem solving in general, the Blueprint overlays the solution process to all the application problems in the first few chapters of the book. As students become more familiar with problem solving, the steps in the Blueprint are streamlined.

Facts from Geometry Many of the important facts from geometry are listed under this heading. In most cases, an example or two accompanies each of the facts to give students a chance to see how topics from geometry are related to the algebra they are learning.

Chapter Openings Each chapter opens with an introduction in which a real-world application is used to stimulate interest in the chapter. These opening applications are expanded on later in the chapter. Most of them are explained using the rule of four: in words, numerically, graphically, and algebraically.

Getting Ready for Class Just before each problem set is a list of four problems under the heading *Getting Ready for Class.* These problems require written responses from students and can be answered by reading the preceding section. They are to be done before students go to class.

Organization of the Problem Sets Six main ideas are incorporated into the problem sets.

1. ***Drill*** There are enough problems in each set to ensure student proficiency in the material.
2. ***Progressive difficulty*** The problems increase in difficulty as the problem set progresses.
3. ***Odd-even similarities*** Each pair of consecutive problems is similar. Since the answers to the odd-numbered problems are listed in the back of the book, the similarity of the odd-even pairs of problems allows your students to check their work on an odd-numbered problem and then to try the similar even-numbered problem.
4. ***Applying the Concepts problems*** Students are always curious about how the algebra they are learning can be applied, but at the same time many of them are apprehensive about attempting application problems. I have found that they are more likely to put some time and effort into trying application problems if they do not have to work an overwhelming number of them at one time and if they work on them every day. For these reasons, I have placed a few application problems in almost every problem set in the book.
5. ***Review problems*** Each problem set, beginning with Chapter 2, contains a few review problems. Where appropriate, the review problems cover material that will be needed in the next section.
6. ***Extending the Concepts problems*** In Part 3, when students are at the intermediate algebra level, many of the problem sets end with a few problems under this heading. These problems are more challenging than those in the problem set, or they are problems that extend some of the topics covered in the section.

End-of-Chapter Retrospective Each chapter ends with the following items that together give a comprehensive reexamination of the chapter.

1. ***Chapter Summary*** Lists all main points from the chapter. In the margin, next to each topic, is an example that illustrates the type of problem associated with the topics being reviewed.
2. ***Chapter Review*** An extensive set of problems that reviews all the main topics in the chapter. The odd-numbered problems in the chapter review can be used as a practice test.

Supplements to the Textbook

Please consult your sales representative, Diana Sulewski for details.

For the Student

Contact your instructor or the math department at Miami Dade College for the most up to date supplements available for your textbook.

Video lessons specific to each section of your book are available at www.mathtv.com through a school specific login page. Simply choose schools from the MathTV website and then select your school from the list.

Acknowledgments

Producing this book was an exciting and challenging project. From the beginning, it was team effort, and there are many people to thank for their contributions.

First on the list is Professor Richard Hochman of Miami Dade College, for the many hours he spent with me, both on the phone and in person, suggesting changes and improvements for the book. Likewise, the faculty in the mathematics department at Miami Dade College made many valuable suggestions, and I appreciate their support for this project and their willingness to take a chance on a new book.

Next, Diana Sulewski, of Cengage Learning, was tireless in her efforts on behalf of this project. She was the catalyst that brought all parties together to make this project possible. No one has taken better care of a visiting author than Diana.

Evelyn Nehme, the Cengage Custom editor does a fantastic job of coordinating the file preparation and printing for each new edition of this book. Her efforts and attention to detail are greatly appreciated.

Devin Christ is the person who has made custom textbooks through our company, XYZ Textbooks, possible. His organization and knowledge of the different file formats, and how they work together allow us to concentrate on writing, rather than typesetting.

Finally we would like to thank our wives, Diane McKeague and Shane McKeague for their encouragement and support.

Pat and Patrick McKeague

Preface to the Student

I often find my students asking themselves the question "Why can't I understand this stuff the first time?" The answer is "You're not expected to." Learning a topic in algebra isn't always accomplished the first time around. There are many instances when you will find yourself reading over new material a number of times before you can begin to work problems. That's just the way things are in algebra. If you don't understand a topic the first time you see it, that doesn't mean there is something wrong with you. Understanding algebra takes time. The process of understanding requires reading the book, studying the examples, working problems, and getting your questions answered.

Here are some questions that are often asked by students starting a beginning algebra class.

How much math do I need to know before taking algebra? You should be able to do the four basic operations (addition, subtraction, multiplication, and division) with whole numbers, fractions, and decimals. Most important is your ability to work with whole numbers. If you are a bit weak at working with fractions because you haven't worked with them in a while, don't be too concerned; we will review fractions as we progress through the book. I have had students who eventually did very well in algebra, even though they were initially unsure of themselves when working with fractions.

What is the best way to study? The best way to study is to study consistently. You must work problems every day. A number of my students spend an hour or so in the morning working problems and reading over new material and then spend another hour in the evening working problems. The students of mine who are most successful in algebra are the ones who find a schedule that works for them and then stick to it. They work problems every day.

If I understand everything that goes on in class, can I take it easy on my homework? Not necessarily. There is a big difference between understanding a problem someone else is working and working the same problem yourself. There is no substitute for working problems yourself. The concepts and properties are understandable to you only if you yourself work problems involving them.

How to Be Successful in Algebra

If you have decided to be successful in algebra, then the following list will be important to you:

1. **If you are in a lecture class, be sure to attend all class sessions on time.** You cannot know exactly what goes on in class unless you are there. Missing class and then expecting to find out what went on from someone else is not the same as being there yourself.
2. **Read the book and work the Practice Problems.** It is best to read the section that will be covered in class beforehand. Reading in advance, even if you do not understand everything you read, is still better than going to class with no idea of what will be discussed. As you read through each section, be sure to work the Practice Problems in the margin of the text. Each Practice Problem is similar to the example with the same number. Look over the

example and then try the corresponding Practice Problem. The answers to the Practice Problems are given on the same page as the problems. If you don't get the correct answer, see if you can rework the problem correctly. If you miss it a second time, check your solution with the solution to the Practice Problem in the back of the book.

3. **Work problems every day and check your answers.** The key to success in mathematics is working problems. The more problems you work, the better you will become at working them. The answers to the odd-numbered problems are given in the back of the book. When you have finished an assignment, be sure to compare your answers with those in the book. If you have made a mistake, find out what it is, and correct it.
4. **Do it on your own.** Don't be misled into thinking someone else's work is your own. Having someone else show you how to work a problem is not the same as working the same problem yourself. It is okay to get help when you are stuck. As a matter of fact, it is a good idea. Just be sure you do the work yourself.
5. **Review every day.** After you have finished the problems your instructor has assigned, take another 15 minutes and review a section you have already completed. The more you review, the longer you will retain the material you have learned.
6. **Don't expect to understand every new topic the first time you see it.** Sometimes you will understand everything you are doing, and sometimes you won't. That's just the way things are in mathematics. Expecting to understand each new topic the first time you see it can lead to disappointment and frustration. The process of understanding algebra takes time. It requires that you read the book, work problems, and get your questions answered.
7. **Spend as much time as it takes for you to master the material.** No set formula exists for the exact amount of time you need to spend on algebra to master it. You will find out as you go along what is or isn't enough time for you. If you end up spending two or more hours on each section in order to master the material there, then that's how much time it takes; trying to get by with less will not work.
8. **Relax.** It's probably not as difficult as you think.

Keep This Book for Reference

I hope you enjoy this book and that you find it to be one of the more readable math books you have used. I encourage you to keep it as a reference after you have finished the courses for which it is required. You will probably have occasion to refer back to the material here from time to time during your college career. If so, it is always easier to find and recall topics in a book you have used than it is in a book you have not used. So, my advice is to keep this book until you have finished all your mathematics classes and all the classes that require mathematics as a prerequisite.

Equations and Inequalities in One Variable

age fotostock/SuperStock

CHAPTER OUTLINE

A recent newspaper article gave the following guideline for college students taking out loans to finance their education: The maximum monthly payment on the amount borrowed should not exceed 8% of their monthly starting salary. In this situation, the maximum monthly payment can be described mathematically with the formula

$$y = 0.08x$$

Using this formula, we can construct a table and a line graph that show the maximum student loan payment that can be made for a variety of starting salaries.

MAXIMUM STUDENT LOAN PAYMENTS

MONTHLY STARTING SALARY	MAXIMUM LOAN PAYMENT
\$2,000	\$160
\$2,500	\$200
\$3,000	\$240
\$3,500	\$280
\$4,000	\$320
\$4,500	\$360
\$5,000	\$400

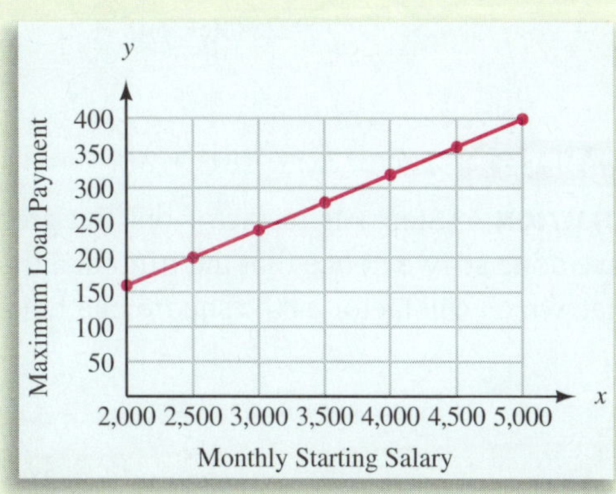

Figure 1

1.1 Review of Factoring

In this section we will review the different methods of factoring that we have presented in the previous chapters in this book. This section is important because it will give you an opportunity to factor a variety of polynomials.

We begin this section by listing the steps that can be used to factor polynomials of any type.

To Factor a Polynomial

Step 1: If the polynomial has a greatest common factor other than 1, then factor out the greatest common factor.

Step 2: If the polynomial has two terms (it is a binomial), then see if it is the difference of two squares or the sum or difference of two cubes, and then factor accordingly. Remember, if it is the sum of two squares it will not factor.

Step 3: If the polynomial has three terms (a trinomial), then it is either a perfect square trinomial, which will factor into the square of a binomial, or it is not a perfect square trinomial, in which case we try to write it as the product of two binomials using the methods developed in this chapter.

Step 4: If the polynomial has more than three terms, then try to factor it by grouping.

Step 5: As a final check, see if any of the factors you have written can be factored further. If you have overlooked a common factor, you can catch it here.

Here are some examples illustrating how we use the steps in our list.

Practice Problems

1. Factor $3x^8 - 27x^6$.

EXAMPLE 1 Factor $2x^5 - 8x^3$.

SOLUTION First we check to see if the greatest common factor is other than 1. Since the greatest common factor is $2x^3$, we begin by factoring it out. Once we have done so, we notice that the binomial that remains is the difference of two squares, which we factor according to the formula $a^2 - b^2 = (a + b)(a - b)$.

$$2x^5 - 8x^3 = 2x^3(x^2 - 4)$$ **Factor out the greatest common factor, $2x^3$**

$$= 2x^3(x + 2)(x - 2)$$ **Factor the difference of two squares**

2. Factor $4x^4 + 40x^3 + 100x^2$.

EXAMPLE 2 Factor $3x^4 - 18x^3 + 27x^2$.

SOLUTION Step 1 is to factor out the greatest common factor $3x^2$. After we have done so, we notice that the trinomial that remains is a perfect square trinomial, which will factor as the square of a binomial.

$$3x^4 - 18x^3 + 27x^2 = 3x^2(x^2 - 6x + 9)$$ **Factor out $3x^2$**

$$= 3x^2(x - 3)^2$$ **$x^2 - 6x + 9$ is the square of $x - 3$**

Answers

1. $3x^6(x + 3)(x - 3)$ 2. $4x^2(x + 5)^2$

EXAMPLE 3 Factor $y^3 + 25y$.

SOLUTION We begin by factoring out the y that is common to both terms. The binomial that remains after we have done so is the sum of two squares, which does not factor, so after the first step, we are finished.

$$y^3 + 25y = y(y^2 + 25)$$

EXAMPLE 4 Factor $6a^2 - 11a + 4$.

SOLUTION Here we have a trinomial that does not have a greatest common factor other than 1. Since it is not a perfect square trinomial, we factor it by trial and error. Without showing all the different possibilities, here is the answer.

$$6a^2 - 11a + 4 = (3a - 4)(2a - 1)$$

EXAMPLE 5 Factor $2x^4 + 16x$.

SOLUTION This binomial has a greatest common factor of $2x$. The binomial that remains after the $2x$ has been factored from each term is the sum of two cubes, which we factor according to the formula $a^3 + b^3 = (a + b)(a^2 - ab + b^2)$.

$$2x^4 + 16x = 2x(x^3 + 8) \quad \textbf{Factor } 2x \textbf{ from each term}$$

$$= 2x(x + 2)(x^2 - 2x + 4) \quad \textbf{The sum of two cubes}$$

EXAMPLE 6 Factor $2ab^5 + 8ab^4 + 2ab^3$.

SOLUTION The greatest common factor is $2ab^3$. We begin by factoring it from each term. After that we find that the trinomial that remains cannot be factored further.

$$2ab^5 + 8ab^4 + 2ab^3 = 2ab^3(b^2 + 4b + 1)$$

EXAMPLE 7 Factor $4x^2 - 6x + 2ax - 3a$.

SOLUTION Our polynomial has four terms, so we factor by grouping.

$$4x^2 - 6x + 2ax - 3a = 2x(2x - 3) + a(2x - 3)$$

$$= (2x - 3)(2x + a)$$

EXAMPLE 8 Factor $12x^4 + 17x^2 + 6$.

SOLUTION This is a trinomial in x^2:

$$12x^4 + 17x^2 + 6 = (4x^2 + 3)(3x^2 + 2)$$

EXAMPLE 9 Factor $2x^2(x - 3) - 5x(x - 3) - 3(x - 3)$.

SOLUTION We begin by factoring out the greatest common factor $(x - 3)$. Then we factor the trinomial that remains.

$$2x^2(x - 3) - 5x(x - 3) - 3(x - 3) = (x - 3)(2x^2 - 5x - 3)$$

$$= (x - 3)(2x + 1)(x - 3)$$

$$= (x - 3)^2(2x + 1)$$

3. Factor $y^4 + 36y^2$.

4. Factor $6x^2 - x - 15$.

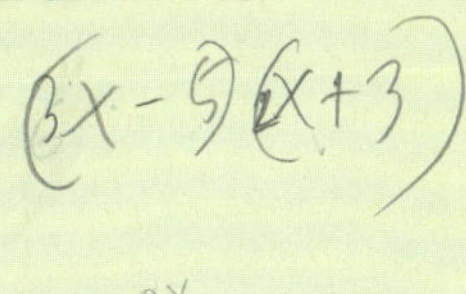

5. Factor $3x^5 - 81x^2$.

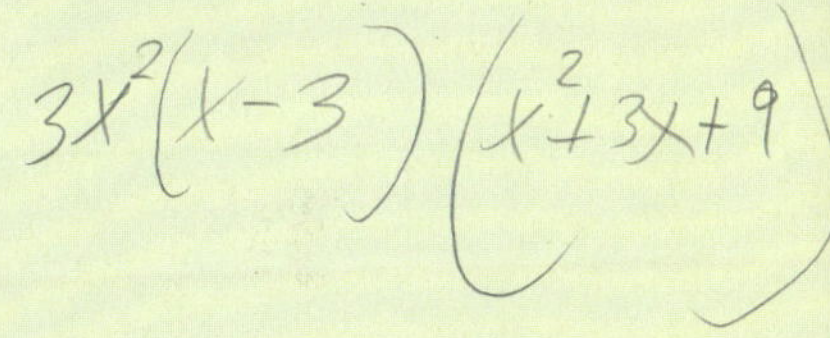

6. Factor $3a^2b^3 + 6a^2b^2 - 3a^2b$.

7. Factor $x^2 - 10x + 25 - b^2$. (*Hint:* Group the first three terms together.)

8. Factor $15x^4 + x^2 - 2$.

9. Factor.
$3x^2(x - 2) - 7x(x - 2) + 2(x - 2)$

Answers

3. $y^2(y^2 + 36)$
4. $(3x - 5)(2x + 3)$
5. $3x^2(x - 3)(x^2 + 3x + 9)$
6. $3a^2b(b^2 + 2b - 1)$
7. $(x - 5 + b)(x - 5 - b)$
8. $(5x^2 + 2)(3x^2 - 1)$
9. $(x - 2)^2(3x - 1)$

Getting Ready for Class

After reading through the preceding section, respond in your own words and in complete sentences.

A. How do you know when you've factored completely?
B. If a polynomial has four terms, what method of factoring should you try?
C. What is the first step in factoring a polynomial?
D. What do we call a polynomial that does not factor?

PROBLEM SET 1.1

The problems below are representative of the type of factoring problems you will see throughout the rest of this book. To make a successful start with intermediate algebra, be sure you are comfortable with all the factoring problems shown here. If you need some additional review, you can find it in Chapter 5.

Factor the following trinomials.

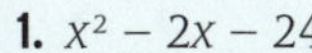

1. $x^2 - 2x - 24$
2. $x^2 + 2x - 24$
3. $x^2 - 5x - 6$
4. $x^2 + 5x - 6$
5. $x^2 - 5x + 6$
6. $x^2 - x - 6$
7. $x^2 - 10x + 25$
8. $4x^2 + 4x + 1$
9. $2x^2 - 5x - 3$
10. $20x^2 - 93x + 34$
11. $21x^2 - 23x + 6$
12. $42x^2 + 23x - 10$

Factor the following binomials as the difference of two squares.

13. $x^2 - 16$
14. $x^2 - 25$
15. $a^2 - 1$
16. $a^2 - 9$
17. $a^2 - 16b^2$
18. $a^2 - 25b^2$
19. $9x^2 - 49$
20. $49x^2 - 144$
21. $16x^4 - 49$
22. $4x^4 - 25$
23. $t^4 - 81$
24. $t^4 - 16$

Factor the following binomials as the sum or difference of two cubes.

25. $a^3 + b^3$
26. $a^3 - b^3$
27. $x^3 - 8$
28. $x^3 + 125$
29. $x^3 + 1$
30. $x^3 - 27$
31. $8x^3 + 1$
32. $27x^3 + 64$

Factor each of the following by first factoring out the greatest common factor and then factoring the polynomial that remains, if possible.

33. $60x^2 - 130x + 60$
34. $90x^2 + 60x - 80$
35. $x^3 + 5x^2 + 6x$
36. $x^3 - 5x^2 - 6x$
37. $2x^3 - 5x^2 - 3x$
38. $6x^3 - 5x^2 - x$
39. $x^3 - 2x^2 - 24x$
40. $x^3 + 2x^2 - 24x$
41. $6x + 24$
42. $3x^2 - 3xy$
43. $100x^2 - 300x$
44. $10x^2 + 100x$
45. $20a^2 - 45$
46. $50a - 2ax^2$
47. $9a^3 - 16a$
48. $16a^3 - 25a$
49. $12y - 2xy - 2x^2y$
50. $6y - 4xy - 2x^2y$

Use factoring by grouping to factor each of the following.

51. $ax + 2x + 3a + 6$
52. $ay + 2y - 4a - 8$
53. $x^2 - 3ax - 2x + 6a$
54. $x^2 - 3ax + 2x - 6a$
55. $x^3 + 2x^2 - 9x - 18$
56. $x^3 + 5x^2 - 4x - 20$
57. $x^3 + 3x^2 - 4x - 12$
58. $x^3 - 3x^2 - 4x + 12$
59. $4x^3 + 12x^2 - 9x - 27$
60. $9x^3 + 18x^2 - 4x - 8$
61. $2x^3 + x^2 - 18x - 9$
62. $3x^3 - x^2 - 12x + 4$

Factor completely, if possible.

63. $4x^2 - 31x - 8$
64. $6x^2 - 55xy + 9y^2$
65. $x^2 + 49$
66. $25 + a^2$
67. $150x^3 + 65x^2 - 280x$
68. $360x^3 - 490x$
69. $24x^2 + 2x - 5$
70. $12x^2 - 49x + 4$
71. $x^6 - 1$
72. $x^6 - 64$
73. $x^2(a + 5) + 6x(a + 5) + 9(a + 5)$
74. $a^2(x - 2) + 4a(x - 2) + 4(x - 2)$
75. $12a^2(x - 7) - 75(x - 7)$
76. $18a^2(2x + 3) - 50(2x + 3)$
77. $r^2 - \frac{1}{9}$
78. $4a^2 + 2a + \frac{1}{4}$

79. $t^2 + 4t + 4 - y^2$

80. $x^2 - 10x + 25 - b^2$

81. $125t^3 + \frac{1}{27}$

82. $\frac{1}{25} + \frac{1}{10}t^2 + \frac{1}{16}t^4$

83. $15t^2 + t - 16$

84. $48x^2 - 74x + 3$

85. $100x^2 - 100x - 600$

86. $100x^2 - 100x - 1200$

87. $4x^3 + 16xy^2$

88. $50 - 2a^2$

89. $25 - 10t + t^2 - x^2$

90. $25 - 20t + 4t^2 - y^2$

91. $30x^2 + 97x + 77$

92. $96a^2 + 44a - 35$

1.2 Review of Solving Equations

This chapter marks the transition from introductory algebra to intermediate algebra. Some of the material here is a review of material we covered earlier, and some of the material here is new. If you cover all the sections in this chapter, you will review all the important points contained in the first six chapters of the book. So, it is a good idea to put some extra time and effort into this chapter to ensure that you get a good start with the rest of the course. Let's begin by reviewing the methods we use to solve equations.

Linear Equations in One Variable

A *linear equation in one variable* is any equation that can be put in the form

$$ax + b = c$$

where a, b, and c are constants and $a \neq 0$. For example, each of the equations

$$5x + 3 = 2 \qquad 2x = 7 \qquad 2x + 5 = 0$$

are linear because they can be put in the form $ax + b = c$. In the first equation, $5x$, 3, and 2 are called *terms* of the equation: $5x$ is a variable term; 3 and 2 are constant terms.

DEFINITION

The **solution set** for an equation is the set of all numbers that, when used in place of the variable, make the equation a true statement.

DEFINITION

Two or more equations with the same solution set are called **equivalent equations.**

EXAMPLE 1 The equations $2x - 5 = 9$, $x - 1 = 6$, and $x = 7$ are all equivalent equations because the solution set for each is {7}.

Practice Problems

1. Show that the equations $3x + 1 = 16$ and $4x - 6 = 14$ both have solution set {5}.

Properties of Equality

The first property states that adding the same quantity to both sides of an equation preserves equality. Or, more importantly, adding the same amount to both sides of an equation *never changes* the solution set. This property is called the *addition property of equality* and is stated in symbols as follows.

Addition Property of Equality

For any three algebraic expressions, A, B, and C,

$$\text{if} \quad A = B$$
$$\text{then} \quad A + C = B + C$$

In words: Adding the same quantity to both sides of an equation will not change the solution set.

Our second new property is called the *multiplication property of equality* and is stated as follows.

Answer

1. See Solutions to Selected Practice Problems.

Note: Because subtraction is defined in terms of addition and division is defined in terms of multiplication, we do not need to introduce separate properties for subtraction and division. The solution set for an equation will never be changed by subtracting the same amount from both sides or by dividing both sides by the same nonzero quantity.

Multiplication Property of Equality

For any three algebraic expressions A, B, and C, where $C \neq 0$,

$$\text{if} \quad A = B$$
$$\text{then} \quad AC = BC$$

In words: Multiplying both sides of an equation by the same nonzero quantity will not change the solution set.

2. Find the solution set for $3a - 3 = -5a + 9$

EXAMPLE 2 Find the solution set for $3a - 5 = -6a + 1$.

SOLUTION To solve for a we must isolate it on one side of the equation. Let's decide to isolate a on the left side by adding $6a$ to both sides of the equation.

$$3a - 5 = -6a + 1$$

$$3a + \mathbf{6a} - 5 = -6a + \mathbf{6a} + 1 \qquad \textbf{Add 6a to both sides}$$

$$9a - 5 = 1$$

$$9a - 5 + \mathbf{5} = 1 + \mathbf{5} \qquad \textbf{Add 5 to both sides}$$

$$9a = 6$$

$$\frac{\mathbf{1}}{\mathbf{9}}(9a) = \frac{\mathbf{1}}{\mathbf{9}}(6) \qquad \textbf{Multiply both sides by } \tfrac{1}{9}$$

$$a = \frac{2}{3} \qquad \tfrac{1}{9}(6) = \tfrac{6}{9} = \tfrac{2}{3}$$

The solution set is $\left\{\frac{2}{3}\right\}$.

Note:
From the previous chapter we know that multiplication by a number and division by its reciprocal always produce the same result. Because of this fact, instead of multiplying each side of our equation in Example 2 by $\frac{1}{9}$, we could just as easily divide each side by 9. If we did so, the last two lines in our solution would look like this:

$$\frac{9a}{9} = \frac{6}{9}$$
$$a = \frac{2}{3}$$

We can check our solution in Example 2 by replacing a in the original equation with $\frac{2}{3}$.

When $\quad a = \frac{2}{3}$

the equation $\quad 3a - 5 = -6a + 1$

becomes $\quad 3\left(\frac{2}{3}\right) - 5 = -6\left(\frac{2}{3}\right) + 1$

$$2 - 5 = -4 + 1$$

$$-3 = -3 \qquad \textbf{A true statement}$$

There will be times when we solve equations and end up with a negative sign in front of the variable. The next example shows how to handle this situation.

3. Solve each equation.

a. $-x = \frac{2}{3}$

b. $-y = -4$

EXAMPLE 3 Solve each equation.

a. $-x = 4$ **b.** $-y = -8$

SOLUTION Neither equation can be considered solved because of the negative sign in front of the variable. To eliminate the negative signs we simply multiply both sides of each equation by -1.

a. $-x = 4$ **b.** $-y = -8$

$\mathbf{-1}(-x) = \mathbf{-1}(4) \qquad \mathbf{-1}(-y) = \mathbf{-1}(-8) \qquad$ **Multiply each side by −1**

$x = -4 \qquad y = 8$

Answers
2. $\left\{\frac{3}{2}\right\}$ 3. a. $x = -\frac{2}{3}$ b. $y = 4$

EXAMPLE 4 Solve $\frac{2}{3}x + \frac{1}{2} = -\frac{3}{8}$.

SOLUTION We can solve this equation by applying our properties and working with fractions, or we can begin by eliminating the fractions. Let's use both methods.

Method 1 Working with the fractions.

$$\frac{2}{3}x + \frac{1}{2} + \left(-\frac{1}{2}\right) = -\frac{3}{8} + \left(-\frac{1}{2}\right) \quad \text{Add } -\tfrac{1}{2} \text{ to each side}$$

$$\frac{2}{3}x = -\frac{7}{8} \quad -\tfrac{3}{8} + \left(-\tfrac{1}{2}\right) = -\tfrac{3}{8} + \left(-\tfrac{4}{8}\right)$$

$$\frac{3}{2}\left(\frac{2}{3}x\right) = \frac{3}{2}\left(-\frac{7}{8}\right) \quad \text{Multiply each side by } \tfrac{3}{2}$$

$$x = -\frac{21}{16}$$

Method 2 Eliminating the fractions in the beginning.

Our original equation has denominators of 3, 2, and 8. The least common denominator, abbreviated LCD, for these three denominators is 24, and it has the property that all three denominators will divide it evenly. If we multiply both sides of our equation by 24, each denominator will divide into 24, and we will be left with an equation that does not contain any denominators other than 1.

$$24\left(\frac{2}{3}x + \frac{1}{2}\right) = 24\left(-\frac{3}{8}\right) \quad \text{Multiply each side by the LCD 24}$$

$$24\left(\frac{2}{3}x\right) + 24\left(\frac{1}{2}\right) = 24\left(-\frac{3}{8}\right) \quad \text{Distributive property on the left side}$$

$$16x + 12 = -9 \quad \text{Multiply}$$

$$16x = -21 \quad \text{Add } -12 \text{ to each side}$$

$$x = -\frac{21}{16} \quad \text{Multiply each side by } \tfrac{1}{16}$$

Check To check our solution, we substitute $x = -\frac{21}{16}$ back into our original equation to obtain

$$\frac{2}{3}\left(-\frac{21}{16}\right) + \frac{1}{2} \stackrel{?}{=} -\frac{3}{8}$$

$$-\frac{7}{8} + \frac{1}{2} \stackrel{?}{=} -\frac{3}{8}$$

$$-\frac{7}{8} + \frac{4}{8} \stackrel{?}{=} -\frac{3}{8}$$

$$-\frac{3}{8} = -\frac{3}{8} \quad \text{A true statement}$$

4. Solve $\frac{3}{5}x + \frac{1}{3} = -\frac{5}{6}$.

Note: We are placing a question mark over the equal sign because we don't know yet if the expression on the left will be equal to the expression on the right.

Answer

4. $-\frac{35}{18}$

5. Solve.
$0.08x + 0.10(8{,}000 - x) = 680$

EXAMPLE 5 Solve the equation $0.06x + 0.05(10{,}000 - x) = 560$.

SOLUTION We can solve the equation in its original form by working with the decimals, or we can eliminate the decimals first by using the multiplication property of equality and solve the resulting equation. Here are both methods.

Method 1 Working with the decimals.

$$0.06x + 0.05(10{,}000 - x) = 560 \quad \text{Original equation}$$

$$0.06x + 0.05(10{,}000) - 0.05x = 560 \quad \text{Distributive property}$$

$$0.01x + 500 = 560 \quad \text{Simplify the left side}$$

$$0.01x + 500 + (\mathbf{-500}) = 560 + (\mathbf{-500}) \quad \text{Add } -500 \text{ to each side}$$

$$0.01x = 60$$

$$\frac{0.01x}{\mathbf{0.01}} = \frac{60}{\mathbf{0.01}} \quad \text{Divide each side by 0.01}$$

$$x = 6{,}000$$

Method 2 Eliminating the decimals in the beginning: To move the decimal point two places to the right in $0.06x$ and 0.05, we multiply each side of the equation by 100.

$$0.06x + 0.05(10{,}000 - x) = 560 \quad \text{Original equation}$$

$$0.06x + 500 - 0.05x = 560 \quad \text{Distributive property}$$

$$\mathbf{100}(0.06x) + \mathbf{100}(500) - \mathbf{100}(0.05x) = \mathbf{100}(560) \quad \text{Multiply each side by 100}$$

$$6x + 50{,}000 - 5x = 56{,}000$$

$$x + 50{,}000 = 56{,}000 \quad \text{Simplify the left side}$$

$$x = 6{,}000 \quad \text{Add } -50{,}000 \text{ to each side}$$

Using either method, the solution to our equation is 6,000.

Check We check our work (to be sure we have not made a mistake in applying the properties or in arithmetic) by substituting 6,000 into our original equation and simplifying each side of the result separately, as the following shows.

$$0.06(\mathbf{6{,}000}) + 0.05(10{,}000 - \mathbf{6{,}000}) \stackrel{?}{=} 560$$

$$0.06(6{,}000) + 0.05(4{,}000) \stackrel{?}{=} 560$$

$$360 + 200 \stackrel{?}{=} 560$$

$$560 = 560 \quad \text{A true statement}$$

Answer
5. 6,000

Here is a list of steps to use as a guideline for solving linear equations in one variable.

Strategy for Solving Linear Equations in One Variable

Step 1: **a.** Use the distributive property to separate terms, if necessary.

b. If fractions are present, consider multiplying both sides by the LCD to eliminate the fractions. If decimals are present, consider multiplying both sides by a power of 10 to clear the equation of decimals.

c. Combine similar terms on each side of the equation.

Step 2: Use the addition property of equality to get all variable terms on one side of the equation and all constant terms on the other side. A **variable term** is a term that contains the variable. A **constant term** is a term that does not contain the variable (the number 3, for example).

Step 3: Use the multiplication property of equality to get the variable by itself on one side of the equation.

Step 4: Check your solution in the original equation to be sure that you have not made a mistake in the solution process.

As you will see as you work through the problems in the problem set, it is not always necessary to use all four steps when solving equations. The number of steps used depends on the equation. In Example 6 there are no fractions or decimals in the original equation, so Step 1b will not be used.

EXAMPLE 6 Solve the equation $8 - 3(4x - 2) + 5x = 35$.

SOLUTION We must begin by distributing the −3 across the quantity $4x - 2$.

Step 1: **a.** $8 - 3(4x - 2) + 5x = 35$ **Original equation**

$8 - 12x + 6 + 5x = 35$ **Distributive property**

c. $-7x + 14 = 35$ **Simplify**

Step 2: $-7x = 21$ **Add −14 to each side**

Step 3: $x = -3$ **Multiply by $-\frac{1}{7}$**

Step 4: When x is replaced by −3 in the original equation, a true statement results. Therefore, −3 is the solution to our equation.

6. Solve for x.
$6 - 2(5x - 1) + 4x = 20$

Note: It would be a mistake to subtract 3 from 8 first because the rule for order of operations indicates we are to do multiplication before subtraction.

Identities and Equations with No Solution

Two special cases are associated with solving linear equations in one variable, each of which is illustrated in the following examples.

EXAMPLE 7 Solve for x: $2(3x - 4) = 3 + 6x$.

SOLUTION Applying the distributive property to the left side gives us

$6x - 8 = 3 + 6x$ **Distributive property**

7. Solve.
$3(5x + 1) = 10 + 15x$

Answer

6. −2

Now, if we add $-6x$ to each side, we are left with the following

$$-8 = 3$$

which is a false statement. This means that there is no solution to our equation. Any number we substitute for x in the original equation will lead to a similar false statement.

8. Solve.
$-4 + 8x = 2(4x - 2)$

EXAMPLE 8 Solve for x: $-15 + 3x = 3(x - 5)$.

SOLUTION We start by applying the distributive property to the right side.

$$-15 + 3x = 3x - 15 \quad \textbf{Distributive property}$$

If we add $-3x$ to each side, we are left with the true statement

$$-15 = -15$$

In this case, our result tells us that any number we use in place of x in the original equation will lead to a true statement. Therefore, all real numbers are solutions to our equation. We say the original equation is an *identity* because the left side is always identically equal to the right side.

Solving Equations by Factoring

Next we will use our knowledge of factoring to solve equations. Most of the equations we will see are *quadratic equations*. Here is the definition of a quadratic equation.

DEFINITION

Any equation that can be written in the form

$$ax^2 + bx + c = 0$$

where a, b, and c are constants and a is not 0 ($a \neq 0$) is called a **quadratic equation.** The form $ax^2 + bx + c = 0$ is called *standard form* for quadratic equations.

Each of the following is a quadratic equation:

$$2x^2 = 5x + 3 \qquad 5x^2 = 75 \qquad 4x^2 - 3x + 2 = 0$$

Note: The third equation is clearly a quadratic equation since it is in standard form. (Notice that a is 4, b is -3, and c is 2.) The first two equations are also quadratic because they could be put in the form $ax^2 + bx + c = 0$ by using the addition property of equality.

Notation For a quadratic equation written in standard form, the first term ax^2 is called the *quadratic term;* the second term bx is the *linear term;* and the last term c is called the *constant term.*

In the past we have noticed that the number 0 is a special number. There is another property of 0 that is the key to solving quadratic equations. It is called the *zero-factor property.*

Zero-Factor Property

For all real numbers r and s,

$$r \cdot s = 0 \quad \text{if and only if} \quad r = 0 \quad \text{or} \quad s = 0 \quad \text{(or both)}$$

Answers

7. No solution
8. All real numbers are solutions.

EXAMPLE 9 Solve $x^2 - 2x - 24 = 0$.

SOLUTION We begin by factoring the left side as $(x - 6)(x + 4)$ and get

$$(x - 6)(x + 4) = 0$$

Now both $(x - 6)$ and $(x + 4)$ represent real numbers. We notice that their product is 0. By the zero-factor property, one or both of them must be 0:

$$x - 6 = 0 \quad \text{or} \quad x + 4 = 0$$

We have used factoring and the zero-factor property to rewrite our original second-degree equation as two first-degree equations connected by the word *or*. Completing the solution, we solve the two first-degree equations:

$$x - 6 = 0 \quad \text{or} \quad x + 4 = 0$$
$$x = 6 \quad \text{or} \quad x = -4$$

We check our solutions in the original equation as follows:

Check $x = 6$	Check $x = -4$
$6^2 - 2(6) - 24 \stackrel{?}{=} 0$	$(-4)^2 - 2(-4) - 24 \stackrel{?}{=} 0$
$36 - 12 - 24 \stackrel{?}{=} 0$	$16 + 8 - 24 \stackrel{?}{=} 0$
$0 = 0$	$0 = 0$

In both cases the result is a true statement, which means that both 6 and -4 are solutions to the original equation.

EXAMPLE 10 Solve $\frac{1}{3}x^3 = \frac{5}{6}x^2 + \frac{1}{2}x$.

SOLUTION We can simplify our work if we clear the equation of fractions. Multiplying both sides by the LCD, 6, we have

$$\mathbf{6} \cdot \frac{1}{3}x^3 = \mathbf{6} \cdot \frac{5}{6}x^2 + \mathbf{6} \cdot \frac{1}{2}x$$
$$2x^3 = 5x^2 + 3x$$

Next we add $-5x^2$ and $-3x$ to each side so that the right side will become 0.

$$2x^3 - 5x^2 - 3x = 0 \qquad \textbf{Standard form}$$

We factor the left side and then use the zero-factor property to set each factor to 0.

$$x(2x^2 - 5x - 3) = 0 \qquad \textbf{Factor out the greatest common factor}$$
$$x(2x + 1)(x - 3) = 0 \qquad \textbf{Continue factoring}$$
$$x = 0 \quad \text{or} \quad 2x + 1 = 0 \quad \text{or} \quad x - 3 = 0 \qquad \textbf{Zero-factor property}$$

Solving each of the resulting equations, we have

$$x = 0 \quad \text{or} \quad x = -\frac{1}{2} \quad \text{or} \quad x = 3$$

To Solve an Equation by Factoring

Step 1: Write the equation in standard form.

Step 2: Factor the left side.

Step 3: Use the zero-factor property to set each factor equal to 0.

Step 4: Solve the resulting linear equations.

Step 5: Check the solution in the original equation.

9. Solve $x^2 - x - 6 = 0$.

Note: What the zero-factor property says in words is that we can't multiply and get 0 without multiplying by 0; that is, if we multiply two numbers and get 0, then one or both of the original two numbers we multiplied must have been 0.

Note: We are placing a question mark over the equal sign because we don't know yet if the expression on the left will be equal to the expression on the right.

10. Solve $\frac{1}{2}x^3 = \frac{5}{6}x^2 + \frac{1}{3}x$.

Answers
9. $-2, 3$ **10.** $0, -\frac{1}{3}, 2$

11. Solve $100x^2 = 500x$.

EXAMPLE 11 Solve $100x^2 = 300x$.

SOLUTION We begin by writing the equation in standard form and factoring:

$$100x^2 = 300x$$

$$100x^2 - 300x = 0 \quad \textbf{Standard form}$$

$$100x(x - 3) = 0 \quad \textbf{Factor}$$

Using the zero-factor property to set each factor to 0, we have

$$100x = 0 \quad \text{or} \quad x - 3 = 0$$

$$x = 0 \quad \text{or} \quad x = 3$$

The two solutions are 0 and 3.

12. Solve $(x + 1)(x + 2) = 12$.

EXAMPLE 12 Solve $(x - 2)(x + 1) = 4$.

SOLUTION We begin by multiplying the two factors on the left side. (Notice that it would be incorrect to set each of the factors on the left side equal to 4. The fact that the product is 4 does not imply that either of the factors must be 4.)

$$(x - 2)(x + 1) = 4$$

$$x^2 - x - 2 = 4 \quad \textbf{Multiply the left side}$$

$$x^2 - x - 6 = 0 \quad \textbf{Standard form}$$

$$(x - 3)(x + 2) = 0 \quad \textbf{Factor}$$

$$x - 3 = 0 \quad \text{or} \quad x + 2 = 0 \quad \textbf{Zero-factor property}$$

$$x = 3 \quad \text{or} \quad x = -2$$

13. Solve $x^3 + 5x^2 - 4x - 20 = 0$.

EXAMPLE 13 Solve for x: $x^3 + 2x^2 - 9x - 18 = 0$.

SOLUTION We start with factoring by grouping.

$$x^3 + 2x^2 - 9x - 18 = 0$$

$$x^2(x + 2) - 9(x + 2) = 0$$

$$(x + 2)(x^2 - 9) = 0$$

$$(x + 2)(x - 3)(x + 3) = 0 \quad \textbf{The difference of two squares}$$

$$x + 2 = 0 \quad \text{or} \quad x - 3 = 0 \quad \text{or} \quad x + 3 = 0 \quad \textbf{Set factors to 0}$$

$$x = -2 \quad \text{or} \quad x = 3 \quad \text{or} \quad x = -3$$

We have three solutions: −2, 3, and −3.

Getting Ready for Class

After reading through the preceding section, respond in your own words and in complete sentences.

A. What is a solution to an equation?
B. What are equivalent equations?
C. Describe how to eliminate fractions in an equation.
D. What is the zero-factor property?

Answers

11. 0, 5 **12.** −5, 2 **13.** −5, −2, 2

PROBLEM SET 1.2

Solve each of the following equations.

1. $2x - 4 = 6$

2. $3x - 5 = 4$

3. $-3 - 4x = 15$

4. $-8 - 5x = -6$

5. $-300y + 100 = 500$

6. $-20y + 80 = 30$

7. $-\frac{3}{5}a + 2 = 8$

8. $-\frac{5}{3}a + 3 = 23$

9. $-x = 2$

10. $-x = \frac{1}{2}$

11. $-a = -\frac{3}{4}$

12. $-a = -5$

13. $7y - 4 = 2y + 11$

14. $8y - 2 = 6y - 10$

15. $5(y + 2) - 4(y + 1) = 3$

16. $6(y - 3) - 5(y + 2) = 8$

17. $x^2 = 64$

18. $x^2 = 36$

19. $5 = 7 - 2(3x - 1) + 4x$

20. $20 = 8 - 5(2x - 3) + 4x$

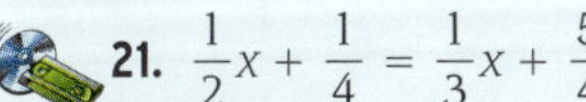

21. $\frac{1}{2}x + \frac{1}{4} = \frac{1}{3}x + \frac{5}{4}$

22. $\frac{2}{3}x - \frac{3}{4} = \frac{1}{6}x + \frac{21}{4}$

23. $x^2 - 5x - 6 = 0$

24. $x^2 + 5x - 6 = 0$

25. $x^3 - 5x^2 + 6x = 0$

26. $x^3 + 5x^2 + 6x = 0$

27. $60x^2 - 130x + 60 = 0$

28. $90x^2 + 60x - 80 = 0$

29. $\frac{1}{5}y^2 - 2 = -\frac{3}{10}y$

30. $\frac{1}{2}y^2 + \frac{5}{3} = \frac{17}{6}y$

31. $-100x = 10x^2$

32. $800x = 100x^2$

33. $(x + 6)(x - 2) = -7$

34. $(x - 7)(x + 5) = -20$

35. $(x + 1)^2 = 3x + 7$

36. $(x + 2)^2 = 9x$

37. $x^3 + 3x^2 - 4x - 12 = 0$

38. $x^3 + 5x^2 - 4x - 20 = 0$

39. $5 - 2x = 3x + 1$

40. $7 - 3x = 8x - 4$

41. $\frac{1}{10}t^2 - \frac{5}{2} = 0$

42. $\frac{2}{7}t^2 - \frac{7}{2} = 0$

43. $7 + 3(x + 2) = 4(x + 1)$

44. $5 + 2(4x - 4) = 3(2x - 1)$

45. $-\frac{2}{5}x + \frac{2}{15} = \frac{2}{3}$

46. $-\frac{1}{6}x + \frac{2}{3} = \frac{1}{4}$

47. $\frac{1}{2}x + \frac{1}{3}x + \frac{1}{4}x = \frac{13}{12}$

48. $\frac{1}{3}x + \frac{1}{4}x + \frac{1}{5}x = \frac{47}{60}$

49. $(2r + 3)(2r - 1) = -(3r + 1)$

50. $(3r + 2)(r - 1) = -(7r - 7)$

51. $9a^3 = 16a$

52. $16a^3 = 25a$

53. $4x^3 + 12x^2 - 9x - 27 = 0$

54. $9x^3 + 18x^2 - 4x - 8 = 0$

55. $\frac{3}{7}x^2 = \frac{1}{14}x$

56. $-\frac{1}{15}x^2 = \frac{2}{5}x$

57. $0.01x - 0.03 = -0.1x^2$

58. $0.02x + 0.45 = 0.08x^2$

59. $24x^3 = 36x^2$

60. $12x^3 = 48x^5$

61. $0.14x + 0.08(10{,}000 - x) = 1220$

62. $0.065x + 0.05(15{,}000 - x) = 870$

63. $\frac{1}{2}x^2 - \frac{9}{8} = 0$

64. $\frac{1}{2}x^2 - \frac{1}{18} = 0$

65. $2x^3 + x^2 - 18x - 9 = 0$

66. $3x^3 - x^2 - 12x + 4 = 0$

Identities and Equations with No Solution

Solve each equation, if possible.

67. $3x - 6 = 3(x + 4)$

68. $7x - 14 = 7(x - 2)$

69. $4y + 2 - 3y + 5 = 3 + y + 4$

70. $7y + 5 - 2y - 3 = 6 + 5y - 4$

71. $2(4t - 1) + 3 = 5t + 4 + 3t$

72. $5(2t - 1) + 1 = 2t - 4 + 8t$
73. $7x - 3(x - 2) = -4(5 - x)$
74. $5x - 2(x + 3) = -3(2 - x)$
75. $7(x + 2) - 4(2x - 1) = 18 - x$
76. $3(2x - 5) - 4(x - 3) = -2(4 - x)$

Applying the Concepts

77. **Cost of a Taxi Ride** The taximeter was invented in 1891 by Wilhelm Bruhn. The city of Chicago charges \$1.80 plus \$0.40 per mile for a taxi ride.
   (a) A woman paid a fare of \$6.60. Write an equation that connects the fare the woman paid, the miles she traveled, n, and the charges the taximeter computes.
   (b) Solve the equation from part (a) to determine how many miles the woman traveled.
78. **Coughs and Earaches** In 1992, twice as many people visited their doctor because of a cough than an earache. The total number of doctor's visits for these two ailments was reported to be 45 million.
   (a) Let x represent the number of earaches reported in 1992, then write an expression using x for the number of coughs reported in 1992.

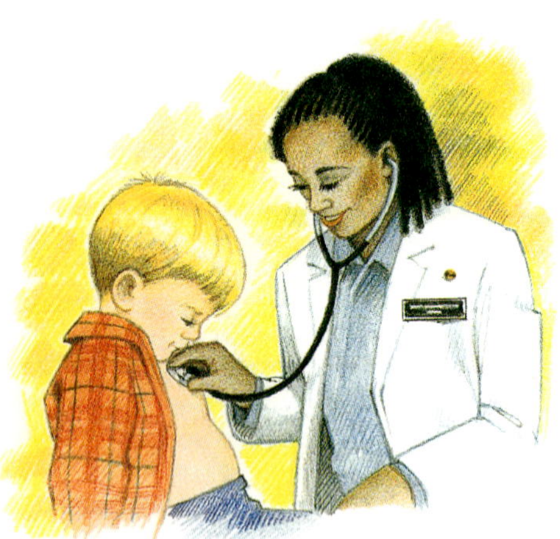

   (b) Write an equation that relates 45 million to the variable x.
   (c) Solve the equation from part (b) to determine the number of people who visited their doctor in 1992 to complain of an earache.
79. **Population Density** In the 1990 census, the population of Puerto Rico was given as 3,522,037, with a population density (people per square mile) of 1,025.
   (a) Let A represent the area of Puerto Rico, and write an equation using the fact that the population is equal to the product of the area and the population density.
   (b) Solve the equation from part (a) for the area of Puerto Rico.
80. **Solving Equations by Trial and Error** Sometimes equations can be solved most easily by trial and error. Solve the following equations by trial and error.
   (a) Find x and y if $x \cdot y + 1 = 36$, and both x and y are prime.
   (b) Find w, t, and z if $w + t + z + 10 = 52$, and w, t, and z are consecutive terms of a Fibonacci sequence.
   (c) Find x and y if $x \neq y$ and $x^y = y^x$.

Review Problems

Simplify each expression as much as possible.

81. $-9 \div \frac{3}{2}$
82. $-\frac{4}{5} \div (-4)$
83. $3 - 7(-6 - 3)$
84. $(3 - 7)(-6 - 3)$
85. $-4(-2)^3 - 5(-3)^2$
86. $4(2 - 5)^3 - 3(4 - 5)^5$
87. $\frac{2(-3) - 5(-6)}{-1 - 2 - 3}$
88. $\frac{4 - 8(3 - 5)}{2 - 4(3 - 5)}$

Extending the Concepts

Solve for x.

89. $\frac{x + 4}{5} - \frac{x + 3}{3} = -\frac{7}{15}$
90. $\frac{x + 1}{7} - \frac{x - 2}{2} = \frac{1}{14}$
91. $\frac{1}{x} - \frac{2}{3} = \frac{2}{x}$
92. $\frac{1}{x} - \frac{3}{5} = \frac{2}{x}$
93. $\frac{x + 3}{2} - \frac{x - 4}{4} = -\frac{1}{8}$
94. $\frac{x - 3}{5} - \frac{x + 1}{10} = -\frac{1}{10}$
95. $\frac{x - 1}{2} - \frac{x + 2}{3} = \frac{x + 3}{6}$
96. $\frac{x + 2}{4} - \frac{x - 1}{3} = -\frac{x + 2}{6}$
97. $x^4 - 13x^2 + 36 = 0$
98. $x^4 - 26x^2 + 25 = 0$

1.3 Equations with Absolute Value

Previously we defined the absolute value of x, $|x|$, to be the distance between x and 0 on the number line. The absolute value of a number measures its distance from 0.

EXAMPLE 1 Solve for x: $|x| = 5$.

SOLUTION Using the definition of absolute value, we can read the equation as, "The distance between x and 0 on the number line is 5." If x is 5 units from 0, then x can be 5 or -5:

$$\text{If } |x| = 5 \quad \text{then} \quad x = 5 \quad \text{or} \quad x = -5$$

In general, then, we can see that any equation of the form $|a| = b$ is equivalent to the equations $a = b$ or $a = -b$, as long as $b > 0$.

EXAMPLE 2 Solve $|2a - 1| = 7$.

SOLUTION We can read this question as "$2a - 1$ is 7 units from 0 on the number line." The quantity $2a - 1$ must be equal to 7 or -7:

$$|2a - 1| = 7$$
$$2a - 1 = 7 \quad \text{or} \quad 2a - 1 = -7$$

We have transformed our absolute value equation into two equations that do not involve absolute value. We can solve each equation separately.

$$\begin{aligned} 2a - 1 &= 7 & \text{or} \quad 2a - 1 &= -7 & \\ 2a &= 8 & \text{or} \quad 2a &= -6 & \textbf{Add 1 to both sides} \\ a &= 4 & \text{or} \quad a &= -3 & \textbf{Multiply by } \tfrac{1}{2} \end{aligned}$$

Our solution set is $\{4, -3\}$.

To check our solutions, we put them into the original absolute value equation:

When	$a = 4$	When	$a = -3$
the equation	$\|2a - 1\| = 7$	the equation	$\|2a - 1\| = 7$
becomes	$\|2(4) - 1\| = 7$	becomes	$\|2(-3) - 1\| = 7$
	$\|7\| = 7$		$\|-7\| = 7$
	$7 = 7$		$7 = 7$

EXAMPLE 3 Solve $\left|\frac{2}{3}x - 3\right| + 5 = 12$.

SOLUTION To use the definition of absolute value to solve this equation, we must isolate the absolute value on the left side of the equal sign. To do so, we add -5 to both sides of the equation to obtain

$$\left|\frac{2}{3}x - 3\right| = 7$$

Now that the equation is in the correct form, we can write

$$\begin{aligned} \frac{2}{3}x - 3 &= 7 & \text{or} \quad \frac{2}{3}x - 3 &= -7 & \\ \frac{2}{3}x &= 10 & \text{or} \quad \frac{2}{3}x &= -4 & \textbf{Add 3 to both sides} \\ x &= 15 & \text{or} \quad x &= -6 & \textbf{Multiply by } \tfrac{3}{2} \end{aligned}$$

The solution set is $\{15, -6\}$.

Practice Problems

1. Solve for x: $|x| = 3$.
2. Solve $|3x - 6| = 9$.
3. Solve $|4x - 3| + 2 = 3$.

Answers

1. $-3, 3$ 2. $-1, 5$ 3. $1, \frac{1}{2}$

4. Solve $|7a - 1| = -2$.

Note: Recall that ∅ is the symbol we use to denote the empty set. When we use it to indicate the solutions to an equation, then we are saying the equation has no solution.

EXAMPLE 4 Solve $|3a - 6| = -4$.

SOLUTION The solution set is ∅ because the left side cannot be negative and the right side is negative. No matter what we try to substitute for the variable a, the quantity $|3a - 6|$ will always be positive or zero. It can never be -4.

Consider the statement $|a| = |b|$. What can we say about a and b? We know they are equal in absolute value. By the definition of absolute value, they are the same distance from 0 on the number line. They must be equal to each other or opposites of each other. In symbols, we write

$$|a| = |b| \iff a = b \quad \text{or} \quad a = -b$$

$\uparrow$	$\uparrow$		$\uparrow$
Equal in absolute value	Equals	or	Opposites

5. Solve $|x + 3| = |x + 8|$.

EXAMPLE 5 Solve $|x - 5| = |x - 7|$.

SOLUTION The quantities $x - 5$ and $x - 7$ must be equal or they must be opposites because their absolute values are equal:

Equals		*Opposites*
$x - 5 = x - 7$	or	$x - 5 = -(x - 7)$
$-5 = -7$		$x - 5 = -x + 7$
No solution here		$2x - 5 = 7$
		$2x = 12$
		$x = 6$

Because the first equation leads to a false statement, it will not give us a solution. (If either of the two equations were to reduce to a true statement, it would mean all real numbers would satisfy the original equation.) In this case, our only solution is $x = 6$.

Getting Ready for Class

After reading through the preceding section, respond in your own words and in complete sentences.

A. Why do some of the equations in this section have two solutions instead of one?

B. Translate $|x| = 6$ into words using the definition of absolute value.

C. Explain in words what the equation $|x - 3| = 4$ means with respect to distance on the number line.

D. When is the statement $|x| = x$ true?

Answers

4. No solution 5. $-\frac{11}{2}$

PROBLEM SET 1.3

Use the definition of absolute value to solve each of the following problems.

1. $|x| = 4$
2. $|x| = 7$
3. $2 = |a|$
4. $5 = |a|$
5. $|x| = -3$
6. $|x| = -4$
7. $|a| + 2 = 3$
8. $|a| - 5 = 2$
9. $|y| + 4 = 3$
10. $|y| + 3 = 1$
11. $|a - 4| = \frac{5}{3}$
12. $|a + 2| = \frac{7}{5}$
13. $\left|\frac{3}{5}a + \frac{1}{2}\right| = 1$
14. $\left|\frac{2}{7}a + \frac{3}{4}\right| = 1$
15. $60 = |20x - 40|$
16. $800 = |400x - 200|$
17. $|2x + 1| = -3$
18. $|2x - 5| = -7$
19. $\left|\frac{3}{4}x - 6\right| = 9$
20. $\left|\frac{4}{5}x + 5\right| = 15$
21. $\left|1 - \frac{1}{2}a\right| = 3$
22. $\left|2 - \frac{1}{3}a\right| = 10$
23. $|2x - 5| = 3$
24. $|3x + 1| = 4$
25. $|4 - 7x| = 5$
26. $|9 - 4x| = 1$
27. $\left|3 - \frac{2}{3}y\right| = 5$
28. $\left|-2 - \frac{3}{4}y\right| = 6$

Solve each equation.

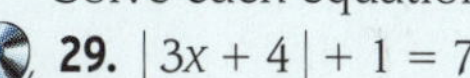

29. $|3x + 4| + 1 = 7$
30. $|5x - 3| - 4 = 3$
31. $|3 - 2y| + 4 = 3$
32. $|8 - 7y| + 9 = 1$
33. $3 + |4t - 1| = 8$
34. $2 + |2t - 6| = 10$
35. $\left|9 - \frac{3}{5}x\right| + 6 = 12$
36. $\left|4 - \frac{2}{7}x\right| + 2 = 14$
37. $5 = \left|\frac{2x}{7} + \frac{4}{7}\right| - 3$
38. $7 = \left|\frac{3x}{5} + \frac{1}{5}\right| + 2$
39. $2 = -8 + \left|4 - \frac{1}{2}y\right|$
40. $1 = -3 + \left|2 - \frac{1}{4}y\right|$
41. $|3(x + 1)| - 4 = -1$
42. $|2(2x + 3)| - 5 = -1$
43. $|1 + 3(2x - 1)| = 5$
44. $|3 + 4(3x + 1)| = 7$
45. $3 = -2 + \left|5 - \frac{2}{3}a\right|$
46. $4 = -1 + \left|6 - \frac{4}{5}a\right|$
47. $6 = |7(k + 3) - 4|$
48. $5 = |6(k - 2) + 1|$

Solve the following equations.

49. $|3a + 1| = |2a - 4|$
50. $|5a + 2| = |4a + 7|$
51. $\left|x - \frac{1}{3}\right| = \left|\frac{1}{2}x + \frac{1}{6}\right|$
52. $\left|\frac{1}{10}x - \frac{1}{2}\right| = \left|\frac{1}{5}x + \frac{1}{10}\right|$
53. $|y - 2| = |y + 3|$
54. $|y - 5| = |y - 4|$
55. $|3x - 1| = |3x + 1|$
56. $|5x - 8| = |5x + 8|$
57. $|0.03 - 0.01x| = |0.04 + 0.05x|$
58. $|0.07 - 0.01x| = |0.08 - 0.02x|$
59. $|x - 2| = |2 - x|$
60. $|x - 4| = |4 - x|$
61. $\left|\frac{x}{5} - 1\right| = \left|1 - \frac{x}{5}\right|$
62. $\left|\frac{x}{3} - 1\right| = \left|1 - \frac{x}{3}\right|$
63. $\left|\frac{2}{3}b - \frac{1}{4}\right| = \left|\frac{1}{6}b + \frac{1}{2}\right|$
64. $\left|-\frac{1}{4}x + 1\right| = \left|\frac{1}{2}x - \frac{1}{3}\right|$
65. $|0.1a - 0.04| = |0.3a + 0.08|$
66. $|-0.4a + 0.6| = |1.3 - 0.2a|$

Applying the Concepts

67. **Amtrak** Amtrak's annual passenger revenue for the years 1985–1995 is modeled approximately by the formula

$$R = -60|x - 11| + 962$$

where R is the annual revenue in millions of dollars and x is the number of years since January 1, 1980 (Association of American Railroads, Washington, DC, *Railroad Facts, Statistics of Railroads of Class 1*, annual). In what years was the passenger revenue \$722 million?

68. **Number of Children** In a given year between 1980 and 1998, the number of U.S. families having three or more children is modeled approximately by

$$N = 88|x - 8.25| + 6405$$

where N is the number of families, in thousands, and x is the number of years since January 1, 1980 (U.S. Census Bureau, *Current Population Reports*).According to the model, in what years were there 7 million U.S. families with three or more children?

69. **Number of Teenagers** The number of U.S. teenagers in the age category "15- to 19-year-olds" during any year between 1980 and 1998 can be modeled approximately by

$$T = 332|x - 11.5| + 17000$$

where T is the number of "15- to 19-year olds" in thousands and x is the number of years since January 1, 1980 (U.S. Census Bureau, *Current Population Reports*). Use the model to determine the year in which the number of 15- to 19-year-olds in the U.S. was 18,700,000 (18,700 thousand).

70. **Corporate Profits** The corporate profits for various U.S. industries vary from year to year. An approximate model for profits of U.S. "communications companies" during a given year between 1990 and 1997 is given by

$$P = -3400|x - 5.5| + 36000$$

where P is the annual profits (in millions of dollars) and x is the number of years since January 1, 1990 (U.S. Bureau of Economic Analysis, Income and Product Accounts of the U.S. (1929–1994), *Survey of Current Business,* September 1998). Use the model to determine the years in which profits of "communication companies" were \$31.5 billion (\$31,500 million).

71. **New Business** For the years 1990–1998, the number of new business starts in the manufacturing industry is modeled approximately by

$$N = 543|x - 5| + 12200$$

where N is the number of new manufacturing business starts in a given year (in thousands) and x is the number of years since January 1, 1990 (The Dun and Bradstreet Corporation, *A Decade of Business Starts,* monthly). In what years were there approximately 13 million (13,000 thousand) new business starts in the manufacturing industry?

72. **Finance Rates** The finance rates from commercial banks for 48-month loans on new automobiles for the years 1985–1995 are modeled approximately by

$$F = 0.6|x - 13.5| + 8$$

where F is the percentage rate (APR) and x is the number of years since January 1, 1980 (Board of Governors of the Federal Reserve System, *Federal Reserve Bulletin,* monthly). Determine the years in which the finance rate (APR) on a new car loan was 9.25%.

Table Building

To obtain a visual representation for expressions that contain absolute value, we can use number sequences. For each of the following sequences, use the formula to construct a table that gives the first five terms in the sequence by substituting 1, 2, 3, 4, and 5 for n in the formula. (Label the first column in the table n, and the second column a_n.) Then use the paired data from the table to construct a scatter diagram.

73. $a_n = |n - 3|$
74. $a_n = |3 - n|$
75. $a_n = |2n - 6|$
76. $a_n = |6 - 2n|$

Review Problems

Multiply.

77. $2x^2(5x^3 + 4x - 3)$
78. $3x^3(7x^2 - 4x - 8)$
79. $(3a - 1)(4a + 5)$
80. $(6a - 3)(2a + 1)$
81. $(x + 3)(x - 3)(x^2 + 9)$
82. $(2x - 3)(4x^2 + 6x + 9)$
83. $(4y - 5)^2$
84. $\left(2y - \frac{1}{2}\right)^2$
85. $(3x + 7)(4y - 2)$
86. $(x + 2a)(2 - 3b)$
87. $(3 - t^2)^2$
88. $(2 - t^3)^2$

One Step Further

Solve each formula for x. (Assume a, b, and c are positive.)

89. $|x - a| = b$
90. $|x + a| - b = 0$
91. $|ax + b| = c$
92. $|ax - b| - c = 0$
93. $\left|\frac{x}{a} + \frac{y}{b}\right| = 1$
94. $\left|\frac{x}{a} + \frac{y}{b}\right| = c$

1.4 Formulas

A *formula* in mathematics is an equation that contains more than one variable. Some formulas are probably already familiar to you—for example, the formula for the area (A) of a rectangle with length l and width w is $A = lw$.

To begin our work with formulas, we will consider some examples in which we are given numerical replacements for all but one of the variables.

EXAMPLE 1 Find y when x is 4 in the formula $3x - 4y = 2$.

SOLUTION We substitute 4 for x in the formula and then solve for y:

When $x = 4$

the formula $3x - 4y = 2$

becomes $3(4) - 4y = 2$

$12 - 4y = 2$ **Multiply 3 and 4**

$-4y = -10$ **Add −12 to each side**

$y = \frac{5}{2}$ **Divide each side by −4**

Note that in the last line of Example 1 we divided each side of the equation by -4. Remember that this is equivalent to multiplying each side of the equation by $-\frac{1}{4}$.

EXAMPLE 2 A store selling art supplies finds that they can sell x sketch pads each week at a price of p dollars each, according to the formula $x = 900 - 300p$. What price should they charge for each sketch pad if they want to sell 525 pads each week?

SOLUTION Here we are given a formula, $x = 900 - 300p$, and asked to find the value of p if x is 525. To do so, we simply substitute 525 for x and solve for p:

When $x = 525$

the formula $x = 900 - 300p$

becomes $525 = 900 - 300p$

$-375 = -300p$ **Add −900 to each side**

$1.25 = p$ **Divide each side by −300**

To sell 525 sketch pads, the store should charge $1.25 for each pad.

Our next example involves a formula that contains the number π. In this book, you can use either 3.14 or $\frac{22}{7}$ as an approximation for π. Generally speaking, if the problem contains decimals, use 3.14 to approximate π.

Practice Problems

1. Find y when x is -3 in $2x - 3y = 6$.

2. Repeat Example 2 if they want to sell 375 pads each week.

Answers

1. -4 2. $1.75

3. Repeat Example 3 using 6 centimeters for the radius of each can. Round your answer to the nearest tenth.

EXAMPLE 3 A company manufacturing coffee cans uses 486.7 square centimeters of material to make each can. Because of the way the cans are to be placed in boxes for shipping, the radius of each can must be 5 centimeters. If the formula for the surface area of a can is $S = \pi r^2 + 2\pi rh$ (the cans are open at the top), find the height of each can. (Use 3.14 as an approximation for π.)

SOLUTION Substituting 486.7 for S, 5 for r, and 3.14 for π into the formula $S = \pi r^2 + 2\pi rh$, we have

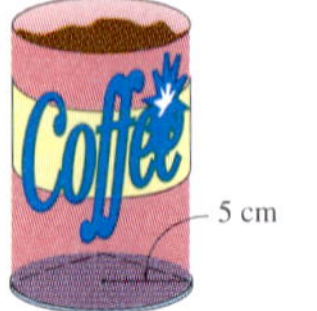

Surface area = 486.7 cm²

$$486.7 = (3.14)(5^2) + 2(3.14)(5)h$$

$$486.7 = 78.5 + 31.4h$$

$$408.2 = 31.4h \qquad \textbf{Add } -78.5 \textbf{ to each side}$$

$$13 = h \qquad \textbf{Divide each side by 31.4}$$

The height of each can is 13 centimeters.

4. Use the formula in Example 4 to find t if $v = 64$ feet/second and $h = 64$ feet.

EXAMPLE 4 If an object is projected into the air with an initial vertical velocity of v feet/second, its height h, in feet, above the ground after t seconds will be given by

$$h = vt - 16t^2$$

Find t if $v = 64$ feet/second and $h = 48$ feet.

SOLUTION Substituting $v = 64$ and $h = 48$ into the preceding formula, we have

$$48 = 64t - 16t^2$$

which is a quadratic equation. We write it in standard form and solve by factoring:

$$16t^2 - 64t + 48 = 0$$

$$t^2 - 4t + 3 = 0 \qquad \textbf{Divide each side by 16}$$

$$(t - 1)(t - 3) = 0$$

$$t - 1 = 0 \quad \text{or} \quad t - 3 = 0$$

$$t = 1 \quad \text{or} \quad t = 3$$

Here is how we interpret our results: If an object is projected upward with an initial vertical velocity of 64 feet/second, it will be 48 feet above the ground after 1 second and after 3 seconds; that is, it passes 48 feet going up and also coming down.

5. Use the information in Example 5 to find the price that should be charged if the weekly revenue is to be $3,600.

EXAMPLE 5 A manufacturer of small portable radios knows that the number of radios she can sell each week is related to the price of the radios by the equation $x = 1{,}300 - 100p$, where x is the number of radios and p is the price per radio. What price should she charge for each radio if she wants the weekly revenue to be $4,000?

SOLUTION The formula for total revenue is $R = xp$. Since we want R in terms of p, we substitute $1{,}300 - 100p$ for x in the equation $R = xp$:

If $\quad R = xp$

and $\quad x = 1{,}300 - 100p$

then $\quad R = (1{,}300 - 100p)p$

Answers
3. 9.9 cm 4. 2 seconds

We want to find p when R is 4,000. Substituting 4,000 for R in the formula gives us

$$4{,}000 = (1{,}300 - 100p)p$$
$$4{,}000 = 1{,}300p - 100p^2$$

which is a quadratic equation. To write it in standard form, we add $100p^2$ and $-1{,}300p$ to each side, giving us

$$100p^2 - 1{,}300p + 4{,}000 = 0$$
$$p^2 - 13p + 40 = 0 \qquad \textbf{Divide each side by 100}$$
$$(p - 5)(p - 8) = 0$$
$$p - 5 = 0 \quad \text{or} \quad p - 8 = 0$$
$$p = 5 \quad \text{or} \quad p = 8$$

If she sells the radios for \$5 each or for \$8 each she will have a weekly revenue of \$4,000.

Facts From Geometry: Formulas for Area and Perimeter

To review, here are the formulas for the area and perimeter of some common geometric objects.

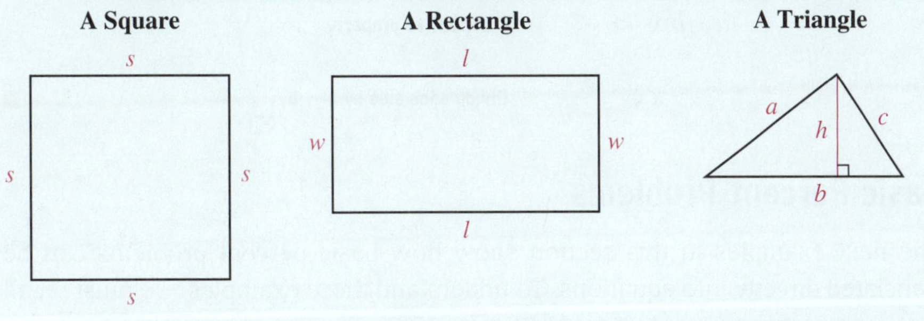

A Square	A Rectangle	A Triangle
Perimeter = $4s$	Perimeter = $2l + 2w$	Perimeter = $a + b + c$
Area = s^2	Area = lw	Area = $\frac{1}{2}bh$

Note: The pink line labeled h in the triangle is its height, or altitude. It extends from the top of the triangle down to the base, meeting the base at an angle of 90°. The altitude of a triangle is always perpendicular to the base. The small square shown where the altitude meets the base is used to indicate that the angle formed is 90°.

The formula for perimeter gives us the distance around the outside of the object along its sides, whereas the formula for area gives us a measure of the amount of surface the object covers.

EXAMPLE 6 Given the formula $P = 2w + 2l$, solve for w.

SOLUTION To solve for w, we must isolate it on one side of the equation. We can accomplish this if we delete the $2l$ term and the coefficient 2 from the right side of the equation.

To begin, we add $-2l$ to both sides:

$$P + (\mathbf{-2l}) = 2w + 2l + (\mathbf{-2l})$$
$$P - 2l = 2w$$

6. Solve $P = 2w + 2l$ for l.

w = width

l = length

Answers

5. \$4 or \$9

To delete the 2 from the right side, we can multiply both sides by $\frac{1}{2}$:

$$\frac{1}{2}(P - 2l) = \frac{1}{2}(2w)$$

$$\frac{P - 2l}{2} = w$$

The two formulas

$$P = 2w + 2l \quad \text{and} \quad w = \frac{P - 2l}{2}$$

give the relationship between P, l, and w. They look different, but they both say the same thing about P, l, and w. The first formula gives P in terms of l and w, and the second formula gives w in terms of P and l.

Note:
We know we are finished solving a formula for a specified variable when that variable appears alone on one side of the equal sign and not on the other.

EXAMPLE 7 Solve for x: $ax - 3 = bx + 5$.

SOLUTION In this example, we must begin by collecting all the variable terms on the left side of the equation and all the constant terms on the other side (just like we did when we were solving linear equations):

$$ax - 3 = bx + 5$$

$$ax - bx - 3 = 5 \quad \text{Add } -bx \text{ to each side}$$

$$ax - bx = 8 \quad \text{Add 3 to each side}$$

At this point we need to apply the distributive property to write the left side as $(a - b)x$. After that, we divide each side by $a - b$:

$$(a - b)x = 8 \quad \text{Distributive property}$$

$$x = \frac{8}{a - b} \quad \text{Divide each side by } a - b$$

7. Solve for x: $ax + 5 = cx + 3$.

Note:
We are applying the distributive property in the same way we applied it when we first learned how to simplify $7x - 4x$. Recall that $7x - 4x = 3x$ because

$$7x - 4x = (7 - 4)x = 3x$$

We are using the same type of reasoning when we write

$$ax - bx = (a - b)x$$

Basic Percent Problems

The next examples in this section show how basic percent problems can be translated directly into equations. To understand these examples, we must recall that percent means "per hundred." That is, 75% is the same as 75/100, 0.75, and, in reduced fraction form, $\frac{3}{4}$. Likewise, the decimal 0.25 is equivalent to 25%. To change a decimal to a percent, we move the decimal point two places to the right and write the % symbol. To change from a percent to a decimal, we drop the % symbol and move the decimal point two places to the left. The table that follows gives some of the most commonly used fractions and decimals and their equivalent percents.

FRACTION	DECIMAL	PERCENT
$\frac{1}{2}$	0.5	50%
$\frac{1}{4}$	0.25	25%
$\frac{3}{4}$	0.75	75%
$\frac{1}{3}$	$0.\overline{3}$	$33\frac{1}{3}\%$
$\frac{2}{3}$	$0.\overline{6}$	$66\frac{2}{3}\%$
$\frac{1}{5}$	0.2	20%
$\frac{2}{5}$	0.4	40%
$\frac{3}{5}$	0.6	60%
$\frac{4}{5}$	0.8	80%

Answers
6. $l = \frac{P - 2w}{2}$ 7. $x = \frac{-2}{a - c}$

EXAMPLE 8 What number is 15% of 63?

SOLUTION To solve a problem like this, we let x = the number in question and then translate the sentence directly into an equation. Here is how it is done:

$$\underbrace{\text{What number}}_{x} \text{ is } 15\% \text{ of } 63?$$

$$x = 0.15 \cdot 63$$
$$= 9.45$$

The number 9.45 is 15% of 63.

8. What number is 25% of 74?

Note: We write 0.15 instead of 15% when we translate the sentence into an equation because we cannot do calculations with the % symbol. Because percent (%) means per hundred, we think of 15% as 15 hundredths, or 0.15.

EXAMPLE 9 What percent of 42 is 21?

SOLUTION We translate the sentence as follows:

$$\underbrace{\text{What percent}}_{x} \text{ of } 42 \text{ is } 21?$$

$$x \cdot 42 = 21$$

Next, we divide each side by 42:

$$x = \frac{21}{42}$$

$$= 0.50 \text{ or } 50\%$$

9. What percent of 84 is 21?

EXAMPLE 10 25 is 40% of what number?

SOLUTION Again, we translate the sentence directly:

$$25 \text{ is } 40\% \text{ of } \underbrace{\text{what number}}_{x}?$$

$$25 = 0.40 \cdot x$$

We solve the equation by dividing both sides by 0.40:

$$\frac{25}{0.40} = \frac{0.40 \cdot x}{0.40}$$

$$62.5 = x$$

25 is 40% of 62.5.

10. 35 is 40% of what number?

Formulas for Number Sequences

In Chapter 1 we used inductive reasoning to find the next number in a sequence of numbers. What we are going to do now is extend our work with number sequences so that we can find any number in the sequence. To begin, we introduce notation that allows us to associate each number in a number sequence with its position in the sequence. To illustrate, consider the sequence of even numbers:

Even numbers: 2, 4, 6, 8, 10, . . .

The number 8 is the fourth number in sequence. Its *position* in the sequence is 4, and its *value* is 8. Here is the sequence of even numbers again, this time written so that the position of each term is noted.

Position: 1, 2, 3, 4, 5, . . .

Value: 2, 4, 6, 8, 10, . . .

The notation we use to indicate the position of a number in a number sequence involves *subscripts.* Subscripts are numbers written to the right and just below

Answers
8. 18.5 9. 25% 10. 87.5

another number or variable. In the expression a_4, the 4 is a subscript. We use subscripts to designate the terms of a sequence this way:

The *first term* of a sequence is designated by a_1.

The *second term* of a sequence is designated by a_2.

The *third term* of a sequence is designated by a_3.

The *fourth term* of a sequence is designated by a_4.

The expression a_n is called the nth term, or the *general term* of the sequence. The general term is used to define the other terms of the sequence; that is, if we are given the formula for the general term a_n, we can find any other term in the sequence. The following examples illustrate the concept.

11. Find the first four terms of the sequence whose general term is given by $a_n = 3n - 2$.

EXAMPLE 11 Find the first four terms of the sequence whose general term is given by $a_n = 2n - 1$.

SOLUTION To find the first, second, third, and fourth terms of this sequence, we simply substitute 1, 2, 3, and 4 for n in the formula $2n - 1$:

If The general term is $a_n = 2n - 1$

then The first term is $a_1 = 2(1) - 1 = 1$

The second term is $a_2 = 2(2) - 1 = 3$

The third term is $a_3 = 2(3) - 1 = 5$

The fourth term is $a_4 = 2(4) - 1 = 7$

The first four terms of this sequence are the odd numbers 1, 3, 5, and 7. The whole sequence can be written as

$$1, 3, 5, \ldots, 2n - 1, \ldots$$

Because each term in this sequence is larger than the preceding term, we say the sequence is an *increasing sequence.*

12. Write the first four terms of the sequence defined by

$$a_n = \frac{n}{n+1}$$

EXAMPLE 12 Write the first four terms of the sequence defined by

$$a_n = \frac{1}{n+1}$$

SOLUTION Replacing n with 1, 2, 3, and 4, we have, respectively, the first four terms:

$$\text{First term} = a_1 = \frac{1}{1+1} = \frac{1}{2}$$

$$\text{Second term} = a_2 = \frac{1}{2+1} = \frac{1}{3}$$

$$\text{Third term} = a_3 = \frac{1}{3+1} = \frac{1}{4}$$

$$\text{Fourth term} = a_4 = \frac{1}{4+1} = \frac{1}{5}$$

The sequence defined by

$$a_n = \frac{1}{n+1}$$

can be written as

$$\frac{1}{2}, \frac{1}{3}, \frac{1}{4}, \ldots, \frac{1}{n+1}, \ldots$$

Because each term in the sequence is smaller than the term preceding it, the sequence is said to be a *decreasing sequence.*

Answers

11. 1, 4, 7, 10

12. $\frac{1}{2}, \frac{2}{3}, \frac{3}{4}, \frac{4}{5}$

USING TECHNOLOGY

Graphing Calculators
Entering, Recalling, and Editing Formulas

Graphing calculators can be useful in evaluating formulas. For example, the formula for the volume of a right circular cylinder is

$$V = \pi r^2 h$$

To find V when $r = 3$ and $h = 4$, we must do three things. First we use the [STO →] key to store 3 in the variable r. Using the same key we store 4 in the variable h. Then we enter the formula V. All three of these things can be done at one time by using the [:] key. Here are the characters that will appear on your calculator screen when you have done the preceding three tasks:

Store 3 in R   Store 4 in H   Enter formula for V

$$3 \rightarrow R : 4 \rightarrow H : \pi R \wedge 2\, H$$

When you press the [ENTER] key, the calculator will display a little more than 113.0973355, which is the volume of a right circular cylinder for which $r = 3$ and $h = 4$, accurate to 10 digits.

If we want to evaluate the same formula for other values of r or h, we can recall it to the screen by pressing [2nd] [ENTER]. (If we have done some other calculations in the meantime, we can repeatedly press the sequence [2nd] [ENTER] until the formula appears on the screen. The Texas Instruments TI-83 will recall as many previous entries as can be stored in its memory.) Once the entry has been recalled, we can use the arrow keys on the calculator to move around the formula to change any part of it that we choose to change. Try it yourself. Recall the formula, and then change the number stored in R to 5, and the number stored in H to 13 (you will have to use the [INS] key to do your editing). When you press [ENTER], you should see a number that represents the volume of the cylinder shown in Example 3.

Getting Ready for Class

After reading through the preceding section, respond in your own words and in complete sentences.

A. What is a formula in mathematics?

B. Create a formula for a sequence and explain how to obtain the first four terms of your sequence.

C. Write a percent problem that can be solved by the equation $30 = 0.25x$.

D. Explain in words the formula for the perimeter of a rectangle.

PROBLEM SET 1.4

Use the formula $3x - 4y = 12$ to find y if

1. x is 0 **2.** x is -2

3. x is 4 **4.** x is -4

Use the formula $y = 2x - 3$ to find x when

5. y is 0 **6.** y is -3

7. y is 5 **8.** y is -5

Use the formula $5x - 3y = -15$ to find y if:

9. $x = 2$ **10.** $x = -3$

11. $x = -\frac{1}{5}$ **12.** $x = 3$

Use the formula $S = 2\pi r^2 + 2\pi rh$, which gives the surface area of a cylinder that is closed at the top and bottom, for the following problems.

13. Find h if $r = 4$ ft and $S = 88\pi$ ft^2.

14. Find h if $r = 8$ ft and $S = 208\pi$ ft^2.

15. Find h if $r = 12$ in. and $S = 1507.2$ in^2. Use $\pi \approx 3.14$.

16. Find h if $r = 9$ in. and $S = 1131\frac{3}{7}$ in^2. Use $\pi \approx \frac{22}{7}$.

17. Find r if $S = 208\pi$ m^2 and $h = 5$ m.

18. Find r if $S = 648\pi$ m^2 and $h = 15$ m.

Solve each of the following formulas for the indicated variable. In parentheses is an indication of what the formula is used for.

19. $A = lw$ for l (Area of a rectangle)

20. $A = \frac{1}{2}bh$ for b (Area of a triangle)

21. $I = prt$ for t (simple interest)

22. $I = prt$ for r **23.** $PV = nRT$ for T

24. $PV = nRT$ for R

25. $y = mx + b$ for x (slope-intercept form)

26. $A = P + Prt$ for t (Interest)

27. $C = \frac{5}{9}(F - 32)$ for F (Temperature conversion)

28. $F = \frac{9}{5}C + 32$ for C

29. $h = vt + 16t^2$ for v (Physics)

30. $h = vt - 16t^2$ for v

31. $A = a + (n - 1)d$ for d (General term of a sequence)

32. $A = a + (n - 1)d$ for n

33. $2x + 3y = 6$ for y

34. $2x - 3y = 6$ for y

35. $-3x + 5y = 15$ for y

36. $-2x - 7y = 14$ for y

37. $2x - 6y + 12 = 0$ for y

38. $7x - 2y - 6 = 0$ for y

39. $ax + 4 = bx + 9$ for x

40. $ax - 5 = cx - 2$ for x

41. $bx - 3 = cx + 5$ for x

42. $bx + 4 = cx - 8$ for x

43. $ax + 3 = cx - 7$ for x

44. $by - 9 = dy + 3$ for y

45. $ax + b = cx + d$ for x

46. $st - h = rt - k$ for t

Solve for y.

47. $\frac{x}{8} + \frac{y}{2} = 1$ **48.** $\frac{x}{7} + \frac{y}{9} = 1$

49. $\frac{x}{5} + \frac{y}{-3} = 1$ **50.** $\frac{x}{16} + \frac{y}{-2} = 1$

Translate each of the following into a linear equation and then solve the equation.

51. What number is 54% of 38?

52. What number is 11% of 67?

53. What percent of 36 is 9?

54. What percent of 50 is 5?

55. 37 is 4% of what number?

56. 8 is 2% of what number?

Write the first five terms of the sequences with the following general terms.

57. $a_n = 3n + 1$ **58.** $a_n = 4n - 1$

59. $a_n = n^2 + 3$ **60.** $a_n = n^3 + 1$

61. $a_n = \frac{n}{n + 3}$ **62.** $a_n = \frac{n}{n + 2}$

63. $a_n = \frac{1}{n^2}$ **64.** $a_n = \frac{1}{n^3}$

65. $a_n = 2^n$ **66.** $a_n = 3^n$

67. $a_n = 1 + \frac{1}{n}$ **68.** $a_n = 1 - \frac{1}{n}$

Applying the Concepts

69. Air Pollution Air pollution in the United States has become an increasing concern in recent years. The formula below models the sulfur dioxide emissions during a given year between 1970 and 1997.

$$E = 30{,}366 - 395x$$

In the formula, E is the emission of sulfur dioxide in thousands of tons, and x is the number of years after 1970. That is, $x = 0$ on January 1, 1970. In what year were sulfur dioxide emissions 22,071 thousand tons? (U.S. Environmental Protection Agency, *National Air Quality and Emissions Trend Report*)

70. **Recycling** The recycling of various waste products has been emphasized by various government agencies during the past 20 years. For the years 1990 to 1997, the relationship between the amount of "paper and paperboard" recycled and the amount of "plastics" recycled is given by the formula

$$y = 19x + 12.8$$

where y is the "paper and paperboard" recycled (in millions of tons), during a given year, and x is the amount of "plastics" recycled in a given year (in millions of tons). During the year that 28 million tons of "paper and paperboard" are recycled, how many tons of "plastics" are recycled? (Franklin Associates, Ltd, Prairie Village, KS. *Characterization of Municipal Waste in the United States: 1998*)

Baseball Earlier, we worked a problem in which we found Rolaids points earned in 1998 for some relief pitchers. Now we are going to do a similar problem, but this time using a formula. The formula $P = 3s + 2w - 2l - 2b$ gives the number of Rolaids points a pitcher earns, where s = saves, w = wins, l = losses, and b = blown saves. Use this formula to complete the following tables.

71. National League

PITCHER, TEAM	W	L	SAVES	BLOWN SAVES	ROLAIDS POINTS
John Franco, New York	0	8	38	8	
Billy Wagner, Houston	4	3	30	5	
Gregg Olson, Arizona	3	4	30	4	
Bob Wickman, Milwaukee	6	9	25	7	

72. American League

PITCHER, TEAM	W	L	SAVES	BLOWN SAVES	ROLAIDS POINTS
Rick Aguilera, Minnesota	4	9	38	11	
Billy Taylor, Oakland	4	9	33	4	
Randy Myers, Toronto	3	4	28	5	
Todd Jones, Detroit	1	4	28	4	

Estimating Vehicle Weight If you can measure the area that the tires on your car contact the ground and know the air pressure in the tires, then you can estimate the weight of your car with the following formula:

$$W = \frac{APN}{2,000}$$

where W is the vehicle's weight in tons, A is the average tire contact area with a hard surface in square inches, P is the air pressure in the tires in pounds per square inch (psi, or lb/in^2), and N is the number of tires.

73. What is the approximate weight of a car if the average tire contact area is a rectangle 6 inches by 5 inches and if the air pressure in the tires is 30 psi?

74. What is the approximate weight of a car if the average tire contact area is a rectangle 5 inches by 4 inches and the tire pressure is 30 psi?

The formula $h = vt - 16t^2$ gives the height h, in feet, of an object projected into the air with an initial vertical velocity v, in feet per second, after t seconds.

75. **Projectile Motion** If an object is propelled upward with an initial velocity of 48 feet per second, at what times will it reach a height of 32 feet above the ground?

76. **Projectile Motion** If an object is propelled upward into the air with an initial velocity of 80 feet per second, at what times will it reach a height of 64 feet above the ground?

77. **Projectile Motion** An object is projected into the air with a velocity of 24 feet per second. At what times will the object be on the ground? (It is on the ground when h is 0).

78. **Projectile Motion** An object is projected into the air with a vertical velocity of 20 feet per second. At what times will the object be on the ground?

79. **Height of a Bullet** A bullet is fired into the air with an initial upward velocity of 80 feet per second from the top of a building 96 feet high. The equation that gives the height of the bullet at any time t is $h = 96 + 80t - 16t^2$. At what times will the bullet be 192 feet in the air?

80. **Height of an Arrow** An arrow is shot into the air with an upward velocity of 48 feet per second from a hill 32 feet high. The equation that gives the height of the arrow at any time t is $h = 32 + 48t - 16t^2$. Find the times at which the arrow will be 64 feet above the ground.
81. **Price and Revenue** A company that manufacturers ink cartridges for printers knows that the number of cartridges x it can sell each week is related to the price per cartridge p by the equation $x = 1{,}200 - 100p$. At what price should it sell the cartridges if it wants the weekly revenue to be \$3,200? (*Remember:* The equation for revenue is $R = xp$.)
82. **Price and Revenue** A company manufacturers diskettes for home computers. It knows from past experience that the number of diskettes x it can sell each day is related to the price per diskette p by the equation $x = 800 - 100p$. At what price should it sell its diskettes if it wants the daily revenue to be \$1,200?
83. **Price and Revenue** The relationship between the number of calculators x a company sells per day and the price of each calculator p is given by the equation $x = 1{,}700 - 100p$. At what price should the calculators be sold if the daily revenue is to be \$7,000?
84. **Price and Revenue** The relationship between the number of pencil sharpeners x a company can sell each week and the price of each sharpener p is given by the equation $x = 1{,}800 - 100p$. At what price should the sharpeners be sold if the weekly revenue is to be \$7,200?

Digital Video The biggest video download of all time was a *Star Wars* movie trailer. The video was compressed so it would be small enough for people to download over the Internet. A formula for estimating the size, in kilobytes, of a compressed video is

$$S = \frac{height \cdot width \cdot fps \cdot time}{35{,}000}$$

where *height* and *width* are in pixels, *fps* is the number of frames per second the video is to play (television plays at 30 fps), and time is given in seconds.

85. Estimate the size in kilobytes of the *Star Wars* trailer that has a height of 480 pixels, a width of 216 pixels, plays at 30 fps, and runs for 150 seconds.
86. Estimate the size in kilobytes of the *Star Wars* trailer that has a height of 320 pixels, a width of 144 pixels, plays at 15 fps, and runs for 150 seconds.

Compound Interest The following formula gives the amount of money A in an account in which P dollars has been invested for t years at an interest rate of r compounded n times a year:

$$A = P\left(1 + \frac{r}{n}\right)^{nt}$$

Use the recall formula function on your calculator to answer the following questions:

87. How much money is in an account in which \$500 has been invested at 5%, compounded quarterly, for 6 years? (I.e., find A when $P = 500$, $r = 0.05$, $n = 4$, and $t = 6$.)
88. How much money is in an account in which \$500 has been invested at 5%, compounded monthly, for 6 years?
89. How much money is in an account in which \$500 has been invested at 5%, compounded daily, for 6 years?
90. How much money is in an account in which \$500 has been invested at 6%, compounded monthly, for 6 years?
91. How much money is in an account in which \$500 has been invested at 7%, compounded monthly, for 6 years?
92. In general, is it more desirable to increase the interest rate or the number of compounding periods to increase the amount of money in the account?

Interest If interest in an investment is compounded just once a year, then the formula given before Problem 87 is simply $A = P(1 + r)^t$.

93. When Benjamin Franklin died in 1790, he left the city of Boston \$5,000. If the city had invested this money at 4% compounded yearly, how much would Franklin's gift have been worth in 2000?

94. How much more would Franklin's gift be worth in 2000 if his gift had been compounded quarterly?

Review Problems

Translate each of the following into symbols.

95. Twice the sum of x and 3
96. The sum of twice x and 3
97. Twice the sum of x and 3 is 16.
98. The sum of twice x and 3 is 16.
99. Five times the difference of x and 3
100. Five times the difference of x and 3 is 10.

101. The sum of $3x$ and 2 is equal to the difference of x and 4.

102. The sum of x and $x + 2$ is 12 more than their difference.

Extending the Concepts

103. Solve for x: $\frac{x}{a} + \frac{y}{b} = 1$

104. Solve for y: $\frac{x}{a} + \frac{y}{b} = 1$

105. Solve for a: $\frac{1}{a} + \frac{1}{b} = \frac{1}{c}$

106. Solve for b: $\frac{1}{a} + \frac{1}{b} = \frac{1}{c}$

Exercise Physiology In exercise physiology, a person's maximum heart rate, in beats per minute, is found by subtracting his age, in years, from 220. So, if A represents your age in years, then your maximum heart rate is

$$M = 220 - A$$

A person's training heart rate, in beats per minute, is her resting heart rate plus 60% of the difference between her maximum heart rate and her resting heart rate. If resting heart rate is R and maximum heart rate is M, then the formula that gives training heart rate is

$$T = R + 0.6(M - R)$$

107. Training Heart Rate Shar is 46 years old. Her daughter, Sara, is 26 years old. If they both have resting heart rate of 60 beats per minute, find the training heart rate for each.

108. Training Heart Rate Shane is 30 years old and has a resting heart rate of 68 beats per minute. Her mother, Carol, is 52 years old and has the same resting heart rate. Find the training heart rate for Shane and for Carol.

1.5 Applications

In this section we use the skills we have developed for solving equations to solve problems written in words. You may find that some of the examples and problems are more realistic than others. Since we are just beginning our work with application problems, even the ones that seem unrealistic are good practice. What is important in this section is the *method* we use to solve application problems, not the applications themselves. The method, or strategy, that we use to solve application problems is called the *Blueprint for Problem Solving.* It is an outline that will overlay the solution process we use on all application problems.

Blueprint for Problem Solving

Step 1: ***Read*** the problem, and then mentally ***list*** the items that are known and the items that are unknown.

Step 2: ***Assign a variable*** to one of the unknown items. (In most cases this will amount to letting x = the item that is asked for in the problem.) Then ***translate*** the other ***information*** in the problem to expressions involving the variable.

Step 3: ***Reread*** the problem, and then ***write an equation,*** using the items and variable listed in steps 1 and 2, that describes the situation.

Step 4: ***Solve the equation*** found in step 3.

Step 5: ***Write your answer*** using a complete sentence.

Step 6: ***Reread*** the problem, and ***check*** your solution with the original words in the problem.

A number of substeps occur within each of the steps in our blueprint. For instance, with steps 1 and 2 it is always a good idea to draw a diagram or picture if it helps you visualize the relationship between the items in the problem.

Practice Problems

1. The width of a rectangle is 10 feet less than 4 times the length. If the perimeter is 12.5 feet, find the length and width.

EXAMPLE 1 The length of a rectangle is 3 inches less than twice the width. The perimeter is 45 inches. Find the length and width.

SOLUTION When working problems that involve geometric figures, a sketch of the figure helps organize and visualize the problem.

Step 1: ***Read and list.***

Known items: The figure is a rectangle. The length is 3 inches less than twice the width. The perimeter is 45 inches.

Unknown items: The length and the width

Step 2: ***Assign a variable and translate information.***

Since the length is given in terms of the width (the length is 3 less than twice the width), we let x = the width of the rectangle. The length is 3 less than twice the width, so it must be $2x - 3$. The diagram in Figure 1 is a visual description of the relationships we have listed so far.

Figure 1

Step 3: ***Reread and write an equation.***

The equation that describes the situation is

$$\begin{array}{ccccc} \text{Twice the length} & + & \text{twice the width} & = & \text{perimeter} \\ 2(2x-3) & + & 2x & = & 45 \end{array}$$

Step 4: ***Solve the equation.***

$$\begin{aligned} 2(2x-3)+2x &= 45 \\ 4x-6+2x &= 45 \\ 6x-6 &= 45 \\ 6x &= 51 \\ x &= 8.5 \end{aligned}$$

Step 5: ***Write the answer.***

The width is 8.5 inches. The length is $2x - 3 = 2(8.5) - 3 = 14$ inches.

Step 6: ***Reread and check.***

If the length is 14 inches and the width is 8.5 inches, then the perimeter must be $2(14) + 2(8.5) = 28 + 17 = 45$ inches. Also, the length, 14, is 3 less than twice the width.

Remember as you read through the steps in the solutions to the examples in this section that step 1 is done mentally. Read the problem and then *mentally* list the items that you know and the items that you don't know. The purpose of step 1 is to give you direction as you begin to work application problems. Finding the solution to an application problem is a process; it doesn't happen all at once. The first step is to read the problem with a purpose in mind. That purpose is to mentally note the items that are known and the items that are unknown.

EXAMPLE 2 In April 1998, Pat bought a new Ford Mustang with a 5.0-liter engine. The total price, which includes the price of the car plus sales tax, was \$17,481.75. If the sales tax rate is 7.25%, what was the price of the car?

SOLUTION

Step 1: ***Read and list.***

Known items: The total price is \$17,481.75. The sales tax rate is 7.25%, which is 0.0725 in decimal form.

Unknown item: The price of the car.

Step 2: ***Assign a variable and translate information.***

If we let x = the price of the car, then to calculate the sales tax, we multiply the price of the car x by the sales tax rate:

$$\begin{aligned} \text{Sales tax} &= (\text{sales tax rate})(\text{price of the car}) \\ &= 0.0725x \end{aligned}$$

Step 3: ***Reread and write an equation.***

$$\begin{array}{ccccc} \text{Car price} & + & \text{sales tax} & = & \text{total price} \\ x & + & 0.0725x & = & 17{,}481.75 \end{array}$$

Step 4: ***Solve the equation.***

$$\begin{aligned} x + 0.0725x &= 17{,}481.75 \\ 1.0725x &= 17{,}481.75 \\ x &= \frac{17{,}481.75}{1.0725} \\ &= 16{,}300.00 \end{aligned}$$

2. If Pat had purchased the convertible version of the Ford Mustang discussed in Example 2, the total price (the price of the car plus sales tax) would have been \$23,466.30. Find the price of the convertible if the sales tax rate is 7.25%.

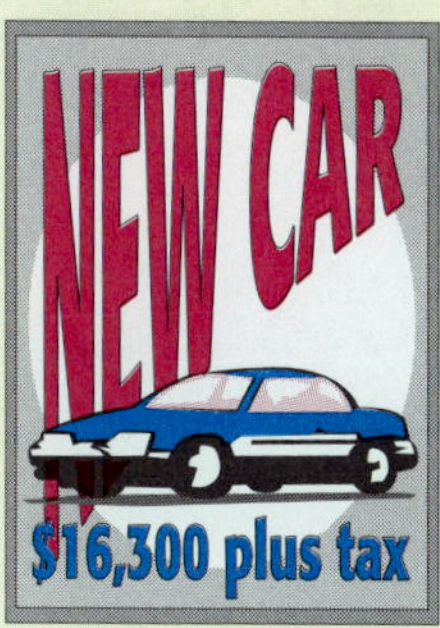

Answer

1. length = 3.25 feet, width = 3 feet

Step 5: ***Write the answer.***
The price of the car is \$16,300.00

Step 6: ***Reread and check.***
The price of the car is \$16,300.00. The tax is 0.0725(16,300) = \$1,181.75. Adding the retail price and the sales tax we have a total bill of \$17,481.75.

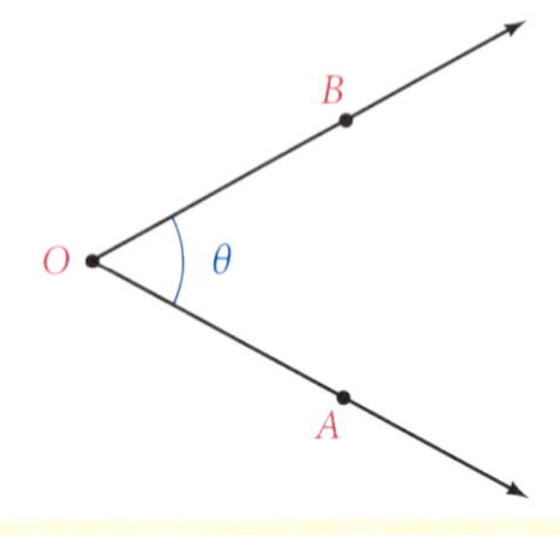

Figure 2

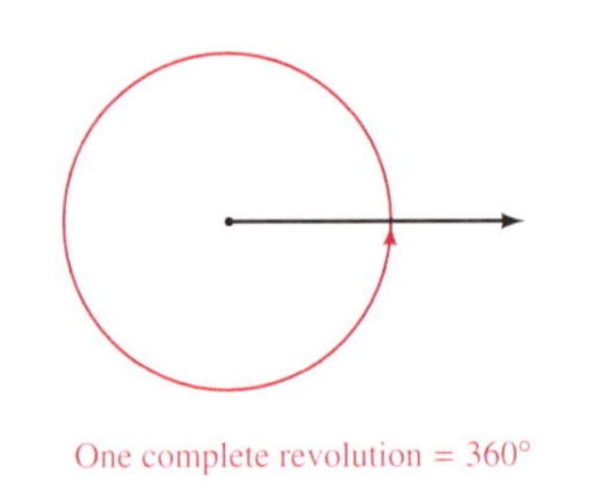

Figure 3

Facts From Geometry: Angles in General

An angle is formed by two rays with the same endpoint. The common endpoint is called the *vertex* of the angle, and the rays are called the *sides* of the angle.

In Figure 2, angle θ (theta) is formed by the two rays *OA* and *OB*. The vertex of θ is *O*. Angle θ is also denoted as angle *AOB*, where the letter associated with the vertex is always the middle letter in the three letters used to denote the angle.

Degree Measure

The angle formed by rotating a ray through one complete revolution about its endpoint (Figure 3) has a measure of 360 degrees, which we write as 360°.

One degree of angle measure, written 1°, is $\frac{1}{360}$ of a complete rotation of a ray about its endpoint; there are 360° in one full rotation. (The number 360 was decided on by early civilizations because it was believed that the earth was at the center of the universe and the sun would rotate once around the earth every 360 days.) Similarly, 180° is half of a complete rotation, and 90° is a quarter of a full rotation. Angles that measure 90° are called *right angles,* and angles that measure 180° are called *straight angles.* If an angle measures between 0° and 90° it is called an *acute angle,* and an angle that measures between 90° and 180° is an *obtuse angle.* Figure 4 illustrates further.

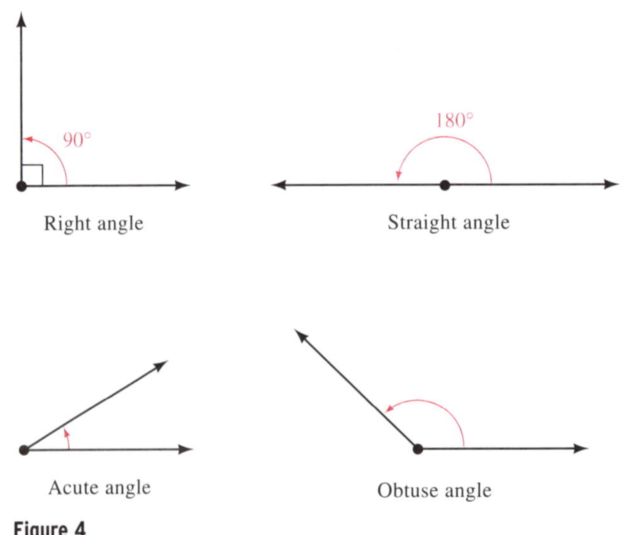

Figure 4

Complementary Angles and Supplementary Angles

If two angles add up to 90°, we call them *complementary angles,* and each is called the *complement* of the other. If two angles have a sum of 180°, we call them *supplementary angles,* and each is called the *supplement* of the other. Figure 5 illustrates the relationship between angles that are complementary and angles that are supplementary.

Answer
2. \$21,880.00

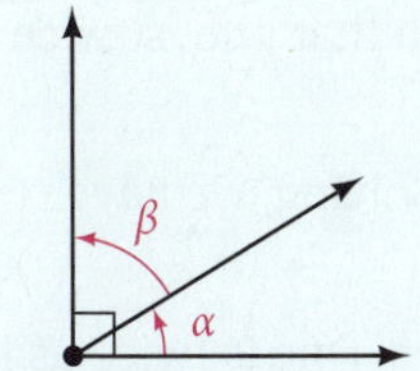

Complementary angles: $\alpha + \beta = 90°$

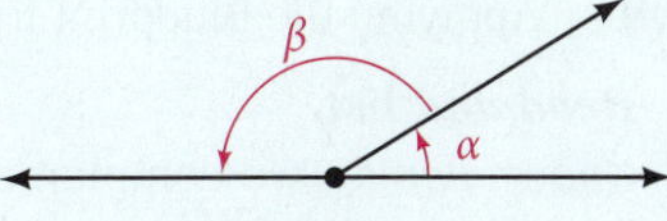

Supplementary angles: $\alpha + \beta = 180°$

Figure 5

Special Triangles

It is not unusual to have the terms we use in mathematics show up in the descriptions of things we find in the world around us. The flag of Puerto Rico shown here is described on the government website as "Five equal horizontal bands of red (top and bottom) alternating with white; a blue isosceles triangle based on the hoist side bears a large white five-pointed star in the center." An *isosceles triangle* as shown here and in Figure 6, is a triangle with two sides of equal length.

Angles A and B in the isosceles triangle in Figure 6 are called the *base angles:* they are the angles opposite the two equal sides. In every isosceles triangle, the base angles are equal.

Isosceles Triangle

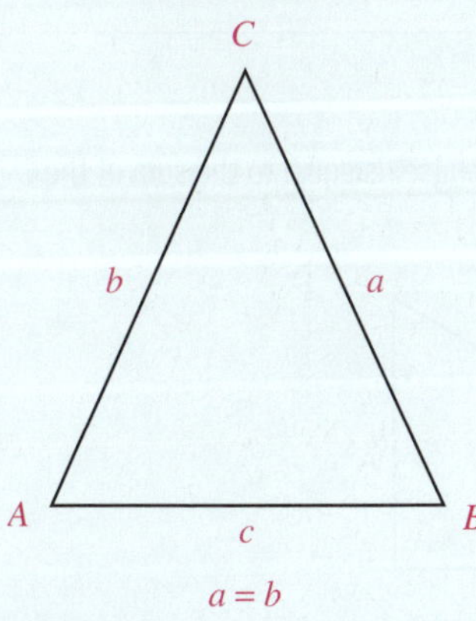

Figure 6

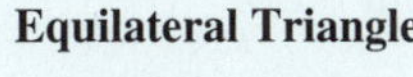

Equilateral Triangle

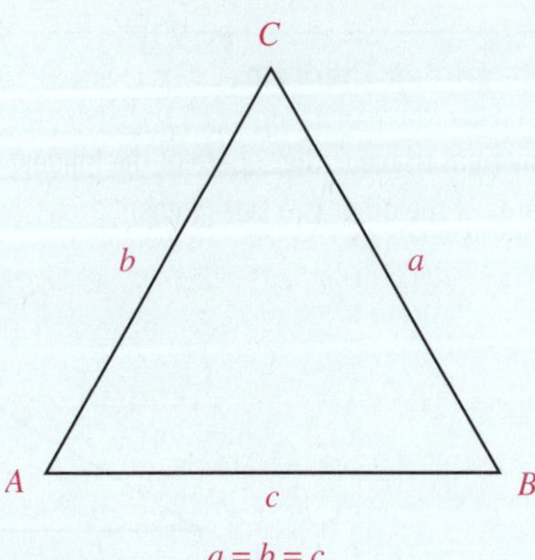

Figure 7

Note: As you can see from Figures 6 and 7, one way to label the important parts of a triangle is to label the vertices with capital letters and the sides with small letters: side a is opposite vertex A, side b is opposite vertex B, and side c is opposite vertex C.

Also, because each vertex is the vertex of one of the angles of the triangle, we refer to the three interior angles as A, B, and C.

Finally, in any triangle, the sum of the interior angles is 180°. For the triangles shown in Figures 6 and 7, the relationship is written

$$A + B + C = 180°$$

An *equilateral triangle* (Figure 7) is a triangle with three sides of equal length. If all three sides in a triangle have the same length, then the three interior angles in the triangle also must be equal. Because the sum of the interior angles in a triangle is always 180°, the three interior angles in any equilateral triangle must be 60°.

3. Two supplementary angles are such that one is five times as large as the other. Find the two angles.

EXAMPLE 3 Two complementary angles are such that one is twice as large as the other. Find the two angles.

SOLUTION Applying the Blueprint for Problem Solving, we have:

Step 1: ***Read and list.***
Known items: Two complementary angles. One is twice as large as the other.
Unknown items: The size of the angles

Step 2: ***Assign a variable and translate information.***
Let x = the smaller angle. The larger angle is twice the smaller so we represent the larger angle with $2x$.

Step 3: ***Reread and write an equation.***
Because the two angles are complementary, their sum is 90. Therefore,

$$x + 2x = 90$$

Step 4: ***Solve the equation.***

$$x + 2x = 90$$
$$3x = 90$$
$$x = 30$$

Step 5: ***Write the answer.***
The smaller angle is 30°, and the larger angle is $2 \cdot 30 = 60°$.

Step 6: ***Reread and check.***
The larger angle is twice the smaller angle, and their sum is 90°.

Another application of quadratic equations involves the Pythagorean theorem, an important theorem from geometry. The theorem gives the relationship between the sides of any right triangle (a triangle with a 90-degree angle). We state it here without proof.

Pythagorean Theorem

In any right triangle, the square of the longest side (hypotenuse) is equal to the sum of the squares of the other two sides (legs).

$$c^2 = a^2 + b^2$$

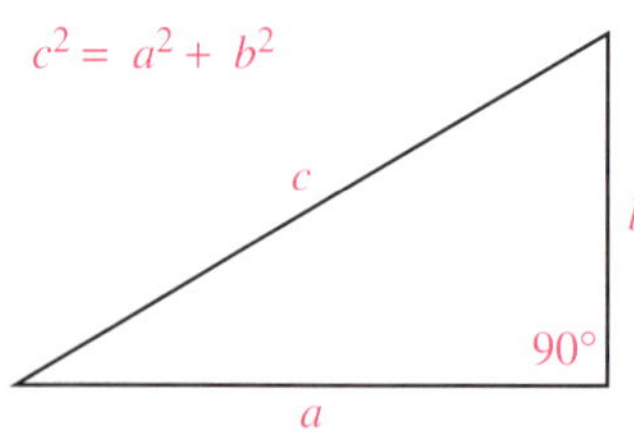

4. The longest side of a right triangle is 4 more than the shortest side. The third side is 2 more than the shortest side. Find the length of each side.

EXAMPLE 4 The lengths of the three sides of a right triangle are given by three consecutive integers. Find the lengths of the three sides.

SOLUTION

Step 1: ***Read and list.***
Known items: A right triangle. The three sides are three consecutive integers.
Unknown items: The three sides

Answer
3. 30° and 150°

Step 2: ***Assign a variable and translate information.***
Let x = first integer (shortest side).
Then $x + 1$ = next consecutive integer
$x + 2$ = last consecutive integer (longest side)

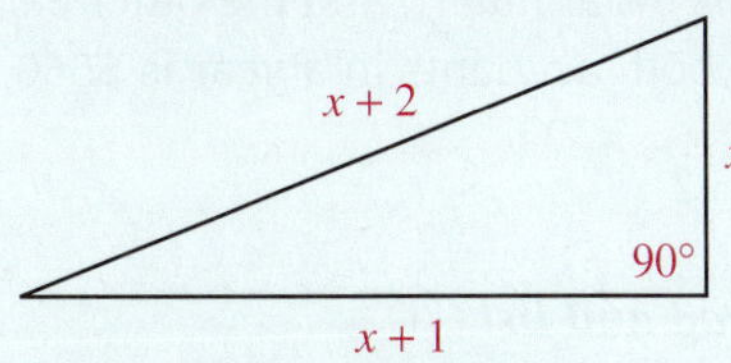

Step 3: ***Reread and write an equation.***
By the Pythagorean theorem, we have

$$(x + 2)^2 = (x + 1)^2 + x^2$$

Step 4: ***Solve the equation.***

$$x^2 + 4x + 4 = x^2 + 2x + 1 + x^2$$
$$x^2 - 2x - 3 = 0$$
$$(x - 3)(x + 1) = 0$$
$$x = 3 \quad \text{or} \quad x = -1$$

Step 5: ***Write the answer.***
Since x is the length of a side in a triangle, it must be a positive number. Therefore, $x = -1$ cannot be used.
The shortest side is 3. The other two sides are 4 and 5.

Step 6: ***Reread and check.***
The three sides are given by consecutive integers. The square of the longest side is equal to the sum of the squares of the two shorter sides.

Suppose we know that the sum of two numbers is 50. If we let x represent one of the two numbers, how can we represent the other? Let's suppose for a moment that x turns out to be 30. Then the other number will be 20 because their sum is 50; that is, if two numbers add up to 50, and one of them is 30, then the other must be $50 - 30 = 20$. Generalizing this to any number x, we see that if two numbers have a sum of 50 and one of the numbers is x, then the other must be $50 - x$. The following table shows some additional examples:

IF TWO NUMBERS HAVE A SUM OF	AND ONE OF THEM IS	THEN THE OTHER MUST BE
50	x	$50 - x$
10	y	$10 - y$
12	n	$12 - n$

Answer
4. 6, 8, 10

Interest Problem

EXAMPLE 5 Suppose a person invests a total of \$10,000 in two accounts. One account earns 5% annually, and the other earns 6% annually. If the total interest earned from both accounts in a year is \$560, how much is invested in each account?

SOLUTION

Step 1: ***Read and list.***

Known items: Two accounts. One pays interest of 5%, and the other pays 6%. The total dollars invested is \$10,000.

Unknown items: The number of dollars invested in each individual account.

Step 2: ***Assign a variable and translate information.***

If we let x equal the amount invested at 6%, then $10{,}000 - x$ is the amount invested at 5%. The total interest earned from both accounts is \$560. The amount of interest earned on x dollars at 6% is $0.06x$, whereas the amount of interest earned on $10{,}000 - x$ dollars at 5% is $0.05(10{,}000 - x)$.

	DOLLARS AT 6%	DOLLARS AT 5%	TOTAL
Number of	x	$10{,}000 - x$	10,000
Interest on	$0.06x$	$0.05(10{,}000 - x)$	560

Step 3: ***Reread and write an equation.***

The last line gives us the equation we are after:

$$0.06x + 0.05(10{,}000 - x) = 560$$

Step 4: ***Solve the equation.***

To make this equation a little easier to solve, we begin by multiplying both sides by 100 to move the decimal point two places to the right:

$$6x + 5(10{,}000 - x) = 56{,}000$$
$$6x + 50{,}000 - 5x = 56{,}000$$
$$x + 50{,}000 = 56{,}000$$
$$x = 6{,}000$$

Step 5: ***Write the answer.***

The amount of money invested at 6% is \$6,000. The amount of money invested at 5% is \$10,000 − \$6,000 = \$4,000.

Step 6: ***Reread and check.***

To check our results, we find the total interest from the two accounts:

The interest on \$6,000 at 6% is 0.06(6,000) = 360
The interest on \$4,000 at 5% is 0.05(4,000) = 200
The total interest = \$560

5. Howard invests a total of \$8,000 in two accounts. One account earns 8% annually, and the other earns 10% annually. If the total interest earned from both accounts in a year is \$680, how much is invested in each account?

Answer

5. \$6,000 at 8%, \$2,000 at 10%

Table Building

We can use our knowledge of formulas to build tables of paired data. As you will see, equations or formulas that contain exactly two variables produce pairs of numbers that can be used to construct tables.

EXAMPLE 6 A piece of string 12 inches long is to be formed into a rectangle. Build a table that gives the length of the rectangle if the width is 1, 2, 3, 4, or 5 inches. Then find the area of each of the rectangles formed.

SOLUTION Because the formula for the perimeter of a rectangle is $P = 2l + 2w$ and our piece of string is 12 inches long, then the formula we will use to find the lengths for the given widths is $12 = 2l + 2w$. To solve this formula for l, we divide each side by 2 and then subtract w. The result is $l = 6 - w$. Table 1 organizes our work so that the formula we use to find l for a given value of w is shown, and we have added a last column to give us the areas of the rectangles formed. The units for the first three columns are inches, and the units for the numbers in the last column are square inches.

Table 1

LENGTH, WIDTH, AND AREA

WIDTH (IN.)	LENGTH (IN.)	AREA (IN.^{2})	
w	$l = 6 - w$	l	$A = lw$
1	$l = 6 - 1$	5	5
2	$l = 6 - 2$	4	8
3	$l = 6 - 3$	3	9
4	$l = 6 - 4$	2	8
5	$l = 6 - 5$	1	5

Figures 8 and 9 show two bar charts constructed from the information in Table 1.

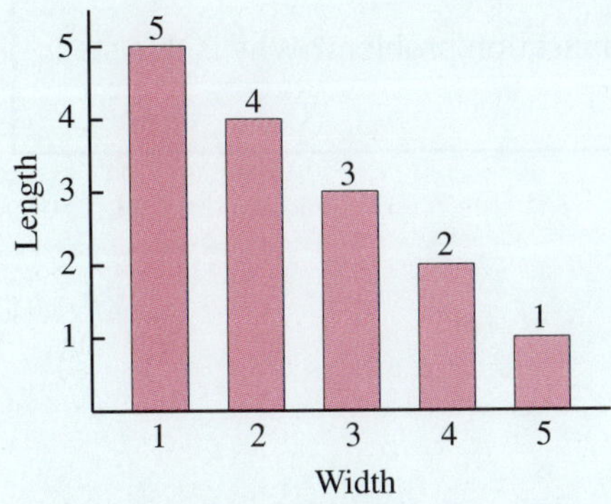

Figure 8 Length and width of rectangles with perimeters fixed at 12 inches.

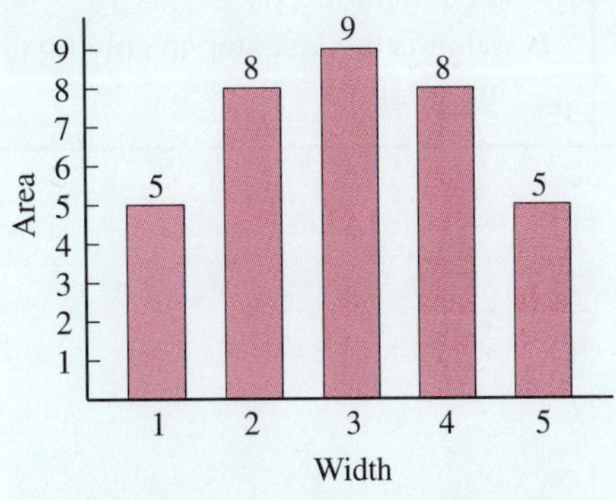

Figure 9 Area and width of rectangles with perimeters fixed at 12 inches.

6. Rework Example 6 if the string is 14 inches long and the rectangle has a width of 1, 2, 3, 4, 5, or 6 inches.

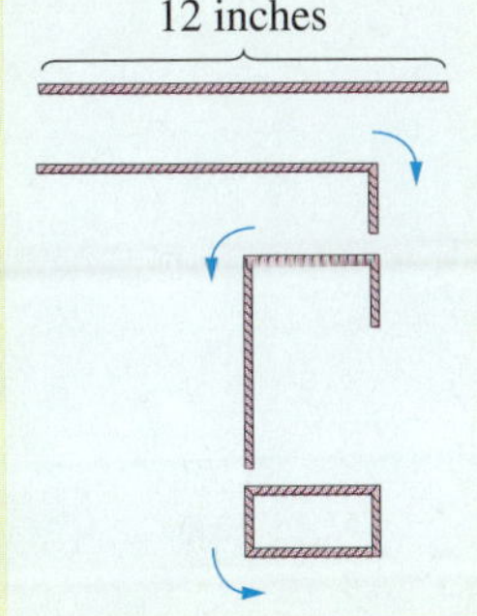

Answer

6. See Solutions to Practice Problems Section

USING TECHNOLOGY

A number of graphing calculators have table-building capabilities. We can let the calculator variable X represent the widths of the rectangles in Example 5. To find the lengths, we set variable Y_1 equal to $6 - X$. The area of each rectangle can be found by setting variable Y_2 equal to $X * Y_1$. To have the calculator produce the table automatically, we use a table minimum of 0 and a table increment of 1. Here is a summary of how the graphing calculator is set up:

Table Setup	*Y Variables Setup*
Table minimum = 0	$Y_1 = 6 - X$
Table increment = 1	$Y_2 = X * Y_1$
Independent variable: Auto	
Dependent variable: Auto	

The table will look like this:

X	Y_1	Y_2
0	6	0
1	5	5
2	4	8
3	3	9
4	2	8
5	1	5
6	0	0

Getting Ready for Class

After reading through the preceding section, respond in your own words and in complete sentences.

A. What is the first step in solving an application problem?

B. What is the biggest obstacle between you and success in solving application problems?

C. Write an application problem for which the solution depends on solving the equation $2x + 2 \cdot 3 = 18$.

D. What is the last step in solving an application problem? Why is this step important?

PROBLEM SET 1.5

Solve each application problem. Be sure to follow the steps in the Blueprint for Problem Solving.

Geometry Problems

1. **Rectangle** A rectangle is twice as long as it is wide. The perimeter is 60 feet. Find the dimensions.
2. **Rectangle** The length of a rectangle is 5 times the width. The perimeter is 48 inches. Find the dimensions.
3. **Square** A square has a perimeter of 28 feet. Find the length of each side.
4. **Square** A square has a perimeter of 36 centimeters. Find the length of each side.
5. **Triangle** A triangle has a perimeter of 23 inches. The medium side is 3 inches more than the shortest side, and the longest side is twice the shortest side. Find the shortest side.
6. **Triangle** The longest side of a triangle is two times the shortest side, while the medium side is 3 meters more than the shortest side. The perimeter is 27 meters. Find the dimensions.

7. **Rectangle** The length of a rectangle is 3 meters less than twice the width. The perimeter is 18 meters. Find the width.
8. **Rectangle** The length of a rectangle is 1 foot more than twice the width. The perimeter is 20 feet. Find the dimensions.
9. **Livestock Pen** A livestock pen is built in the shape of a rectangle that is twice as long as it is wide. The perimeter is 48 feet. If the material used to build the pen is \$1.75 per foot for the longer side and \$2.25 per foot for the shorter sides (the shorter sides have gates, which increase the cost per foot), find the cost to build the pen.

10. **Garden** A garden is in the shape of a square with a perimeter of 42 feet. The garden is surrounded by two fences. One fence is around the perimeter of the garden, whereas the second fence is 3 feet from the first fence, as Figure 10 indicates. If the material used to build the two fences is \$1.28 per foot, what was the total cost of the fences?

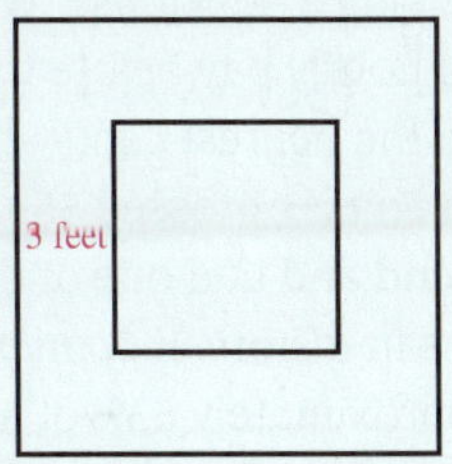

Figure 10

Percent Problems

11. **Truck Price** Suppose the total price, including sales tax, of a used Ford F-150 pickup truck is \$10,039.43. If the sales tax rate is 7.5%, what was the price of the truck? Round your answer to the nearest cent.
12. **Money** Shane returned from a trip to Las Vegas with \$300.00, which was 50% more money than he had at the beginning of the trip. How much money did Shane have at the beginning of his trip?
13. **Items Sold** Every item in the Just a Dollar store is priced at \$1.00. When Mary Jo opens the store, there is \$125.50 in the cash register. When she counts the money in the cash register at the end of the day, the total is \$1,058.60. If the sales tax rate is 8.5%, how many items were sold that day?
14. **TV Show Ratings** During one ratings period in 1993, it was estimated that 5.4 million people watched *The Tonight Show* with Jay Leno. If the number of people watching Jay Leno was 1.8% more than the number of viewers watching *Nightline* with Ted Koppel, approximately how many viewers did Ted Koppel have? Round your answer to the nearest tenth of a million viewers.
15. **Workforce** On Monday, August 16, 1993, Continental Airlines announced it was laying off 2,500 workers, or 6% of its workforce, by December 31 of that year. How many workers did Continental Airlines have on Friday, August 13, 1993?

16. **Textbook Price** Suppose a college bookstore buys a textbook from a publishing company and then marks up the price they paid for the book 33% and sells it to a student at the marked-up price. If the student pays \$45.00 for the textbook, what did the bookstore pay for it? Round your answer to the nearest cent.

17. **Monthly Salary** An accountant earns \$3,440 per month after receiving a 5.5% raise. What was the accountant's monthly income before the raise? Round your answer to the nearest cent.

18. **Hourly Wage** A sheet metal worker earns \$26.80 per hour after receiving a 4.5% raise. What was the sheet metal worker's hourly pay before the raise? Round your answer to the nearest cent.

19. **Movies** *Batman Forever* grossed \$52.8 million on its opening weekend and had one of the most successful movie launches in history. If *Batman Forever* accounted for approximately 53% of all box office receipts that weekend, what were the total box office receipts?

20. **Movies** The two top money-making films from 1990–1995 were *Jurassic Park* and *Lion King*. If together these two films grossed \$651 million and *Lion King* grossed \$25 million less than *Jurassic Park*, what were the gross earnings for *Jurassic Park?*

21. **Loans** Upon graduation in 1996, 52% of bachelor's degree recipients from public institutions and 25% of associate's degree recipients from community colleges had federal loan debt. This represents a significant increase from 1993 in the percent of students incurring debt. In addition, the average per-student loan total upon graduation significantly increased from 1993 to 1996.
   (a) The average per-student debt at graduation from community college increased from \$4,380 to \$5,450. Find the percent of increase.
   (b) In 1993, the average per-student debt for private school students was \$10,214. By 1996, this average had risen by 40%. Find the 1996 average debt per student to the nearest dollar.
   (c) The per-student debt averaged \$11,950 at public institutions in 1996, which represented a 61.48% increase over the average in 1993. What was the average debt per student (rounded to the nearest dollar) in 1993?

22. **Loans** In 1997, 5.4 million students obtained federal loans. Currently, the default rate on student loans has been reduced to approximately 11%. If this rate holds, how many of the students who borrowed in 1997 will default on their loan (with serious consequences)?

23. **Fat Content in Milk** I was reading the information on the milk carton in Figure 11 at breakfast one morning when I was working on this book. According to the carton, this milk contains 70% less fat than whole milk. The nutrition label on the other side of the carton states that one serving of this milk contains 2.5 grams of fat. How many grams of fat are in an equivalent serving of whole milk?

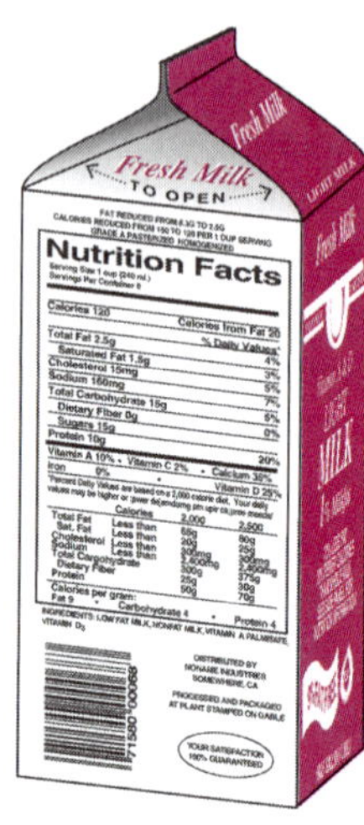

Figure 11

24. **Calories in Milk** The information on the milk carton in Figure 11 states that it contains 20% fewer calories than whole milk. If one serving of this milk contains 120 calories, how many calories are in an equivalent serving of whole milk?

More Geometry Problems

25. **Angles** Two supplementary angles are such that one is eight times larger than the other. Find the two angles.

26. **Angles** Two complementary angles are such that one is five times larger than the other. Find the two angles.

27. **Angles** One angle is 12° less than four times another. Find the measure of each angle if
   (a) they are complements of each other.
   (b) they are supplements of each other.

28. **Angles** One angle is 4° more than three times another. Find the measure of each angle if
   (a) they are complements of each other.
   (b) they are supplements of each other.

29. **Triangles** A triangle is such that the largest angle is three times the smallest angle. The third angle is 9° less than the largest angle. Find the measure of each angle.

30. **Triangles** The smallest angle in a triangle is half of the largest angle. The third angle is 15° less than the largest angle. Find the measure of all three angles.

31. **Triangles** The smallest angle in a triangle is one-third of the largest angle. The third angle is 10° more than the smallest angle. Find the measure of all three angles.
32. **Triangles** The third angle in an isosceles triangle is half as large as each of the two base angles. Find the measure of each angle.
33. **Isosceles Triangles** The third angle in an isosceles triangle is 8° more than twice as large as each of the two base angles. Find the measure of each angle.
34. **Isosceles Triangles** The third angle in an isosceles triangle is 4° more than one-fifth of each of the two base angles. Find the measure of each angle.
35. **Right Triangle** A 25-foot ladder is leaning against a building. The base of the ladder is 7 feet from the side of the building. How high does the ladder reach along the side of the building?

36. **Right Triangle** Noreen wants to place a 13-foot ramp against the side of her house so the top of the ramp rests on a ledge that is 5 feet above the ground. How far will the base of the ramp be from the house?
37. **Right Triangle** The lengths of the three sides of a right triangle are given by three consecutive even integers. Find the lengths of the three sides.
38. **Right Triangle** The longest side of a right triangle is 3 less than twice the shortest side. The third side measures 12 inches. Find the length of the shortest side.
39. **Rectangle** The length of a rectangle is 2 feet more than three times the width. If the area is 16 square feet, find the width and the length.
40. **Rectangle** The length of a rectangle is 4 yards more than twice the width. If the area is 70 square yards, find the width and the length.
41. **Rectangle** Professor Clark is building a rectangular dog run for his dog. He would like the length to be 1 foot longer than twice the width. If he has enough cement to make a foundation that is 66 square feet, find the length and width of the dog run.
42. **Rectangle** Professor Parke is building a rectangular flower garden. She would like the length to be 2 feet less than four times the width. If she has enough topsoil to cover 72 square feet, find the length and width of the flower garden.
43. **Triangle** When viewed from the front, the roof of a house in the neighborhood of Miami Dade College is in the shape of a triangle, with an area of 100 square feet. The height of the triangle is one foot more than three times the base. Find the base and height of the roof.
44. **Triangle** Professor Garcia is designing a triangular reflecting pool that covers an area of 135 square feet. The height of the triangle is three feet less than three times the base. Find the base and height of the reflecting pool.
45. **Triangle** The base of a triangle is 2 inches more than four times the height. If the area is 36 square inches, find the base and the height.
46. **Triangle** The height of a triangle is 4 feet less than twice the base. If the area is 48 square feet, find the base and the height.

Interest Problems

47. **Investing** A woman has a total of $9,000 to invest. She invests part of the money in an account that pays 8% per year and the rest in an account that pays 9% per year. If the interest earned in the first year is $750, how much did she invest in each account?

	DOLLARS AT 8%	DOLLARS AT 9%	TOTAL
Number of			
Interest on			

48. **Investing** A man invests $12,000 in two accounts. If one account pays 10% per year and the other pays 7% per year, how much was invested in each account if the total interest earned in the first year was $960?

	DOLLARS AT 10%	DOLLARS AT 7%	TOTAL
Number of			
Interest on			

49. **Investing** A total of $15,000 is invested in two accounts. One of the accounts earns 12% per year, and the other earns 10% per year. If the total interest earned in the first year is $1,600, how much was invested in each account?
50. **Investing** A total of $11,000 is invested in two accounts. One of the two accounts pays 9% per year, and the other account pays 11% per year. If the total interest paid in the first year is $1,150, how much was invested in each account?
51. **Investing** Stacey has a total of $6,000 in two accounts. The total amount of interest she earns from both accounts in the first year is $500. If one of the accounts

earns 8% interest per year and the other earns 9% interest per year, how much did she invest in each account?

52. **Investing** Travis has a total of \$6,000 invested in two accounts. The total amount of interest he earns from the accounts in the first year is \$410. If one account pays 6% per year and the other pays 8% per year, how much did he invest in each account?

Miscellaneous Problems

53. **Tickets** Tickets for the father-and-son breakfast were \$2.00 for fathers and \$1.50 for sons. If a total of 75 tickets were sold for \$127.50, how many fathers and how many sons attended the breakfast?
54. **Tickets** A Girl Scout troop sells 62 tickets to their mother-and-daughter dinner, for a total of \$216. If the tickets cost \$4.00 for mothers and \$3.00 for daughters, how many of each ticket did they sell?
55. **Sales Tax** A woman owns a small, cash-only business in a state that requires her to charge 6% sales tax on each item she sells. At the beginning of the day, she has \$250 in the cash register. At the end of the day, she has \$1,204 in the register. How much money should she send to the state government for the sales tax she collected?
56. **Sales Tax** A store is located in a state that requires 6% tax on all items sold. If the store brings in a total of \$3,392 in one day, how much of that total was sales tax?
57. **Long-Distance Phone Calls** Patrick goes away to college. The first week he is away from home he calls his girlfriend, using his parents' telephone credit card, and talks for a long time. The telephone company charges 40 cents for the first minute and 30 cents for each additional minute, and then adds on a 50-cent service charge for using the credit card. If his parents receive a bill for \$13.80 for Patrick's call, how long did he talk?
58. **Long-Distance Phone Calls** A person makes a long distance person-to-person call to Santa Barbara, California. The telephone company charges 41 cents for the first minute and 32 cents for each additional minute. Because the call is person-to-person, there is also a service charge of \$3.00. If the cost of the call is \$6.29, how many minutes did the person talk?

Problems 59–62 may be solved using a graphing calculator.

Table Building

59. **Livestock Pen** A farmer buys 48 feet of fencing material to build a rectangular livestock pen. Fill in the second column of the table to find the length of the pen if the width is 2, 4, 6, 8, 10, or 12 feet. Then find the area of each of the pens formed.

W	*L*	*A*
2		
4		
6		
8		
10		
12		

60. **Art Supplies** A store selling art supplies finds they can sell x sketch pads each week at a price of p dollars each according to the formula $x = 900 - 300p$. Use this formula to build a table that gives the number of pads sold each week if the price per pad is \$0.50, \$1.00, \$1.50, \$2.00, or \$2.50.
61. **Model Rocket** A small rocket is projected straight up into the air with a velocity of 128 feet per second. The formula that gives the height h of the rocket t seconds after it is launched is

$$h = -16t^2 + 128t$$

Use this formula to find the height of the rocket after 1, 2, 3, 4, 5, and 6 seconds.

TIME (SECONDS)	HEIGHT (FEET)
1	
2	
3	
4	
5	
6	

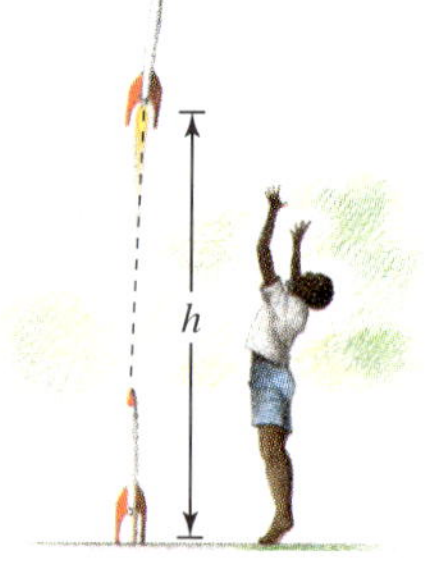

62. **Cellular Phone** If you were to buy a cellular phone in California and access it through GTE Wireless, you would have a choice of rates. One rate card lists a flat rate of \$19.95 per month plus \$0.38 for each minute you use the phone. Write an equation in two variables that will let you calculate the monthly charge for talking x minutes. Then build a table that shows the cost for talking 10, 15, 20, 25, or 30 minutes in one month.

Maximum Heart Rate In exercise physiology, a person's maximum heart rate, in beats per minute, is found by subtracting his age in years from 220. So, if A represents your age in years, then your maximum heart rate is

$$M = 220 - A$$

Use this formula to complete the following tables.

63.

AGE (YEARS)	MAXIMUM HEART RATE (BEATS PER MINUTE)
18	
19	
20	
21	
22	
23	

64.

AGE (YEARS)	MAXIMUM HEART RATE (BEATS PER MINUTE)
15	
20	
25	
30	
35	
40	

Training Heart Rate A person's training heart rate, in beats per minute, is his resting heart rate plus 60% of the difference between his maximum heart rate and his resting heart rate. If resting heart rate is R and maximum heart rate is M, then the formula that gives training heart rate is

$$T = R + 0.6(M - R)$$

Use this formula along with the results of Problems 63 and 64 to fill in the following two tables.

65. For a 20-year-old person

RESTING HEART RATE (BEATS PER MINUTE)	TRAINING HEART RATE (BEATS PER MINUTE)
60	
62	
64	
68	
70	
72	

66. For a 40-year-old person

RESTING HEART RATE (BEATS PER MINUTE)	TRAINING HEART RATE (BEATS PER MINUTE)
60	
62	
64	
68	
70	
72	

67. Length of String Use a graphing calculator or a spreadsheet program to produce Table 1 from this section. Start with a width of 0, but increase the width by 0.5 inch each time. What is the largest area produced by this table?

68. Length of String Use a graphing calculator or a spreadsheet program to reproduce Table 1 from this section. This time, start with a width of 0 and increase the width by 0.25 inch each time. What is the largest area produced by this table?

69. Investing If you invest $100 in an account that earns 6% annual interest, compounded monthly, the amount of money in that account t months later is given by the formula $A = 100(1.005)^t$. Use a graphing calculator or a spreadsheet program to produce a table that gives the amount of money in the account each month for the first 2 years it is on deposit. (On your graphing calculator, let X represent t, and Y_1 represent A. The minimum X is 0 and the table increment is 1.)

70. Volume of a Balloon A spherical balloon is being inflated at a constant rate so that the radius increases by 1 centimeter every second. Because the balloon is a sphere, its volume when the radius is r is given by the formula $V = \frac{4}{3}\pi r^3$. Construct a table that gives the volume of the balloon every second, starting when the radius is 0 centimeters and ending when the radius is 10 centimeters. (Round to the nearest whole number.)

Review Problems

For the set $\{-4, -\frac{2}{5}, 0, 2, \sqrt{5}, 3, \pi\}$, name all the:

71. Integers

72. Irrational numbers

73. Rational numbers

74. Whole numbers

If $A = \{3, 5, 7, 9\}$ and $B = \{4, 5, 6, 7\}$, find:

75. $A \cup B$

76. $A \cap B$

77. $\{x \mid x \in A \text{ and } x > 5\}$

78. $\{x \mid x \in B \text{ and } x \leq 5\}$

1.6 Linear Inequalities in One Variable

The instrument panel on most cars includes a temperature gauge. The one shown below indicates that the normal operating temperature for the engine is from 50°F to 270°F.

We can represent the same situation with an inequality by writing $50 \le F \le 270$, where F is the temperature in degrees Fahrenheit. This inequality is a compound inequality. In this section we present the notation and definitions associated with compound inequalities.

Let's begin by reviewing. Here is an example that show's how we solved inequalities in Chapter 2.

Practice Problems

1. Solve $4x - 2 > 3x + 4$ and graph the solution.

EXAMPLE 1 Solve $3x + 3 < 2x - 1$, and graph the solution.

SOLUTION We use the addition property for inequalities to write all the variable terms on one side and all constant terms on the other side:

$$3x + 3 < 2x - 1$$
$$3x + (\mathbf{-2x}) + 3 < 2x + (\mathbf{-2x}) - 1 \quad \text{Add } -2x \text{ to each side}$$
$$x + 3 < -1$$
$$x + 3 + (\mathbf{-3}) < -1 + (\mathbf{-3}) \quad \text{Add } -3 \text{ to each side}$$
$$x < -4$$

The solution set is all real numbers that are less than -4. To show this we can use set notation and write

$$\{x \mid x < -4\}$$

We can graph the solution set on the number line using an open circle at -4 to show that -4 is not part of the solution set. This is the format you used when graphing inequalities in Chapter 2.

Here is an equivalent graph that uses a parenthesis opening left, instead of an open circle, to represent the end point of the graph.

This graph gives rise to the following notation, called *interval notation,* that is an alternative way to write the solution set.

$$(-\infty, -4)$$

Note:
The English mathematician John Wallis (1616–1703) was the first person to use the ∞ symbol to represent infinity. When we encounter the interval $(3, \infty)$, we read it as "the interval from 3 to infinity," and we mean the set of real numbers that are greater than three. Likewise, the interval $(-\infty, -4)$ is read "the interval from negative infinity to -4," which is all real numbers less than -4.

The preceding expression indicates that the solution set is all real numbers from negative infinity up to, but not including, -4.

We have three equivalent representations for the solution set to our original inequality. Here are all three together.

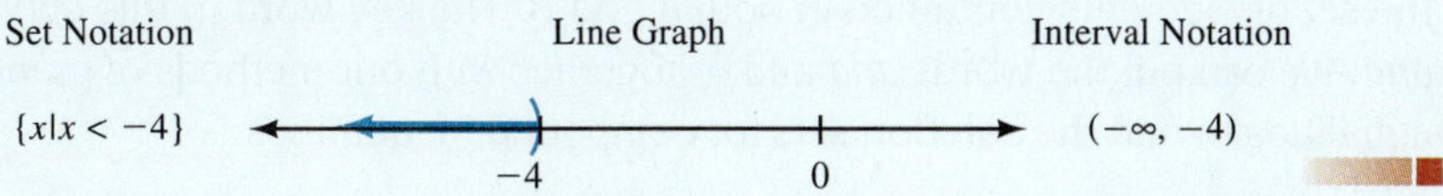

Interval Notation and Graphing

The table below shows the connection between inequalities, interval notation, and number line graphs. We have included the graphs with open and closed circles for those of you who have used this type of graph previously. In this book, we will continue to show our graphs using the parentheses/brackets method.

Inequality notation	Interval notation	Graph using parentheses/brackets	Graph using open and closed circles
$x < 2$	$(-\infty, 2)$	0 2	0 2
$x \leq 2$	$(-\infty, 2]$	0 2	0 2
$x \geq -3$	$(-3, \infty)$	−3 0	−3 0
$x > -3$	$(-3, \infty)$	−3 0	−3 0

Recall that the multiplication property for inequalities allows us to multiply both sides of an inequality by any nonzero number we choose. If the number we multiply by happens to be *negative,* then we *must also reverse* the direction of the inequality.

EXAMPLE 2 Find the solution set for $-2y - 3 \leq 7$.

SOLUTION We begin by adding 3 to each side of the inequality:

$$-2y - 3 \leq 7$$

$$-2y \leq 10 \qquad \textbf{Add 3 to both sides}$$

$$-\frac{1}{2}(-2y) \geq -\frac{1}{2}(10) \qquad \textbf{Multiply by } -\tfrac{1}{2} \textbf{ and reverse the direction of the inequality symbol}$$

$$y \geq -5$$

The solution set is all real numbers that are greater than or equal to -5. Below are three equivalent ways to represent this solution set.

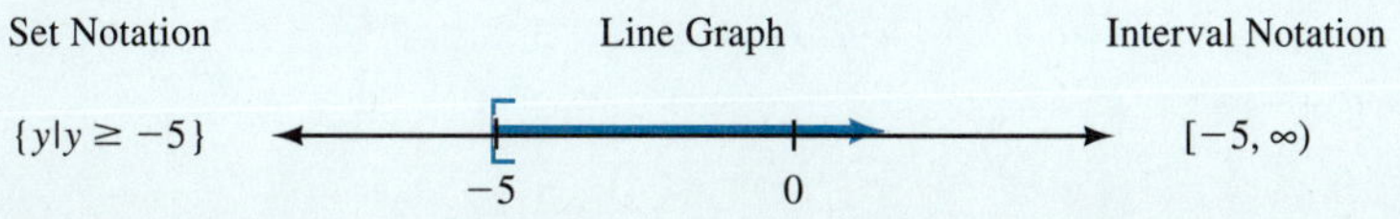

Notice how a bracket is used with interval notation to show that -5 is part of the solution set.

2. Find the solution set for
$-3y - 2 < 7$

Answers

1-6. See Solutions to Selected Practice Problems for this section.

Compound Inequalities

The *union* of two sets A and B is the set of all elements that are in A or in B. The word *or* is the key word in the definition. The *intersection* of two sets A and B is the set of elements contained in both A and B. The key word in this definition is *and*. We can put the words *and* and *or* together with our methods of graphing inequalities to find the solution sets for compound inequalities.

DEFINITION

A **compound inequality** is two or more inequalities connected by the word *and* or *or*.

3. Graph $x < -3$ or $x \geq 2$.

EXAMPLE 3 Graph the solution set for the compound inequality

$$x < -1 \qquad \text{or} \qquad x \geq 3$$

SOLUTION Graphing each inequality separately, we have

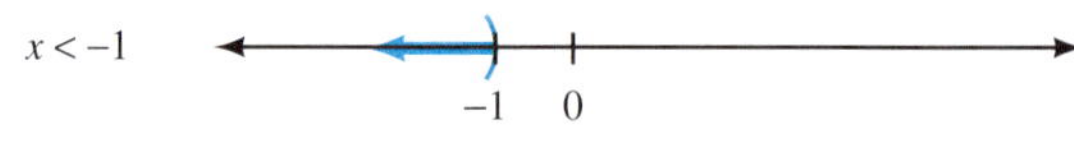

x ≥ 3
0
3

Because the two inequalities are connected by *or*, we want to graph their union; that is, we graph all points that are on either the first graph or the second graph. Essentially, we put the two graphs together on the same number line.

To represent this set of numbers with interval notation we use two intervals connected with the symbol for the union of two sets. Here is the equivalent set of numbers described with interval notation:

$$(-\infty, -1) \cup [3, \infty)$$

4. Graph $x > -3$ and $x < 2$.

EXAMPLE 4 Graph the solution set for the compound inequality

$$x > -2 \qquad \text{and} \qquad x < 3$$

SOLUTION Graphing each inequality separately, we have

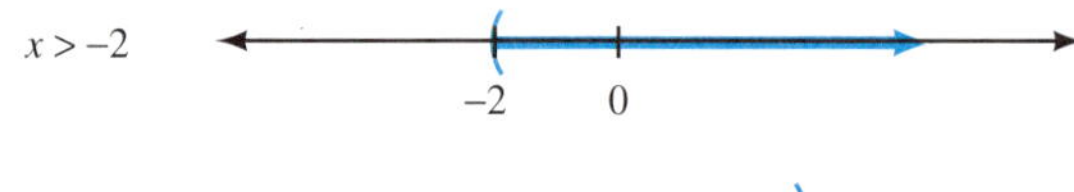

Because the two inequalities are connected by the word *and,* we will graph their intersection, which consists of all points that are common to both graphs; that is, we graph the region where the two graphs overlap.

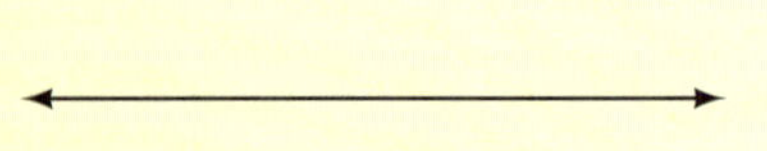

This graph corresponds to the interval $(-2, 3)$ which is called an *open interval* because neither endpoint is included in the interval.

Notation: Sometimes compound inequalities that use the word *and* as the connecting word can be written in a shorter form. For example, the compound inequality $-3 \le x$ and $x \le 4$ can be written $-3 \le x \le 4$. The word *and* does not appear when an inequality is written in this form. It is implied. Inequalities of the form $-3 \le x \le 4$ are called *continued inequalities.* This new notation is useful because it takes fewer symbols to write it. The graph of $-3 \le x \le 4$ is

The corresponding interval is $[-3,4]$, which is a *closed interval* because both endpoints are included in the inteval.

The table that follows shows the connection between number line graphs for a variety of continued inequalities. Again, we have included the graphs with open and closed circles for those of you who have used this type of graph previously. Remember, however, that in this book we will be using the parentheses/brackets method of graphing.

Interval Notation and Graphing

The table below shows the connection between interval notation and number line graphs for a variety of continued inequalities. Again, we have included the graphs with open and closed circles for those of you who have used this type of graph previously.

Inequality notation	Interval notation	Graph using parentheses/brackets	Graph using open and closed circles
$-3 < x < 2$	$(-3, 2)$	−3 2	−3 2
$-3 \le x \le 2$	$[-3, 2]$	−3 2	−3 2
$-3 \le x < 2$	$[-3, 2)$	−3 2	−3 2
$-3 < x \le 2$	$(-3, 2]$	−3 2	−3 2

EXAMPLE 5 Solve and graph $-3 \le 2x - 5 \le 3$.

SOLUTION We can extend our properties for addition and multiplication to cover this situation. If we add a number to the middle expression, we must add the same number to the outside expressions. If we multiply the center expression by a number, we must do the same to the outside expressions, remembering to reverse the direction of the inequality symbols if we multiply by a negative number. We begin by adding 5 to all three parts of the inequality:

$$-3 \le 2x - 5 \le 3$$

$$2 \le 2x \le 8 \quad \textbf{Add 5 to all three members}$$

$$1 \le x \le 4 \quad \textbf{Multiply through by } \tfrac{1}{2}$$

5. Solve and graph the solution.
$-7 \le 2x + 1 \le 7$

Here are three ways to write this solution set:

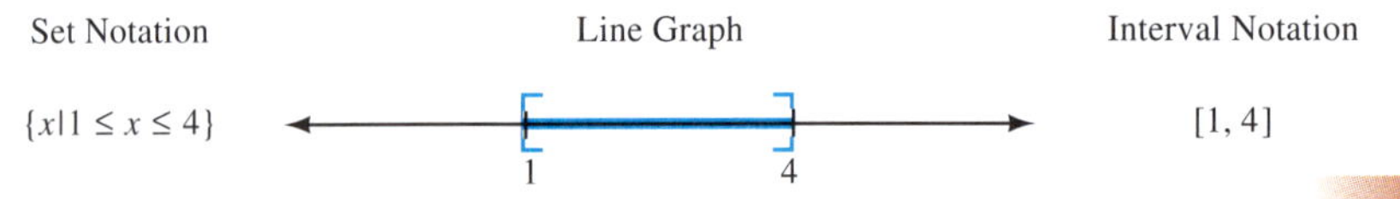

6. Graph the solution for the compound inequality
$3t - 6 \le -3 \quad \text{or} \quad 3t - 6 \ge 3$

EXAMPLE 6 Solve the compound inequality.

$$3t + 7 \le -4 \quad \text{or} \quad 3t + 7 \ge 4$$

SOLUTION We solve each half of the compound inequality separately, then we graph the solution set:

$$\begin{aligned} 3t + 7 &\le -4 & \text{or} \quad 3t + 7 &\ge 4 & \\ 3t &\le -11 & \text{or} \quad 3t &\ge -3 & \textbf{Add } -7 \\ t &\le -\frac{11}{3} & \text{or} \quad t &\ge -1 & \textbf{Multiply by } \tfrac{1}{3} \end{aligned}$$

The solution set can be written in any of the following ways:

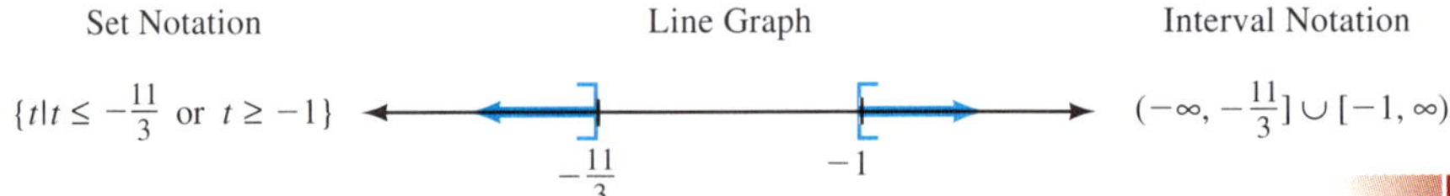

Getting Ready for Class

After reading through the preceding section, respond in your own words and in complete sentences.

A. What is a compound inequality?

B. Explain the shorthand notation that can be used to write two inequalities connected by the word *and*.

C. Write two inequalities connected by the word *and* that together are equivalent to $-1 < x < 2$.

D. Explain in words how you would graph the compound inequality

$$x < 2 \text{ or } x > -3$$

PROBLEM SET 1.6

Solve each of the following inequalities, and graph each solution.

1. $2x \le 3$
2. $5x \ge -115$
3. $\frac{1}{2}x > 2$
4. $\frac{1}{3}x > 4$
5. $-5x \le 25$
6. $-7x \ge 35$
7. $-\frac{3}{2}x > -6$
8. $-\frac{2}{3}x < -8$
9. $-12 \le 2x$
10. $-20 \ge 4x$
11. $-1 \ge -\frac{1}{4}x$
12. $-1 \le -\frac{1}{5}x$

Solve each of the following inequalities, and graph each solution.

13. $-3x + 1 > 10$
14. $-2x - 5 \le 15$
15. $\frac{1}{2} - \frac{m}{12} \le \frac{7}{12}$
16. $\frac{1}{2} - \frac{m}{10} > -\frac{1}{5}$

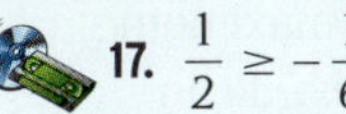

17. $\frac{1}{2} \ge -\frac{1}{6} - \frac{2}{9}x$
18. $\frac{9}{5} > -\frac{1}{5} - \frac{1}{2}x$
19. $-40 \le 30 - 20y$
20. $-20 > 50 - 30y$
21. $\frac{2}{3}x - 3 < 1$
22. $\frac{3}{4}x - 2 > 7$
23. $10 - \frac{1}{2}y \le 36$
24. $8 - \frac{1}{3}y \ge 20$
25. $4 - \frac{1}{2}x < \frac{2}{3}x - 5$
26. $5 - \frac{1}{3}x > \frac{1}{4}x + 2$
27. $0.03x - 0.4 \le 0.08x + 1.2$
28. $2.0 - 0.7x < 1.3 - 0.3x$
29. $3 - \frac{x}{5} < 5 - \frac{x}{4}$
30. $-2 + \frac{x}{3} \ge \frac{x}{2} - 5$

Simplify each side first, then solve the following inequalities. Write your answers with interval notation.

31. $2(3y + 1) \le -10$
32. $3(2y - 4) > 0$
33. $-(a + 1) - 4a \le 2a - 8$
34. $-(a - 2) - 5a \le 3a + 7$
35. $\frac{1}{3}t - \frac{1}{2}(5 - t) < 0$
36. $\frac{1}{4}t - \frac{1}{3}(2t - 5) < 0$
37. $-2 \le 5 - 7(2a + 3)$
38. $1 < 3 - 4(3a - 1)$
39. $-\frac{1}{3}(x + 5) \le -\frac{2}{9}(x - 1)$
40. $-\frac{1}{2}(2x + 1) \le -\frac{3}{8}(x + 2)$
41. $5(x - 2) - 7(x + 1) \le -4x + 3$
42. $-3(1 - 2x) - 3(x - 4) < -3 - 4x$
43. $\frac{2}{3}x - \frac{1}{3}(4x - 5) < 1$
44. $\frac{1}{4}x - \frac{1}{2}(3x + 1) \ge 2$
45. $5(y + 3) + 4 < 6y - 1 - 5y$
46. $4(y - 1) + 2 \ge 3y + 8 - 2y$

Solve the following continued inequalities. Use both a line graph and interval notation to write each solution set.

47. $-2 \le m - 5 \le 2$
48. $-3 \le m + 1 \le 3$

49. $-60 < 20a + 20 < 60$
50. $-60 < 50a - 40 < 60$
51. $0.5 \le 0.3a - 0.7 \le 1.1$
52. $0.1 \le 0.4a + 0.1 \le 0.3$
53. $3 < \frac{1}{2}x + 5 < 6$
54. $5 < \frac{1}{4}x + 1 < 9$
55. $4 < 6 + \frac{2}{3}x < 8$
56. $3 < 7 + \frac{4}{5}x < 15$

Graph the solution sets for the following compound inequalities. Then write each solution set using interval notation.

57. $x + 5 \le -2$ or $x + 5 \ge 2$
58. $3x + 2 < -3$ or $3x + 2 > 3$
59. $5y + 1 \le -4$ or $5y + 1 \ge 4$
60. $7y - 5 \le -2$ or $7y - 5 \ge 2$
61. $2x + 5 < 3x - 1$ or $x - 4 > 2x + 6$
62. $3x - 1 > 2x + 4$ or $5x - 2 < 3x + 4$
63. $3x + 1 < -8$ or $-2x + 1 \le -3$
64. $2x - 5 \le -1$ or $-3x - 6 < -15$
65. $4x - 1 \le 2x - 9$ or $-3x + 1 < 6 - x$
66. $-3x - 7 < -5x - 12$ or $-9x + 2 < x + 12$

Translate each of the following phrases into an equivalent inequality statement.

67. x is greater than -2 and at most 4
68. x is less than 9 and at least -3
69. x is less than -4 or at least 1
70. x is at most 1 or more than 6

Applying the Concepts

Art Supplies A store selling art supplies finds that they can sell x sketch pads each week at a price of p dollars each, according to the formula $x = 900 - 300p$. What price should they charge if they want to sell

71. At least 300 pads each week?
72. More than 600 pads each week?
73. Less than 525 pads each week?
74. At most 375 pads each week?

Temperature Range Each of the following temperature ranges is in degrees Fahrenheit. Use the formula $F = \frac{9}{5}C + 32$ to find the corresponding temperature range in degrees Celsius.

75. 95° to 113°

76. 68° to 86°

77. −13° to 14°

78. −4° to 23°

79. Fuel Efficiency The fuel efficiency (mpg rating) for cars has been increasing steadily since 1980. The formula for a car's fuel efficiency for a given year between 1980 and 1996 is

$$E = 0.36x + 15.9$$

where E is miles per gallon and x is the number of years since January 1, 1980 (U.S. Federal Highway Administration. *Highway Statistics,* annual).

(a) In what years was the average fuel efficiency for cars less than 17 mpg?

(b) In what years was the average fuel efficiency for cars more than 20 mpg?

80. Amtrak The average number of passengers carried by Amtrak declined each year for the years 1990–1996 (American Association of Railroads, Washington, DC, *Railroad Facts, Statistics of Railroads of Class 1*). The linear model for the number of passengers carried each year by Amtrak is given by $P = 22{,}419 - 399x$, where P is the number of passengers in millions, and x is the number of years since January 1, 1990. In what years did Amtrak have more than 20.5 billion (20,500 million) passengers?

81. Insurance Costs The average cost of automobile insurance in New Jersey for a given year, 1995–1997, is given by the model, $I = 56.5x + 740$, where I is the average annual cost of automobile insurance, in dollars, and x is the number of years since January 1, 1990 (National Association of Insurance Commissioners, Kansas City, KS, *State Average Expenditures and Premiums for Personal Automobile Insurance*).

(a) In what year will the average annual cost for auto insurance in New Jersey exceed $1,300?

(b) For the period 1995–1997, the average annual cost for auto insurance in Texas is modeled by $T = 14.5x + 639$, where T is the annual average cost in dollars and x is the number of years since January 1, 1990. In what years will the average annual cost for auto insurance in Texas be less than half of the average annual cost in New Jersey?

82. Health Insurance Income The revenues from health insurance premiums collected by life insurance companies in the United States has been increasing since 1980 (American Council of Life Insurance, *Life Insurance Fact Book,* annual). A model for this growth during 1980–1997 is

$$P = 4x + 23$$

where P is the annual health insurance premiums collected in billions of dollars, and x is the number of years since January 1, 1980. In what years did the insurance companies collect between $60 and $80 billion in health insurance premiums each year?

83. Polling Error The October 24, 1998, issue of *Chicago Tribune* reported the results of a telephone poll of 1,099 registered Illinois voters for the office of governor. This poll indicated that 49% would vote for the Republican, George Ryan, and only 34% would vote for his Democratic opponent, Glen Poshard. The margin of error of this poll was given as ±3%.

(a) Based on the results of this poll, write an inequality for the minimum and maximum percentage of the vote, R, that George Ryan can expect to receive, and an inequality for the minimum and maximum percentage of the vote, P, that Glen Poshard can expect.

(b) Based on the results of this poll, what is the minimum winning margin that George Ryan can expect?

(c) Is George Ryan "guaranteed" to win this election? Explain.

84. Photocopying Money Federal law states that it is legal to photocopy United States money if the photocopy is not color, and the copy is smaller than three fourths of the actual size, or greater than one and one-half times the actual size. A $10 bill measures approximately 6 inches by $2\frac{1}{2}$ inches. What are the maximum and minimum size photo reproductions of this bill that may be legally made from a black and white photocopy machine? Express your answer as an inequality.

85. Sound Frequencies Nearly every animal can emit and hear sound waves. A dog can emit sound waves between 452 and 1,080 vibrations per second, but hear from 15 to 50,000 vibrations per second. Humans can emit sound waves from 85 to 1,100 vibrations per second and hear between 20 and 20,000 vibrations per second.

(a) Write as an inequality the number of vibrations per second, D, that a dog can hear but a human cannot.

(b) Write as an inequality the number of vibrations per second, H, that a human can emit but a dog cannot.

(c) Refer to your answer for part (a). Explain what this means in terms of the ability to hear sound for a dog versus a human.

86. **Cell Phone Cost** Suppose you have a cellular phone for which you have committed to spend less than \$30 on each monthly bill. Your rate plan costs \$20 a month and includes 30 free minutes. Each additional minute costs \$0.40. Set up and solve an inequality to determine how many minutes you may use each month to keep your bill under \$30.

87. **Student Loan** When considering how much debt to incur in student loans, you learn that it is wise to keep your student loan payment to 8% or less of your starting monthly income. Suppose you anticipate a starting annual salary of \$24,000. Set up and solve an inequality that represents the amount of monthly debt for student loans that would be considered manageable.

88. **Student Loan** Suppose you have already borrowed \$30,000 in student loans, which you will pay back at \$368 a month for 10 years. Set up and solve an inequality that determines how much you should make per month as your starting salary to keep your debt at 8% or less of your monthly income.

Review Problems

Factor completely.

89. $6x^4y - 9xy^4 + 18x^3y^3$
90. $x^2y + 3xy + 4x + 12$
91. $x^2 - 5x + 6$
92. $2x^3 + 4x^2 - 30x$
93. $4x^2 - 20x + 25$
94. $2xb^3 - 8x^3b$
95. $x^4 - 16$
96. $8x^3 - 125$
97. $x^2 - 12x + 36 - y^2$
98. $49 - 25x^2$
99. $a^2 + 49$
100. $x^2 - 3x - 3$

One Step Further

Assume that a, b and c are positive and solve each formula for x.

101. $ax + b < c$
102. $\frac{x}{a} + \frac{y}{b} < 1$
103. $-c < ax + b < c$
104. $-1 < \frac{ax + b}{c} < 1$

Assume that $0 < a < b < c < d$ and solve each formula for x.

105. $ax + b < cx + d$
106. $cx + b < ax + d$
107. $-d < \frac{ax + b}{c} < d$
108. $\frac{x}{a} + \frac{1}{b} < \frac{x}{c}$
109. $\frac{x}{a} + \frac{1}{b} < \frac{x}{c} + \frac{1}{d}$
110. $\frac{3}{a} - \frac{x}{b} \leq \frac{5}{c} - \frac{x}{d}$

CHAPTER 1 SUMMARY

Examples

1. Here are some binomials that have been factored this way.

$x^2 + 6x + 9 = (x + 3)^2$

$x^2 - 6x + 9 = (x - 3)^2$

$x^2 - 9 = (x + 3)(x - 3)$

$x^3 - 27 = (x - 3)(x^2 + 3x + 9)$

$x^3 + 27 = (x + 3)(x^2 - 3x + 9)$

2. Factor completely.
   a. $3x^3 - 6x^2 = 3x^2(x - 2)$
   b. $x^2 - 9 = (x + 3)(x - 3)$
      $x^3 - 8 = (x - 2)(x^2 + 2x + 4)$
      $x^3 + 27 = (x + 3)(x^2 - 3x + 9)$
   c. $x^2 - 6x + 9 = (x - 3)^2$
      $6x^2 - 7x - 5 = (2x + 1)(3x - 5)$
   d. $x^2 + ax + bx + ab$
      $= x(x + a) + b(x + a)$
      $= (x + a)(x + b)$

3. Solve: $2(x + 3) = 10$.

$2x + 6 = 10$

$2x + 6 + (\mathbf{-6}) = 10 + (\mathbf{-6})$

$2x = 4$

$\frac{\mathbf{1}}{\mathbf{2}}(2x) = \frac{\mathbf{1}}{\mathbf{2}}(4)$

$x = 2$

4. Solve: $x^2 - 5x = -6$.

$x^2 - 5x + 6 = 0$

$(x - 3)(x - 2) = 0$

$x - 3 = 0$ or $x - 2 = 0$

$x = 3$ or $x = 2$

Special Factoring [1.1]

$a^2 + 2ab + b^2 = (a + b)^2$	**Perfect square trinomials**
$a^2 - 2ab + b^2 = (a - b)^2$	
$a^2 - b^2 = (a - b)(a + b)$	**Difference of two squares**
$a^3 - b^3 = (a - b)(a^2 + ab + b^2)$	**Difference of two cubes**
$a^3 + b^3 = (a + b)(a^2 - ab + b^2)$	**Sum of two cubes**

To Factor Polynomials in General [1.1]

Step 1: If the polynomial has a greatest common factor other than 1, then factor out the greatest common factor.

Step 2: If the polynomial has two terms (it is a binomial), then see if it is the difference of two squares, or the sum or difference of two cubes, and then factor accordingly. Remember, if it is the sum of two squares it will not factor.

Step 3: If the polynomial has three terms (a trinomial), then it is either a perfect square trinomial, which will factor into the square of a binomial, or it is not a perfect square trinomial, in which case you use one of the methods developed in Section 5.5.

Step 4: If the polynomial has more than three terms, then try to factor it by grouping.

Step 5: As a final check, see if any of the factors you have written can be factored further. If you have overlooked a common factor, you can catch it here.

Strategy for Solving Linear Equations in One Variable [1.2]

Step 1a: Use the distributive property to separate terms, if necessary.

1b: If fractions are present, consider multiplying both sides by the LCD to eliminate the fractions. If decimals are present, consider multiplying both sides by a power of 10 to clear the equation of decimals.

1c: Combine similar terms on each side of the equation.

Step 2: Use the addition property of equality to get all variable terms on one side of the equation and all constant terms on the other side. A variable term is a term that contains the variable (for example, $5x$). A constant term is a term that does not contain the variable (the number 3, for example.)

Step 3: Use the multiplication property of equality to get x (that is, $1x$) by itself on one side of the equation.

Step 4: Check your solution in the original equation to be sure that you have not made a mistake in the solution process.

To Solve a Quadratic Equation by Factoring [1.2]

Step 1: Write the equation in standard form.

Step 2: Factor the left side.

Step 3: Use the zero-factor property to set each factor equal to 0.

Step 4: Solve the resulting linear equations.

Step 5: Check the solutions in the original equation.

Absolute Value Equations [1.3]

To solve an equation that involves absolute value, we isolate the absolute value on one side of the equation and then rewrite the absolute value equation as two separate equations that do not involve absolute value. In general, if b is a positive real number, then

$$|a| = b \quad \text{is equivalent to} \quad a = b \quad \text{or} \quad a = -b$$

5. To solve

$$|2x - 1| + 2 = 7$$

we first isolate the absolute value on the left side by adding -2 to each side to obtain

$$|2x - 1| = 5$$

$$\begin{aligned} 2x - 1 &= 5 & \text{or} && 2x - 1 &= -5 \\ 2x &= 6 & \text{or} && 2x &= -4 \\ x &= 3 & \text{or} && x &= -2 \end{aligned}$$

Formulas [1.4]

A *formula* in algebra is an equation involving more than one variable. To solve a formula for one of its variables, simply isolate that variable on one side of the equation.

6. Solve for w:

$$\begin{aligned} P &= 2l + 2w \\ P - 2l &= 2w \\ \frac{P - 2l}{2} &= w \end{aligned}$$

Blueprint for Problem Solving [1.5]

Step 1: ***Read*** the problem, and then mentally ***list*** the items that are known and the items that are unknown.

Step 2: ***Assign a variable*** to one of the unknown items. (In most cases this will amount to letting x = the item that is asked for in the problem.) Then ***translate*** the other ***information*** in the problem to expressions involving the variable.

Step 3: ***Reread*** the problem, and then ***write an equation,*** using the items and variables listed in steps 1 and 2, that describes the situation.

Step 4: ***Solve the equation*** found in step 3.

Step 5: ***Write*** your ***answer*** using a complete sentence.

Step 6: ***Reread*** the problem, and ***check*** your solution with the original words in the problem.

7. The perimeter of a rectangle is 32 inches. If the length is 3 times the width, find the dimensions.

Step 1: This step is done mentally.

Step 2: Let x = the width. Then the length is $3x$.

Step 3: The perimeter is 32; therefore

$$2x + 2(3x) = 32$$

Step 4:

$$\begin{aligned} 8x &= 32 \\ x &= 4 \end{aligned}$$

Step 5: The width is 4 inches. The length is $3(4) = 12$ inches.

Step 6: The perimeter is $2(4) + 2(12)$, which is 32. The length is 3 times the width.

Compound Inequalities [1.6]

Two inequalities connected by the word *and* or *or* form a compound inequality. If the connecting word is *or,* we graph all points that are on either graph. If the connecting word is *and,* we graph only those points that are common to both graphs. The inequality $-2 \le x \le 3$ is equivalent to the compound inequality $-2 \le x$ and $x \le 3$.

8. $x < -3$ or $x > 1$

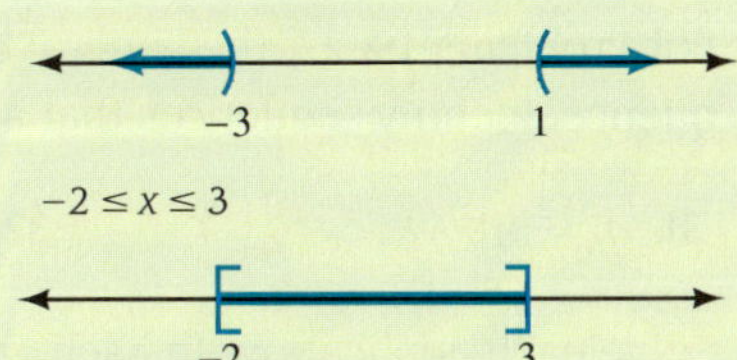

$-2 \le x \le 3$

CHAPTER 1 REVIEW

The problems below form a comprehensive review of the material in this chapter. They can be used to study for exams. If you would like to take a practice test on this chapter, you can use the odd-numbered problems. Give yourself an hour and work as many of the odd-numbered problems as possible. When you are finished, or when an hour has passed, check your answers with the answers in the back of the book. You can use the even-numbered problems for a second practice test.

Factor completely. [1.1]

1. $x^4 - 16$

2. $3a^4 + 18a^2 + 27$

3. $a^3 - 8$

4. $5x^3 + 30x^2y + 45xy^2$

5. $3a^3b - 27ab^3$

6. $x^2 - 10x + 25 - y^2$

7. $36 - 25a^2$

8. $x^3 + 4x^2 - 9x - 36$

Solve each equation [1.2]

9. $4x - 2 = 7x + 7$

10. $\frac{3y}{4} - \frac{1}{2} + \frac{3y}{2} = 2 - y$

11. $8 - 3(2t + 1) = 5(t + 2)$

12. $6 + 4(1 - 3t) = -3(t - 4) + 2$

13. $2x^2 - 5x = 12$

14. $81a^2 = 1$

15. $(x - 2)(x - 3) = 2$

16. $9x^3 + 18x^2 - 4x - 8 = 0$

Solve each equation. [1.3]

17. $|x - 3| = 1$

18. $|x - 2| = 3$

19. $|2y - 3| = 5$

20. $|3y - 2| = 7$

21. $|4x - 3| + 2 = 11$

22. $|6x - 2| + 4 = 16$

23. $\left|\frac{7}{3} - \frac{x}{3}\right| + \frac{4}{3} = 2$

24. $\left|\frac{1}{2} - x\right| = \left|x + \frac{1}{2}\right|$

Substitute the given values in each formula and then solve for the variable that does not have a numerical replacement. [1.4]

25. $P = 2b + 2h$: $P = 40$, $b = 3$

26. $A = P + Prt$: $A = 2{,}000$, $P = 1{,}000$, $r = 0.05$

Solve each formula for the indicated variable. [1.4]

27. $I = prt$ for p

28. $y = mx + b$ for x

29. $4x - 3y = 12$ for y

30. $d = vt + 16t^2$ for v

For Problems 19–22, the general term of a sequence is given. Use the general term to find the first four terms of each sequence. Identify sequences that are increasing or decreasing. [1.4]

31. $a_n = 3n$

32. $a_n = n + 3$

33. $a_n = (-2)^n$

34. $a_n = \frac{1}{n}$

Solve each application. In each case, be sure to show the equation that describes the situation. [1.5]

35. Geometry The length of a rectangle is 3 times the width. The perimeter is 32 feet. Find the length and width.

36. Geometry The three sides of a triangle are given by three consecutive integers. If the perimeter is 12 meters, find the length of each side.

37. Brick Laying The formula $N = 7 \cdot L \cdot H$ gives the number of standard bricks needed in a wall L feet long and H feet high and is called the *brick-layer's formula.*

(a) How many bricks would be required in a wall 45 fcct long by 12 feet high?

(b) An 8-foot-high wall is to be built from a load of 35,000 bricks. What length wall can be built?

38. Salary A teacher has a salary of $25,920 for her second year on the job. If this is 4.2% more than her first-year salary, how much did she earn her first year?

Solve each inequality. Write your answer using interval notation. [1.6]

39. $-8a > -4$

40. $6 - a \geq -2$

41. $\frac{3}{4}x + 1 \leq 10$

42. $800 - 200x < 1{,}000$

43. $\frac{1}{3} \leq \frac{1}{6}x \leq 1$

44. $-0.01 \leq 0.02x - 0.01 \leq 0.01$

45. $5t + 1 \leq 3t - 2$ or $-7t \leq -21$

46. $3(x + 1) < 2(x + 2)$ or $2(x - 1) \geq x + 2$

Equations and Inequalities in Two Variables

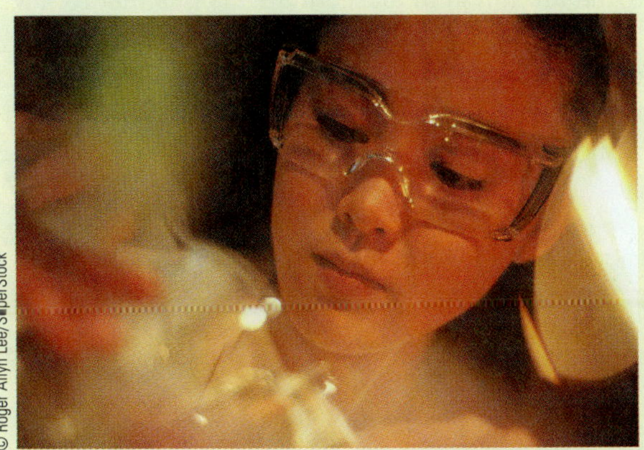

A student is heating water in a chemistry lab. As the water heats, she records the temperature readings from two thermometers, one giving temperature in degrees Fahrenheit and the other in degrees Celsius. The table below shows some of the data she collects. Figure 1 is a scatter diagram that gives a visual representation of the data in the table.

CORRESPONDING TEMPERATURES

IN DEGREES FAHRENHEIT	IN DEGREES CELSIUS
77	25
95	35
167	75
212	100

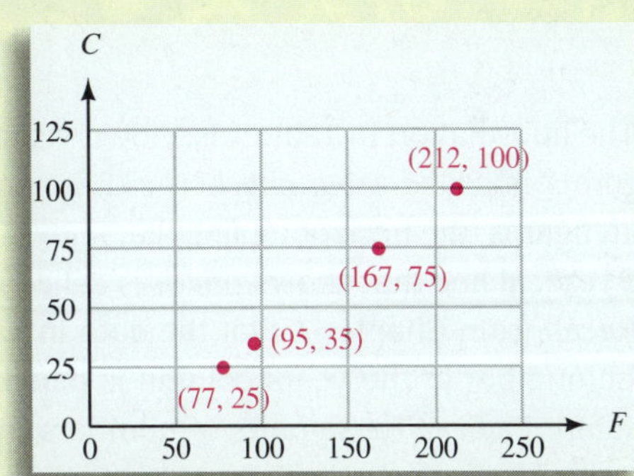

Figure 1

The exact relationship between the Fahrenheit and Celsius temperature scales is given by the formula

$$C = \frac{5}{9}(F - 32)$$

We have three ways to describe the relationship between the two temperature scales: a table, a graph, and an equation. But, most important to us, we don't need to accept this formula on faith, because the material we will present in this chapter gives us the ability to derive the formula from the data in the table above.

CHAPTER OUTLINE

2.1 Paired Data and the Rectangular Coordinate System

In this section, we begin our work with the visual component of algebra. We are setting the stage for two important topics: *functions* and *graphs*. Both topics have a wide variety of applications and are found in all the math classes that follow this one.

Table 1 gives the net price of a popular intermediate algebra text at the beginning of each year in which a new edition was published. (The net price is the price the bookstore pays for the book, not the price you pay for it.)

Table 1

PRICE OF A TEXTBOOK

EDITION	YEAR PUBLISHED	NET PRICE ($)
First	1982	16.95
Second	1985	23.50
Third	1989	30.50
Fourth	1993	39.25
Fifth	1997	47.50
Sixth	2001	55.00

The information in Table 1 is shown visually in Figures 1 and 2. The diagram in Figure 1 is called a *bar chart*. The diagram in Figure 2 is called a *line graph*. In both figures, the horizontal line that shows years is called the *horizontal axis*, and the vertical line that shows prices is called the *vertical axis*.

Recall from Chapter 1 that the data in Table 1 are called *paired data* because each number in the years column is paired with a specific number in the price column. Each of these pairs of numbers from Table 1 is associated with one of the solid bars in the bar chart (Figure 1) and one of the dots in the line graph (Figure 2).

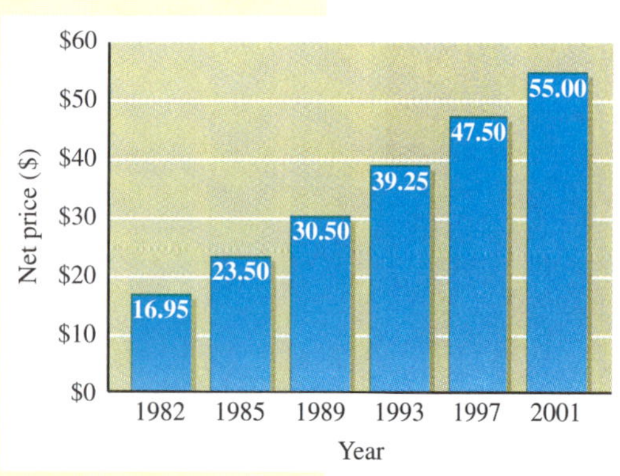

Figure 1

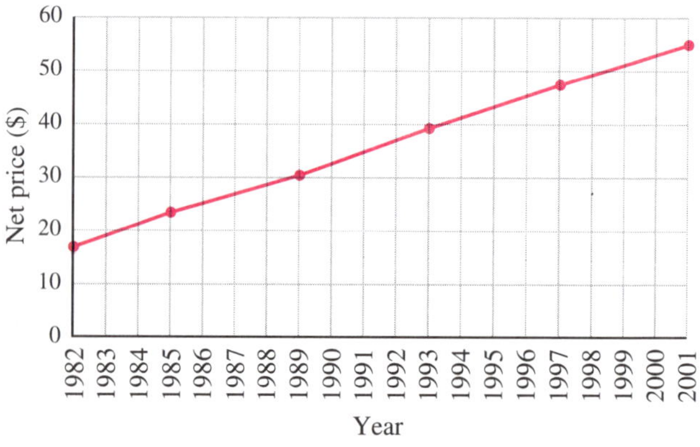

Figure 2

USING TECHNOLOGY

Spreadsheet Programs

When I put together the manuscript for this book, I used a spreadsheet program to draw the bar chart and line graph shown in Figures 1 and 2.

Figure 3 shows what the screen of my computer looked like when I was preparing Figure 1. The bar chart was drawn by the computer from the data in the table. This was just one of many ways I could have chosen to display the data.

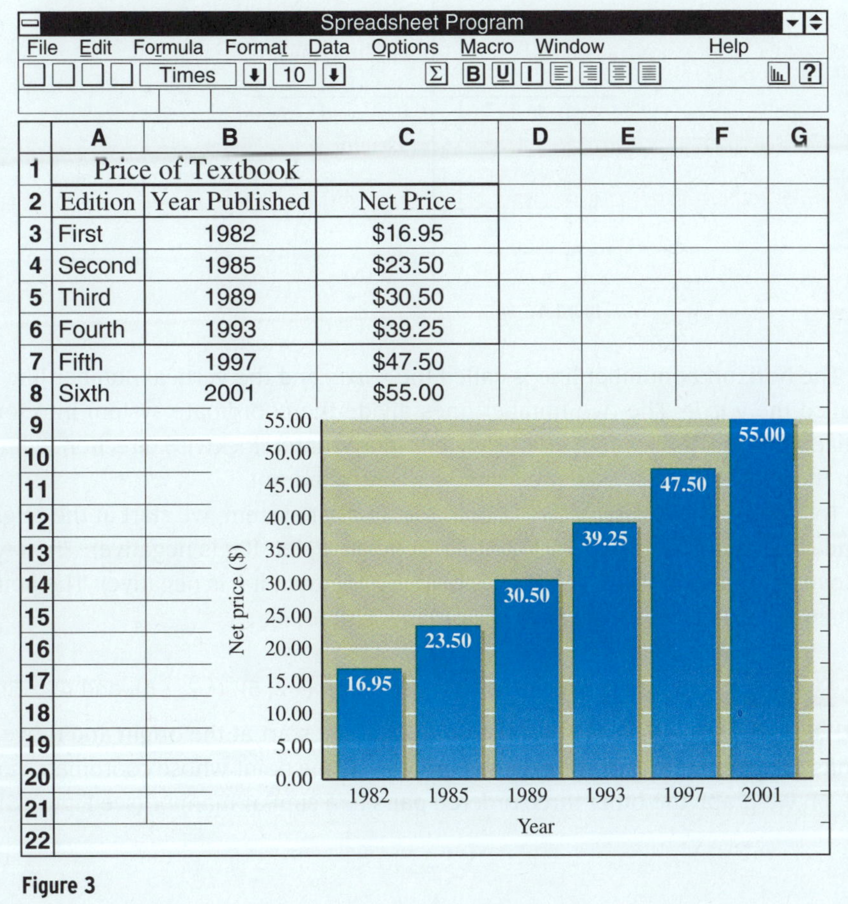

	A	B	C
1	Price of Textbook		
2	Edition	Year Published	Net Price
3	First	1982	$16.95
4	Second	1985	$23.50
5	Third	1989	$30.50
6	Fourth	1993	$39.25
7	Fifth	1997	$47.50
8	Sixth	2001	$55.00

Figure 3

Ordered Pairs

Paired data play an important role in equations that contain two variables. Working with these equations is easier if we standardize the terminology and notation associated with paired data. So here is a definition that will do just that.

DEFINITION

A pair of numbers enclosed in parentheses and separated by a comma, such as $(-2, 1)$, is called an **ordered pair** of numbers. The first number in the pair is called the ***x*-coordinate** of the ordered pair; the second number is called the ***y*-coordinate.** For the ordered pair $(-2, 1)$, the x-coordinate is -2 and the y-coordinate is 1.

To standardize the way in which we display paired data visually, we use a rectangular coordinate system. A *rectangular coordinate system* is made by drawing two real number lines at right angles to each other. The two number lines, called *axes,* cross each other at 0. This point is called the *origin.* Positive directions are to the right and up. Negative directions are down and to the left. The rectangular coordinate system is shown in Figure 4.

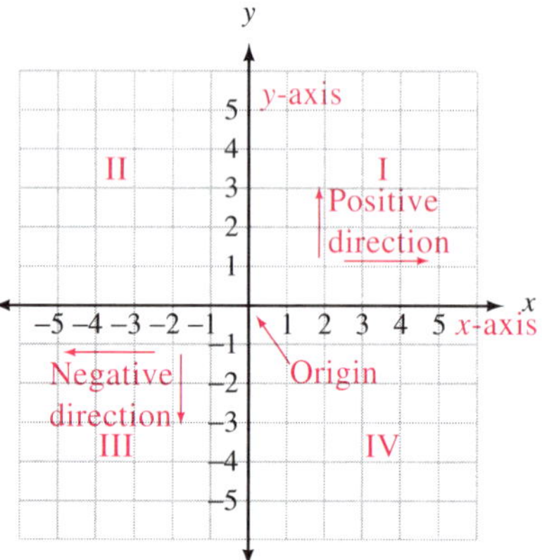

Figure 4

Note: A rectangular coordinate system allows us to connect algebra and geometry by associating geometric shapes (the curves shown in the diagrams) with algebraic equations. The French philosopher and mathematician René Descartes (1596–1650) usually is credited with the invention of the rectangular coordinate system, which often is referred to as the *Cartesian coordinate system* in his honor. As a philosopher, Descartes is responsible for the statement, "I think, therefore, I am." Until Descartes invented his coordinate system in 1637, algebra and geometry were treated as separate subjects.

The horizontal number line is called the *x-axis* and the vertical number line is called the *y-axis.* The two number lines divide the coordinate system into four quadrants, which we number I through IV in a counterclockwise direction. Points on the axes are not considered as being in any quadrant.

To graph the ordered pair (a, b) on a rectangular system, we start at the origin and move a units right or left (right if a is positive, left if a is negative). Then we move b units up or down (up if b is positive and down if b is negative). The point where we end up is the graph of the ordered pair (a, b).

Practice Problems

1. Plot the ordered pairs (1, 3), (−1, 3), (−1, −3), and (1, −3).

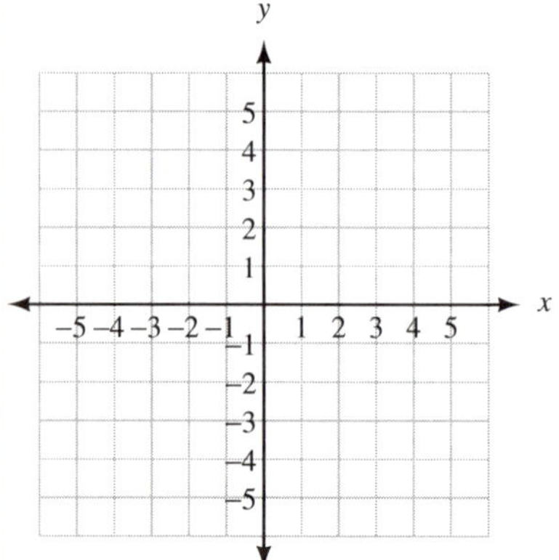

EXAMPLE 1 Plot (graph) the ordered pairs (2, 5), (−2, 5), (−2, −5), and (2, −5).

SOLUTION To graph the ordered pair (2, 5), we start at the origin and move 2 units to the right, then 5 units up. We are now at the point whose coordinates are (2, 5). We graph the other three ordered pairs in a similar manner (see Figure 5).

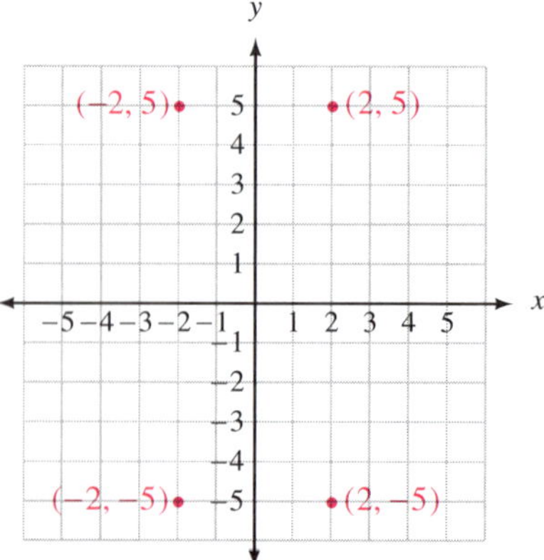

Figure 5

Note: From Example 1 we see that any point in quadrant I has both its *x*- and *y*-coordinates positive (+, +). Points in quadrant II have negative *x*-coordinates and positive *y*-coordinates (−, +). In quadrant III both coordinates are negative (−, −). In quadrant IV the form is (+, −).

Answers

See Solutions to Selected Practice Problems for all answers in this section.

EXAMPLE 2 Graph the ordered pairs $(1, -3)$, $(\frac{1}{2}, 2)$, $(3, 0)$, $(0, -2)$, $(-1, 0)$, and $(0, 5)$.

SOLUTION

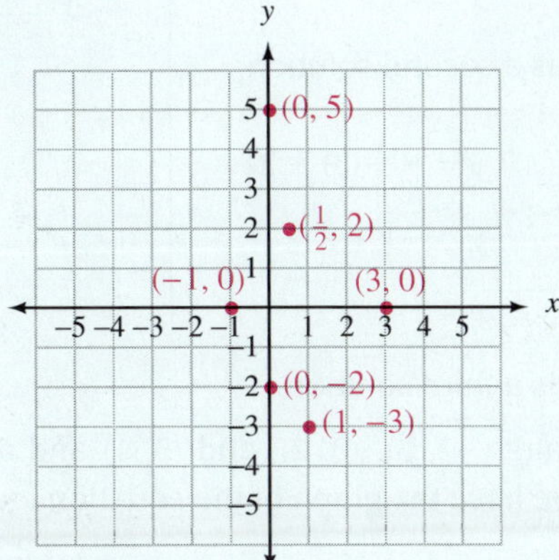

Figure 6

From Figure 6 we see that any point on the x-axis has a y-coordinate of 0 (it has no vertical displacement), and any point on the y-axis has an x-coordinate of 0 (no horizontal displacement).

2. Plot the ordered pairs $(4, -2)$, $(-\frac{1}{2}, 3)$, $(1, 0)$, $(0, -5)$, $(-6, 0)$, and $(0, 4)$.

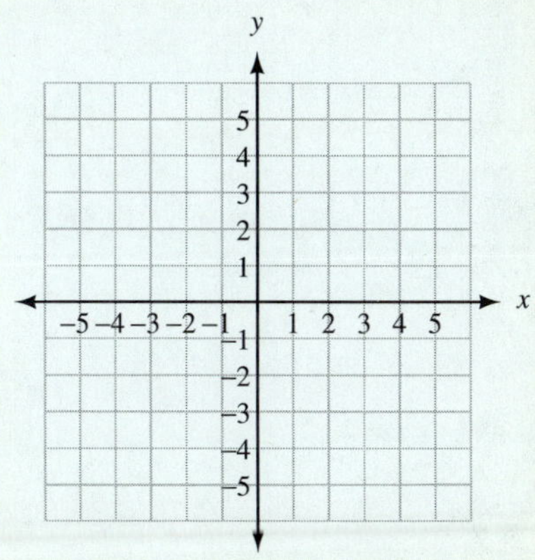

Linear Equations

We can plot a single point from an ordered pair, but to draw a line, we need two points or an equation in two variables.

> **DEFINITION**
>
> Any equation that can be put in the form $ax + by = c$, where a, b, and c are real numbers and a and b are not both 0, is called a **linear equation in two variables.** The graph of any equation of this form is a straight line (that is why these equations are called "linear"). The form $ax + by = c$ is called **standard form.**

To graph a linear equation in two variables, we simply graph its solution set; that is, we draw a line through all the points whose coordinates satisfy the equation.

EXAMPLE 3 Graph the equation $y = -\frac{1}{3}x + 2$.

SOLUTION We need to find three ordered pairs that satisfy the equation. To do so, we can let x equal any numbers we choose and find corresponding values of y. But since every value of x we substitute into the equation is going to be multiplied by $-\frac{1}{3}$, let's use numbers for x that are divisible by 3, like -3, 0, and 3. That way, when we multiply them by $-\frac{1}{3}$, the result will be an integer.

Let $x = -3$; $\quad y = -\dfrac{1}{3}(-3) + 2$

$$y = 1 + 2$$

$$y = 3$$

The ordered pair $(-3, 3)$ is one solution.

3. Graph $y = \frac{1}{2}x + 3$.

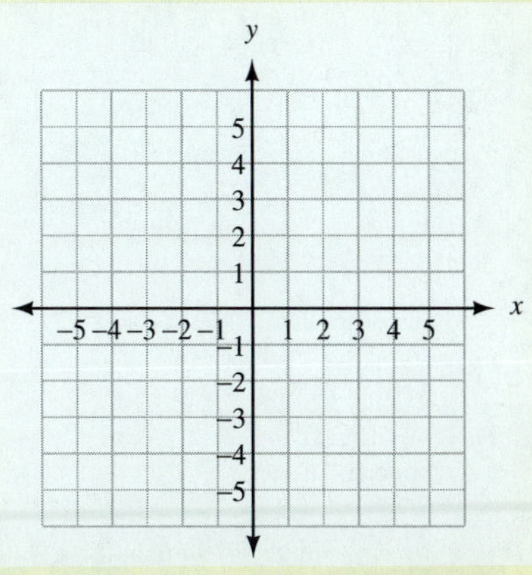

Let $x = 0$; $\quad y = -\frac{1}{3}(0) + 2$

$$y = 0 + 2$$

$$y = 2$$

The ordered pair (0, 2) is a second solution.

Let $x = 3$; $\quad y = -\frac{1}{3}(3) + 2$

$$y = -1 + 2$$

$$y = 1$$

The ordered pair (3, 1) is a third solution.

In table form

x	y
−3	3
0	2
3	1

Plotting the ordered pairs (−3, 3), (0, 2), and (3, 1) and drawing a straight line through their graphs, we have the graph of the equation $y = -\frac{1}{3}x + 2$, as shown in Figure 7.

Note: It takes only two points to determine a straight line. We have included a third point for "insurance." If all three points do not line up in a straight line, we have made a mistake.

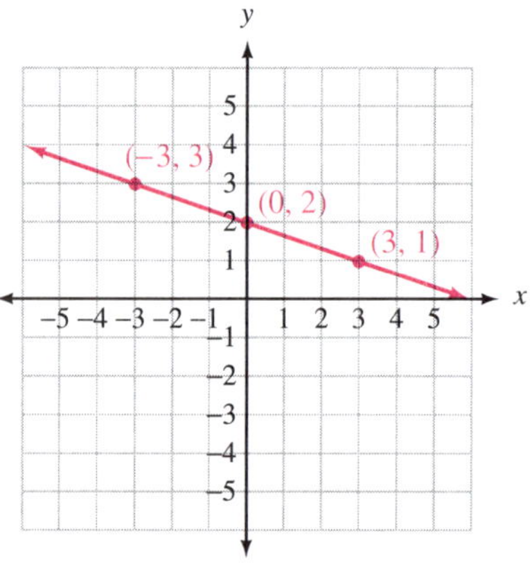

Figure 7

Example 3 illustrates again the connection between algebra and geometry that we mentioned earlier in this section. Descartes' rectangular coordinate system allows us to associate the equation $y = -\frac{1}{3}x + 2$ (an algebraic concept) with a specific straight line (a geometric concept). The study of the relationship between equations in algebra and their associated geometric figures is called *analytic geometry*.

Intercepts

Two important points on the graph of a straight line, if they exist, are the points where the graph crosses the axes.

DEFINITION

The **x-intercept** of the graph of an equation is the x-coordinate of the point where the graph crosses the x-axis. The **y-intercept** is defined similarly.

Because any point on the x-axis has a y-coordinate of 0, we can find the x-intercept by letting $y = 0$ and solving the equation for x. We find the y-intercept by letting $x = 0$ and solving for y.

EXAMPLE 4 Find the x- and y-intercepts for $2x + 3y = 6$; then graph the solution set.

SOLUTION To find the y-intercept we let $x = 0$.

When $x = 0$
we have $2(0) + 3y = 6$
$3y = 6$
$y = 2$

The y-intercept is 2, and the graph crosses the y-axis at the point (0, 2).

When $y = 0$
we have $2x + 3(0) = 6$
$2x = 6$
$x = 3$

The x-intercept is 3, so the graph crosses the x-axis at the point (3, 0). We use these results to graph the solution set for $2x + 3y = 6$. The graph is shown in Figure 8.

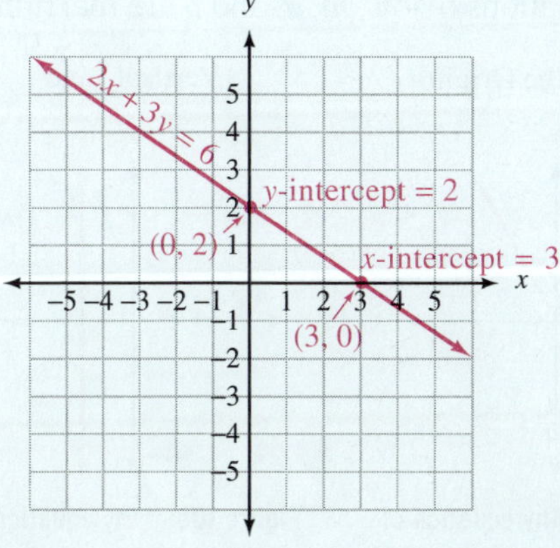

Figure 8

EXAMPLE 5 Graph each of the following lines.

(a) $y = \frac{1}{2}x$ **(b)** $x = 3$ **(c)** $y = -2$

SOLUTION

(a) The line $y = \frac{1}{2}x$ passes through the origin because (0, 0) satisfies the equation. To sketch the graph we need at least one more point on the line. When x is 2, we obtain the point (2, 1), and when x is -4, we obtain the point $(-4, -2)$. The graph of $y = \frac{1}{2}x$ is shown in Figure 9a.

(b) The line $x = 3$ is the set of all points whose x-coordinate is 3. The variable y does not appear in the equation, so the y-coordinate can be any number. Note that we can write our equation as a linear equation in two variables by writing it as $x + 0y = 3$. Because the product of 0 and y will always be 0, y can be any number. The graph of $x = 3$ is the vertical line shown in Figure 9b.

(c) The line $y = -2$ is the set of all points whose y-coordinate is -2. The variable x does not appear in the equation, so the x-coordinate can be any number. Again, we can write our equation as a linear equation in two variables by writing it as $0x + y = -2$. Because the product of 0 and x will always be 0, x can be any number. The graph of $y = -2$ is the horizontal line shown in Figure 9c.

4. Find the x- and y-intercepts for $2x - 3y = 6$, then graph the solution set.

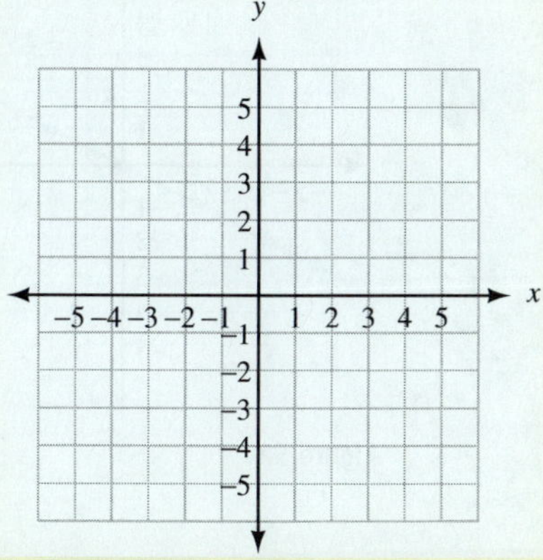

Note: Graphing straight lines by finding the intercepts works best when the coefficients of x and y are factors of the constant term.

5. Graph the line $x = -1$ and the line $y = 4$.

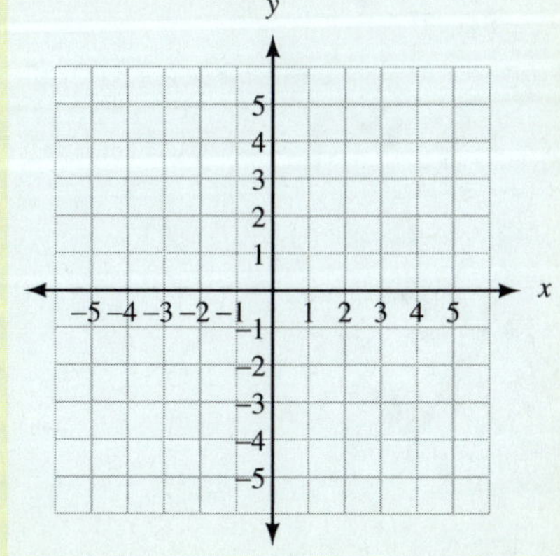

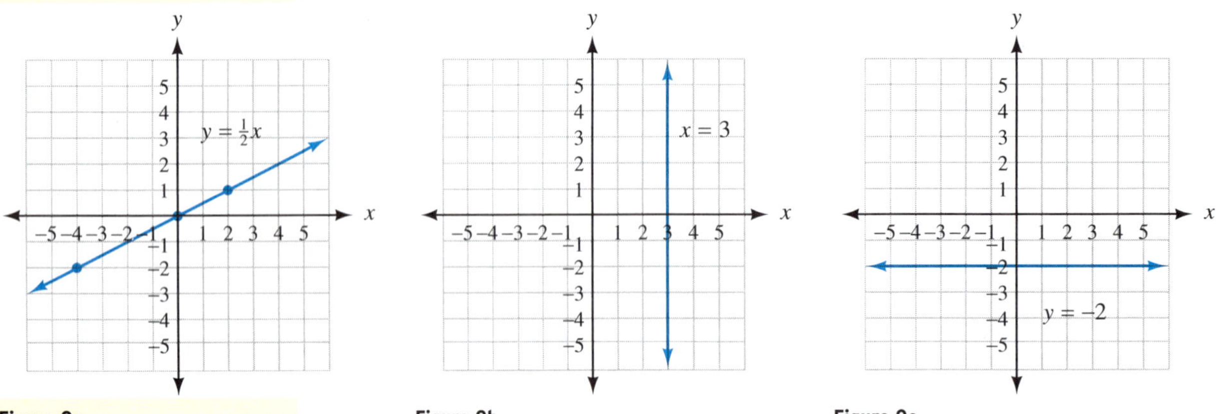

Figure 9a Figure 9b Figure 9c

Facts From Geometry: Special Equations and Their Graphs

For the equations below, m, a, and b are real numbers.

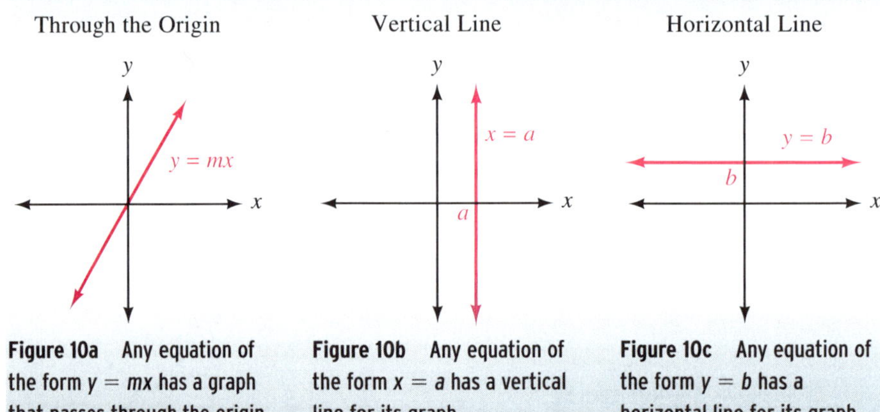

Figure 10a Any equation of the form $y = mx$ has a graph that passes through the origin.

Figure 10b Any equation of the form $x = a$ has a vertical line for its graph.

Figure 10c Any equation of the form $y = b$ has a horizontal line for its graph.

USING TECHNOLOGY

Graphing Calculators and Computer Graphing Programs

A variety of computer programs and graphing calculators are currently available to help us graph equations and then obtain information from those graphs much faster than we could with paper and pencil. We will not give instructions for all the available calculators. Most of the instructions we give are generic in form. You will have to use the manual that came with your calculator to find the specific instructions for your calculator.

Graphing with Trace and Zoom

All graphing calculators have the ability to graph a function and then trace over the points on the graph, giving their coordinates. Furthermore, all graphing calculators can zoom in and out on a graph that has been drawn. To graph a linear equation on a graphing calculator, we first set the graph window. Most calculators call the smallest value of x Xmin and the largest value of x Xmax. The counterpart values of y are Ymin and Ymax. We will use the notation

Window: X from -5 to 4, Y from -3 to 2

to stand for a window in which

Xmin $= -5$ Ymin $= -3$
Xmax $= 4$ Ymax $= 2$

Set your calculator with the following window:

Window: X from −10 to 10, Y from −10 to 10

Graph the equation Y = −X + 8. On the TI-82/83, you use the [Y=] key to enter the equation; you enter a negative sign with the [(−)] key, and a subtraction sign with the [−] key. The graph will be similar to the one shown in Figure 11.

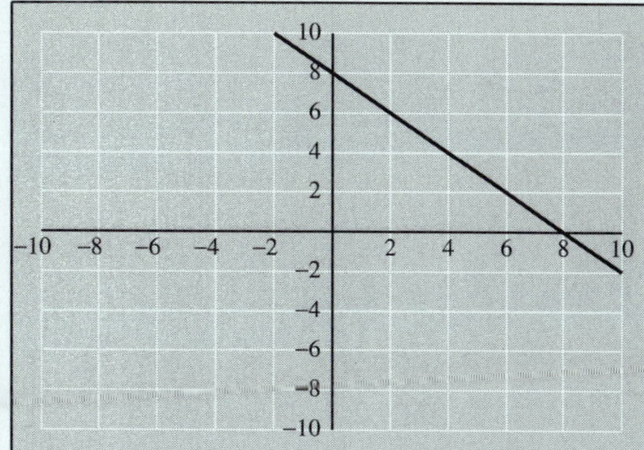

Figure 11

Use the Trace feature of your calculator to name three points on the graph. Next, use the Zoom feature of your calculator to zoom out so your window is twice as large.

Solving for *y* First

To graph the equation from Example 4, $2x + 3y = 6$, on a graphing calculator, you must first solve it for y. When you do so, you will get $y = -\frac{2}{3}x + 2$, which you enter into your calculator as Y = −(2/3)X + 2. Graph this equation in the window described here, and compare your results with the graph in Figure 8.

Window: X from −6 to 6, Y from −6 to 6

Hint on Tracing

If you are going to use the Trace feature and you want the x-coordinates to be exact numbers, set your window so that the range of X inputs is a multiple of the number of horizontal pixels on your calculator screen. On the TI-82/83, the screen has 94 pixels across. Here are a few convenient trace windows:

X from −4.7 to 4.7	To trace to the nearest tenth
X from −47 to 47	To trace to the nearest integer
X from 0 to 9.4	To trace to the nearest tenth
X from 0 to 94	To trace to the nearest integer
X from −94 to 94	To trace to the nearest even integer

Getting Ready for Class

After reading through the preceding section, respond in your own words and in complete sentences.

A. Explain how you would construct a rectangular coordinate system from two real number lines.

B. Explain in words how you would graph the ordered pair (2, −3).

C. How can you tell if an ordered pair is a solution to the equation $y = 2x - 5$?

D. If you were looking for solutions to the equation $y = \frac{1}{3}x + 5$, why would it be easier to substitute 6 for x than to substitute 5 for x?

PROBLEM SET 2.1

Graph each of the following ordered pairs on a rectangular coordinate system.

1. **(a)** $(1, 2)$
(b) $(-1, -2)$
(c) $(5, 0)$
(d) $(0, 2)$
(e) $(-5, -5)$
(f) $\left(\frac{1}{2}, 2\right)$

2. **(a)** $(-1, 2)$
(b) $(1, -2)$
(c) $(0, -3)$
(d) $(4, 0)$
(e) $(-4, -1)$
(f) $\left(3, \frac{1}{4}\right)$

Give the coordinates of each point.

3.

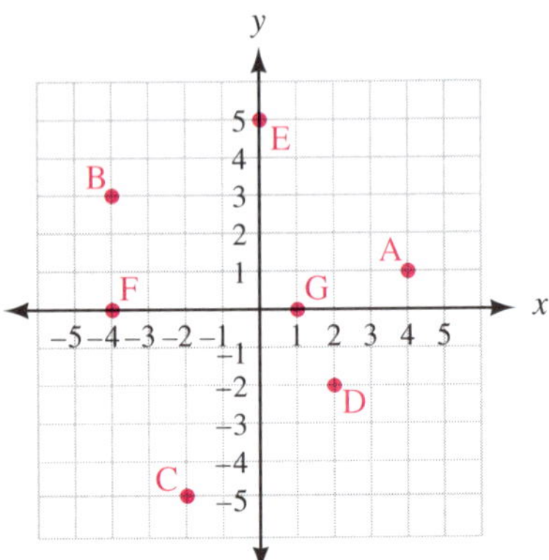

4.

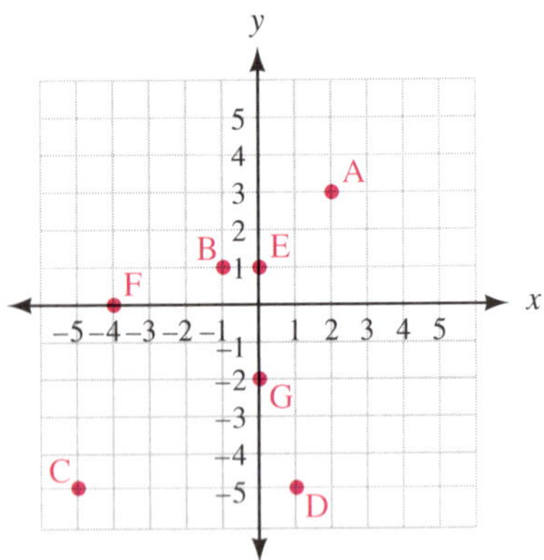

Graph each of the following linear equations by first finding the intercepts.

5. $2x - 3y = 6$
6. $y - 2x = 4$
7. $4x - 5y = 20$
8. $5x - 3y - 15 = 0$
9. $y = 2x + 3$
10. $y = 3x - 2$
11. $-3x + 2y = 12$
12. $5x - 7y = -35$
13. $6x - 5y - 20 = 0$
14. $-4x - 6y + 15 = 0$
15. $y = 3x \quad 5$
16. $y = -4x + 1$
17. $\frac{x}{2} + \frac{y}{3} = 1$
18. $\frac{x}{4} - \frac{y}{7} = 1$

19. Which of the following tables could be produced from the equation $y = 2x - 6$?

(a)

x	y
0	6
1	4
2	2
3	0

(b)

x	y
0	−6
1	−4
2	−2
3	0

(c)

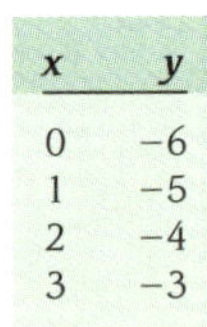

x	y
0	−6
1	−5
2	−4
3	−3

20. Which of the following tables could be produced from the equation $3x - 5y = 15$?

(a)

x	y
0	5
−3	0
10	3

(b)

x	y
0	−3
5	0
10	3

(c)

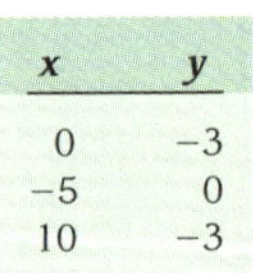

x	y
0	−3
−5	0
10	−3

Graph each of the following straight lines.

21. $y = \frac{1}{3}x$
22. $y = \frac{1}{2}x$
23. $-2x + y = -3$
24. $-3x + y = -2$
25. $5x - 2y = 5$
26. $2x - 3y = 3$
27. $y = -\frac{1}{2}x$
28. $y = \frac{4}{3}x$
29. $y = \frac{2}{3}x - 4$
30. $y = -\frac{3}{4}x + 5$
31. $y = \frac{2}{3}x + \frac{2}{3}$
32. $y = -\frac{3}{4}x + \frac{3}{2}$
33. $-3x + 4y = 6$
34. $2x - 3y = 6$
35. $\frac{x}{-3} + \frac{y}{2} = 1$
36. $\frac{x}{6} - \frac{y}{2} = 1$

37. The graph shown here is the graph of which of the following equations?
(a) $3x - 2y = 6$
(b) $2x - 3y = 6$
(c) $2x + 3y = 6$

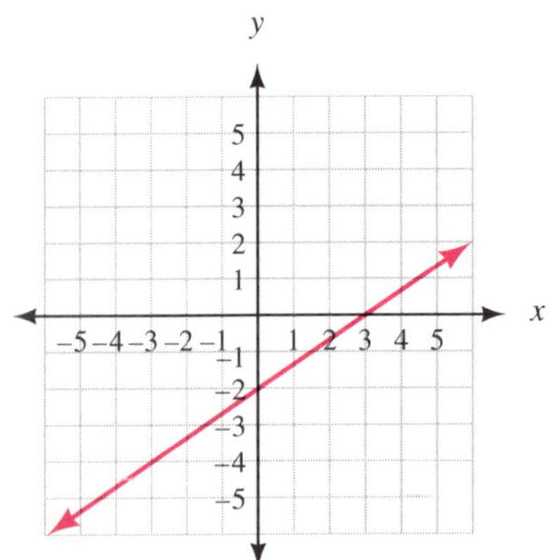

38. The graph shown here is the graph of which of the following equations?

(a) $3x - 2y = 8$

(b) $2x - 3y = 8$

(c) $2x + 3y = 8$

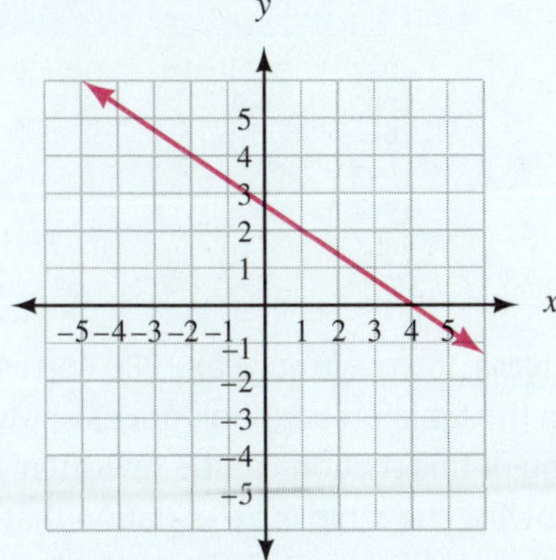

39. Graph each of the following lines.

(a) $y = \frac{1}{3}x$

(b) $x = 2$

(c) $y = 3$

40. Graph each of the following lines.

(a) $y = \frac{1}{2}x$

(b) $x = 5$

(c) $y = -2$

41. Graph each of the following lines.

(a) $y = 2x$

(b) $x = -3$

(c) $y = 2$

42. Graph each of the following lines.

(a) $y = 3x$

(b) $x = -2$

(c) $y = 4$

43. Graph each of the following lines.

(a) $y = -\frac{1}{2}x$

(b) $x = 4$

(c) $y = -3$

44. Graph each of the following lines.

(a) $y = -\frac{1}{3}x$

(b) $x = 1$

(c) $y = -5$

45. The ordered pairs that satisfy the equation $y = 3x$ all have the form $(x, 3x)$ because y is always 3 times x. Graph all ordered pairs of the form $(x, 3x)$.

46. Graph all ordered pairs of the form $(x, -3x)$.

47. Replace each question mark with the symbol > or <.

(a) In the first quadrant, x ? 0, and y ? 0.

(b) In the second quadrant, x ? 0, and y ? 0.

(c) In the third quadrant, x ? 0, and y ? 0.

(d) In the fourth quadrant, x ? 0, and y ? 0.

48. Replace each question mark with the symbol > or <.

(a) In the first quadrant, the product $x \cdot y$? 0.

(b) In the second quadrant, the product $x \cdot y$? 0.

(c) In the third quadrant, the product $x \cdot y$? 0.

(d) In the fourth quadrant, the product $x \cdot y$? 0.

Applying the Concepts

For each of the following applied problems, first build a table of values. Make sure each input is accompanied by an output.

49. Cost of a Phone Call If the cost of a long-distance phone call is 50¢ for the first minute and 25¢ for each additional minute, then the total cost y (in cents) of a call that goes x minutes past the first minute is $y = 25x + 50$. Let 1 unit on the x-axis equal 1 minute and 1 unit on the y-axis equal 25¢, and graph this equation.

50. Cost of a Taxi Ride If the cost of a taxi ride in Las Vegas is \$1.50 for the first mile and \$0.50 for each tenth of a mile after the first mile, then the total cost y (in cents) of a ride that goes x tenths of a mile after the first mile is $y = 50x + 150$. Let each unit on the x-axis equal one-tenth of a mile and each unit on the y-axis equal \$0.50, and graph this equation.

51. Demand Equation A company that manufactures typewriter ribbons knows that the number of ribbons they can sell each week, x, is related to the price p of each ribbon by the equation $x + 100p = 1{,}200$. Note any restrictions on the variables; then graph the relationship between x and p, with the values of x along the horizontal axis and the values of p on the vertical axis.

52. Demand Equation A company that manufacturers diskettes for home computers finds that they can sell x diskettes each day at p dollars per diskette according to the equation $x + 100p = 800$. Note any restrictions on the variables; then graph the relationship between x and p.

53. A Snail's Pace A snail is climbing straight up a brick wall at a constant rate. It takes the snail 4 hours to climb up 6 feet. After climbing for 4 hours, the snail rests for 4 hours, during which time it slides 2 feet down the wall. If the snail starts at the bottom of the wall and repeats this climbing up/sliding down process continuously, construct a table that gives the snail's height above the ground every 4 hours, starting at 0 and ending at 24 hours. Then construct a line graph from the information in the table.

54. Air Temperature A pilot checks the weather conditions before flying and finds that the air temperature drops 3.5°F every 1,000 feet. If the air temperature is 41°F when the plane reaches 10,000 feet, construct a table that gives the air temperature every 2,000 feet, starting at 10,000 feet and ending at 20,000 feet. Then construct a line graph from the information in the table.

55. Accidents at Work Between 1930 and the mid-1990s, the number of accident-related deaths at the workplace in the United States was decreasing in a linear fashion. If D represents the annual number of accident-related deaths and y represents the year, then

$$D = 325{,}870 - 159y$$

(a) Each successive year saw a reduction of how many additional accident-related deaths?

(b) How many accident-related deaths occurred in 1930?

(c) How many accident-related deaths occurred in 1975?

(d) Can this pattern continue to the year 2050? Explain why or why not.

56. Disposable Income Between 1970 and 1990, inclusive, the disposable income in the United States for each person increased significantly and may be described by the equation

$$I = 4{,}000 + 600x$$

where I is the disposable income and x is the number of years since January 1, 1970.

(a) What are the restrictions on x?

(b) What was the disposable income in 1972?

(c) By how much did the disposable income increase over any 4-year period between 1970 and 1990?

57. Price of a Textbook The table on textbook prices shown at the beginning of this section is repeated here. Below the table is an equation that approximates the pairs of numbers in the table.

YEAR OF NEW EDITION	NET PRICE ($)
x	y
1982	16.95
1985	23.50
1989	30.50
1993	39.25
1997	47.50
2001	55.00

Approximating Equation: $y = 2.022x - 3{,}991$

(a) Use your calculator to graph the equation using the following window:

Window: X from 1981 to 2001, Y from 0 to 60

(b) Trace along the graph to find the value of y that corresponds to each of the following values of x:

FROM APPROXIMATING EQUATION	
x	y
1982	
1985	
1989	
1993	
1997	
2001	

58. Price of a Textbook If we let the year 1980 correspond to $x = 0$ then the table on textbook prices shown at the beginning of this section can be rewritten as shown. Following the table is an equation that approximates the pairs of numbers in the table.

YEAR OF NEW EDITION	IF $x = 0$ AT 1980	NET PRICE ($)
	x	y
1982	2	16.95
1985	5	23.50
1989	9	30.50
1993	13	39.25
1997	17	47.50
2001	21	55.00

Approximating Equation: $y = 2.022x + 12.94$

(a) Use your calculator to graph the equation using the following window:

Window: X from 0 to 23.5, Y from 0 to 60

(b) Trace along the graph to find the value of y that corresponds to each of the following values of x:

YEAR OF NEW EDITION	FROM APPROXIMATING EQUATION	
	x	y
1982	2	
1985	5	
1989	9	
1993	13	
1997	17	
2001	21	

Review Problems

Solve each equation:

59. $5x - 4 = -3x + 12$

60. $\frac{1}{2} - \frac{y}{5} = -\frac{9}{10} + \frac{y}{2}$

61. $2x^2 - 5x = 12$
62. $8x^2 = 14x - 5$
63. $\frac{1}{2} - \frac{1}{8}(3t - 4) = -\frac{7}{8}t$
64. $3(5t - 1) - (3 - 2t) = 5t - 8$
65. $4t^2 + 12t = 0$
66. $(x - 3)(x - 5) = 3$

Extending the Concepts

Each of the following equations is a graph that is not a straight line. Set up a table of values, then sketch the graph of the equation.

67. $y = x^2$
68. $y = x^2 - 3$
69. $y = |x|$
70. $y = |x + 2|$
71. $y = x^3$
72. $y = x^3 + 2$

Find the x- and y-intercepts for each equation. Your answers will contain the constants a, b, and c.

73. $ax + by = c$
74. $ax - by = c$
75. $\frac{x}{a} + \frac{y}{b} = 1$
76. $y = ax + b$

Spreadsheets and Drag Racing Jim Rizzoli lives in San Luis Obispo, California. He owns and operates an alcohol-fueled dragster. The information in the table below was recorded by a computer in the dragster during one of his races at the 1993 Winternationals. It shows the time and speed of the dragster, along with the distance traveled past the starting line and the front axle RPM (revolutions per minute). Use a spreadsheet program to work these problems.

TIME (sec)	SPEED (mph)	DISTANCE TRAVELED (ft)	FRONT AXLE RPM
0	0.0	0	0
1	72.7	69	1,107
2	129.9	231	1,978
3	162.8	439	2,486
4	192.2	728	2,919
5	212.4	1,000	3,233
6	228.1	1,373	3,473

77. Construct a bar chart that shows the relationship between time and distance.
78. Construct a line graph that shows the relationship between time and front axle RPM.
79. Construct a line graph that shows the relationship between speed and distance traveled.
80. Construct a line graph that shows the relationship between speed and front axle RPM. Does your line graph suggest that the relationship between speed and front axle RPM is a linear one?

2.2 The Slope of a Line

A highway sign tells us we are approaching a 6% downgrade. As we drive down this hill, each 100 feet we travel horizontally is accompanied by a 6-foot drop in elevation.

Highway sign Mathematical model

In mathematics we say the slope of the highway is $-0.06 = -\frac{6}{100} = -\frac{3}{50}$. The *slope* is the ratio of the vertical change to the accompanying horizontal change.

In defining the slope of a straight line, we are looking for a number to associate with a straight line that does two things. First, we want the slope of a line to measure the "steepness" of the line; that is, in comparing two lines, the slope of the steeper line should have the larger numerical value. Second, we want a line that *rises* going from left to right to have a *positive* slope. We want a line that *falls* going from left to right to have a *negative* slope. (A line that neither rises nor falls going from left to right must, therefore, have 0 slope.)

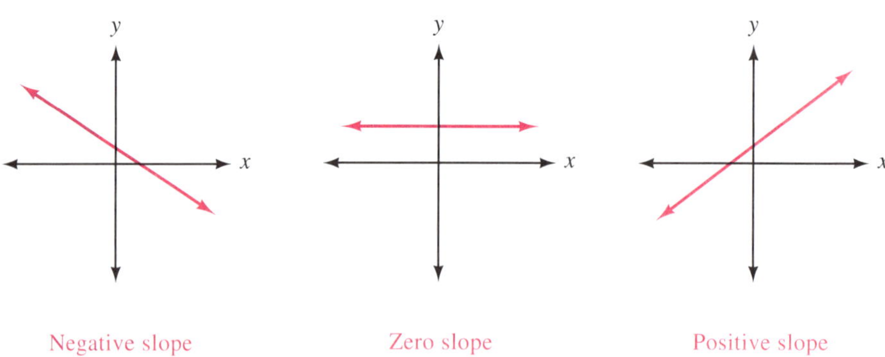

Negative slope Zero slope Positive slope

Geometrically, we can define the *slope* of a line as the ratio of the vertical change to the horizontal change encountered when moving from one point to another on the line. The vertical change is sometimes called the *rise*. The horizontal change is called the *run*.

EXAMPLE 1 Find the slope of the line $y = 2x - 3$.

SOLUTION To use our geometric definition, we first graph $y = 2x - 3$ (Figure 1). We then pick any two convenient points and find the ratio of rise to run. By convenient points we mean points with integer coordinates. If we let $x = 2$ in the equation, then $y = 1$. Likewise if we let $x = 4$, then y is 5.

Practice Problems

1. Graph the line $y = 3x - 5$ and then find the slope.

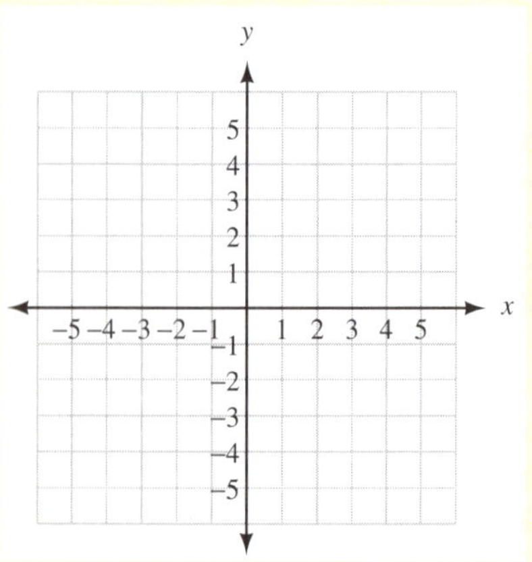

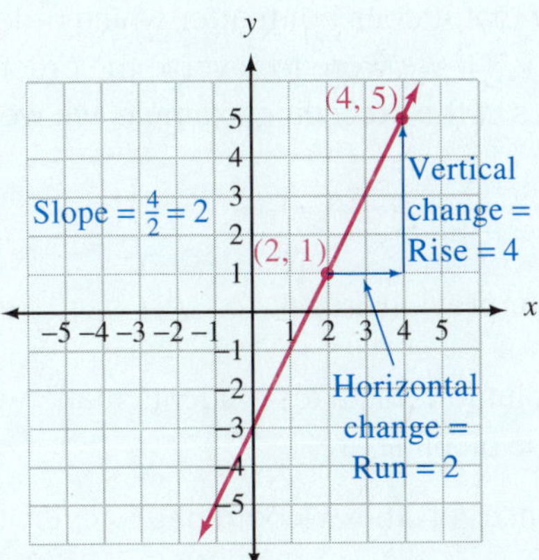

Figure 1

Our line has a slope of 2.

Notice that we can measure the vertical change (rise) by subtracting the y-coordinates of the two points shown in Figure 1: $5 - 1 = 4$. The horizontal change (run) is the difference of the x-coordinates: $4 - 2 = 2$. This gives us a second way of defining the slope of a line.

DEFINITION

The **slope** of the line between two points (x_1, y_1) and (x_2, y_2) is given by

$$\text{Slope} = m = \frac{\text{Rise}}{\text{Run}} = \frac{y_2 - y_1}{x_2 - x_1}$$

↑ Geometric form ↑ Algebraic form

EXAMPLE 2 Find the slope of the line through $(-2, -3)$ and $(-5, 1)$.

SOLUTION

$$m = \frac{y_2 - y_1}{x_2 - x_1} = \frac{1 - (-3)}{-5 - (-2)} = \frac{4}{-3} = -\frac{4}{3}$$

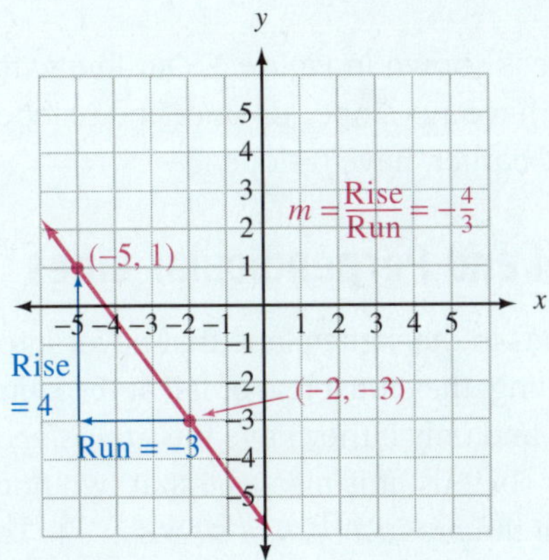

Figure 2

Looking at the graph of the line between the two points (Figure 2), we can see our geometric approach does not conflict with our algebraic approach.

2. Find the slope of the line through (3, 4) and (1, −2).

Answer

1. 3

We should note here that it does not matter which ordered pair we call (x_1, y_1) and which we call (x_2, y_2). If we were to reverse the order of subtraction of both the x- and y-coordinates in the preceding example, we would have

$$m = \frac{-3 - 1}{-2 - (-5)} = \frac{-4}{3} = -\frac{4}{3}$$

which is the same as our previous result.

Note: The two most common mistakes students make when first working with the formula for the slope of a line are

1. Putting the difference of the x-coordinates over the difference of the y-coordinates.
2. Subtracting in one order in the numerator and then subtracting in the opposite order in the denominator. You would make this mistake in Example 2 if you wrote $1 - (-3)$ in the numerator and then $-2 - (-5)$ in the denominator.

3. Find the slope of the line through $(2, -3)$ and $(-1, -3)$.

EXAMPLE 3 Find the slope of the line containing $(3, -1)$ and $(3, 4)$.

SOLUTION Using the definition for slope, we have

$$m = \frac{-1 - 4}{3 - 3} = \frac{-5}{0}$$

The expression $\frac{-5}{0}$ is undefined; that is, there is no real number to associate with it. In this case, we say the line has *no slope* or *undefined slope*.

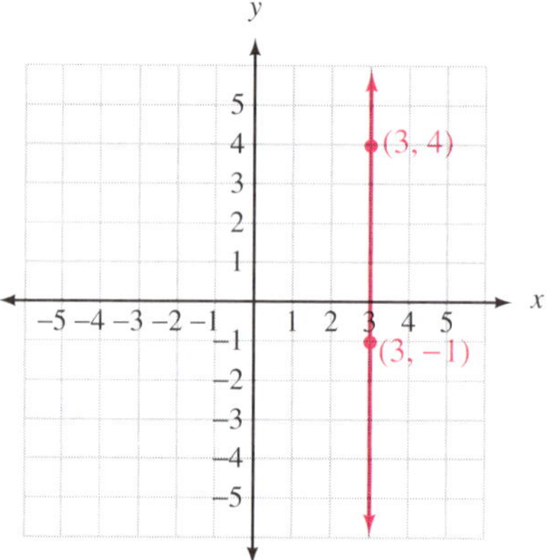

Figure 3

The graph of our line is shown in Figure 3. Our line with no slope is a vertical line. All vertical lines have no slope (or undefined slope). (And all horizontal lines, as we mentioned earlier, have 0 slope.)

Slopes of Parallel and Perpendicular Lines

In geometry we call lines in the same plane that never intersect parallel. For two lines to be nonintersecting, they must rise or fall at the same rate. In other words, two lines are *parallel* if and only if they have the *same slope.*

Although it is not as obvious, it is also true that two nonvertical lines are *perpendicular* if and only if the *product of their slopes is* -1. This is the same as saying their slopes are negative reciprocals.

Answers
2. 3 **3.** 0

We can state these facts with symbols as follows: If line l_1 has slope m_1 and line l_2 has slope m_2, then

$$l_1 \text{ is parallel to } l_2 \Leftrightarrow m_1 = m_2$$

and

$$l_1 \text{ is perpendicular to } l_2 \Leftrightarrow m_1 \cdot m_2 = -1$$

$$\left(\text{or } m_1 = \frac{-1}{m_2}\right)$$

For example, if a line has a slope of $\frac{2}{3}$, then any line parallel to it has a slope of $\frac{2}{3}$. Any line perpendicular to it has a slope of $-\frac{3}{2}$ (the negative reciprocal of $\frac{2}{3}$).

Although we cannot give a formal proof of the relationship between the slopes of perpendicular lines at this level of mathematics, we can offer some justification for the relationship. Figure 4 shows the graphs of two lines. One of the lines has a slope of $\frac{2}{3}$; the other has a slope of $-\frac{3}{2}$. As you can see, the lines are perpendicular.

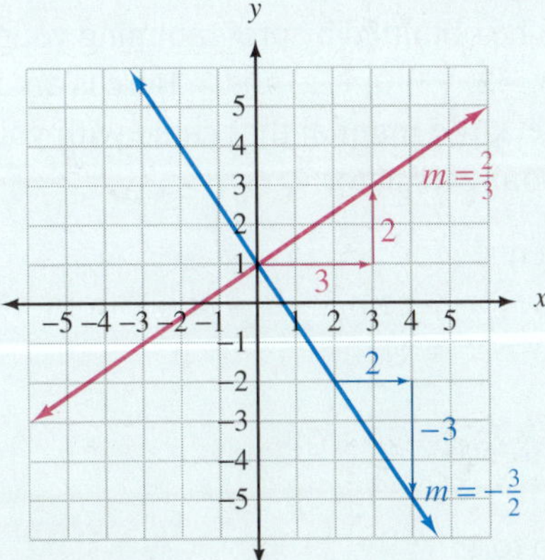

Figure 4

USING TECHNOLOGY

Families of Curves

We can use a graphing calculator to investigate the effects of the numbers a and b on the graph of $y = ax + b$. To see how the number b affects the graph, we can hold a constant and let b vary. Doing so will give us a *family* of curves. Suppose we set $a = 1$ and then let b take on integer values from -3 to 3. The equations we obtain are

$$y = x - 3$$
$$y = x - 2$$
$$y = x - 1$$
$$y = x$$
$$y = x + 1$$
$$y = x + 2$$
$$y = x + 3$$

We will give three methods of graphing this set of equations on a graphing calculator.

Method 1: Y-Variables List

To use the Y-variables list, enter each equation at one of the Y variables, set the graph window, then graph. The calculator will graph the equations in order, starting with Y_1 and ending with Y_7. Following is the Y-variables list, an appropriate window, and a sample of the type of graph obtained (Figure 5).

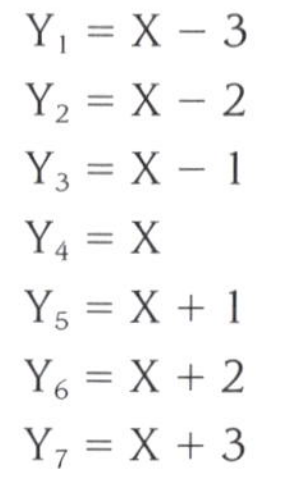
$Y_1 = X - 3$
$Y_2 = X - 2$
$Y_3 = X - 1$
$Y_4 = X$
$Y_5 = X + 1$
$Y_6 = X + 2$
$Y_7 = X + 3$

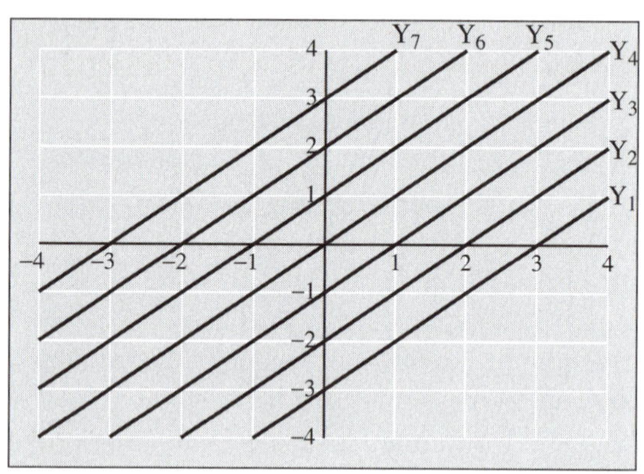

Figure 5

Window: X from −4 to 4, Y from −4 to 4

Method 2: Programming

The same result can be obtained by programming your calculator to graph $y = x + b$ for $b = -3, -2, -1, 0, 1, 2,$ and 3. Here is an outline of a program that will do this. Check the manual that came with your calculator to find the commands for your calculator.

Step 1: Clear screen
Step 2: Set window for X from −4 to 4 and Y from −4 to 4
Step 3: $-3 \rightarrow B$
Step 4: Label 1
Step 5: Graph $Y = X + B$
Step 6: $B + 1 \rightarrow B$
Step 7: If $B < 4$, go to 1
Step 8: End

Method 3: Using Lists

On the TI-82/83 you can set Y_1 as follows

$$Y_1 = X + \{-3, -2, -1, 0, 1, 2, 3\}$$

When you press GRAPH, the calculator will graph each line from $y = x + (-3)$ to $y = x + 3$.

Each of the three methods will produce graphs similar to those in Figure 5.

Getting Ready for Class

After reading through the preceding section, respond in your own words and in complete sentences.

A. If you were looking at a graph that described the performance of a stock you had purchased, why would it be better if the slope of the line were positive, rather than negative?
B. Describe the behavior of a line with a negative slope.
C. Would you rather climb a hill with a slope of $\frac{1}{2}$ or a slope of 3? Explain why.
D. Describe how to obtain the slope of a line if you know the coordinates of two points on the line.

PROBLEM SET 2.2

Find the slope of each of the following lines from the given graph.

1.

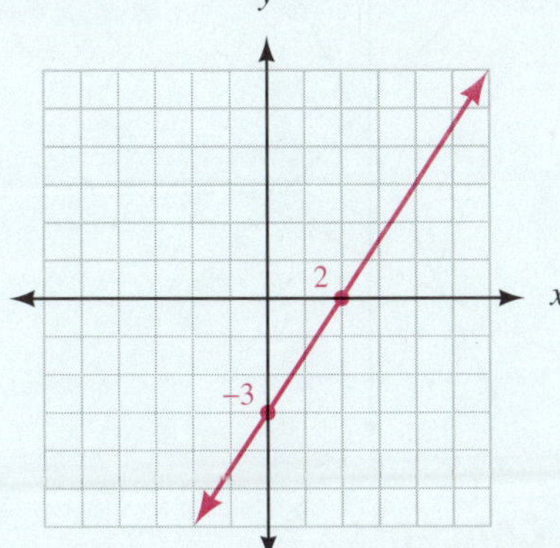

2.

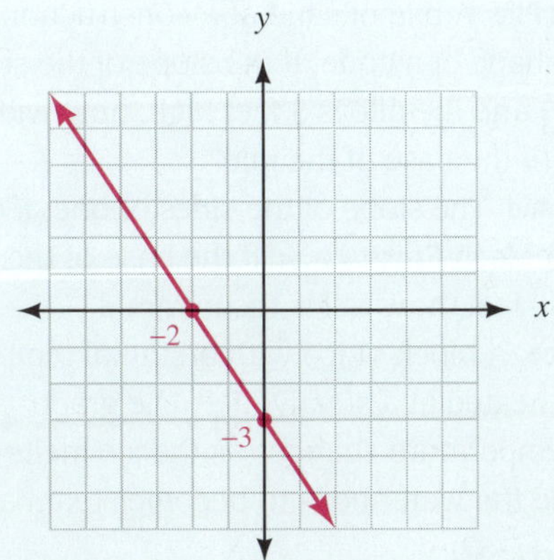

3.

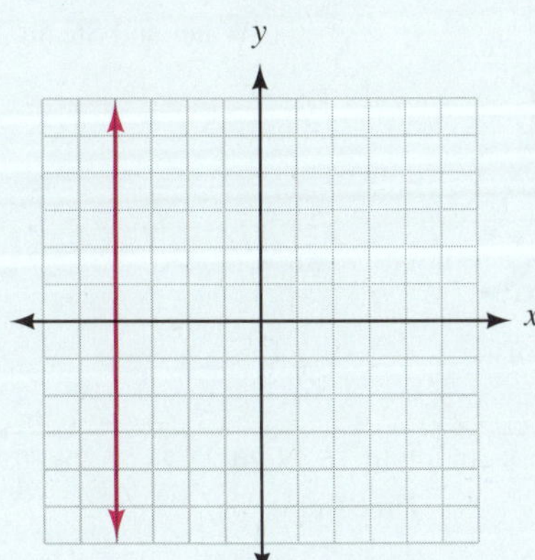

4.

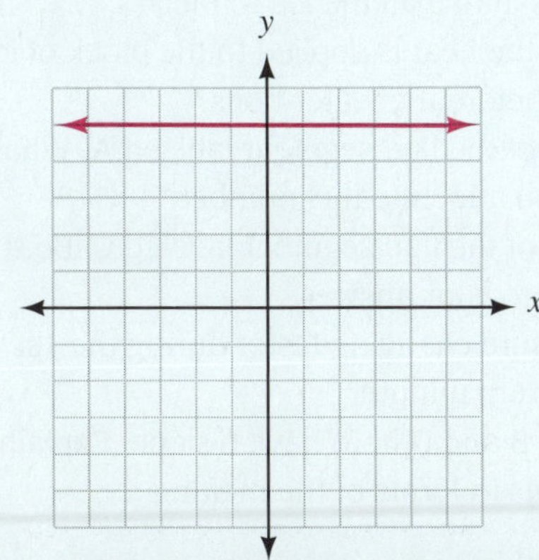

5.

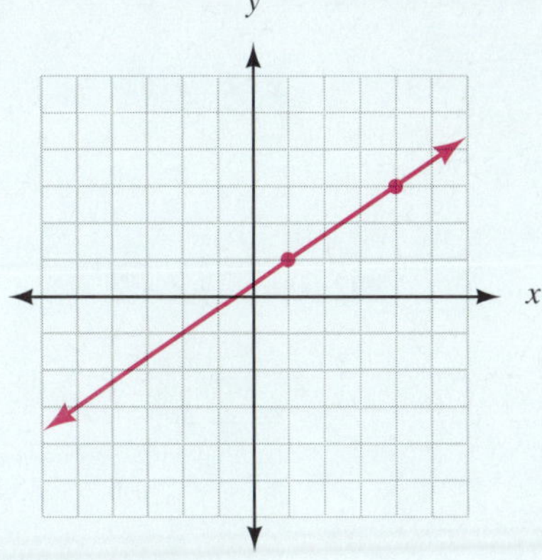

6.

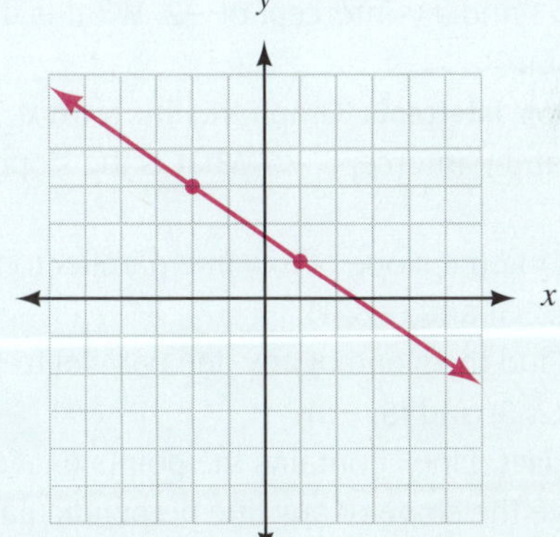

Find the slope of the line through the following pairs of points. Then, plot each pair of points, draw a line through them, and indicate the rise and run in the graph in the manner shown in Example 2.

7. (2, 1) and (4, 4)
8. (3, 1) and (5, 4)
9. (1, 4) and (5, 2)
10. (1, 3) and (5, 2)
11. (1, −3) and (4, 2)
12. (2, −3) and (5, 2)
13. (2, −4) and (5, −9)
14. (−3, 2) and (−1, 6)
15. (−3, 5) and (1, −1)
16. (−2, −1) and (3, −5)
17. (−4, 6) and (2, 6)
18. (2, −3) and (2, 7)
19. $(a, -3)$ and $(a, 5)$
20. $(x, 2y)$ and $(4x, 8y)$

Solve for the indicated variable if the line through the two given points has the given slope.

21. $(a, 3)$ and $(2, 6)$, $m = -1$
22. $(a, -2)$ and $(4, -6)$, $m = -3$
23. $(2, b)$ and $(-1, 4b)$, $m = -2$
24. $(-4, y)$ and $(-1, 6y)$, $m = 2$
25. $(2, 4)$ and (x, x^2), $m = 5$
26. $(3, 9)$ and (x, x^2), $m = -2$
27. $(1, 3)$ and $(x, 2x^2 + 1)$, $m = -6$
28. $(3, 7)$ and $(x, x^2 - 2)$, $m = -4$

For each of the equations in Problems 29–32, complete the table, and then use the results to find the slope of the graph of the equation.

29. $2x + 3y = 6$

x	y
0	
	0

30. $3x - 2y = 6$

x	y
0	
	0

31. $y = \frac{2}{3}x - 5$

x	y
0	
3	

32. $y = -\frac{3}{4}x + 2$

x	y
0	
4	

33. Finding Slope from Intercepts Graph the line that has an x-intercept of 3 and a y-intercept of -2. What is the slope of this line?

34. Finding Slope from Intercepts Graph the line with x-intercept -4 and y-intercept -2. What is the slope of this line?

35. Parallel Lines Find the slope of any line parallel to the line through (2, 3) and (−8, 1).

36. Parallel Lines Find the slope of any line parallel to the line through (2, 5) and (5, −3).

37. Perpendicular Lines Line l contains the points (5, −6) and (5, 2). Give the slope of any line perpendicular to l.

38. Perpendicular Lines Line l contains the points (3, 4) and (−3, 1). Give the slope of any line perpendicular to l.

39. Parallel Lines Line L contains the points (−2, 1) and (4, −5). Find the slope of any line parallel to L.

40. Parallel Lines Line L contains the points (3, −4) and (−2, −6). Find the slope of any line parallel to L.

41. Perpendicular Lines Line L contains the points (−2, −5) and (1, −3). Find the slope of any line perpendicular to L.

42. Perpendicular Lines Line L contains the points (6, −3) and (−2, 7). Find the slope of any line perpendicular to L.

43. Determine if each of the following tables could represent ordered pairs from an equation of a line.

(a)

x	y
0	5
1	7
2	9
3	11

(b)

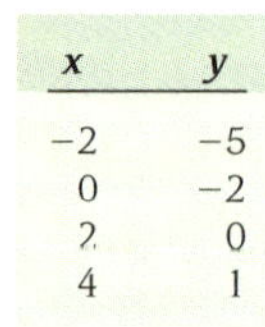

x	y
−2	−5
0	−2
2	0
4	1

44. The following lines have slope 2, $\frac{1}{2}$, 0, and -1. Match each line to its slope value.

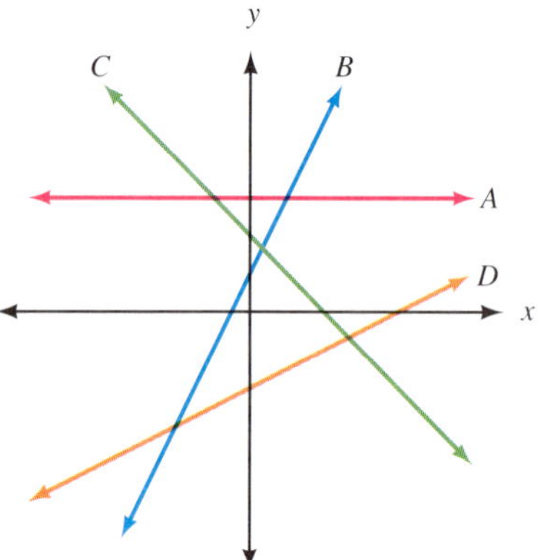

Applying the Concepts

45. Slope of a Sand Pile A pile of sand at a construction site is in the shape of a cone. If the slope of the side of the pile is $\frac{2}{3}$ and the pile is 8 feet high, how wide is the diameter of the base of the pile?

46. Slope of a Pyramid The slope of the sides of one of the ancient pyramids in Egypt is $\frac{13}{10}$. If the base of the pyramid is 750 feet, how tall is the pyramid?

Heating a Block of Ice A block of ice with an initial temperature of −20°C is heated at a steady rate. The graph shows how the temperature changes as the ice melts to become water and the water boils to become steam and water. (Problems 47–52)

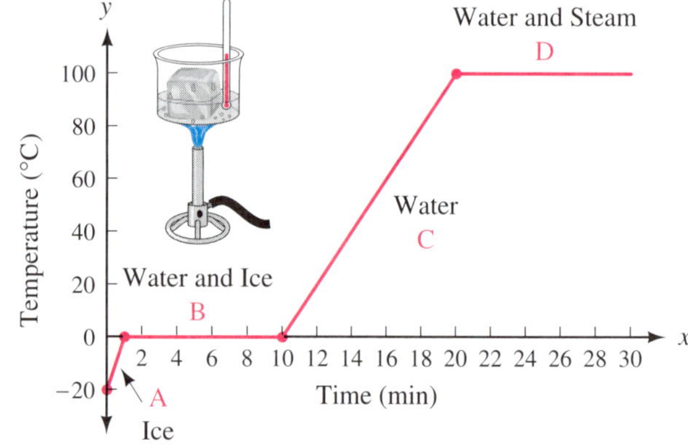

47. How long does it take all the ice to melt?

48. From the time the heat is applied to the block of ice, how long is it before the water boils?

49. Find the slope of the line segment labeled A. What units would you attach to this number?

50. Find the slope of the line segment labeled C. Be sure to attach units to your answer.

51. Is the temperature changing faster during the 1st minute or the 16th minute?

52. Line segments B and D both have 0 slope. Explain what this means in terms of the melting ice.

Value of a Used Car The 1998 edition of a popular consumer car price book gives the values shown in the table for Volkswagen Jettas in good condition. The line graph was drawn from the information in the table. Use this information to solve Problems 53–56.

YEAR	AGE IN 1998	VALUE ($)
1994	4	8,525
1995	3	9,575
1996	2	11,950
1997	1	13,200
1998	0	15,250

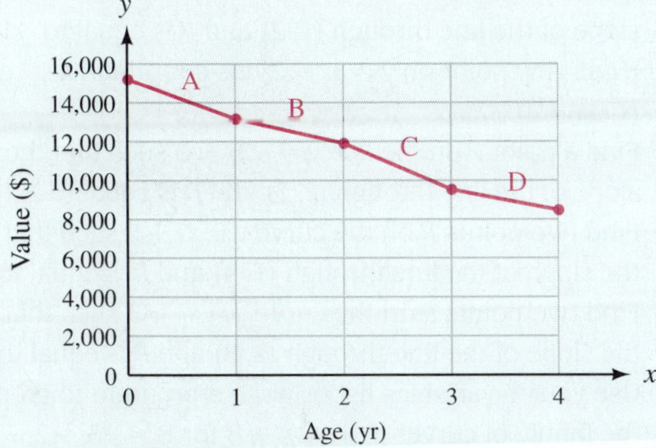

53. Find the slope of the line segment labeled B. What units should you attach to this number?

54. Find the slope of line segment C. Be sure to include units with your answer.

55. From the graph, does the value of the car decrease more from 1 to 2 years, or from 2 to 3 years?

56. From the graph, does the value of the car decrease more from 2 to 3 years, or from 3 to 4 years?

57. Slope of a Highway A sign at the top of the Cuesta Grade, outside of San Luis Obispo, reads "7% downgrade next 3 miles." The diagram below is a model of the Cuesta Grade that takes into account the information on that sign.

(a) At point B, the graph crosses the y-axis at 1,106 feet. How far is it from the origin to point A?

(b) What is the slope of the Cuesta Grade?

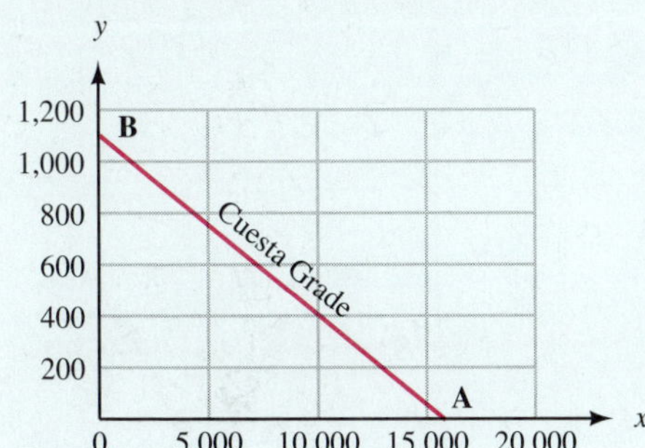

58. Graph Reading A fraternity is planning a fund-raising concert at the Veterans Memorial Building, which can accommodate a maximum of 300 ticket holders. Tickets are sold at $10 each. Their expenses are $455 for the band and $125 rent for the Veterans building. If 150 or more tickets are sold, the fraternity is required to hire a security guard for $200. The following diagram can be used to find the profit they will make by selling various numbers of tickets. As you can see, if no one buys a ticket, their profit will be −$580, the amount they will pay for the band and for rent.

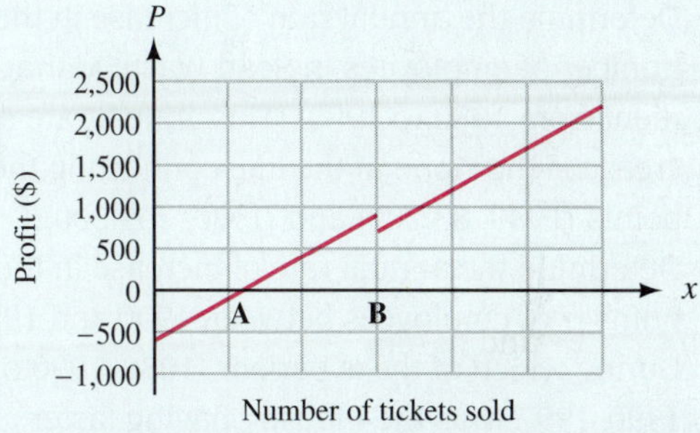

(a) Both line segments have the same slope. What is that slope?

(b) Point A is the break-even point. It corresponds to the number of tickets they must sell so that their income is equal to their expenses. How many tickets is this?

(c) How many tickets does point B represent?

(d) How many tickets must they sell to make a profit of $500?

59. Baseball Salaries The average salary of a major league baseball player in 1967 was $25,000, and in 1975 it was $40,000. What is the slope of the line that connects these two points; that is, what is the average yearly salary increase for this period?

The average salary of a major league baseball player in 1990 was $600,000, and in 1994 it was $1,300,000. What is the slope of the line that connects these two points; that is, what is the average yearly salary increase for this period?

60. Blood Pressure The average systolic blood pressure for a 26-year-old is 113 and for a 40-year-old is 120.

(a) Using the age as the horizontal axis and the blood pressure as the vertical axis, plot this blood pressure data. Draw a line connecting the points.

(b) Determine the slope of the line connecting the two points from part (a).

(c) What does the slope from part (b) represent in terms of blood pressure and age?

61. **Waste Management** The data table for number of employees in the Solid Waste Management field during certain years between 1980 and 1997 is given below.

SOLID WASTE MANAGEMENT	
YEAR	NUMBER OF EMPLOYEES
1980	83,200
1990	209,500
1995	234,400
1997	249,900

Source: Environmental Business International, Inc., San Diego, CA, *Environmental Business Journal.*

(a) Determine the annual rate of increase in the number of employees in Solid Waste Management from 1980 to 1990. [This annual rate of increase is the slope of the line connecting the points (1980, 83,200) and (1990, 209,500).]

(b) Determine the annual rate of increase in the number of employees between 1990 and 1997.

(c) During which of these periods, 1980–1990 or 1990–1997, was the industry having faster growth?

62. **Administrative Salaries** The average annual salary for a superintendent of a U.S. public school system for the years 1980–1998 can be estimated by the linear equation

$$S = 3514.4x + 40772$$

where S is the average annual salary in dollars, and x is the number of years after 1980. (That is, $x = 0$ on January 1, 1980.)

(a) Use the model to predict the average annual salary of a superintendent in 1985 and in 1997.

(b) Write your answers from part (a) as ordered pairs and then find the slope of the line between them.

(c) Does the number for slope of the line appear in the equation? If so, where?

Review Problems

63. If $3x + 2y = 12$, find y when x is 4.
64. If $y = 3x - 1$, find x when y is 0.
65. Solve the formula $3x + 2y = 12$ for y.
66. Solve the formula $y = 3x - 1$ for x.
67. Solve the formula $A = P + Prt$ for t.
68. Solve the formula $S = \pi r^2 + 2\pi rh$ for h.

Extending the Concepts

69. Find the slope of the line passing through the two points $(2, 4)$ and (x, x^2).
70. Find the slope of the line passing through the two points (x, x^2) and $(x + 2, (x + 2)^2)$.
71. Find a point P on the line $y = 3x - 2$ such that the slope of the line through $(1, 2)$ and P is equal to -1. [*Hint:* Any point on $y = 3x - 2$ has coordinates $(x, 3x - 2)$.]
72. Find a point P on the line $y = -2x + 3$ such that the slope of the line through $(2, 3)$ and P is equal to 2.
73. Find two points P on the curve $y = x^2 + 2$ such that the slope of the line through $(1, 4)$ and P is equal to 2.
74. Find two points P on the curve $y = x^2 - 3$ such that the slope of the line through $(2, 3)$ and P is equal to 3.
75. Use your Y-variables list or write a program to graph the family of curves $Y = -2X + B$ for $B = -3, -2, -1, 0, 1, 2$, and 3.
76. Use your Y-variables list or write a program to graph the family of curves $Y = 2X + B$ for $B = -3, -2, -1, 0, 1, 2$, and 3.
77. Use your Y-variables list or write a program to graph the family of curves $Y = AX$ for $A = -3, -2, -1, 0, 1, 2$, and 3.
78. Use your Y-variables list or write a program to graph the family of curves $Y = AX$ for $A = \frac{1}{4}, \frac{1}{3}, \frac{1}{2}, 1, 2$, and 3.
79. Use your Y-variables list or write a program to graph the family of curves $Y = AX + 2$ for $A = -3, -2, -1, 0, 1, 2$, and 3.
80. Use your Y-variables list or write a program to graph the family of curves $Y = AX - 2$ for $A = \frac{1}{4}, \frac{1}{3}, \frac{1}{2}, 1, 2$, and 3.

2.3 The Equation of a Line

The table and illustrations show some corresponding temperatures on the Fahrenheit and Celsius temperature scales. For example, water freezes at 32°F and 0°C, and boils at 212°F and 100°C.

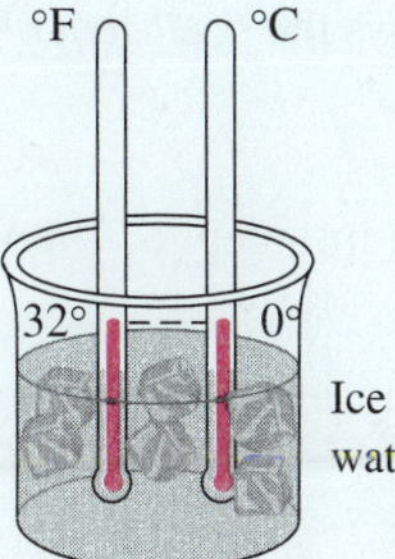

DEGREES CELSIUS	DEGREES FAHRENHEIT
0	32
25	77
50	122
75	167
100	212

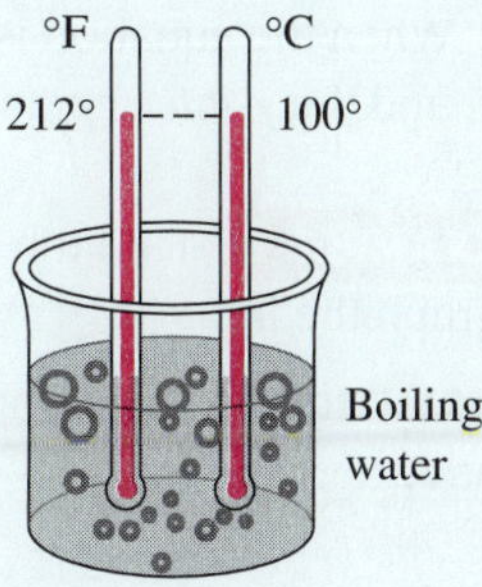

If we plot all the points in the table using the x-axis for temperatures on the Celsius scale and the y-axis for temperatures on the Fahrenheit scale, we see that they line up in a straight line (Figure 1). This means that a linear equation in two variables will give a perfect description of the relationship between the two scales. That equation is

$$F = \frac{9}{5}C + 32$$

The techniques we use to find the equation of a line from a set of points is what this section is all about.

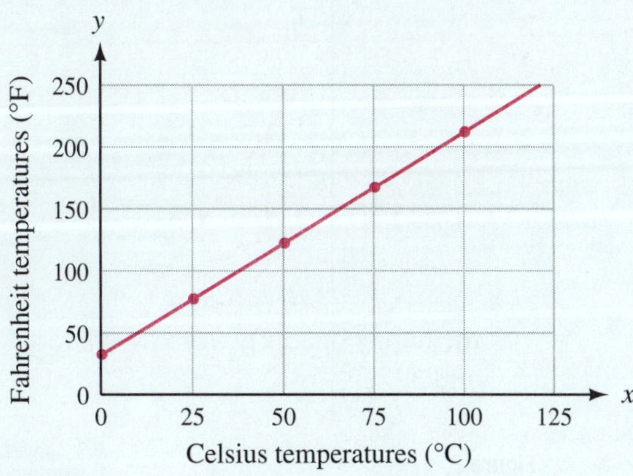

Figure 1

Suppose line l has slope m and y-intercept b. What is the equation of l? Because the y-intercept is b, we know the point $(0, b)$ is on the line. If (x, y) is any other point on l, then using the definition for slope, we have

$$\frac{y-b}{x-0} = m \qquad \textbf{Definition of slope}$$

$$y - b = mx \qquad \textbf{Multiply both sides by } x$$

$$y = mx + b \qquad \textbf{Add } b \textbf{ to both sides}$$

This last equation is known as the *slope-intercept form* of the equation of a straight line.

Slope-Intercept Form of the Equation of a Line

The equation of any line with slope m and y-intercept b is given by

$$y = mx + b$$

Slope ↗ (m) ↑ y-intercept (b)

When the equation is in this form, the *slope* of the line is always the *coefficent of* x, and the *y-intercept* is always the *constant term.*

Practice Problems

1. Find the equation of the line with slope $\frac{2}{3}$ and y-intercept 1. Then graph the line.

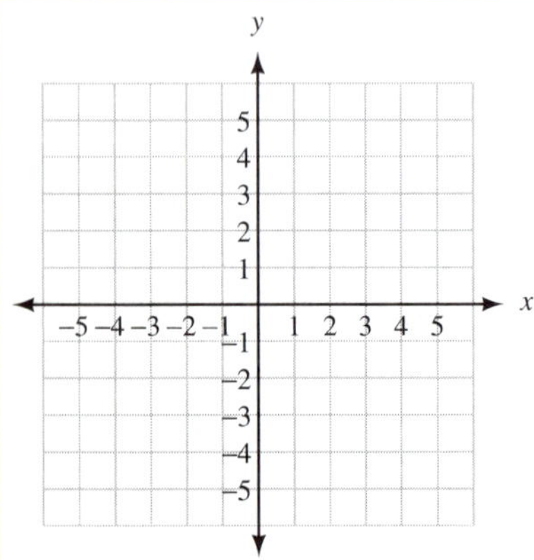

EXAMPLE 1 Find the equation of the line with slope $-\frac{4}{3}$ and y-intercept 5. Then graph the line.

SOLUTION Substituting $m = -\frac{4}{3}$ and $b = 5$ into the equation $y = mx + b$, we have

$$y = -\frac{4}{3}x + 5$$

Finding the equation from the slope and y-intercept is just that easy. If the slope is m and the y-intercept is b, then the equation is always $y = mx + b$. Now, let's graph the line.

Because the y-intercept is 5, the graph goes through the point (0, 5). To find a second point on the graph, we start at (0, 5) and move 4 units down (that's a rise of −4) and 3 units to the right (a run of 3). The point we end up at is (3, 1). Drawing a line that passes through (0, 5) and (3, 1), we have the graph of our equation. (Note that we could also let the rise = 4 and the run = −3 and obtain the same graph.) The graph is shown in Figure 2.

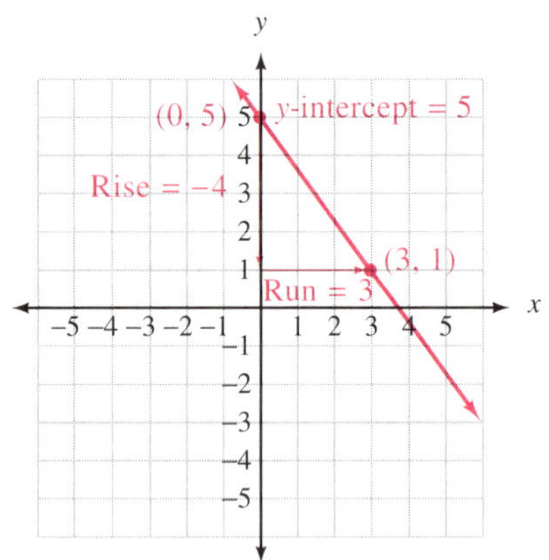

Figure 2

2. Give the slope and y-intercept for the line $4x - 5y = 7$.

EXAMPLE 2 Give the slope and y-intercept for the line $2x - 3y = 5$.

SOLUTION To use the slope-intercept form, we must solve the equation for y in terms of x.

$$2x - 3y = 5$$

$$-3y = -2x + 5 \quad \textbf{Add } -2x \textbf{ to both sides}$$

$$y = \frac{2}{3}x - \frac{5}{3} \quad \textbf{Divide by } -3$$

The last equation has the form $y = mx + b$. The slope must be $m = \frac{2}{3}$, and the y-intercept is $b = -\frac{5}{3}$.

Answers

1. See Solutions to Selected Practice Problems.
2. $m = \frac{4}{5}$, $b = -\frac{7}{5}$

EXAMPLE 3 Graph the equation $2x + 3y = 6$ using the slope and y-intercept.

SOLUTION Although we could graph this equation by finding ordered pairs that are solutions to the equation and drawing a line through their graphs, it is sometimes easier to graph a line using the slope-intercept form of the equation.

Solving the equation for y, we have,

$$2x + 3y = 6$$

$$3y = -2x + 6 \quad \text{Add } -2x \text{ to both sides}$$

$$y = -\frac{2}{3}x + 2 \quad \text{Divide by 3}$$

The slope is $m = -\frac{2}{3}$ and the y-intercept is $b = 2$. Therefore, the point (0, 2) is on the graph and the ratio of rise to run going from (0, 2) to any other point on the line is $-\frac{2}{3}$. If we start at (0, 2) and move 2 units up (that's a rise of 2) and 3 units to the left (a run of −3), we will be at another point on the graph. (We could also go down 2 units and right 3 units and still be assured of ending up at another point on the line because $\frac{2}{-3}$ is the same as $\frac{-2}{3}$.)

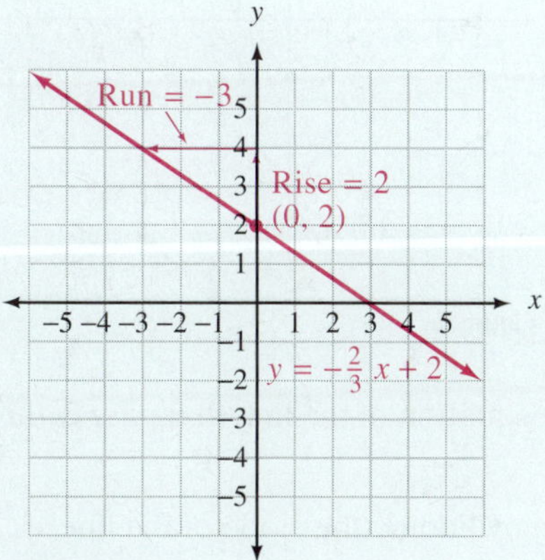

Figure 3

A second useful form of the equation of a straight line is the point-slope form. Let line l contain the point (x_1, y_1) and have slope m. If (x, y) is any other point on l, then by the definition of slope we have

$$\frac{y - y_1}{x - x_1} = m$$

Multiplying both sides by $(x - x_1)$ gives us

$$(x - x_1) \cdot \frac{y - y_1}{x - x_1} = m(x - x_1)$$

$$y - y_1 = m(x - x_1)$$

This last equation is known as the *point-slope form* of the equation of a straight line.

Point-Slope Form of the Equation of a Line

The equation of the line through (x_1, y_1) with slope m is given by

$$y - y_1 = m(x - x_1)$$

This form of the equation of a straight line is used to find the equation of a line, either given one point on the line and the slope, or given two points on the line.

3. Graph the line $-3x + 2y = -6$ on the following coordinate system. Use the method shown in Example 3.

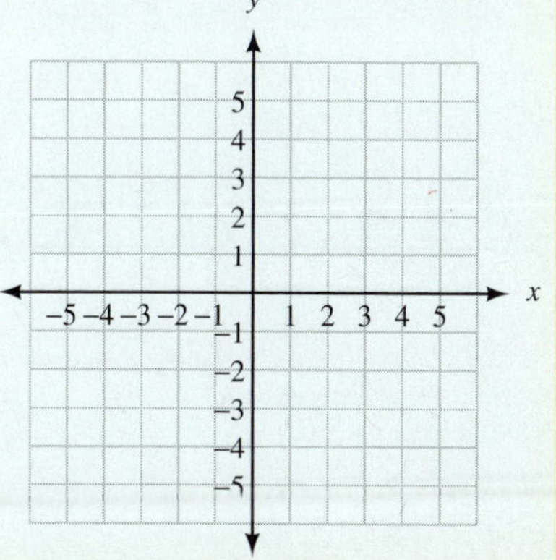

Note:
As we mentioned earlier, the rectangular coordinate system is the tool we use to connect algebra and geometry. Example 3 illustrates this connection, as do the many other examples in this chapter. In Example 3, Descartes' rectangular coordinate system allows us to associate the equation $2x + 3y = 6$ (an algebraic concept) with the straight line (a geometric concept) shown in Figure 3.

Answer
3. See Solutions to Selected Practice Problems.

4. Find the equation of the line with slope 3 that contains the point (−1, 2).

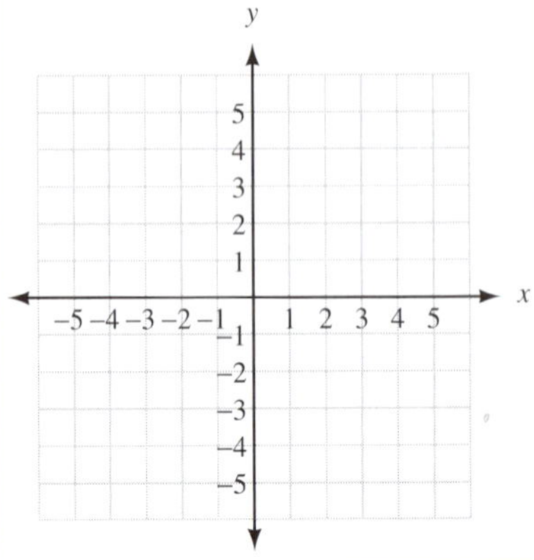

EXAMPLE 4 Find the equation of the line with slope −2 that contains the point (−4, 3). Write the answer in slope-intercept form.

SOLUTION

Using	$(x_1, y_1) = (-4, 3)$ and $m = -2$	
in	$y - y_1 = m(x - x_1)$	**Point-slope form**
gives us	$y - 3 = -2(x + 4)$	***Note:*** $x - (-4) = x + 4$
	$y - 3 = -2x - 8$	**Multiply out right side**
	$y = -2x - 5$	**Add 3 to each side**

Figure 4 is the graph of the line that contains (−4, 3) and has a slope of −2. Notice that the y-intercept on the graph matches that of the equation we found.

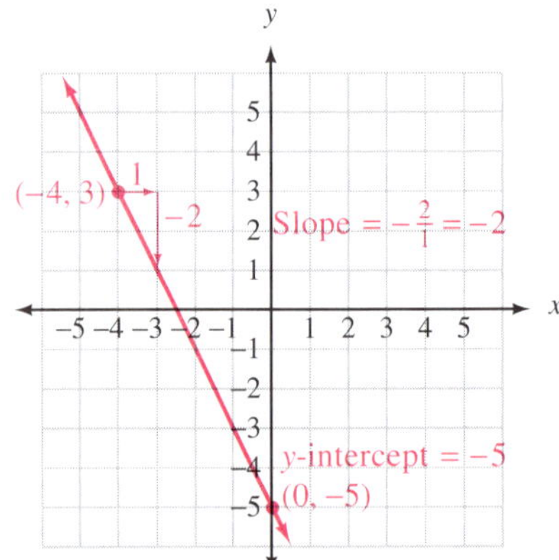

Figure 4

5. Find the equation of the line through (2, 5), and (6, −3).

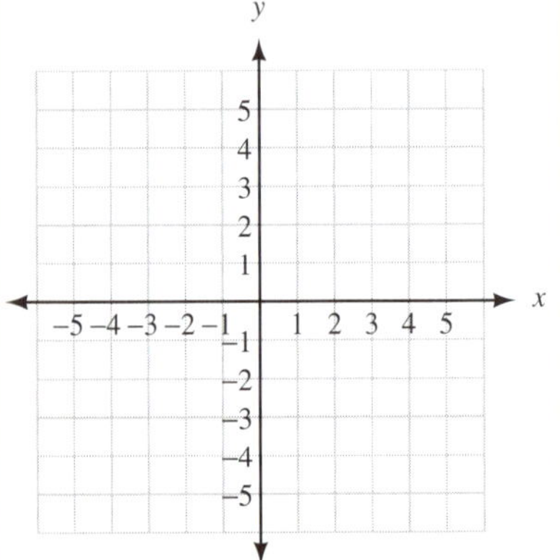

EXAMPLE 5 Find the equation of the line that passes through the points (−3, 3) and (3, −1).

SOLUTION We begin by finding the slope of the line:

$$m = \frac{3 - (-1)}{-3 - 3} = \frac{4}{-6} = -\frac{2}{3}$$

Using $(x_1, y_1) = (3, -1)$ and $m = -\frac{2}{3}$ in $y - y_1 = m(x - x_1)$ yields

$$y + 1 = -\frac{2}{3}(x - 3)$$

$$y + 1 = -\frac{2}{3}x + 2 \qquad \textbf{Multiply out right side}$$

$$y = -\frac{2}{3}x + 1 \qquad \textbf{Add −1 to each side}$$

Figure 5 shows the graph of the line that passes through the points (−3, 3) and (3, −1). As you can see, the slope and y-intercept are $-\frac{2}{3}$ and 1, respectively.

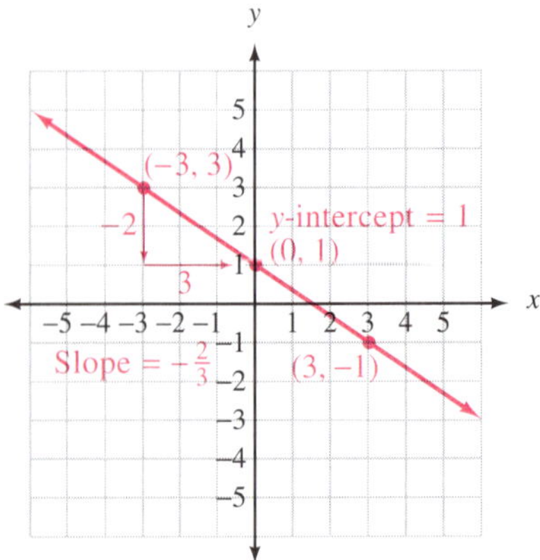

Figure 5

Note: In Example 5 we could have used the point (−3, 3) instead of (3, −1) and obtained the same equation; that is, using $(x_1, y_1) = (-3, 3)$ and $m = -\frac{2}{3}$ in $y - y_1 = m(x - x_1)$ gives us

$$y - 3 = -\frac{2}{3}(x + 3)$$

$$y - 3 = -\frac{2}{3}x - 2$$

$$y = -\frac{2}{3}x + 1$$

which is the same result we obtained using (3, −1).

Answers

4. $y = 3x + 5$ **5.** $y = -2x + 9$

The last form of the equation of a line that we will consider in this section is called the standard form. It is used mainly to write equations in a form that is free of fractions and is easy to compare with other equations.

Standard Form for the Equation of a Line

If a, b, and c are integers, then the equation of a line is in standard form when it has the form

$$ax + by = c$$

If we were to write the equation

$$y = -\frac{2}{3}x + 1$$

in standard form, we would first multiply both sides by 3 to obtain

$$3y = -2x + 3$$

Then we would add $2x$ to each side, yielding

$$2x + 3y = 3$$

which is a linear equation in standard form.

EXAMPLE 6 Give the equation of the line through $(-1, 4)$ whose graph is perpendicular to the graph of $2x - y = -3$. Write the answer in standard form.

SOLUTION To find the slope of $2x - y = -3$, we solve for y:

$$2x - y = -3$$

$$y = 2x + 3$$

The slope of this line is 2. The line we are interested in is perpendicular to the line with slope 2 and must, therefore, have a slope of $-\frac{1}{2}$.

Using $(x_1, y_1) = (-1, 4)$ and $m = -\frac{1}{2}$, we have

$$y - y_1 = m(x - x_1)$$

$$y - 4 = -\frac{1}{2}(x + 1)$$

Because we want our answer in standard form, we multiply each side by 2.

$$2y - 8 = -1(x + 1)$$

$$2y - 8 = -x - 1$$

$$x + 2y - 8 = -1$$

$$x + 2y = 7$$

The last equation is in standard form.

6. Find the equation of the line through (3, 2) that is perpendicular to the graph of $3x - y = 2$. Write your answer in standard form.

As a final note, the summary reminds us that all horizontal lines have equations of the form $y = b$ and slopes of 0. Because they cross the y-axis at b, the y-intercept is b; there is no x-intercept. Vertical lines have no slope and equations of the form $x = a$. Each will have an x-intercept at a and no y-intercept. Finally, equations of the form $y = mx$ have graphs that pass through the origin. The slope is always m and both the x-intercept and the y-intercept are 0.

Answer

6. $x + 3y = 9$

Facts From Geometry: Special Equations and Their Graphs, Slopes, and Intercepts

For the equations below, m, a, and b are real numbers.

Through the Origin	*Vertical Line*	*Horizontal Line*
Equation: $y = mx$	Equation: $x = a$	Equation: $y = b$
Slope $= m$	No slope	Slope $= 0$
x-intercept $= 0$	x-intercept $= a$	No x-intercept
y-intercept $= 0$	No y-intercept	y-intercept $= b$

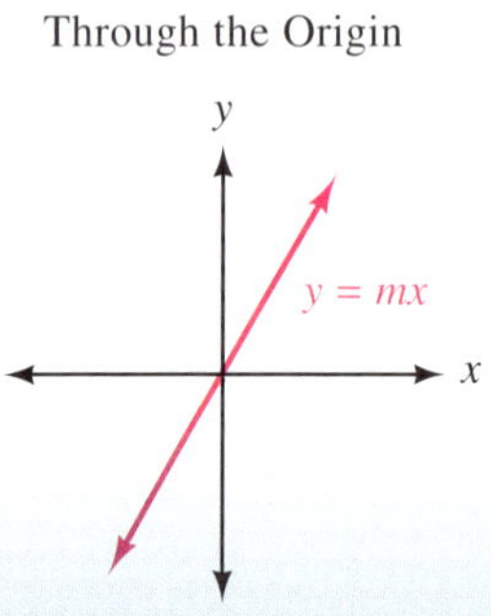

Through the Origin

Figure 6a

Vertical Line

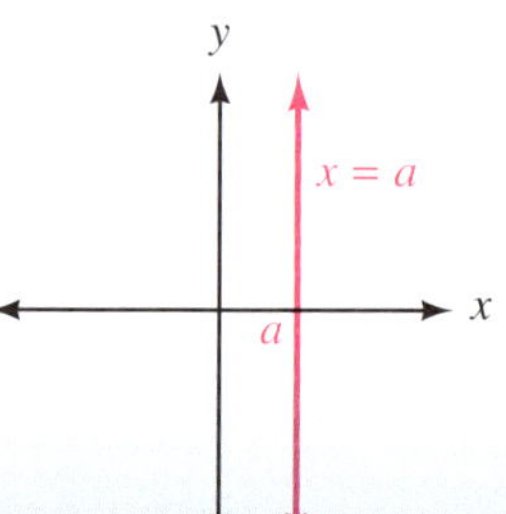

Figure 6b

Horizontal Line

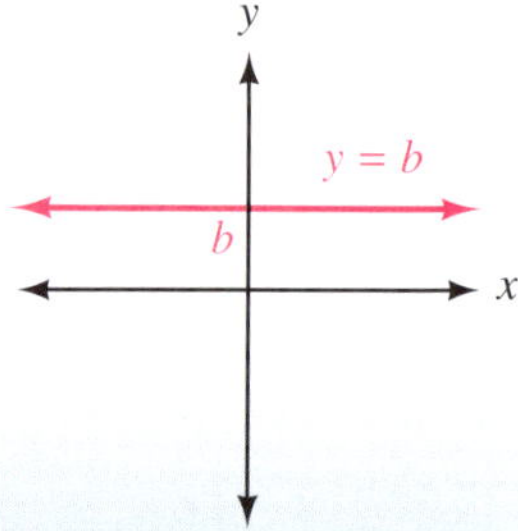

Figure 6c

USING TECHNOLOGY

Graphing Calculators

One advantage of using a graphing calculator to graph lines is that a calculator does not care whether the equation has been simplified or not. To illustrate, in Example 5 we found that the equation of the line with slope $-\frac{2}{3}$ that passes through the point $(3, -1)$ is

$$y + 1 = -\frac{2}{3}(x - 3)$$

Normally, to graph this equation we would simplify it first. With a graphing calculator, we add -1 to each side and enter the equation this way:

$$Y_1 = -(2/3)(X - 3) - 1$$

No simplification is necessary. We can graph the equation in this form, and the graph will be the same as the simplified form of the equation, which is $y = -\frac{2}{3}x + 1$. To convince yourself that this is true, graph both the simplified form for the equation and the unsimplified form in the same window. As you will see, the two graphs coincide.

PROBLEM SET 2.3

Give the equation of the line with the following slope and y-intercept.

1. $m = -4, b = -3$
2. $m = -6, b = \frac{4}{3}$
3. $m = -\frac{2}{3}, b = 0$
4. $m = 0, b = \frac{3}{4}$
5. $m = -\frac{2}{3}, b = \frac{1}{4}$
6. $m = \frac{5}{12}, b = -\frac{3}{2}$

Find the slope of a line (a) parallel and (b) perpendicular to the given line.

7. $y = 3x - 4$
8. $y = -4x + 1$
9. $3x + y = -2$
10. $2x - y = -4$
11. $2x + 5y = -11$
12. $3x - 5y = -4$

Give the slope and y-intercept for each of the following equations. Sketch the graph using the slope and y-intercept. Give the slope of any line perpendicular to the given line.

13. $y = 3x - 2$
14. $y = 2x + 3$
15. $2x - 3y = 12$
16. $3x - 2y = 12$
17. $4x + 5y = 20$
18. $5x - 4y = 20$

For each of the following lines, name the slope and y-intercept. Then write the equation of the line in slope-intercept form.

19.

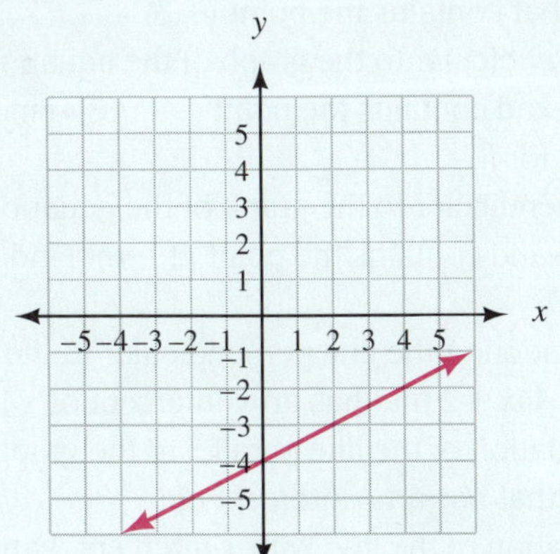

20.

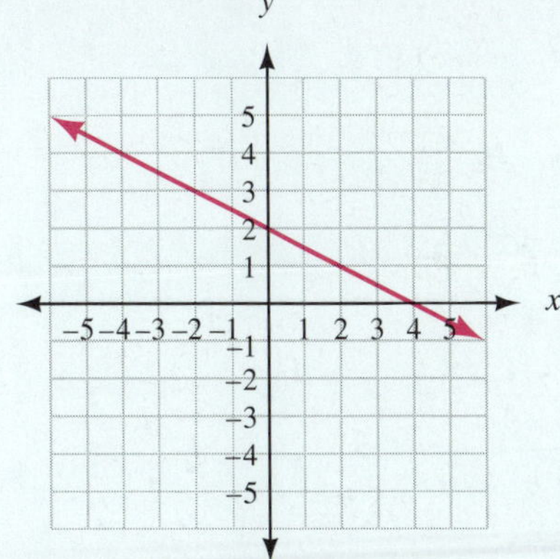

21.

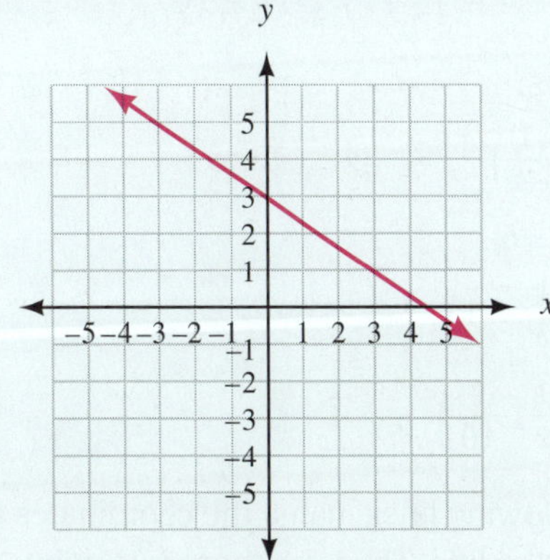

22.

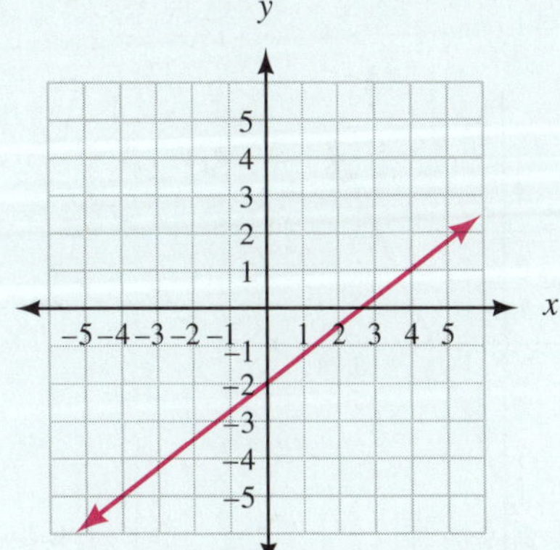

For each of the following problems, the slope and one point on the line are given. In each case, find the equation of that line. (Write the equation for each line in slope-intercept form.)

23. $(-2, -5)$; $m = 2$
24. $(-1, -5)$; $m = 2$
25. $(-4, 1)$; $m = -\frac{1}{2}$

26. $(-2, 1);\ m = -\dfrac{1}{2}$

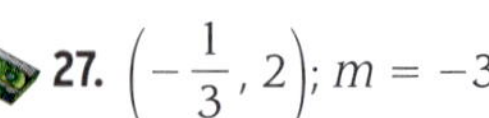

27. $\left(-\dfrac{1}{3}, 2\right);\ m = -3$

28. $\left(-\dfrac{2}{3}, 5\right);\ m = -3$

29. $(-4, 2),\ m = \dfrac{2}{3}$

30. $(3, -4),\ m = -\dfrac{1}{3}$

31. $(-5, -2),\ m = -\dfrac{1}{4}$

32. $(-4, -3),\ m = \dfrac{1}{6}$

Find the equation of the line that passes through each pair of points. Write your answers in standard form.

33. $(3, -2), (-2, 1)$

34. $(-4, 1), (-2, -5)$

35. $\left(-2, \dfrac{1}{2}\right), \left(-4, \dfrac{1}{3}\right)$

36. $(-6, -2), (-3, -6)$

37. $\left(\dfrac{1}{3}, -\dfrac{1}{5}\right), \left(-\dfrac{1}{3}, -1\right)$

38. $\left(-\dfrac{1}{2}, -\dfrac{1}{2}\right), \left(\dfrac{1}{2}, \dfrac{1}{10}\right)$

For each of the following lines, name the coordinates of any two points on the line. Then use those two points to find the equation of the line.

39.

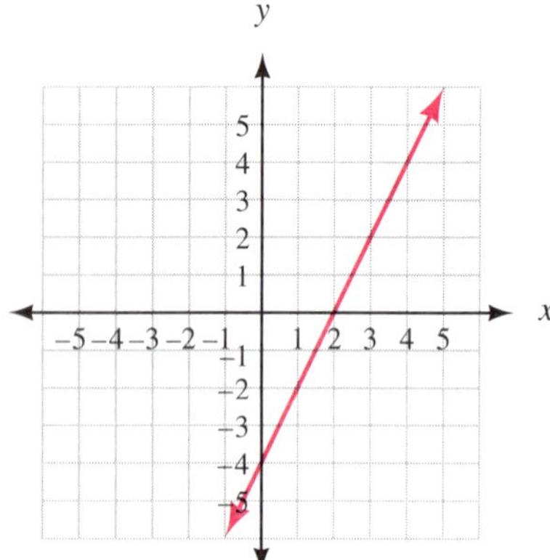

40.

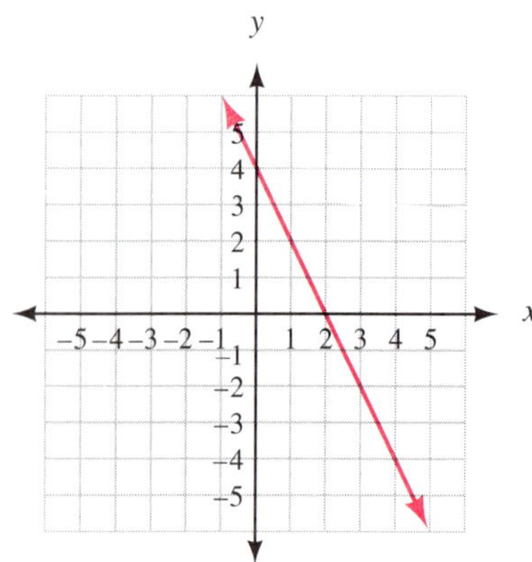

41.

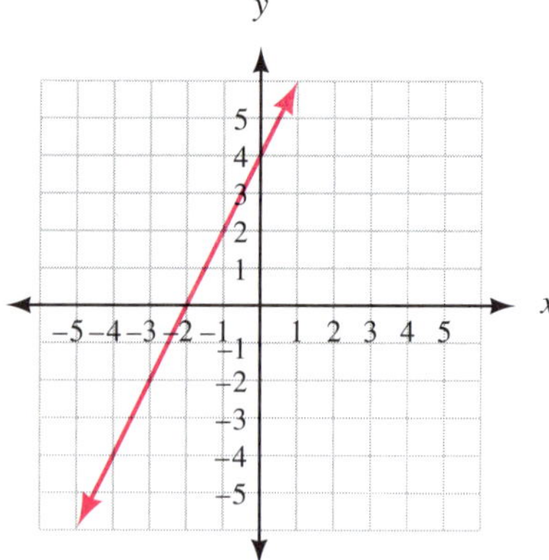

42.

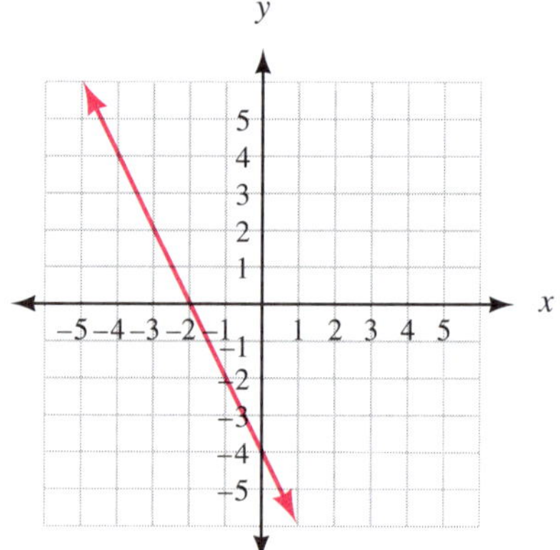

43. Graph each of the following lines. In each case, name the slope, the x-intercept, and the y-intercept.

(a) $y = \dfrac{1}{2}x$ **(b)** $x = 3$ **(c)** $y = -2$

44. Graph each of the following lines. In each case, name the slope, the x-intercept, and the y-intercept.

(a) $y = -2x$ **(b)** $x = 2$ **(c)** $y = -4$

45. Find the equation of the line parallel to the graph of $3x - y = 5$ that contains the point $(-1, 4)$.

46. Find the equation of the line parallel to the graph of $2x - 4y = 5$ that contains the point $(0, 3)$.

47. Line l is perpendicular to the graph of the equation $2x - 5y = 10$ and contains the point $(-4, -3)$. Find the equation for l.

48. Line l is perpendicular to the graph of the equation $-3x - 5y = 2$ and contains the point $(2, -6)$. Find the equation for l.

49. Give the equation of the line perpendicular to the graph of $y = -4x + 2$ that has an x-intercept of -1.

50. Write the equation of the line parallel to the graph of $7x - 2y = 14$ that has an x-intercept of 5.

51. Find the equation of the line with x-intercept 3 and y-intercept 2.

52. Find the equation of the line with x-intercept 2 and y-intercept 3.

Applying the Concepts

53. Deriving the Temperature Equation The table below resembles the table from the introduction to this section. The rows of the table give us ordered pairs (C, F).

DEGREES CELSIUS	DEGREES FAHRENHEIT
C	F
0	32
25	77
50	122
75	167
100	212

(a) Use any two of the ordered pairs from the table to derive the equation $F = \frac{9}{5}C + 32$.

(b) Use the equation from part (a) to find the Fahrenheit temperature that corresponds to a Celsius temperature of 30°.

54. Maximum Heart Rate The table below gives the maximum heart rate for adults 30, 40, 50, and 60 years old. Each row of the table gives us an ordered pair (A, M).

AGE (YEARS)	MAXIMUM HEART RATE (BEATS PER MINUTE)
A	M
30	190
40	180
50	170
60	160

(a) Use any two of the ordered pairs from the table to derive the equation $M = 220 - A$, which gives the maximum heart rate M for an adult whose age is A.

(b) Use the equation from part (a) to find the maximum heart rate for a 25-year-old adult.

55. Oxygen Consumption and Exercise A person's oxygen consumption (in liters per minute) is linearly related to that person's heart rate (in beats per minute). Suppose that at 98 beats per minute a person consumes 1 liter of oxygen per minute, and after exercising with a heart rate of 155 beats per minute consumes 1.5 liters of oxygen per minute.

(a) Find an equation that relates oxygen consumption to heart rate.

(b) What restrictions seem reasonable on values of the independent variable?

56. Exercise Heart Rate In an aerobics, class, the instructor indicates that her students' exercise heart rate is 60% of their maximum heart rate, where maximum heart rate is 220 minus their age.

(a) Determine the equation that gives exercise heart rate E in terms of age A.

(b) Use the equation to find the exercise heart rate of a 22-year-old student.

(c) Sketch the graph of the equation for students from 18 to 80 years of age.

57. AIDS Cases The number of AIDS cases in the United States from 1984 through 1988 increased in a linear relationship. In 1984, there were 3,000 known cases, and in 1988, the number of cases had risen to 20,000.

(a) Write the equation of the line that relates the number of AIDS cases to the year for the years 1984 through 1988.

(b) Use the equation found in part (a) to determine how many AIDS cases were reported in 1986.

58. Automobile Expenses The variable cost for operating an automobile is considered to be the costs for "gas and oil," maintenance, and tires. The variable cost (in cents per mile) for operating an automobile in the United States during a given year between 1990 and 1997 is given in the table that follows.

YEAR	x	VARIABLE COST C (IN CENTS PER MILE)
1990	0	8.4
1992	2	9.1
1993	3	9.3
1994	4	9.2
1995	5	10.0
1996	6	10.1
1997	7	10.8

Source: American Automobile Manufacturers Association, Inc., Detroit, MI, *Motor Vehicle Facts and Figures,* annual.

(a) Let x be the number of years since January 1, 1990, and interpret the table data as ordered pairs, (x, C). Find an equation for the line through the points (3, 9.3) and (5, 10).

(b) Interpret the "slope" of this line in terms of change in variable cost per year.

(c) Use the equation as the "predicting equation" for the variable cost of operating an auto during a year between 1990 and 1997. Find the predicted variable cost for 1992, 1993, 1999, and 2002, from the equation instead of the table.

(d) Are you confident using this equation to predict for the years 1999 and 2002? Explain.

59. Motor Vehicle Accidents The table below gives data indicating the number of deaths (within 30 days of the accident) per year from motor vehicle accidents. The data were collected for the years 1972–1997.

FATAL MOTOR VEHICLE ACCIDENTS IN THE UNITED STATES

YEAR	x	NUMBER N (IN THOUSANDS)
1972	2	54.6
1980	10	51.1
1985	15	43.8
1990	20	44.6
1994	24	40.7
1996	26	41.9

Source: U.S. National Highway Traffic Administration, Unpublished data from Fatal Accident Reporting System.

(a) Draw a scatter diagram for these data (or use the scatter plot function on your graphing calculator). Use ordered pairs (x, N), where x is the number of years since January 1, 1970.

(b) Find the equation for the line through the points (10, 51.1) and (24, 40.7), then draw this line on the scatter diagram from part (a).

(c) Use the equation from part (b) to predict the number of deaths resulting from motor vehicle accidents in the years 1975 and 2000.

(d) Do these estimates in part (c) appear to be reasonable? Why or why not?

60. Motor Vehicle Accidents Use the information in Problem 59 to work the following problems.

(a) Find the equation for the line through the points (2, 54.6) and (26, 41.9). Draw the line through these points.

(b) Use this equation to predict the number of deaths resulting from motor vehicle accidents in the years 1975 and 2000.

Review Problems

61. The length of a rectangle is 3 inches more than 4 times the width. The perimeter is 56 inches. Find the length and width.

62. One angle is 10 degrees less than 4 times another. Find the measure of each angle if

(a) The two angles are complementary.

(b) The two angles are supplementary.

63. The cash register in a candy shop contains \$66 at the beginning of the day. At the end of the day, it contains \$732.50. If the sales tax rate is 7.5%, how much of the total is sales tax?

64. The third angle in an isosceles triangle is 20 degrees less than twice as large as each of the two base angles. Find the measure of each angle.

Extending the Concepts

The midpoint M of two points (x_1, y_1) and (x_2, y_2) is defined to be the average of each of their coordinates, so

$$M = \left(\frac{x_1 + x_2}{2}, \frac{y_1 + y_2}{2}\right)$$

For example, the midpoint of $(-2, 3)$ and $(6, 8)$ is given by

$$\left(\frac{-2+6}{2}, \frac{3+8}{2}\right) = \left(2, \frac{11}{2}\right)$$

For each given pair of points, find the equation of the line perpendicular to the line through these points that passes through their midpoint. Answer using slope-intercept form.

NOTE This line is called the perpendicular bisector of the line segment connecting the two points.

65. (1, 4) and (7, 8)

66. $(-2, 1)$ and (6, 7)

67. $(-5, 1)$ and $(-1, 4)$

68. $(-6, -2)$ and (2, 1)

Write each equation in slope-intercept form. Then name the slope, the y-intercept, and the x-intercept.

69. $\frac{x}{2} + \frac{y}{3} = 1$

70. $\frac{x}{5} + \frac{y}{4} = 1$

71. $\frac{x}{-2} + \frac{y}{3} = 1$

72. $\frac{x}{2} + \frac{y}{-3} = 1$

73. When a linear equation is written in the form

$$\frac{x}{a} + \frac{y}{b} = 1$$

it is said to be in *two-intercept form.* Find the x-intercept, the y-intercept, and the slope of this line.

2.4 Linear Inequalities in Two Variables

A small movie theater holds 100 people. The owner charges more for adults than for children, so it is important to know the different combinations of adults and children that can be seated at one time. The shaded region in Figure 1 contains all the seating combinations. The line $x + y = 100$ shows the combinations for a full theater: The y-intercept corresponds to a theater full of adults, and the x-intercept corresponds to a theater full of children. In the shaded region below the line $x + y = 100$ are the combinations that occur if the theater is not full.

Shaded regions like the one shown in Figure 1 are produced by linear inequalities in two variables, which is the topic of this section.

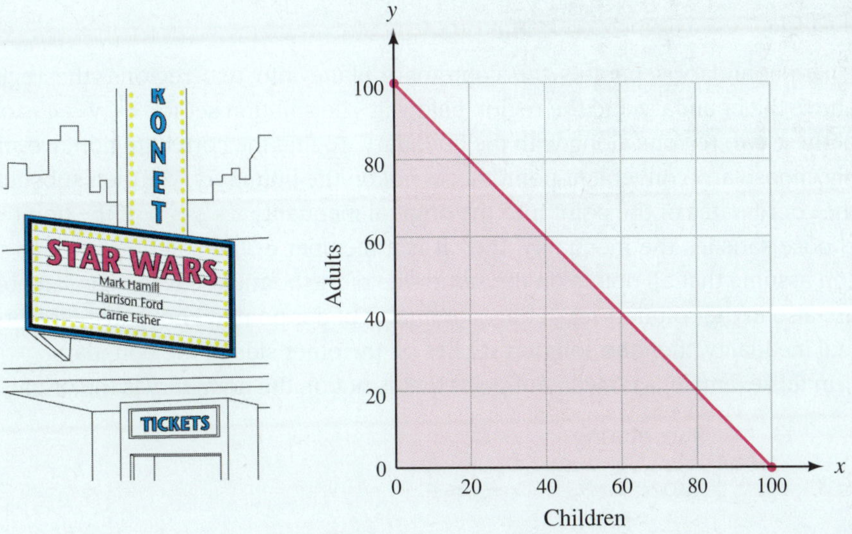

Figure 1

A *linear inequality in two variables* is any expression that can be put in the form

$$ax + by < c$$

where a, b, and c are real numbers (a and b not both 0). The inequality symbol can be any one of the following four: $<, \le, >, \ge$.

Some examples of linear inequalities are

$$2x + 3y < 6 \qquad y \ge 2x + 1 \qquad x - y \le 0$$

Although not all of these examples have the form $ax + by < c$, each one can be put in that form.

The solution set for a linear inequality is a *section of the coordinate plane.* The *boundary* for the section is found by replacing the inequality symbol with an equal sign and graphing the resulting equation. The boundary is included in the solution set (and represented with a *solid line*) if the inequality symbol used originally is $\le$ or $\ge$. The boundary is not included (and is represented with a *broken line*) if the original symbol is $<$ or $>$.

EXAMPLE 1 Graph the solution set for $x + y \le 4$.

SOLUTION The boundary for the graph is the graph of $x + y = 4$. The boundary is included in the solution set because the inequality symbol is $\le$.

Practice Problems

1. Graph the solution set for $x - y \ge 3$. Follow Example 1 carefully. First graph the boundary. Then shade in the correct region after testing a point not on the boundary.

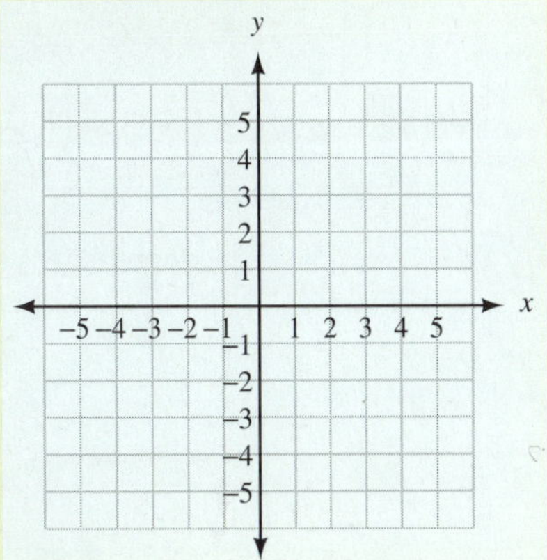

Answers

See Solutions to Selected Practice Problems for all answers in this section.

Figure 2 is the graph of the boundary:

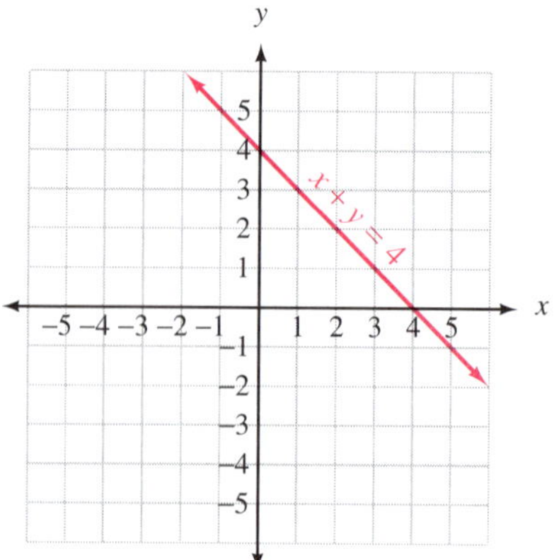

Figure 2

The boundary separates the coordinate plane into two regions: the region above the boundary and the region below it. The solution set for $x + y \le 4$ is one of these two regions along with the boundary. To find the correct region, we simply choose any convenient point that is *not* on the boundary. We then substitute the coordinates of the point into the original inequality $x + y \le 4$. If the point we choose satisfies the inequality, then it is a member of the solution set, and we can assume that all points on the same side of the boundary as the chosen point are also in the solution set. If the coordinates of our point do not satisfy the original inequality, then the solution set lies on the other side of the boundary.

In this example, a convenient point that is not on the boundary is the origin.

Substituting	$(0, 0)$	
into	$x + y \le 4$	
gives us	$0 + 0 \le 4$	
	$0 \le 4$	**A true statement**

Because the origin is a solution to the inequality $x + y \le 4$ and the origin is below the boundary, all other points below the boundary are also solutions.

Figure 3 is the graph of $x + y \le 4$.

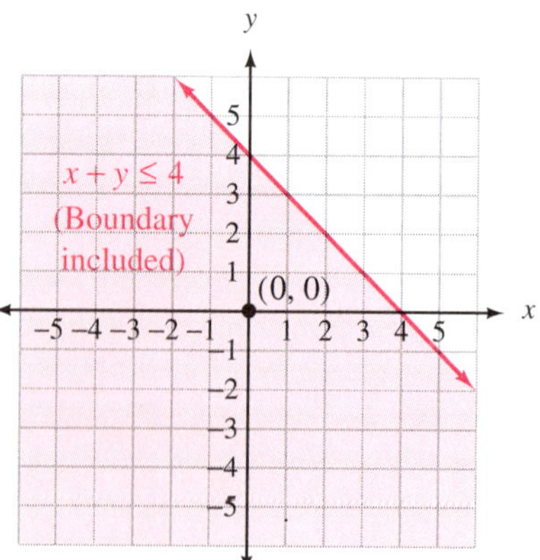

Figure 3

The region above the boundary is described by the inequality $x + y > 4$.

Here is a list of steps to follow when graphing the solution set for linear inequalities in two variables.

To Graph a Linear Inequality in Two Variables

Step 1: Replace the inequality symbol with an equal sign. The resulting equation represents the boundary for the solution set.

Step 2: Graph the boundary found in step 1 using a *solid line* if the boundary is included in the solution set (that is, if the original inequality symbol was either $\leq$ or $\geq$). Use a *broken line* to graph the boundary if it is *not* included in the solution set. (It is not included if the original inequality was either $<$ or $>$.)

Step 3: Choose any convenient point not on the boundary and substitute the coordinates into the *original* inequality. If the resulting statement is *true*, the graph lies on the *same* side of the boundary as the chosen point. If the resulting statement is *false*, the solution set lies on the *opposite* side of the boundary.

EXAMPLE 2 Graph the solution set for $y < 2x - 3$.

SOLUTION The boundary is the graph of $y = 2x - 3$, a line with slope 2 and y-intercept -3. The boundary is not included because the original inequality symbol is $<$. We therefore use a broken line to represent the boundary in Figure 4.

2. Graph the solution set for $y < \frac{1}{2}x + 3$

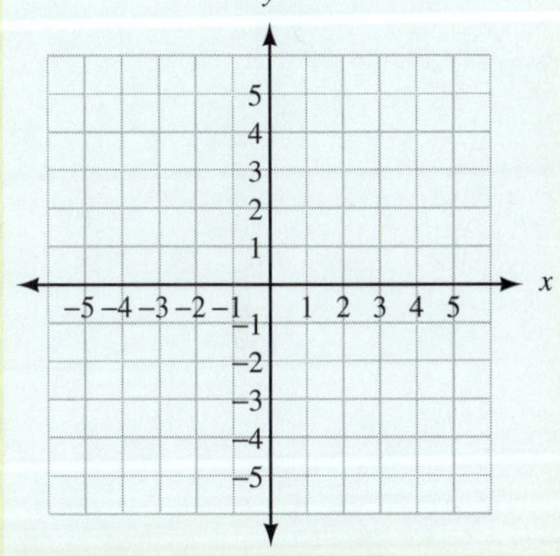

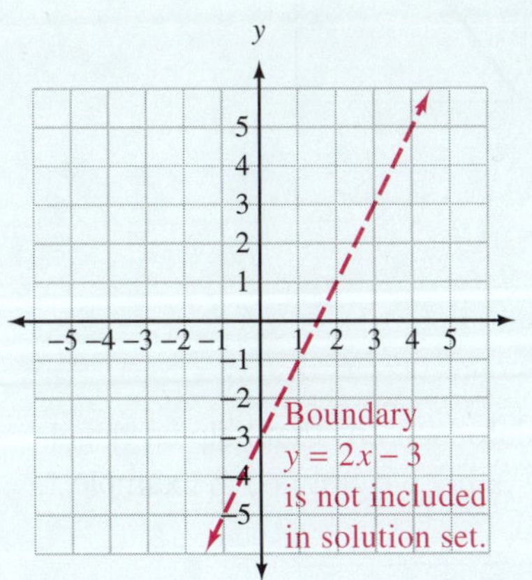

Figure 4

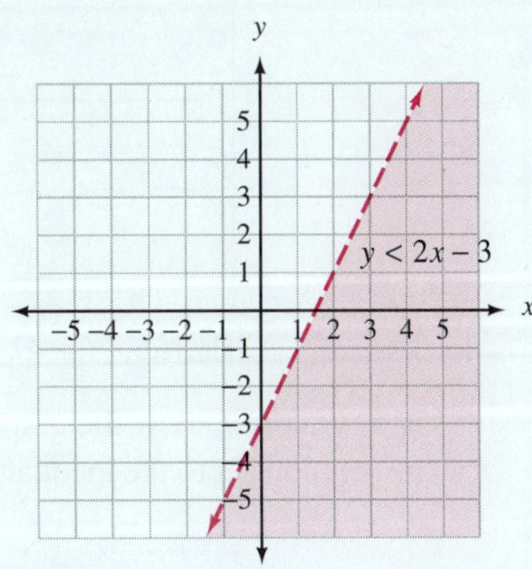

Figure 5

A convenient test point is again the origin:

Using	$(0, 0)$	
in	$y < 2x - 3$	
we have	$0 < 2(0) - 3$	
	$0 < -3$	**A false statement**

Because our test point gives us a false statement and it lies above the boundary, the solution set must lie on the other side of the boundary (Figure 5).

USING TECHNOLOGY

Graphing Calculators

Most graphing calculators have a Shade command that allows a portion of a graphing screen to be shaded. With this command we can visualize the solution sets to linear inequalities in two variables. Because most graphing calculators cannot draw a dotted line, however, we are not actually "graphing" the solution set, only visualizing it.

Strategy for Visualizing a Linear Inequality in Two Variables on a Graphing Calculator

Step 1: Solve the inequality for y.

Step 2: Replace the inequality symbol with an equal sign. The resulting equation represents the boundary for the solution set.

Step 3: Graph the equation in an appropriate viewing window.

Step 4: Use the Shade command to indicate the solution set:

For inequalities having the $<$ or $\leq$ sign, use Shade(Xmin, Y_1).

For inequalities having the $>$ or $\geq$ sign, use Shade(Y_1, Xmax).

Note: On the TI-83, step 4 can be done by manipulating the icons in the left column in the list of Y variables.

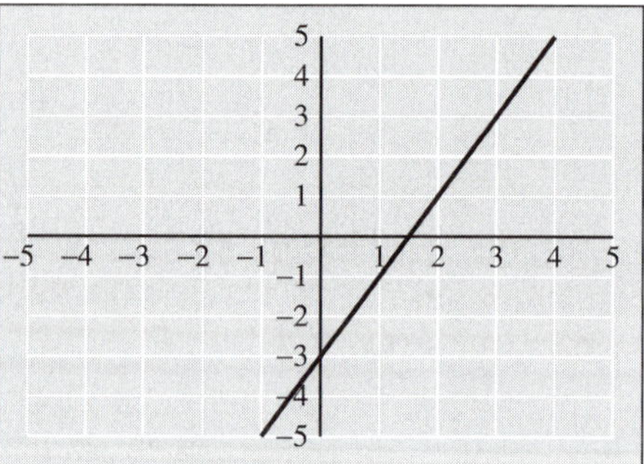

Figure 6 $Y_1 = 2X - 3$

Figures 6 and 7 show the graphing calculator screens that help us visualize the solution set to the inequality $y < 2x - 3$ that we graphed in Example 2.

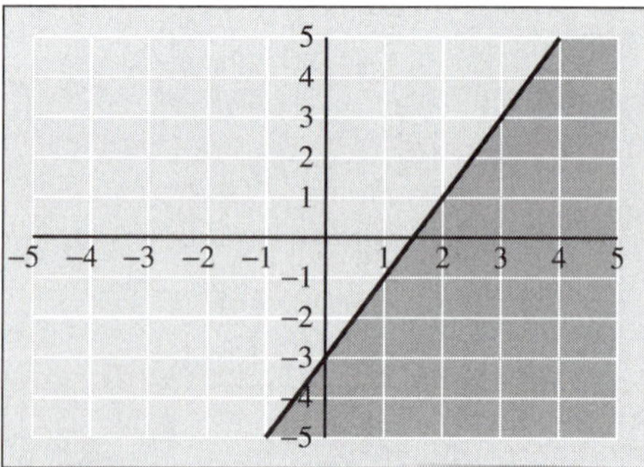

Figure 7 Shade (Xmin, Y_1)

Windows: X from -5 to 5, Y from -5 to 5

EXAMPLE 3 Graph the solution set for $x \le 5$.

SOLUTION The boundary is $x = 5$, which is a vertical line. All points in Figure 8 to the left of the boundary have x-coordinates less than 5 and all points to the right have x-coordinates greater than 5.

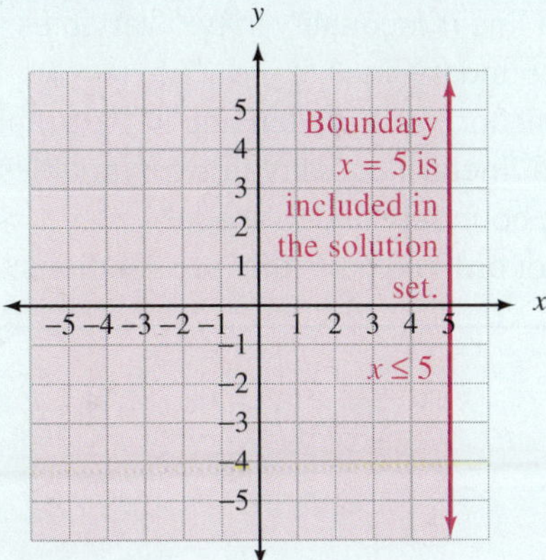

Figure 8

EXAMPLE 4 Graph the solution set for $y > \frac{1}{4}x$.

SOLUTION The boundary is the line $y = \frac{1}{4}x$, which has a slope of $\frac{1}{4}$ and passes through the origin. The graph of the boundary line is shown in Figure 9. Since the boundary passes through the origin, we cannot use the origin as our test point. Remember, the test point cannot be on the boundary line. Let's use the point $(0, -4)$ as our test point. It lies below the boundary line. When we substitute the coordinates into our original inequality, the result is the false statement $-4 > 0$. This tells us that the solution set is on the other side of the boundary line. The solution set for our original inequality is shown in Figure 10.

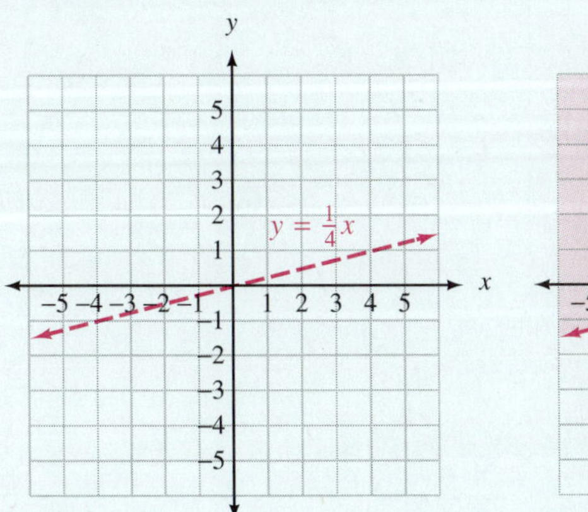

Figure 9

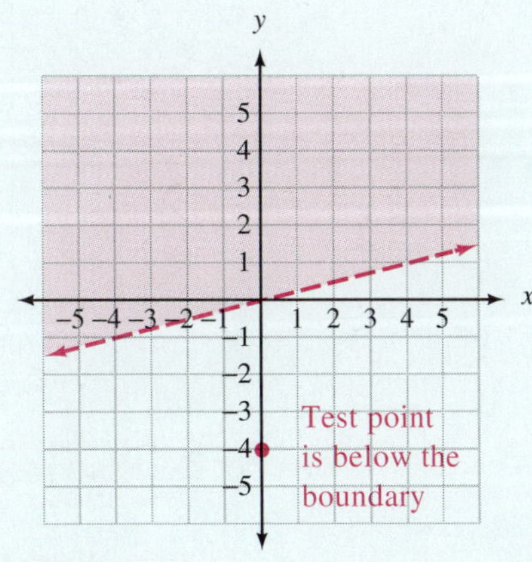

Figure 10

3. Graph $y > -2$.

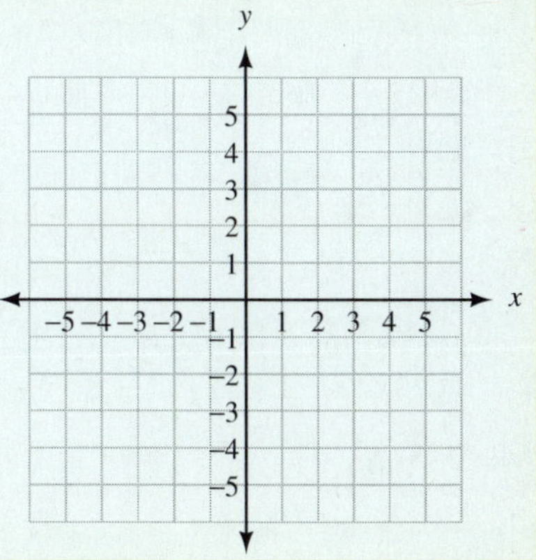

4. Graph the solution set for $y > -\frac{2}{3}x$.

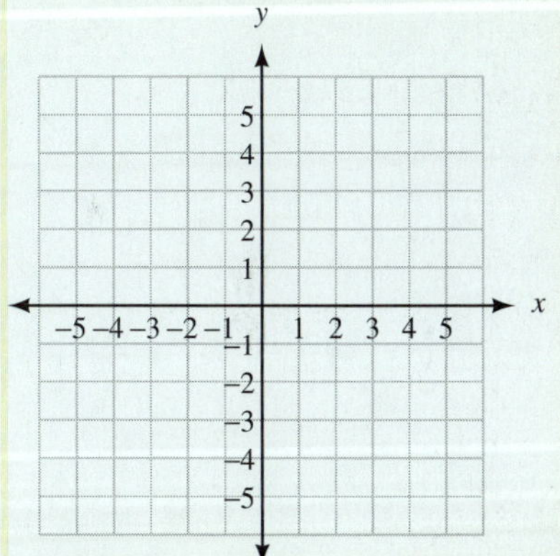

Getting Ready for Class

After reading through the preceding section, respond in your own words and in complete sentences.

A. When graphing a linear inequality in two variables, how do you find the equation of the boundary line?

B. What is the significance of a broken line in the graph of an inequality?

C. When graphing a linear inequality in two variables, how do you know which side of the boundary line to shade?

D. Describe the set of ordered pairs that are solutions to $x + y < 6$.

PROBLEM SET 2.4

Graph the solution set for each of the following.

1. $x + y < 5$
2. $x + y \leq 5$
3. $x - y \geq -3$
4. $x - y > -3$
5. $2x + 3y < 6$
6. $2x - 3y > -6$
7. $-x + 2y > -4$
8. $-x - 2y < 4$
9. $2x + y < 5$
10. $2x + y < -5$
11. $y < 2x - 1$
12. $y \leq 2x - 1$
13. $3x - 4y < 12$
14. $-2x + 3y < 6$
15. $-5x + 2y \leq 10$
16. $4x - 2y \leq 8$

For each graph shown here, name the linear inequality in two variables that is represented by the shaded region.

17.

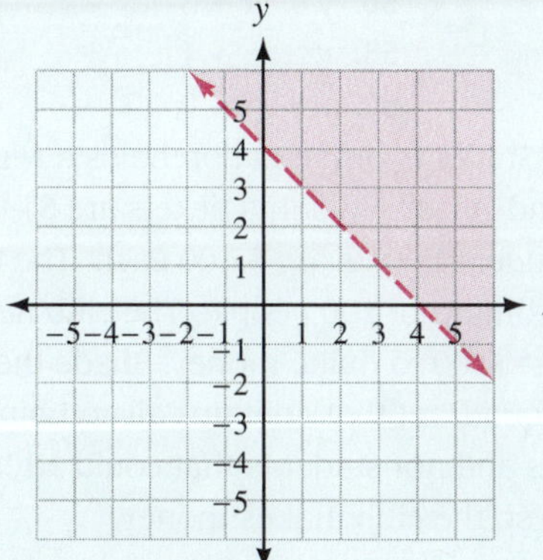

18.

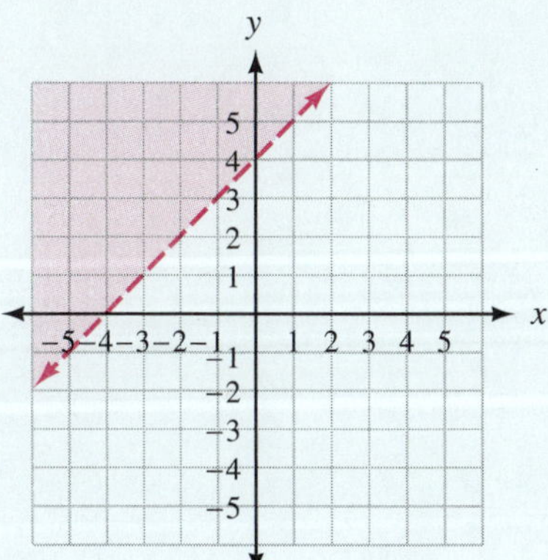

19.

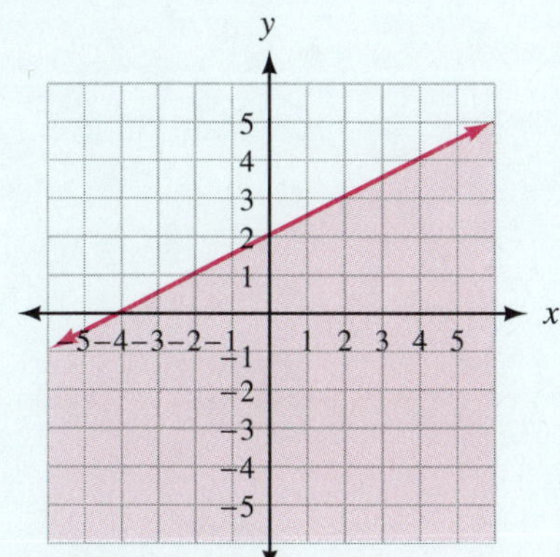

20.

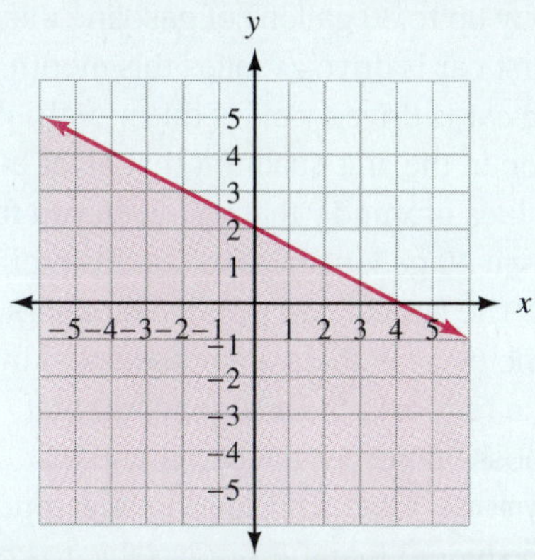

Graph each inequality.

21. $x \geq 3$
22. $x > -2$
23. $y \leq 4$
24. $y > -5$
25. $y < 2x$
26. $y > -3x$
27. $y \geq \frac{1}{2}x$
28. $y \leq \frac{1}{3}x$
29. $y \geq \frac{3}{4}x - 2$
30. $y > -\frac{2}{3}x + 3$
31. $\frac{x}{3} + \frac{y}{2} > 1$
32. $\frac{x}{5} + \frac{y}{4} < 1$
33. $\frac{x}{3} - \frac{y}{2} > 1$
34. $-\frac{x}{4} - \frac{y}{3} > 1$
35. $y \leq -\frac{2}{3}x$
36. $y \geq \frac{1}{4}x$
37. $5x - 3y < 0$
38. $2x + 3y > 0$
39. $\frac{x}{4} + \frac{y}{5} \leq 1$
40. $\frac{x}{2} + \frac{y}{3} < 1$

Applying the Concepts

41. **Number of People in a Dance Club** A dance club holds a maximum of 200 people. The club charges one price for students and a higher price for nonstudents. If the number of students in the club at any time is x and the number of nonstudents is y, shade the region in the first quadrant that contains all combinations of students and nonstudents that are in the club at any time.
42. **Many Perimeters** Suppose you have 500 feet of fencing that you will use to build a rectangular livestock pen. Let x represent the length of the pen and y represent the width. Shade the region in the first quadrant that contains all possible values of x and y that will give you a rectangle from 500 feet of fencing. (You don't have to use all of the fencing, so the perimeter of the pen could be less than 500 feet.)

43. **Gas Mileage** You have two cars. The first car travels an average of 12 miles on a gallon of gasoline, and the second averages 22 miles per gallon. Suppose you can afford to buy up to 30 gallons of gasoline this month. If the first car is driven x miles this month, and the second car is driven y miles this month, shade the region in the first quadrant that gives all the possible values of x and y that will keep you from buying more than 30 gallons of gasoline this month.
44. **Number Problem** The sum of two positive numbers is at most 20. If the two numbers are represented by x and y, shade the region in the first quadrant that shows all the possibilities for the two numbers.
45. **Student Loan Payments** When considering how much debt to incur in student loans, it is advisable to keep your student loan payment after graduation to 8% or less of your starting monthly income. Let x represent your starting monthly salary and let y represent your monthly student loan payment, and write an inequality that describes this situation. Shade the region in the first quadrant that is the solution to your inequality.
46. **Student Loan Payments** During your college career you anticipate borrowing money two times: once as an undergraduate student and once as a graduate student. You anticipate a starting monthly salary after graduate school of approximately $3,500. You are advised to keep your total in student loan payments to 8% or less of your starting monthly salary. If x represents the monthly payment on the first loan and y represents the monthly payment on the second loan, shade the region in the first quadrant representing all combinations of possible monthly payments on the two loans that will total 8% or less of your starting monthly salary.

Review Problems

Solve each of the following inequalities.

47. $\frac{1}{3} + \frac{y}{5} \leq \frac{26}{15}$
48. $-\frac{1}{3} \geq \frac{1}{6} - \frac{y}{2}$
49. $5t - 4 > 3t - 8$
50. $-3(t - 2) < 6 - 5(t + 1)$
51. $-9 < -4 + 5t < 6$
52. $-3 < 2t + 1 < 3$

Extending the Concepts

Graph each inequality.

53. $y < |x + 2|$
54. $y > |x - 2|$
55. $y > |x - 3|$
56. $y < |x + 3|$
57. $y \leq |x - 1|$
58. $y \geq |x + 1|$
59. $y < x^2$
60. $y > x^2 - 2$
61. The Associated Students organization holds a *Night at the Movies* fund-raiser. Students tickets are $1.00 each and nonstudent tickets are $2.00 each. The theater holds a maximum of 200 people. The club needs to collect at least $100 to make money. Shade the region in the first quadrant that contains all combinations of students and nonstudents that could attend the movie night so the club makes money.

2.5 Introduction to Functions

The ad shown in the margin appeared in the help wanted section of the local newspaper the day I was writing this section of the book. We can use the information in the ad to start an informal discussion of our next topic: functions.

312 Help Wanted

YARD PERSON
Full-time 40 hrs. with weekend work required. Cleaning & loading trucks. $7.50/hr. Valid CDL with clean record & drug screen required. Submit current MVR to KCI, 225 Suburban Rd., SLO. 805-555-3304.

An Informal Look at Functions

To begin with, suppose you have a job that pays $7.50 per hour and that you work anywhere from 0 to 40 hours per week. The amount of money you make in one week depends on the number of hours you work that week. In mathematics we say that your weekly earnings are a *function* of the number of hours you work. If we let the variable x represent hours and the variable y represent the money you make, then the relationship between x and y can be written as

$$y = 7.5x \quad \text{for} \quad 0 \le x \le 40$$

EXAMPLE 1 Construct a table and graph for the function

$$y = 7.5x \quad \text{for} \quad 0 \le x \le 40$$

Practice Problems

1. Construct a table and graph for
$y = 8x, \quad 0 \le x \le 40$

x	y
0	
10	
20	
30	
40	

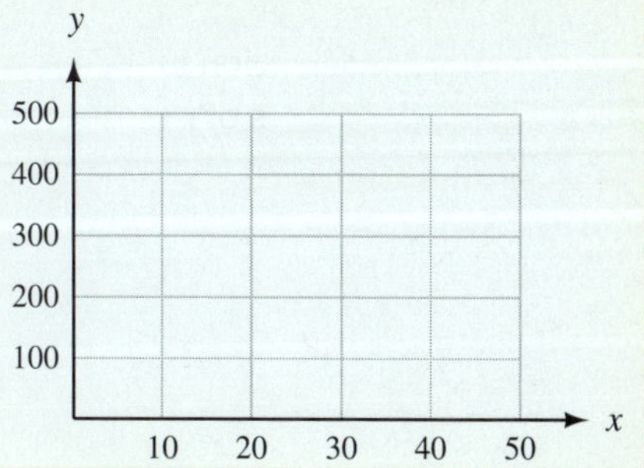

SOLUTION Table 1 gives some of the paired data that satisfy the equation $y = 7.5x$. Figure 1 is the graph of the equation with the restriction $0 \le x \le 40$.

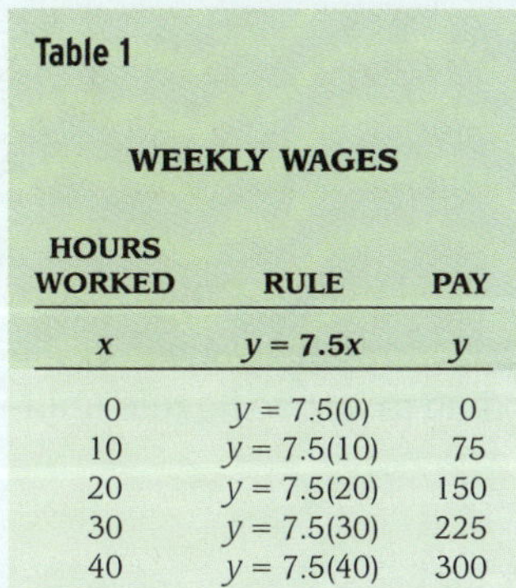

Table 1

WEEKLY WAGES

HOURS WORKED	RULE	PAY
x	$y = 7.5x$	y
0	$y = 7.5(0)$	0
10	$y = 7.5(10)$	75
20	$y = 7.5(20)$	150
30	$y = 7.5(30)$	225
40	$y = 7.5(40)$	300

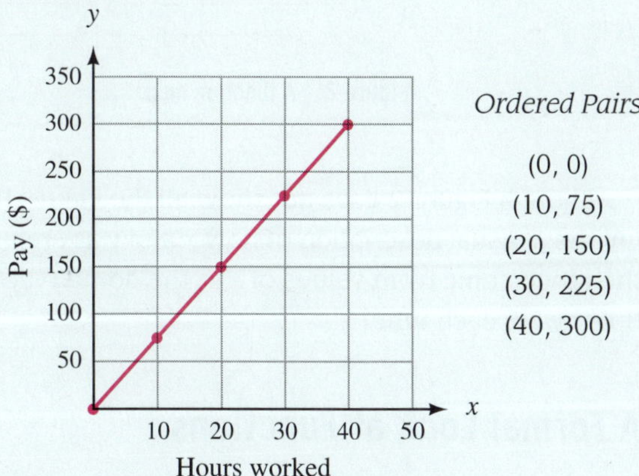

Figure 1 Weekly wages at $7.50 per hour.

The equation $y = 7.5x$ with the restriction $0 \le x \le 40$, Table 1, and Figure 1 are three ways to describe the same relationship between the number of hours you work in 1 week and your gross pay for that week. In all three, we *input* values of x, and then use the function rule to *output* values of y.

Domain and Range of a Function

We began this discussion by saying that the number of hours worked during the week was from 0 to 40, so these are the values that x can assume. From the line graph in Figure 1, we see that the values of y range from 0 to 300. We call the

Answer

1. See Solutions to Selected Practice Problems.

complete set of values that x can assume the *domain* of the function. The values that are assigned to y are called the *range* of the function.

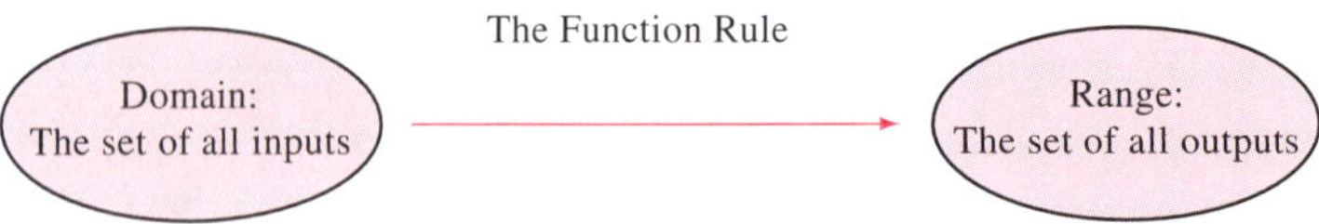

2. State the domain and range for $y = 8x, \quad 0 \le x \le 40$

EXAMPLE 2 State the domain and range for the function

$$y = 7.5x, \quad 0 \le x \le 40$$

SOLUTION From the previous discussion we have

$$\text{Domain} = \{x \mid 0 \le x \le 40\}$$

$$\text{Range} = \{y \mid 0 \le y \le 300\}$$

Function Maps

Another way to visualize the relationship between x and y is with the diagram in Figure 2, which we call a *function map:*

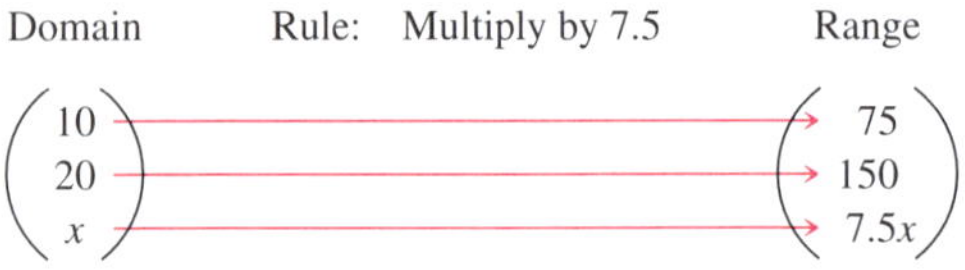

Figure 2 **A function map.**

Although Figure 2 does not show all the values that x and y can assume, it does give us a visual description of how x and y are related. It shows that values of y in the range come from values of x in the domain according to a specific rule (multiply by 7.5 each time).

A Formal Look at Functions

What is apparent from the preceding discussion is that we are working with paired data. The solutions to the equation $y = 7.5x$ are pairs of numbers; the points on the line graph in Figure 1 come from paired data; and the diagram in Figure 2 pairs numbers in the domain with numbers in the range. We are now ready for the formal definition of a function.

DEFINITION

A **function** is a rule that pairs each element in one set, called the **domain,** with exactly one element from a second set, called the **range.**

In other words, a function is a rule for which each input is paired with exactly one output.

Answer

2. Domain = $\{x \mid 0 \le x \le 40\}$
Range = $\{y \mid 0 \le y \le 320\}$

Functions as Ordered Pairs

The function rule $y = 7.5x$ from Example 1 produces ordered pairs of numbers (x, y). The same thing happens with all functions: The function rule produces ordered pairs of numbers. We use this result to write an alternative definition for a function.

> **ALTERNATIVE DEFINITION**
>
> A **function** is a set of ordered pairs in which no two different ordered pairs have the same first coordinate. The set of all first coordinates is called the **domain** of the function. The set of all second coordinates is called the **range** of the function.

The restriction on first coordinates in the alternative definition keeps us from assigning a number in the domain to more than one number in the range.

A Relationship That Is Not a Function

You may be wondering if any sets of paired data fail to qualify as functions. The answer is yes, as the next example reveals.

EXAMPLE 3 Table 2 shows the prices of used Ford Mustangs that were listed in the local newspaper. The diagram in Figure 3 is called a *scatter diagram.* It gives a visual representation of the data in Table 2. Why are these data not a function?

3. With respect to Table 2 and Figure 3, how many outputs are paired with each of the following inputs?
 (a) 1996
 (b) 1994

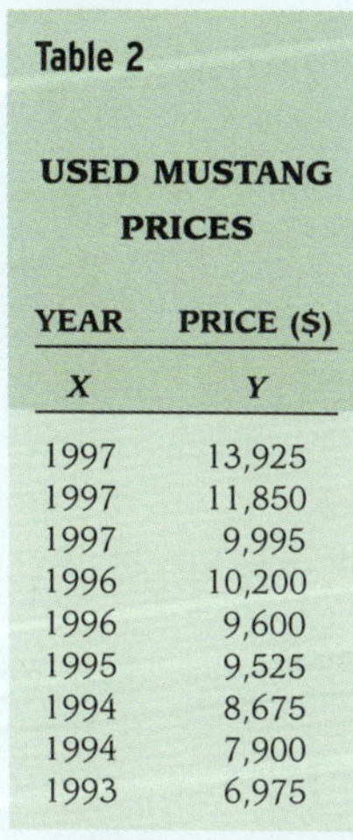

Table 2

USED MUSTANG PRICES

YEAR *X*	PRICE ($) *Y*
1997	13,925
1997	11,850
1997	9,995
1996	10,200
1996	9,600
1995	9,525
1994	8,675
1994	7,900
1993	6,975

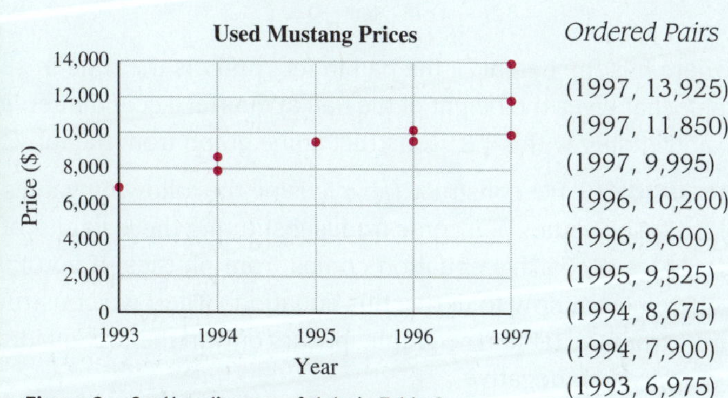

Figure 3 Scatter diagram of data in Table 2

SOLUTION In Table 2, the year 1997 is paired with three different prices: \$13,925, \$11,850, and \$9,995. That is enough to disqualify the data from belonging to a function. For a set of paired data to be considered a function, each number in the domain must be paired with exactly one number in the range. ■

Still, there is a relationship between the first coordinates and second coordinates in the used-car data. It is not a function relationship, but it is a relationship. To classify all relationships specified by ordered pairs, whether they are functions or not, we include the following two definitions.

> **DEFINITION**
>
> A **relation** is a rule that pairs each element in one set, called the **domain,** with one or more elements from a second set, called the **range.**

Answer

3. **(a)** 2 **(b)** 2

ALTERNATIVE DEFINITION

A **relation** is a set of ordered pairs. The set of all first coordinates is the **domain** of the relation. The set of all second coordinates is the **range** of the relation.

Here are some facts that will help clarify the distinction between relations and functions.

1. Any rule that assigns numbers from one set to numbers in another set is a relation. If that rule makes the assignment so that no input has more than one output, then it is also a function.
2. Any set of ordered pairs is a relation. If none of the first coordinates of those ordered pairs is repeated, the set of ordered pairs is also a function.
3. Every function is a relation.
4. Not every relation is a function.

Graphing Relations and Functions

To give ourselves a wider perspective on functions and relations, we consider some equations whose graphs are not straight lines.

EXAMPLE 4 Kendra is tossing a softball into the air with an underhand motion. The distance of the ball above her hand at any time is given by the function

$$h = 32t - 16t^2 \quad \text{for} \quad 0 \le t \le 2$$

where h is the height of the ball in feet and t is the time in seconds. Construct a table that gives the height of the ball at quarter-second intervals, starting with $t = 0$ and ending with $t = 2$. Construct a line graph from the table.

SOLUTION We construct Table 3 using the following values of t: $0, \frac{1}{4}, \frac{1}{2}, \frac{3}{4}, 1, \frac{5}{4}, \frac{3}{2}, \frac{7}{4}, 2$. The values of h come from substituting these values of t into the equation $h = 32t - 16t^2$. (This equation comes from physics. If you take a physics class, you will learn how to derive this equation.) Then we construct the graph in Figure 4 from the table. The graph appears only in the first quadrant because neither t nor h can be negative.

Table 3

TOSSING A SOFTBALL INTO THE AIR

TIME (SEC) t	FUNCTION RULE $h = 32t - 16t^2$	DISTANCE (FT) h
0	$h = 32(0) - 16(0)^2 = 0 - 0 = 0$	0
$\frac{1}{4}$	$h = 32(\frac{1}{4}) - 16(\frac{1}{4})^2 = 8 - 1 = 7$	7
$\frac{1}{2}$	$h = 32(\frac{1}{2}) - 16(\frac{1}{2})^2 = 16 - 4 = 12$	12
$\frac{3}{4}$	$h = 32(\frac{3}{4}) - 16(\frac{3}{4})^2 = 24 - 9 = 15$	15
1	$h = 32(1) - 16(1)^2 = 32 - 16 = 16$	16
$\frac{5}{4}$	$h = 32(\frac{5}{4}) - 16(\frac{5}{4})^2 = 40 - 25 = 15$	15
$\frac{3}{2}$	$h = 32(\frac{3}{2}) - 16(\frac{3}{2})^2 = 48 - 36 = 12$	12
$\frac{7}{4}$	$h = 32(\frac{7}{4}) - 16(\frac{7}{4})^2 = 56 - 49 = 7$	7
2	$h = 32(2) - 16(2)^2 = 64 - 64 = 0$	0

4. Kendra is tossing a softball into the air so that the distance h the ball is above her hand t seconds after she begins the toss is given by
$h = 48t - 16t^2, \quad 0 \le t \le 3$
Construct a table and line graph for this function.

t	h
0	
$\frac{1}{2}$	
1	
$\frac{3}{2}$	
2	
$\frac{5}{2}$	
3	

Answer

4. See Solutions to Selected Practice Problems.

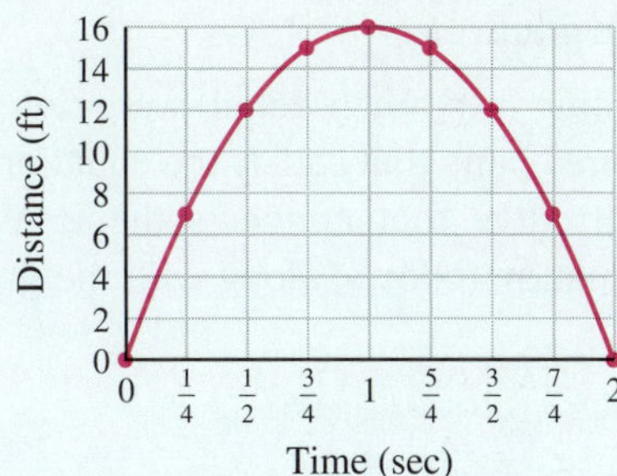

Figure 4

Here is a summary of what we know about functions as it applies to this example: We input values of t and output values of h according to the function rule

$$h = 32t - 16t^2 \quad \text{for} \quad 0 \le t \le 2$$

The domain is given by the inequality that follows the equation; it is

$$\text{Domain} = \{t \mid 0 \le t \le 2\}$$

The range is the set of all outputs that are possible by substituting the values of t from the domain into the equation. From our table and graph, it seems that the range is

$$\text{Range} = \{h \mid 0 \le h \le 16\}$$

USING TECHNOLOGY

More About Example 4

Most graphing calculators can easily produce the information in Table 3. Simply set Y_1 equal to $32X - 16X^2$. Then set up the table so it starts at 0 and increases by an increment of 0.25 each time. (On a TI-82/83, use the TBLSET key to set up the table.)

Table Setup

Table minimum = 0
Table increment = .25
Dependent variable: Auto
Independent variable: Auto

Y Variables Setup

$Y_1 = 32X - 16X^2$

The table will look like this:

X	Y1
0	0
.25	7
.5	12
.75	15
1	16
1.25	15
1.5	12

5. Use the equation

$$x = y^2 - 4$$

to fill in the following table. Then use the table to sketch the graph

x	y
	−3
	−2
	−1
	0
	1
	2
	3

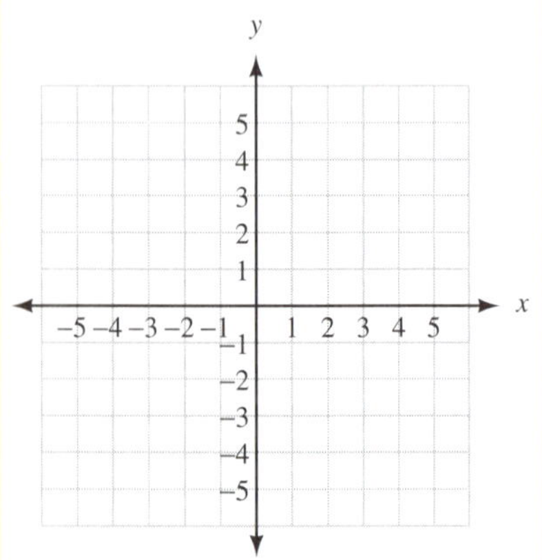

Answer

5. See Solutions to Selected Practice Problems.

EXAMPLE 5 Sketch the graph of $x = y^2$.

SOLUTION Without going into much detail, we graph the equation $x = y^2$ by finding a number of ordered pairs that satisfy the equation, plotting these points, then drawing a smooth curve that connects them. A table of values for x and y that satisfy the equation follows, along with the graph of $x = y^2$ shown in Figure 5.

x	y
0	0
1	1
1	−1
4	2
4	−2
9	3
9	−3

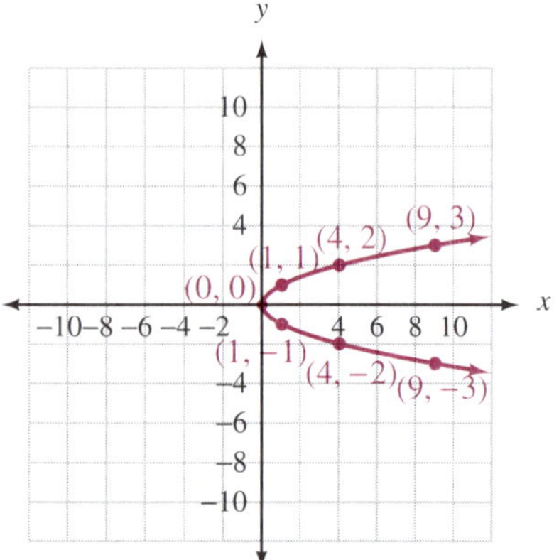

Figure 5

As you can see from looking at the table and the graph in Figure 5, several ordered pairs whose graphs lie on the curve have repeated first coordinates. For instance, (1, 1) and (1, −1), (4, 2) and (4, −2), and (9, 3) and (9, −3). The graph is therefore not the graph of a function.

Vertical Line Test

Look back at the scatter diagram for used Mustang prices shown in Figure 3. Notice that some of the points on the diagram lie above and below each other along vertical lines. This is an indication that the data do not constitute a function. Two data points that lie on the same vertical line must have come from two ordered pairs with the same first coordinates.

Now, look at the graph shown in Figure 5. The reason this graph is the graph of a relation, but not of a function, is that some points on the graph have the same first coordinates—for example, the points (4, 2) and (4, −2). Furthermore, any time two points on a graph have the same first coordinates, those points must lie on a vertical line. [To convince yourself, connect the points (4, 2) and (4, −2) with a straight line. You will see that it must be a vertical line.] This allows us to write the following test that uses the graph to determine whether a relation is also a function.

Vertical Line Test

If a vertical line crosses the graph of a relation in more than one place, the relation cannot be a function. If no vertical line can be found that crosses a graph in more than one place, then the graph is the graph of a function.

If we look back to the graph of $h = 32t - 16t^2$ as shown in Figure 4, we see that no vertical line can be found that crosses this graph in more than one place. The graph shown in Figure 4 is therefore the graph of a function.

EXAMPLE 6 Graph $y = |x|$. Use the graph to determine whether we have the graph of a function. State the domain and range.

SOLUTION We let x take on values of $-4, -3, -2, -1, 0, 1, 2, 3$, and 4. The corresponding values of y are shown in the table. The graph is shown in Figure 6.

x	y
−4	4
−3	3
−2	2
−1	1
0	0
1	1
2	2
3	3
4	4

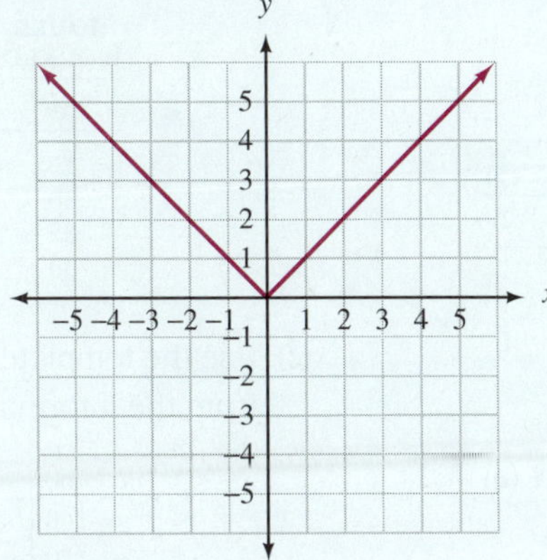

Figure 6

Because no vertical line can be found that crosses the graph in more than one place, $y = |x|$ is a function. The domain is all real numbers. The range is $\{y \mid y \geq 0\}$.

Getting Ready for Class

After reading through the preceding section, respond in your own words and in complete sentences.

A. What is a function?
B. What is the vertical line test?
C. Is every line the graph of a function? Explain.
D. Which variable is usually associated with the domain of a function?

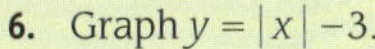

6. Graph $y = |x| - 3$.

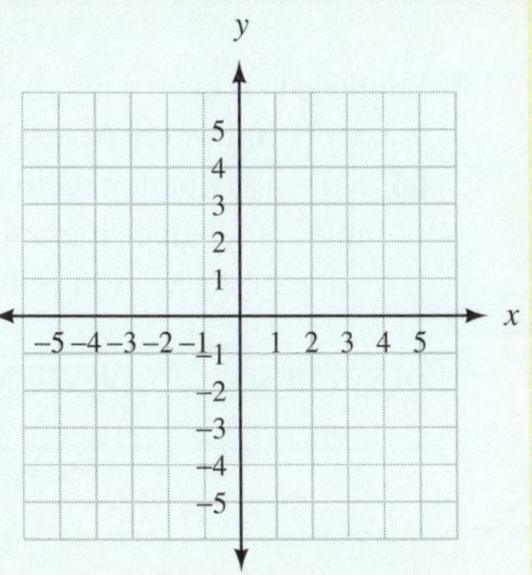

Answer

6. See Solutions to Selected Practice Problems.

PROBLEM SET 2.5

1. Suppose you have a job that pays $8.50 per hour and you work anywhere from 10 to 40 hours per week.
   (a) Write an equation, with a restriction on the variable x, that gives the amount of money, y, you will earn for working x hours in one week.

HOURS WORKED x	FUNCTION RULE	GROSS PAY ($) y
10		
20		
30		
40		

   (b) Use the function rule you have written in part (a) to complete the table.
   (c) Use the template below to construct a line graph from the information in the table.

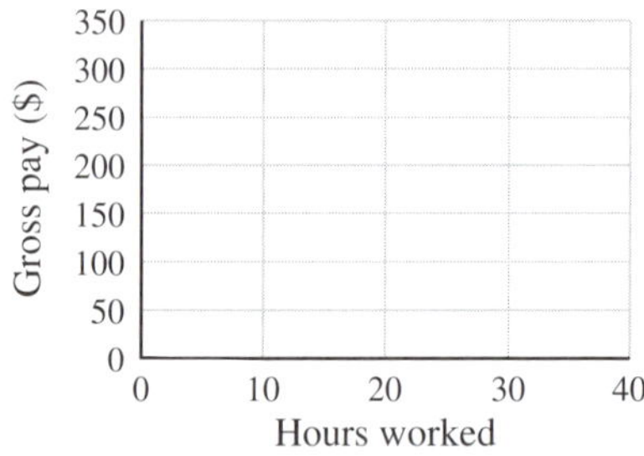

   (d) State the domain and range of this function.
   (e) What is the minimum amount you can earn in a week with this job? What is the maximum amount?

2. The ad shown here was in the local newspaper. Suppose you are hired for the job described in the ad.

312 Help Wanted

ESPRESSO BAR OPERATOR
Must be dependable, honest, service-oriented. Coffee exp desired. 15–30 hrs per wk. $5.25/hr. Start 5/31. Apply in person: Espresso Yourself, Central Coast Mall. Deadline 5/23.

   (a) If x is the number of hours you work per week and y is your weekly gross pay, write the equation for y. (Be sure to include any restrictions on the variable x that are given in the ad.)
   (b) Use the function rule you have written in part (a) to complete the table.

HOURS WORKED x	FUNCTION RULE	GROSS PAY ($) y
15		
20		
25		
30		

   (c) Use the template below to construct a line graph from the information in the table.

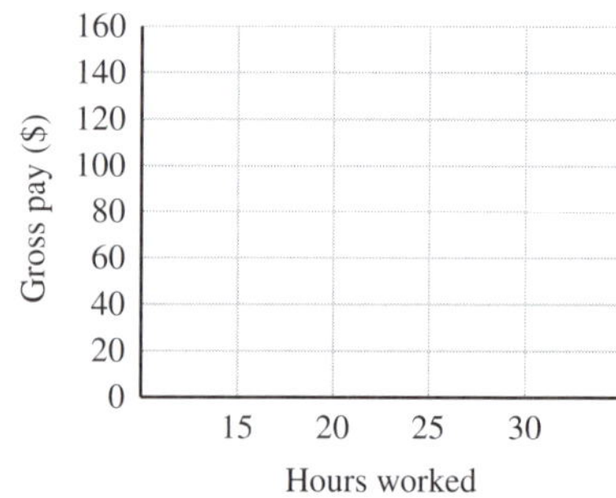

   (d) State the domain and range of this function.
   (e) What is the minimum amount you can earn in a week with this job? What is the maximum amount?

For each of the following relations, give the domain and range, and indicate which are also functions.

3. $\{(2, 5), (3, 4), (1, 4), (0, 6)\}$
4. $\{(0, 4), (1, 6), (2, 4), (1, 5)\}$
5. $\{(a, 3), (b, 4), (c, 3), (d, 5)\}$
6. $\{(a, 5), (b, 5), (c, 4), (d, 5)\}$
7. $\{(a, 1), (a, 2), (a, 3), (a, 4)\}$
8. $\{(a, 1), (b, 1), (c, 1), (d, 1)\}$

State whether each of the following graphs represents a function.

9.

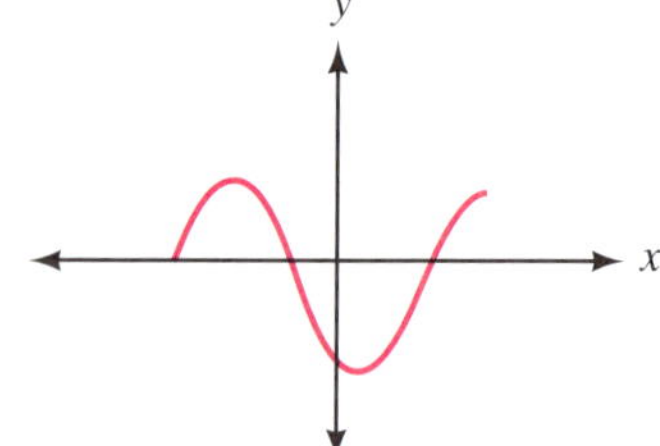

10.

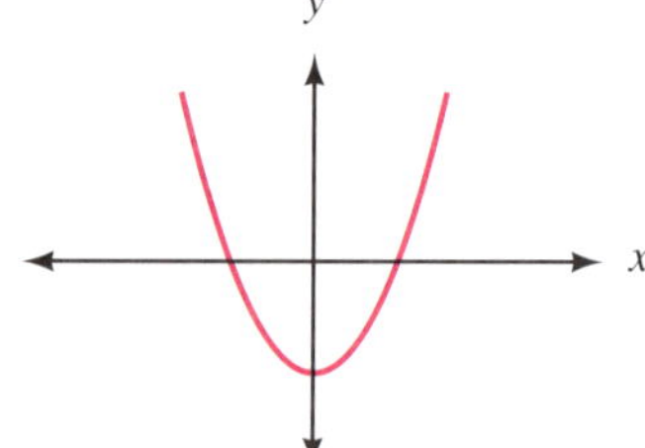

11.

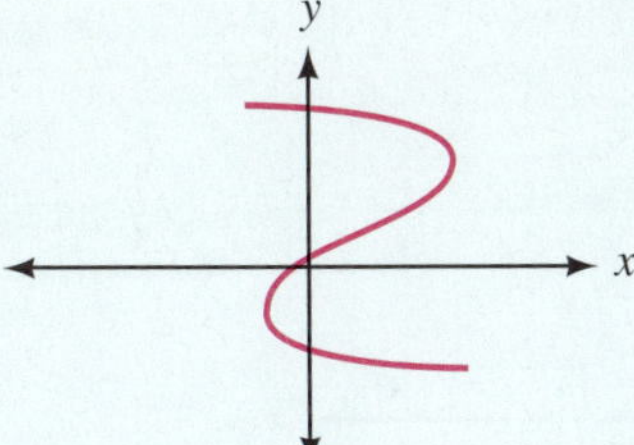

12.

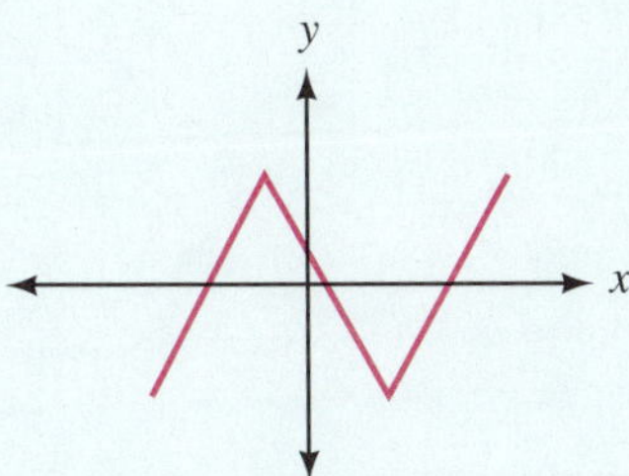

13.

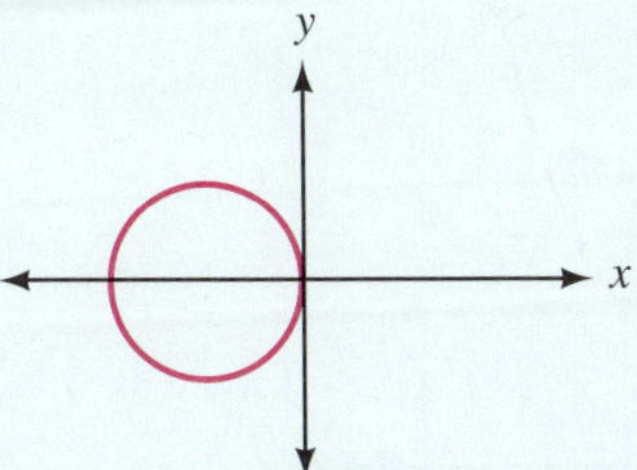

14.

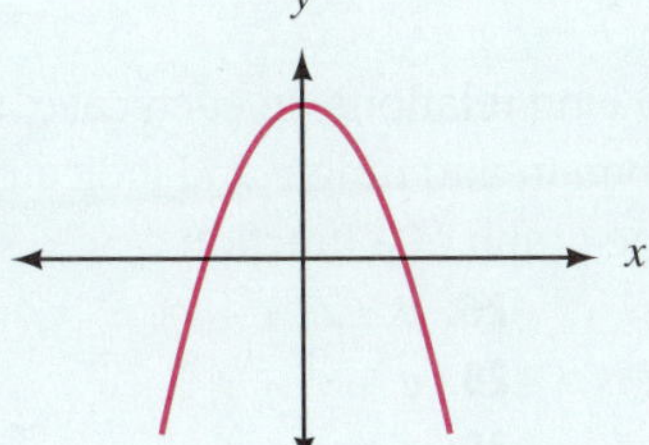

15.

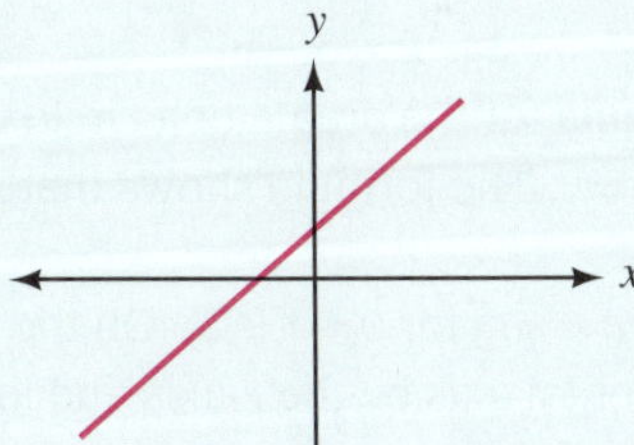

16.

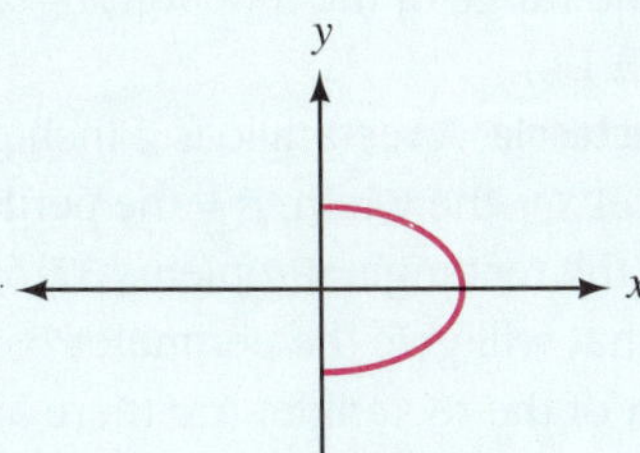

17.

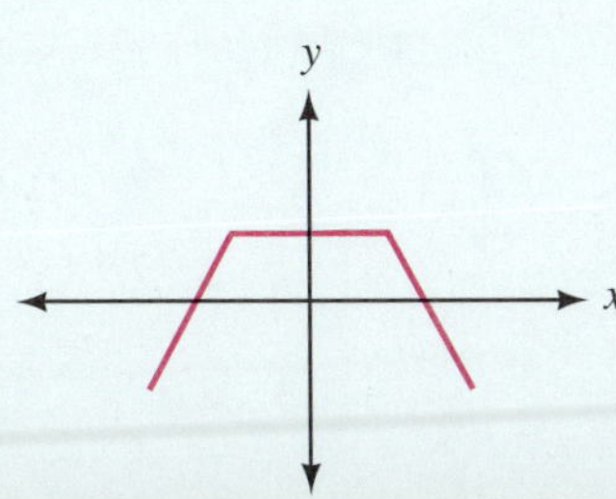

18.

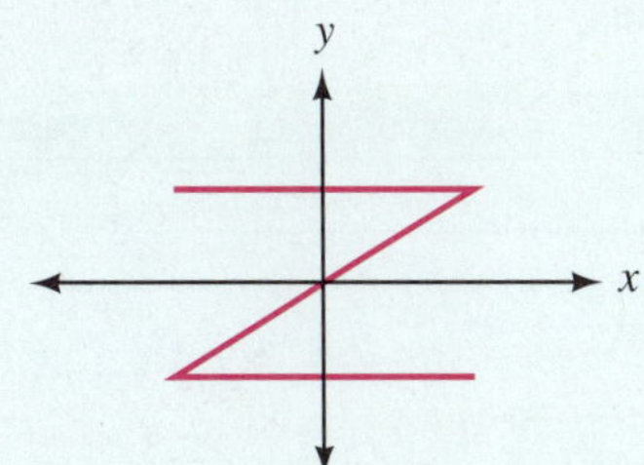

19. Tossing a Coin Hali tosses a quarter into the air with an underhand motion. The distance the quarter is above her hand at any time is given by the function

$$h = 16t - 16t^2 \quad \text{for } 0 \le t \le 1$$

where h is the height of the quarter in feet, and t is the time in seconds.

(a) Fill in the table.

TIME (sec) t	FUNCTION RULE $h = 16t - 16t^2$	DISTANCE (ft) h
0		
0.1		
0.2		
0.3		
0.4		
0.5		
0.6		
0.7		
0.8		
0.9		
1		

(b) State the domain and range of this function.

(c) Use the data from the table to graph the function.

20. Intensity of Light The following formula gives the intensity of light that falls on a surface at various distances from a 100-watt light bulb:

$$I = \frac{120}{d^2} \quad \text{for} \quad d > 0$$

where I is the intensity of light (in lumens per square foot) and d is the distance (in feet) from the light bulb to the surface.

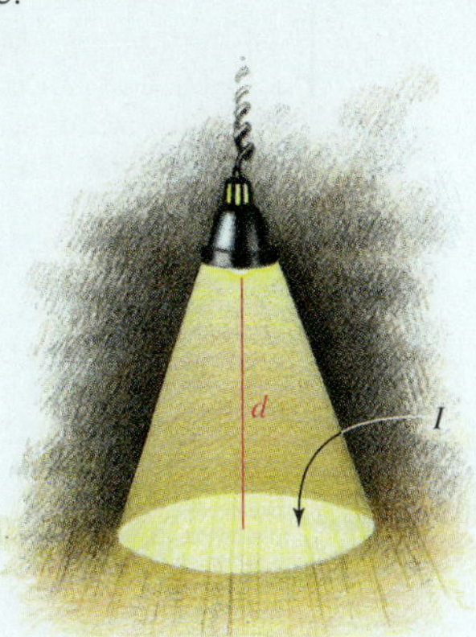

(a) Fill in the table.

DISTANCE (ft)	FUNCTION RULE	INTENSITY
d		I
1		
2		
3		
4		
5		
6		

(b) Use the data from the table in part (a) to construct a line graph of the function.

Determine the domain and range of the following functions. Assume the *entire* function is shown.

21.

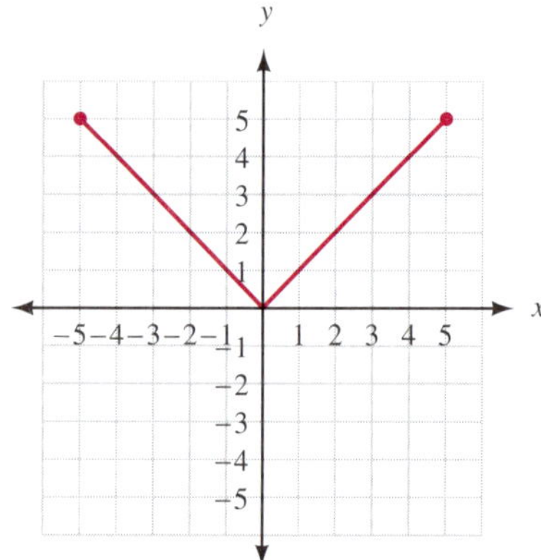

22.

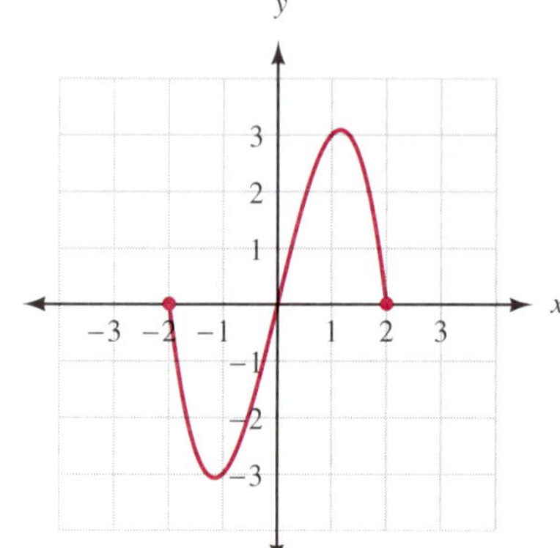

23.

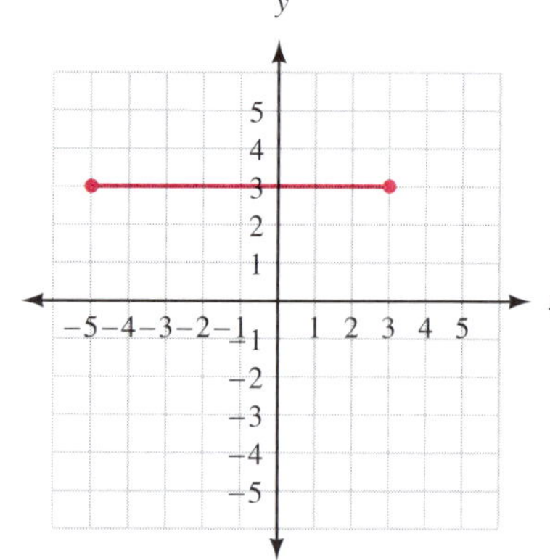

24.

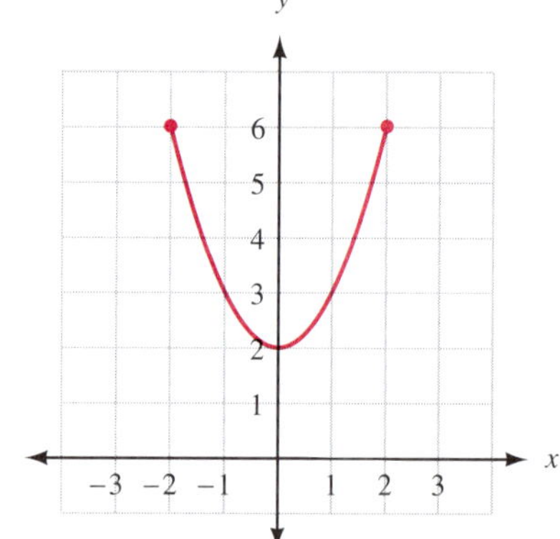

Graph each of the following relations. In each case, use the graph to find the domain and range, and indicate whether the graph is the graph of a function.

25. $y = x^2 - 1$ **26.** $y = x^2 + 1$

27. $y = x^2 + 4$ **28.** $y = x^2 - 9$

29. $x = y^2 - 1$ **30.** $x = y^2 + 1$

31. $y = (x + 2)^2$ **32.** $y = (x - 3)^2$

33. $x = (y + 1)^2$ **34.** $x = 3 - y^2$

Area of a Circle The formula for the area A of a circle with radius r is given by $A = \pi r^2$. The formula shows that A is a function of r.

35. Graph the function $A = \pi r^2$ for $0 \le r \le 3$. (On the graph, let the horizontal axis be the r-axis and let the vertical axis be the A-axis.)

36. State the domain and range of the function $A = \pi r^2$, $0 \le r \le 3$. (Use $\pi \approx 3.14$.)

Area and Perimeter of a Rectangle A rectangle is 2 inches longer than it is wide. Let x = the width, P = the perimeter, and A = the area of the rectangle (Problems 37–40).

37. Write an equation that will give the perimeter P in terms of the width x of the rectangle. Are there any restrictions on the values that x can assume?

38. Graph the relationship between P and x.

39. Write an equation that will give the area A in terms of the width x of the rectangle.

Are there any restrictions on the values that x can assume?

40. Graph the relationship between A and x.

41. Tossing a Ball A ball is thrown straight up into the air from ground level. The relationship between the height h of the ball at any time t is illustrated by the following graph:

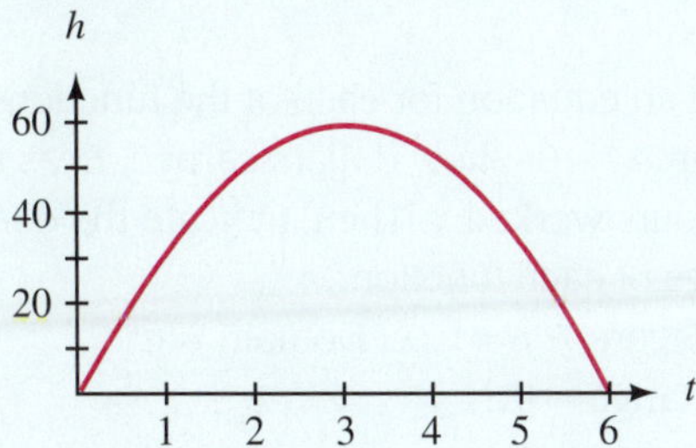

The horizontal axis represents time t, and the vertical axis represents height h.

(a) Is this graph the graph of a function?

(b) State the domain and range.

(c) At what time does the ball reach its maximum height?

(d) What is the maximum height of the ball?

(e) At what time does the ball hit the ground?

Applying the Concepts

42. Construction Statistics The function below gives the relationship between the total annual payroll for the U.S. construction industry in a given year between 1980 and 1996 (U.S. Census, *County Business Patterns,* annual).

$$\{(x, P) \mid P = 5.12x + 76.3\}$$

where P is the annual payroll in billions of dollars and x is the number of years after 1980; that is, $x = 0$ on January 1, 1980 and $0 < x < 16$.

(a) Complete the table below for the given model.

YEAR	x	PAYROLL (P) IN BILLIONS OF DOLLARS
1980	0	76.3
1985	5	101.9
1990	10	127.5
1995	15	153.1
1996	16	158.2

(b) Plot the information in the table as ordered pairs (x, P) on a rectangle grid, and draw the graph of this function. Is the graph a line or a curve?

(c) Give the domain and range of this function.

43. Death Rate Data from 1920–1997 tracking the death rate per 100,000 population from "major cardiovascular/renal" diseases give the following function as a model to estimate death rate per 100,000 population during a given year of the time period above (1900–1970: U.S. Public Health Service, *Vital Statistics of the U.S.,* annual; 1971–1997: U.S. National Center for Health Statistics, *Vital Statistics of the U.S.,* annual).

$$Y = -0.11x^2 + 12.9x + 142$$

where Y is the death rate per 100,000 and x is the number of years since January 1, 1900, and $20 < x < 97$.

(a) Complete the table below for this function.

YEAR	x	DEATH RATE PER 100,000 POPULATION
1930	30	430
1940	40	482
1950	50	512
1960	60	520
1970	70	506
1980	80	470
1990	90	412
1995	95	375

(b) Write the information in the table as ordered pairs (x, Y), and plot the corresponding points on a rectangular grid. Draw a curve through the points.

(c) Determine the maximum point on the curve. Interpret the meaning of this maximum point in terms of the "problem scenario."

(d) Find the domain and range of this function.

44. Agricultural Land The total number of acres used for agricultural purposes in the United States from 1900 to 1990 is approximated by the function

$$Y = -7|x - 55| + 1202$$

where Y is the number of acres (in millions) and x is the number of years since January 1, 1900 (U.S. Department of Agriculture, National Agricultural Statistics Service).

(a) Use the function to estimate the number of acres being used for agricultural purposes in 1900, 1920, 1940, 1955, 1970, 1985, and 1990.

(b) Plot the ordered pairs (x, Y) found in part (a) and sketch the graph of this function.

(c) Does this function have a maximum? If so, what is it?

(d) Interpret the maximum function value in terms of the relationship that the function models.

(e) Determine the domain and range for this function.

(f) Use the function to predict the number of acres used for agricultural purposes in the year 2001.

45. **Profits** Match each of the following statements to the appropriate graph indicated by Figures 7–10.
   (a) Sarah works 25 hours in a week to earn \$250.
   (b) Justin works 35 hours in a week to earn \$560.
   (c) Rosemary works 30 hours in a week to earn \$360.
   (d) Marcus works 40 hours in a week to earn \$320.

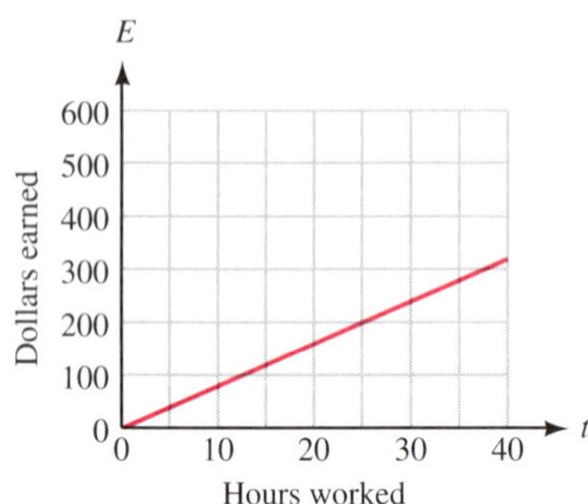

Figure 7

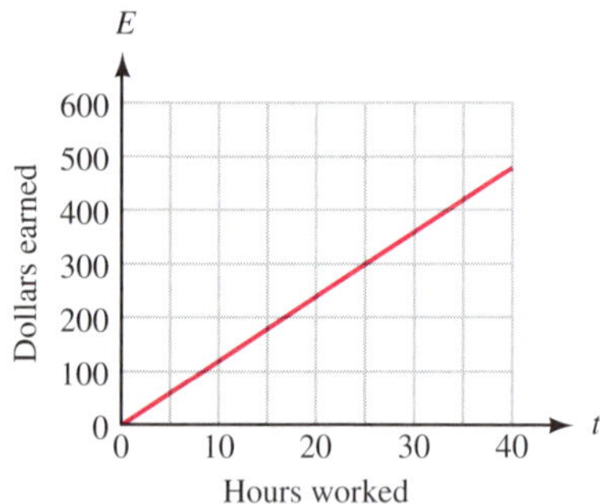

Figure 8

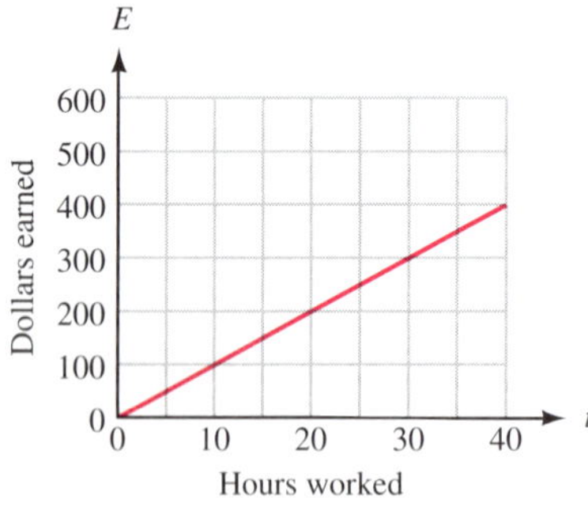

Figure 9

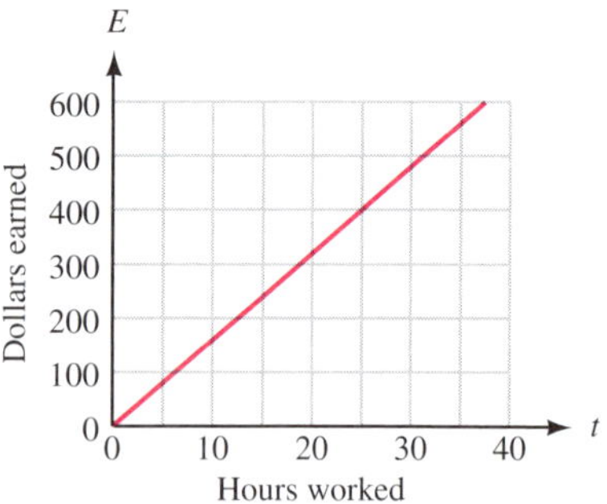

Figure 10

46. Find an equation for each of the functions shown in Figures 7–10. Show dollars earned, E, as a function of hours worked, t. Then, indicate the domain and range of each function.
   (a) Figure 7: $E =$ ____ Domain = $\{t \mid$; Range = $\{E \mid$ $\}$
   (b) Figure 8: $E =$ ____ Domain = $\{t \mid$; Range = $\{E \mid$ $\}$
   (c) Figure 9: $E =$ ____ Domain = $\{t \mid$; Range = $\{E \mid$ $\}$

Review Problems

For the equation $y = 3x - 2$:

47. Find y if x is 4.
48. Find y if x is 0.
49. Find y if x is -4.
50. Find y if x is -2.

For the equation $y = x^2 - 3$:

51. Find y if x is 2.
52. Find y if x is -2.
53. Find y if x is 0.
54. Find y if x is -4.

Extending the Concepts

Graph each of the following relations. In each case, use the graph to find the domain and range, and indicate whether the graph is the graph of a function.

55. $y = 5 - |x|$
56. $y = |x| - 3$
57. $x = |y| + 3$
58. $x = 2 - |y|$
59. $|x| + |y| = 4$
60. $2|x| + |y| = 6$

2.6 Function Notation

Let's return to the discussion that introduced us to functions. If a job pays \$7.50 per hour for working from 0 to 40 hours a week, then the amount of money y earned in 1 week is a function of the number of hours worked, x. The exact relationship between x and y is written

$$y = 7.5x \quad \text{for} \quad 0 \le x \le 40$$

Because the amount of money earned y depends on the number of hours worked x, we call y the *dependent variable* and x the *independent variable.* Furthermore, if we let f represent all the ordered pairs produced by the equation, then we can write

$$f = \{(x, y) \mid y = 7.5x \text{ and } 0 \le x \le 40\}$$

Once we have named a function with a letter, we can use an alternative notation to represent the dependent variable y. The alternative notation for y is $f(x)$. It is read "f of x" and can be used instead of the variable y when working with functions. The notation y and the notation $f(x)$ are equivalent—that is,

$$y = 7.5x \Leftrightarrow f(x) = 7.5x$$

When we use the notation $f(x)$ we are using *function notation.* The benefit of using function notation is that we can write more information with fewer symbols than we can by using just the variable y. For example, asking how much money a person will make for working 20 hours is simply a matter of asking for $f(20)$. Without function notation, we would have to say "find the value of y that corresponds to a value of $x = 20$." To illustrate further, using the variable y, we can say "y is 150 when x is 20." Using the notation $f(x)$, we simply say "$f(20) = 150$." Each expression indicates that you will earn \$150 for working 20 hours.

EXAMPLE 1 If $f(x) = 7.5x$, find $f(0)$, $f(10)$, and $f(20)$.

SOLUTION To find $f(0)$ we substitute 0 for x in the expression $7.5x$ and simplify. We find $f(10)$ and $f(20)$ in a similar manner—by substitution.

$$\begin{aligned} \text{If} \quad f(x) &= 7.5x \\ \text{then} \quad f(\mathbf{0}) &= 7.5(\mathbf{0}) = 0 \\ f(\mathbf{10}) &= 7.5(\mathbf{10}) = 75 \\ f(\mathbf{20}) &= 7.5(\mathbf{20}) = 150 \end{aligned}$$

If we changed the example in the discussion that opened this section so that the hourly wage was \$6.50 per hour, we would have a new equation to work with:

$$y = 6.5x \quad \text{for} \quad 0 \le x \le 40$$

Suppose we name this new function with the letter g. Then

$$g = \{(x, y) \mid y = 6.5x \text{ and } 0 \le x \le 40\}$$

and

$$g(x) = 6.5x$$

If we want to talk about both functions in the same discussion, having two different letters, f and g, makes it easy to distinguish between them. For example,

Practice Problems

1. If $f(x) = 8x$, find
   (a) $f(0)$
   (b) $f(5)$
   (c) $f(10.5)$

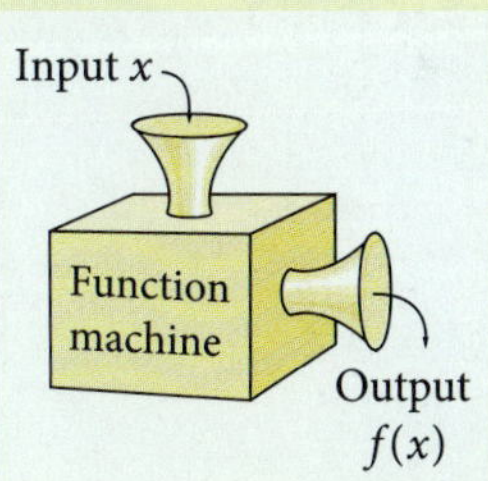

Some students like to think of functions as machines. Values of x are put into the machine, which transforms them into values of $f(x)$, which then are output by the machine.

Answer

1. **(a)** 0 **(b)** 40 **(c)** 84

because $f(x) = 7.5x$ and $g(x) = 6.5x$, asking how much money a person makes for working 20 hours is simply a matter of asking for $f(20)$ or $g(20)$, avoiding any confusion over which hourly wage we are talking about.

The diagrams shown in Figure 1 further illustrate the similarities and differences between the two functions we have been discussing.

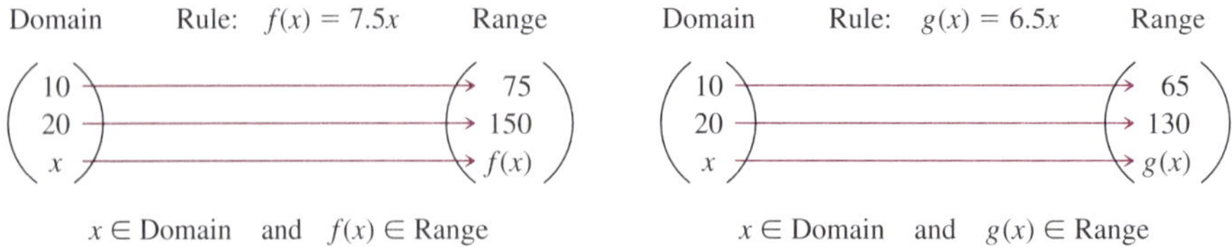

Figure 1 Function maps

Function Notation and Graphs

We can visualize the relationship between x and $f(x)$ or $g(x)$ on the graphs of the two functions. Figure 2 shows the graph of $f(x) = 7.5x$ along with two additional line segments. The horizontal line segment corresponds to $x = 20$, and the vertical line segment corresponds to $f(20)$. Figure 3 shows the graph of $g(x) = 6.5x$ along with the horizontal line segment that corresponds to $x = 20$, and the vertical line segment that corresponds to $g(20)$. (Note that the domain in each case is restricted to $0 \le x \le 40$.)

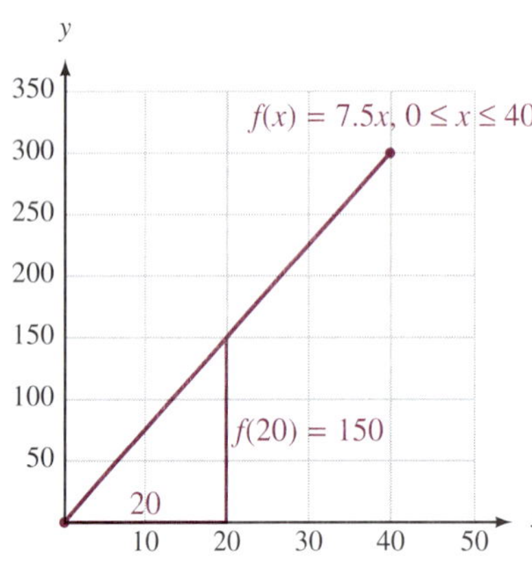

Figure 2

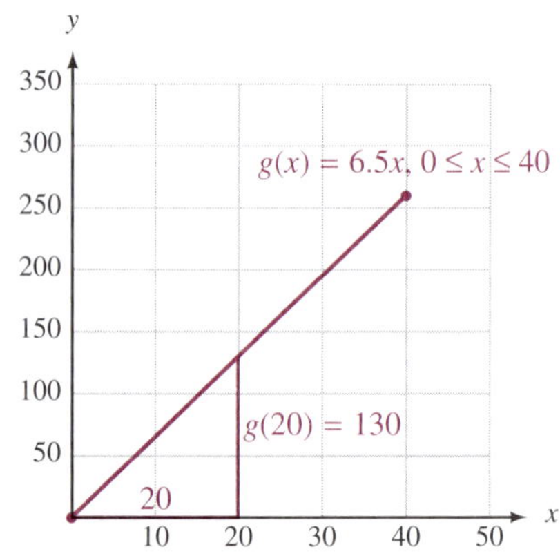

Figure 3

Using Function Notation

The remaining examples in this section show a variety of ways to use and interpret function notation.

2. When Lorena runs a mile in t minutes, then her average speed in feet per seconds is given by $s(t) = \frac{88}{t}$, $t > 0$
   (a) Find $s(8)$ and explain what it means.
   (b) Find $s(11)$ and explain what it means.

EXAMPLE 2 If it takes Lorena t minutes to run a mile, then her average speed $s(t)$ in miles per hour is given by the formula

$$s(t) = \frac{60}{t} \quad \text{for} \quad t > 0$$

Find $s(10)$ and $s(8)$, and then explain what they mean.

SOLUTION To find $s(10)$, we substitute 10 for t in the equation and simplify:

$$s(\mathbf{10}) = \frac{60}{\mathbf{10}} = 6$$

In words: When Lorena runs a mile in 10 minutes, her average speed is 6 miles per hour.

We calculate $s(8)$ by substituting 8 for t in the equation. Doing so gives us

$$s(\mathbf{8}) = \frac{60}{\mathbf{8}} = 7.5$$

In words: Running a mile in 8 minutes is running at a rate of 7.5 miles per hour.

EXAMPLE 3 A painting is purchased as an investment for \$125. If its value increases continuously so that it doubles every 5 years, then its value is given by the function

$$V(t) = 125 \cdot 2^{t/5} \quad \text{for} \quad t \geq 0$$

where t is the number of years since the painting was purchased, and $V(t)$ is its value (in dollars) at time t. Find $V(5)$ and $V(10)$, and explain what they mean.

SOLUTION The expression $V(5)$ is the value of the painting when $t = 5$ (5 years after it is purchased). We calculate $V(5)$ by substituting 5 for t in the equation $V(t) = 125 \cdot 2^{t/5}$. Here is our work:

$$V(\mathbf{5}) = 125 \cdot 2^{\mathbf{5}/5} = 125 \cdot 2^1 = 125 \cdot 2 = 250$$

In words: After 5 years, the painting is worth \$250.

The expression $V(10)$ is the value of the painting after 10 years. To find this number, we substitute 10 for t in the equation:

$$V(\mathbf{10}) = 125 \cdot 2^{\mathbf{10}/5} = 125 \cdot 2^2 = 125 \cdot 4 = 500$$

In words: The value of the painting 10 years after it is purchased is \$500.

EXAMPLE 4 A balloon has the shape of a sphere with a radius of 3 inches. Use the following formulas to find the volume and surface area of the balloon.

$$V(r) = \frac{4}{3}\pi r^3 \qquad S(r) = 4\pi r^2$$

SOLUTION As you can see, we have used function notation to write the two formulas for volume and surface area because each quantity is a function of the radius. To find these quantities when the radius is 3 inches, we evaluate $V(3)$ and $S(3)$:

$V(\mathbf{3}) = \frac{4}{3}\pi \mathbf{3}^3 = \frac{4}{3}\pi 27 = 36\pi$ cubic inches, or 113 cubic inches **To the nearest whole number**

$S(\mathbf{3}) = 4\pi \mathbf{3}^2 = 36\pi$ square inches, or 113 square inches **To the nearest whole number**

The fact that $V(3) = 36\pi$ means that the ordered pair $(3, 36\pi)$ belongs to the function V. Likewise, the fact that $S(3) = 36\pi$ tells us that the ordered pair $(3, 36\pi)$ is a member of function S.

We can generalize the discussion at the end of Example 4 this way:

$$(a, b) \in f \quad \text{if and only if} \quad f(a) = b$$

3. A medication has a half-life of 5 days. If the concentration of the medication in a patient's system is 80 ng/mL, and the patient stops taking it, then t days later the concentration will be

$$C(t) = 80\left(\frac{1}{2}\right)^{t/5}$$

Find each of the following, and explain what they mean.

(a) $C(5)$

(b) $C(10)$

4. The following formulas give the circumference and area of a circle with a radius of r. Use the formulas to find the circumference and area of a circular plate if the radius is 5 inches.

$$C(r) = 2\pi r$$
$$A(r) = \pi r^2$$

Answers

2. **(a)** $s(8) = 11$; runs a mile in 8 minutes, average speed is 11 feet per second. **(b)** $s(11) = 8$; runs a mile in 11 minutes, average speed is 8 feet per second.
3. **(a)** $C(5) = 40$ ng/mL; after 5 days the concentration is 40 ng per mL. **(b)** $C(10) = 20$ ng/mL; after 10 days the concentration is 20 ng per mL.
4. $C(5) = 10\pi \approx 31.4$ inches
$A(5) = 25\pi \approx 78.5$ in^2

USING TECHNOLOGY

More About Example 4

If we look back at Example 4, we see that when the radius of a sphere is 3, the numerical values of the volume and surface area are equal. How unusual is this? Are there other values of r for which $V(r)$ and $S(r)$ are equal? We can answer this question by looking at the graphs of both V and S.

To graph the function $V(r) = \frac{4}{3}\pi r^3$, set $Y_1 = 4\pi X^3/3$. To graph $S(r) = 4\pi r^2$, set $Y_2 = 4\pi X^2$. Graph the two functions in each of the following windows:

Window 1: X from −4 to 4, Y from −2 to 10

Window 2: X from 0 to 4, Y from 0 to 50

Window 3: X from 0 to 4, Y from 0 to 150

Then use the Trace and Zoom features of your calculator to locate the point in the first quadrant where the two graphs intersect. How do the coordinates of this point compare with the results in Example 4?

5. If $f(x) = 4x^2 - 3$, find
(a) $f(0)$
(b) $f(3)$
(c) $f(-2)$

EXAMPLE 5 If $f(x) = 3x^2 + 2x - 1$, find $f(0)$, $f(3)$, and $f(-2)$.

SOLUTION Because $f(x) = 3x^2 + 2x - 1$, we have

$$f(\mathbf{0}) = 3(\mathbf{0})^2 + 2(\mathbf{0}) - 1 \quad = 0 + 0 - 1 = -1$$

$$f(\mathbf{3}) = 3(\mathbf{3})^2 + 2(\mathbf{3}) - 1 \quad = 27 + 6 - 1 = 32$$

$$f(\mathbf{-2}) = 3(\mathbf{-2})^2 + 2(\mathbf{-2}) - 1 = 12 - 4 - 1 = 7$$

In Example 5, the function f is defined by the equation $f(x) = 3x^2 + 2x - 1$. We could just as easily have said $y = 3x^2 + 2x - 1$; that is, $y = f(x)$. Saying $f(-2) = 7$ is exactly the same as saying y is 7 when x is −2.

6. If $f(x) = 2x + 1$ and $g(x) = x^2 - 3$, find
(a) $f(5)$
(b) $g(5)$
(c) $f(-2)$
(d) $g(-2)$
(e) $f(a)$
(f) $g(a)$

EXAMPLE 6 If $f(x) = 4x - 1$ and $g(x) = x^2 + 2$, then

$f(\mathbf{5}) = 4(\mathbf{5}) - 1 = 19$ and $g(\mathbf{5}) = \mathbf{5}^2 + 2 = 27$

$f(\mathbf{-2}) = 4(\mathbf{-2}) - 1 = -9$ and $g(\mathbf{-2}) = (\mathbf{-2})^2 + 2 = 6$

$f(\mathbf{0}) = 4(\mathbf{0}) - 1 = -1$ and $g(\mathbf{0}) = \mathbf{0}^2 + 2 = 2$

$f(\boldsymbol{z}) = 4\boldsymbol{z} - 1$ and $g(\boldsymbol{z}) = \boldsymbol{z}^2 + 2$

$f(\boldsymbol{a}) = 4\boldsymbol{a} - 1$ and $g(\boldsymbol{a}) = \boldsymbol{a}^2 + 2$

Answers
5. **(a)** −3 **(b)** 33 **(c)** 13
6. **(a)** 11 **(b)** 22 **(c)** −3 **(d)** 1 **(e)** $2a + 1$ **(f)** $a^2 - 3$

USING TECHNOLOGY

More About Example 6

Most graphing calculators can use tables to evaluate functions. To work Example 6 using a graphing calculator table, set Y_1 equal to $4X - 1$ and Y_2 equal to $X^2 + 2$. Then set the independent variable in the table to Ask instead of Auto. Go to your table and input 5, −2, and 0. Under Y_1 in the table, you will find $f(5)$, $f(-2)$, and $f(0)$. Under Y_2, you will find $g(5)$, $g(-2)$, and $g(0)$.

Table Setup

Table minimum = 0
Table increment = 1
Independent variable: Ask
Dependent variable: Ask

Y Variables Setup

$Y_1 = 4X - 1$
$Y_2 = X^2 + 2$

The table will look like this:

X	Y1	Y2
5	19	27
−2	−9	6
0	−1	2

Although the calculator asks us for a table increment, the increment doesn't matter since we are inputting the X values ourselves.

EXAMPLE 7 If the function f is given by

$$f = \{(-2, 0), (3, -1), (2, 4), (7, 5)\}$$

then $f(-2) = 0$, $f(3) = -1$, $f(2) = 4$, and $f(7) = 5$.

EXAMPLE 8 If $f(x) = 2x^2$ and $g(x) = 3x - 1$, find

(a) $f[g(2)]$ **(b)** $g[f(2)]$

SOLUTION The expression $f[g(2)]$ is read "f of g of 2."

(a) Since $g(2) = 3(2) - 1 = 5$,

$$f[g(2)] = f(5) = 2(5)^2 = 50$$

(b) Since $f(2) = 2(2)^2 = 8$,

$$g[f(2)] = g(8) = 3(8) - 1 = 23$$

7. If $f = \{(-4, 1), (2, -3), (7, 9)\}$, find
(a) $f(-4)$
(b) $f(2)$
(c) $f(7)$

8. If $f(x) = 3x^2$ and $g(x) = 4x + 1$, find
(a) $f[g(2)]$
(b) $g[f(2)]$

Getting Ready for Class

After reading through the preceding section, respond in your own words and in complete sentences.

A. Explain what you are calculating when you find $f(2)$ for a given function f.

B. If $s(t) = \frac{60}{t}$, how do you find $s(10)$?

C. If $f(2) = 3$ for a function f, what is the relationship between the numbers 2 and 3 and the graph of f?

D. If $f(6) = 0$ for a particular function f, then you can immediately graph one of the intercepts. Explain.

Answers

7. **(a)** 1 **(b)** −3 **(c)** 9
8. **(a)** 243 **(b)** 49

PROBLEM SET 2.6

Let $f(x) = 2x - 5$ and $g(x) = x^2 + 3x + 4$. Evaluate the following.

1. $f(2)$
2. $f(3)$
3. $f(-3)$
4. $g(-2)$
5. $g(-1)$
6. $f(-4)$
7. $g(-3)$
8. $g(2)$
9. $g(a)$
10. $f(a)$
11. $f(a + 6)$
12. $g(a + 6)$

Let $f(x) = 3x^2 - 4x + 1$ and $g(x) = 2x - 1$. Evaluate the following.

13. $f(0)$
14. $g(0)$
15. $g(-4)$
16. $f(1)$
17. $f(-1)$
18. $g(-1)$
19. $g\left(\frac{1}{2}\right)$
20. $g\left(\frac{1}{4}\right)$
21. $f(a)$
22. $g(a)$
23. $f(a + 2)$
24. $g(a + 2)$

If $f = \{(1, 4), (-2, 0), (3, \frac{1}{2}), (\pi, 0)\}$ and $g = \{(1, 1), (-2, 2), (\frac{1}{2}, 0)\}$, find each of the following values of f and g.

25. $f(1)$
26. $g(1)$
27. $g\left(\frac{1}{2}\right)$
28. $f(3)$
29. $g(-2)$
30. $f(\pi)$

Let $f(x) = x^2 - 2x$ and $g(x) = 5x - 4$. Evaluate the following.

31. $f(-4)$
32. $g(-3)$
33. $f(-2) + g(-1)$
34. $f(-1) + g(-2)$
35. $2f(x) - 3g(x)$
36. $f(x) - g(x^2)$
37. $f[g(3)]$
38. $g[f(3)]$

Let $f(x) = \dfrac{1}{x + 3}$ and $g(x) = \dfrac{1}{x} + 1$. Evaluate the following.

39. $f\left(\frac{1}{3}\right)$
40. $g\left(\frac{1}{3}\right)$
41. $f\left(-\frac{1}{2}\right)$
42. $g\left(-\frac{1}{2}\right)$
43. $f(-3)$
44. $g(0)$
45. For the function $f(x) = x^2 - 4$, evaluate each of the following expressions.
    (a) $f(a) - 3$
    (b) $f(a - 3)$
    (c) $f(x) + 2$
    (d) $f(x + 2)$
    (e) $f(a + b)$
    (f) $f(x + h)$
46. For the function $f(x) = 3x^2$, evaluate each of the following expressions.
    (a) $f(a) - 2$
    (b) $f(a - 2)$
    (c) $f(x) + 5$
    (d) $f(x + 5)$
    (e) $f(a + b)$
    (f) $f(x + h)$
47. Graph the function $f(x) = \frac{1}{2}x + 2$. Then draw and label the line segments that represent $x = 4$ and $f(4)$.
48. Graph the function $f(x) = -\frac{1}{2}x + 6$. Then draw and label the line segments that represent $x = 4$ and $f(4)$.
49. For the function $f(x) = \frac{1}{2}x + 2$, find the value of x for which $f(x) = x$.
50. For the function $f(x) = -\frac{1}{2}x + 6$, find the value of x for which $f(x) = x$.
51. Graph the function $f(x) = x^2$. Then draw and label the line segments that represent $x = 1$ and $f(1)$, $x = 2$ and $f(2)$ and, finally, $x = 3$ and $f(3)$.
52. Graph the function $f(x) = x^2 - 2$. Then draw and label the line segments that represent $x = 2$ and $f(2)$ and the line segments corresponding to $x = 3$ and $f(3)$.

Applying the Concepts

53. **Investing in Art** A painting is purchased as an investment for \$150. If its value increases continuously so that it doubles every 3 years, then its value is given by the function

$$V(t) = 150 \cdot 2^{t/3} \quad \text{for} \quad t \geq 0$$

where t is the number of years since the painting was purchased, and $V(t)$ is its value (in dollars) at time t. Find $V(3)$ and $V(6)$, and then explain what they mean.

54. **Average Speed** If it takes Minke t minutes to run a mile, then her average speed $s(t)$, in miles per hour, is given by the formula

$$s(t) = \frac{60}{t} \quad \text{for} \quad t > 0$$

Find $s(4)$ and $s(5)$, and then explain what they mean.

55. **Dimensions of a Rectangle**
    (a) The length of a rectangle is 3 inches more than twice the width. Let x represent the width of the rectangle and $P(x)$ represent the perimeter of the rectangle. Use function notation to write the relationship between x and $P(x)$, noting any restrictions on the variable x.
    (b) The length of a rectangle is 3 inches more than twice the width. Let x represent the width of the rectangle and $A(x)$ represent the area of the rectangle. Use function notation to write the relationship between x and $A(x)$, noting any restrictions on the variable x.

56. **Cost of a Phone Call** Suppose a phone company charges 33¢ for the first minute and 24¢ for each additional minute to place a long-distance call between 5 P.M. and 11 P.M. If x is the number of additional minutes and $f(x)$ is the cost of the call, then $f(x) = 24x + 33$.
    (a) How much does it cost to talk for 10 minutes?
    (b) What does $f(5)$ represent in this problem?
    (c) If a call costs $1.29, how long was it?

57. **Life Expectancy** The life expectancy (from birth) for males in the U.S. is modeled by the function $L(x) = 0.21x + 53.4$, where x is the number of years since January 1, 1900 (U.S. National Center for Health Statistics, *Vital Statistics of the United States,* annual).
    (a) Explain how to use this function to find the life expectancy for U.S. men in 1950, 1960, 1975, and 1995.
    (b) Find the life expectancy for U.S. men for the years given in part (a)
    (c) In approximately what year did a U.S. male's life expectancy reach 72.1 years?

58. **Motor Vehicle Accidents** The function below gives the death rate per 100,000 population from motor vehicle accidents in the U.S. during a given year between 1915 and 1997:

$$G(x) = -0.008x^2 + 0.96x - 2.2$$

where x is the number of years since January 1, 1990 (U.S. National Center for Health Statistics, *Vital Statistics of the United States,* annual; U.S. Census Bureau, *1990 Census of Population and Housing*).
    (a) Find $G(35)$, $G(50)$, $G(70)$, and $G(85)$. Explain what these function values mean in terms of death rate from motor vehicle accidents.
    (b) The 1950 population of the United States was approximately 150,700,000. Determine the number of deaths from motor vehicle accidents in 1950.
    (c) The U.S. population in 1985 was approximately 237,900,000. How many people died as a result of motor vehicle accidents in 1985?

59. **Cotton Production** The annual cotton production in the United States in the twentieth century is modeled by the function

$$F(x) = 0.00005x^4 + 0.068x^3 - 9.16x^2 + 303.6x + 10,010$$

where $F(x)$ is the number of cotton bales (in thousands) and x is the number of years since January 1, 1900. (U.S. Department of Agriculture, National Agricultural Statistics Service.)
    (a) Find $F(10)$, $F(25)$, $F(40)$, $F(65)$, $F(80)$, and $F(95)$. Explain the meaning of each of these function values.
    (b) Use the function to predict U.S. cotton production in the year 2001.

60. **Farming Statistics** The average size of a U.S. farm has changed dramatically during the twentieth century. The invention and improvement of farm equipment allows farmers to work larger farms. Economic changes have also had a great impact on the farming industry. The function below gives the average size of a farm in a given year between 1900 and 1998.

$$S(x) = -0.0014x^3 + 0.235x^2 - 6.83x + 170$$

where $S(x)$ is the average farm size (in acres) and x is the number of years since January 1, 1900. (U.S. Department of Agriculture, National Agricultural Statistics Service.)
    (a) Explain how to use the function to find the average farm size in 1900, 1925, 1950, and 1975.
    (b) What is the average farm size in each of the years in part (a)?
    (c) Predict the average farm size in the year 2002.

Straight-Line Depreciation Straight-line depreciation is an accounting method used to help spread the cost of new equipment over a number of years. It takes into account both the cost when new and the salvage value, which is the value of the equipment at the time it gets replaced.

61. **Value of a Copy Machine** The function $V(t) = -3,300t + 18,000$, where V is value and t is time in years, can be used to find the value of a large copy machine during the first 5 years of use.
    (a) What is the value of the copier after 3 years and 9 months?
    (b) What is the salvage value of this copier if it is replaced after 5 years?
    (c) State the domain of this function.

(d) Sketch the graph of this function.

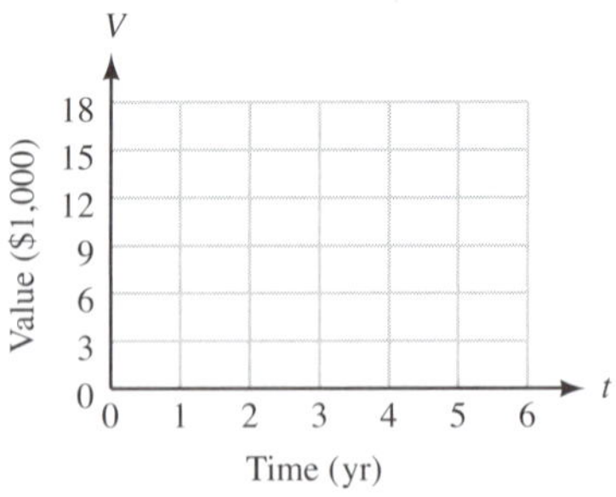

(e) What is the range of this function?

(f) After how many years will the copier be worth only $10,000?

PhotoDisc/Getty Images

62. Value of a Forklift The function $V = -16{,}500t + 125{,}000$, where V is value and t is time in years, can be used to find the value of an electric forklift during the first 6 years of use.

(a) What is the value of the forklift after 2 years and 3 months?

(b) What is the salvage value of this forklift if it is replaced after 6 years?

(c) State the domain of this function.

(d) Sketch the graph of this function.

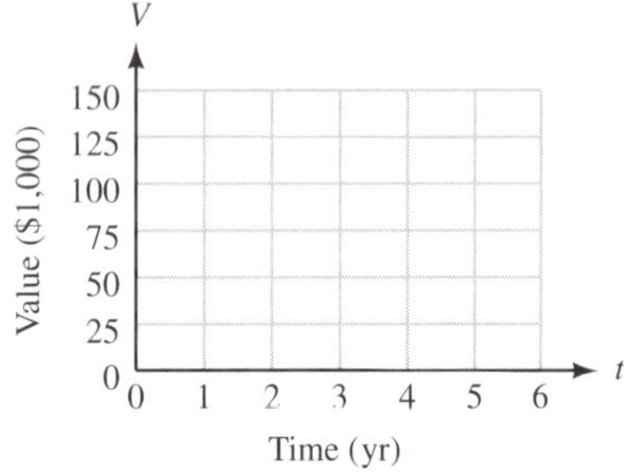

(e) What is the range of this function?

(f) After how many years will the forklift be worth only $45,000?

© Michael Grecco/Getty Images

63. Step Function Figure 4 shows the graph of the step function C that was used to calculate the first-class postage on a letter weighing x ounces in 1997. Use this graph to answer questions (a) through (d).

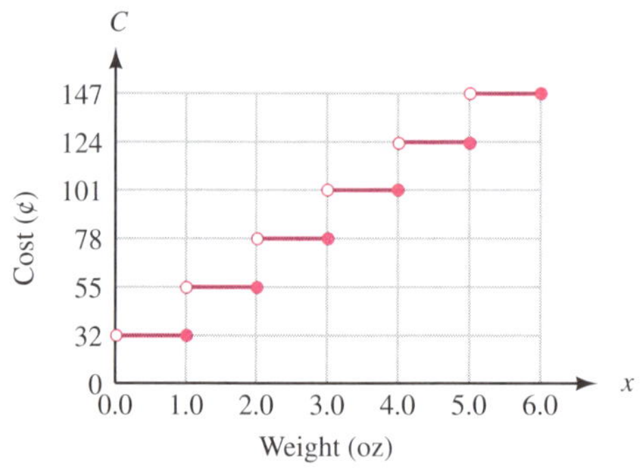

Figure 4 The graph of $C(x)$

(a) Fill in the following table:

Weight (ounces)	0.6	1.0	1.1	2.5	3.0	4.8	5.0	5.3
Cost (cents)								

(b) If a letter cost 78 cents to mail, how much does it weigh? State your answer in words. State your answer as an inequality.

(c) If the entire function is shown in Figure 4, state the domain.

(d) State the range of the function shown in Figure 4.

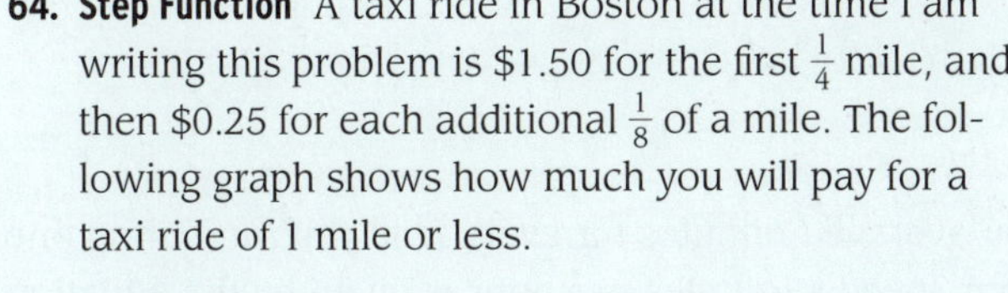

64. Step Function A taxi ride in Boston at the time I am writing this problem is $1.50 for the first $\frac{1}{4}$ mile, and then $0.25 for each additional $\frac{1}{8}$ of a mile. The following graph shows how much you will pay for a taxi ride of 1 mile or less.

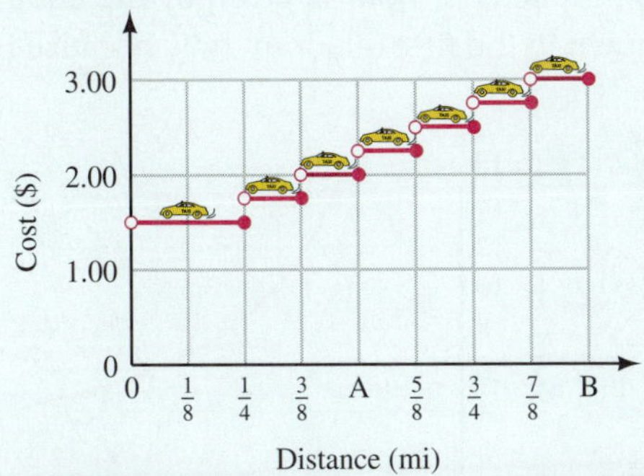

Figure 5

(a) What is the most you will pay for this taxi ride?

(b) How much does it cost to ride the taxi for $\frac{8}{10}$ of a mile?

(c) Find the values of A and B on the horizontal axis.

(d) If a taxi ride costs $2.50, what distance was the ride?

(e) If the complete function is shown in Figure 5, find the domain and range of the function.

Review Problems

Solve each equation.

65. $|3x - 5| = 7$ **66.** $|0.04 - 0.03x| = 0.02$

67. $|4y + 2| - 8 = -2$ **68.** $4 - |3 - 2y| = -5$

69. $5 + |6t + 2| = 3$ **70.** $7 + |3 - \frac{3}{4}t| = 10$

Extending the Concepts

The graphs of two functions are shown in Figure 6 and 7. Use the graphs to find the following.

71. **(a)** $f(2)$ **(b)** $f(-4)$

(c) $g(0)$ **(d)** $g(3)$

72. **(a)** $g(2) - f(2)$ **(b)** $f(1) + g(1)$

(c) $f[g(3)]$ **(d)** $g[f(3)]$

The fixed points of a function f are the values of x such that $f(x) = x$. Find the fixed points of each function.

73. $f(x) = x^2$ **74.** $f(x) = 3x - 5$

75. $f(x) = \frac{1}{2}x - 3$ **76.** $f(x) = x^2 - 6$

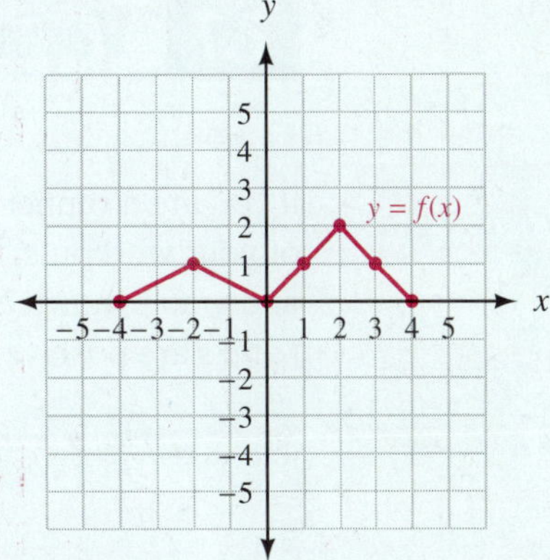

Figure 6

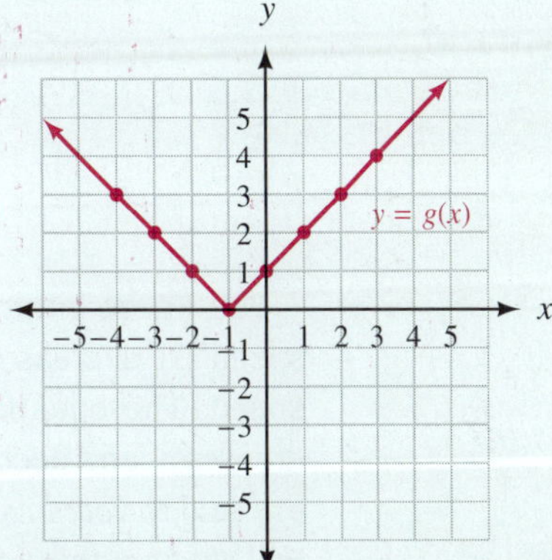

Figure 7

The average rate of change of a function f on the interval $a \le x \le b$ is defined by the ratio $[f(b) - f(a)]/(b - a)$. Find the average rate of change of each function on the interval $1 \le x \le 3$.

77. $f(x) = 2x - 3$ **78.** $f(x) = x^2 + 3$

79. $f(x) = \frac{1}{x}$ **80.** $f(x) = \frac{1}{x^2}$

The difference quotient of a function f at $x = a$ is defined to be the ratio $[f(a + h) - f(a)]/h$. Find the difference quotient of each function at $a = 2$.

81. $f(x) = x^2$ **82.** $f(x) = \frac{1}{2}x - 3$

83. $f(x) = \frac{1}{x}$ **84.** $f(x) = x^2 - 3x$

2.7 Variation

If you are a runner and you average t minutes for every mile you run during one of your workouts, then your speed s in miles per hour is given by the equation and graph shown here. Figure 1 is shown in the first quadrant only because both t and s are positive.

$$s = \frac{60}{t}$$

INPUT t	OUTPUT s
4	15
6	10
8	7.5
10	6
12	5
14	4.3

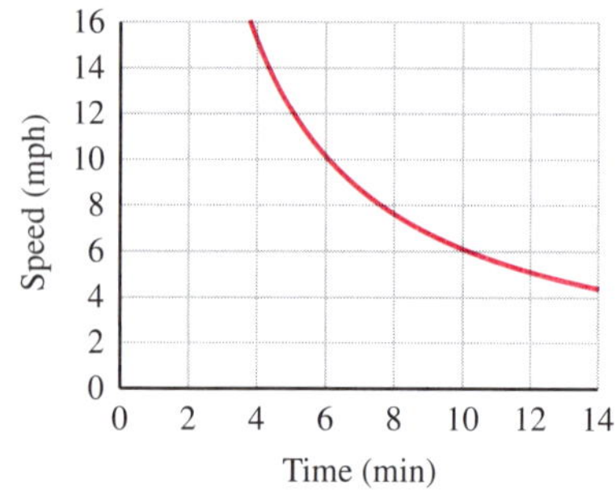

Figure 1

You know intuitively that as your average time per mile (t) increases, your speed (s) decreases. Likewise, lowering your time per mile will increase your speed. The equation and Figure 1 also show this to be true: increasing t decreases s, and decreasing t increases s. Quantities that are connected in this way are said to *vary inversely* with each other. Inverse variation is one of the topics we will study in this section.

There are two main types of variation: *direct variation* and *inverse variation*. Variation problems are most common in the sciences, particularly in chemistry and physics.

Direct Variation

When we say the variable y *varies directly* with the variable x, we mean that the relationship can be written in symbols as $y = Kx$, where K is a nonzero constant called the *constant of variation* (or *proportionality constant*). Another way of saying y varies directly with x is to say y is *directly proportional* to x.

Study the following list. It gives the mathematical equivalent of some direct variation statements.

ENGLISH PHRASE	ALGEBRAIC EQUATION
y varies directly with x	$y = Kx$
s varies directly with the square of t	$s = Kt^2$
y is directly proportional to the cube of z	$y = Kz^3$
u is directly proportional to the square root of v	$u = K\sqrt{v}$

Practice Problems

1. y varies directly with x. If y is 24 when x is 8, find y when x is 2.

EXAMPLE 1 y varies directly with x. If y is 15 when x is 5, find y when x is 7.

SOLUTION The first sentence gives us the general relationship between x and y. The equation equivalent to the statement "y varies directly with x" is

$$y = Kx$$

The first part of the second sentence in our example gives us the information necessary to evaluate the constant K:

When	$y = 15$
and	$x = 5$
the equation	$y = Kx$
becomes	$15 = K \cdot 5$
or	$K = 3$

The equation now can be written specifically as

$$y = 3x$$

Letting $x = 7$, we have

$$y = 3 \cdot 7$$

$$y = 21$$

The graph of the direct variation statement $y = 3x$ is a line with slope 3 that passes through the origin.

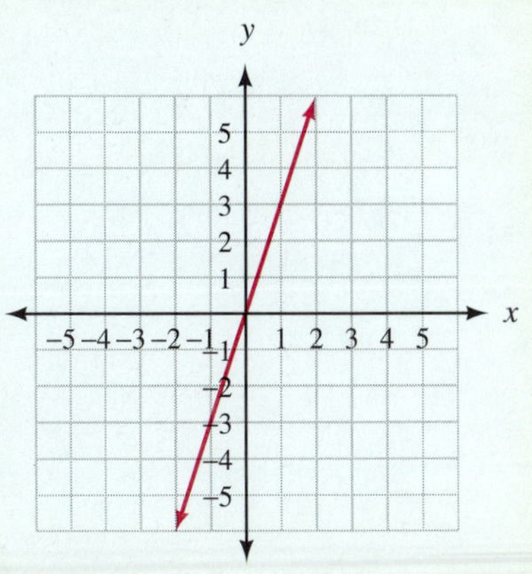

EXAMPLE 2 A skydiver jumps from a plane. Like any object that falls toward Earth, the distance the skydiver falls is directly proportional to the square of the time he has been falling until he reaches his terminal velocity. If the skydiver falls 64 feet in the first 2 seconds of the jump, then

(a) How far will he have fallen after 3.5 seconds?

(b) Graph the relationship between distance and time.

(c) How long will it take him to fall 256 feet?

SOLUTION We let t represent the time the skydiver has been falling, then we can let $d(t)$ represent the distance he has fallen.

(a) Because $d(t)$ is directly proportional to the square of t, we have the general function that describes this situation:

$$d(t) = Kt^2$$

Next, we use the fact that $d(2) = 64$ to find K.

$$64 = K(2^2)$$

$$K = 16$$

The specific equation that describes this situation is

$$d(t) = 16t^2$$

To find how far a skydiver will fall after 3.5 seconds, we find $d(3.5)$:

$$d(3.5) = 16(3.5^2)$$

$$d(3.5) = 196$$

A skydiver will fall 196 feet after 3.5 seconds.

(b) To graph this equation, we use a table:

INPUT t	OUTPUT $d(t)$
0	0
1	16
2	64
3	144
4	256
5	400

2. Use the information in Example 2 to find how far the skydiver will fall in 2.5 seconds.

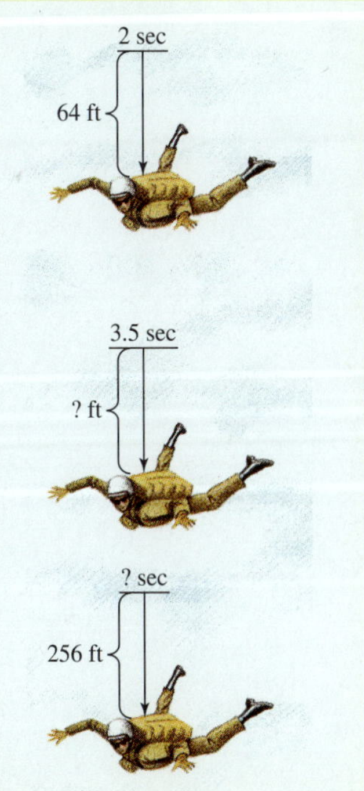

Answer

1. 6

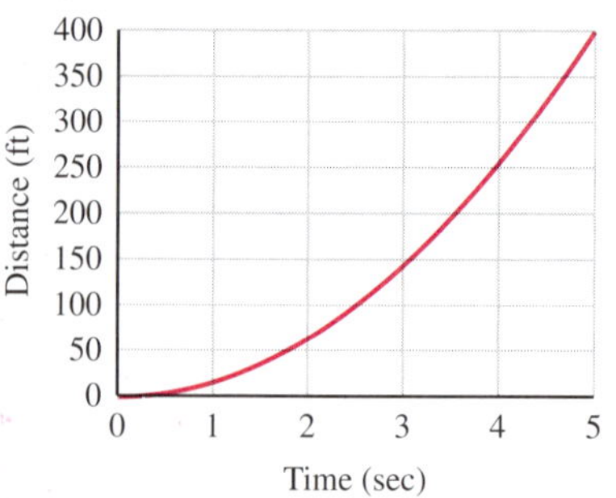

Figure 2

(c) From the table or the graph (Figure 2), we see that it will take 4 seconds for the skydiver to fall 256 feet.

Inverse Variation

Running From the introduction to this section, we know that the relationship between the number of minutes (t) it takes a person to run a mile and his or her average speed in miles per hour (s) can be described with the following equation, table, and Figure 3.

$$s = \frac{60}{t}$$

INPUT t	OUTPUT s
4	15
6	10
8	7.5
10	6
12	5
14	4.3

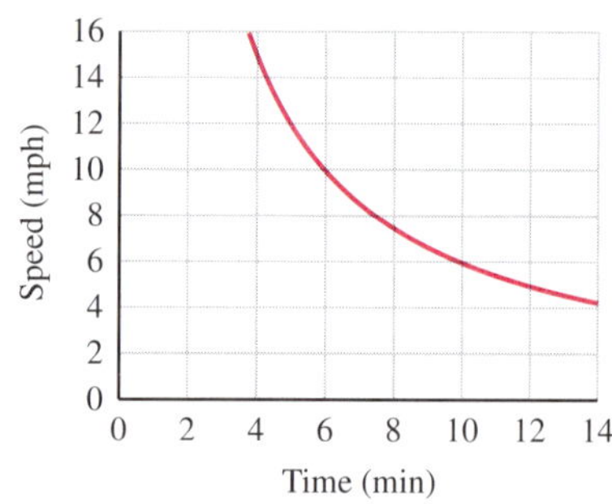

Figure 3

If t decreases, then s will increase, and if t increases, then s will decrease. The variable s is *inversely proportional* to the variable t. In this case, the *constant of proportionality* is 60.

Photography If you are familiar with the terminology and mechanics associated with photography, you know that the *f*-stop for a particular lens will increase as the aperture (the maximum diameter of the opening of the lens) decreases. In mathematics we say that *f*-stop and aperture vary inversely with each other. The diagram illustrates this relationship.

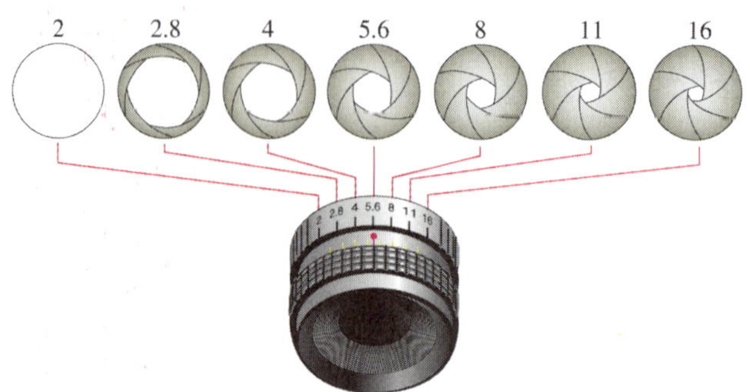

Answer
2. 100 feet

If f is the f-stop and d is the aperture, then their relationship can be written

$$f = \frac{K}{d}$$

In this case, K is the constant of proportionality. (Those of you familiar with photography know that K is also the focal length of the camera lens.)

In General We generalize this discussion of inverse variation as follows: If y varies inversely with x, then

$$y = K\frac{1}{x} \quad \text{or} \quad y = \frac{K}{x}$$

We can also say y is inversely proportional to x. The constant K is again called the constant of variation or proportionality constant.

ENGLISH PHRASE	ALGEBRAIC EQUATION
y is inversely proportional to x	$y = \frac{K}{x}$
s varies inversely with the square of t	$s = \frac{K}{t^2}$
y is inversely proportional to x^4	$y = \frac{K}{x^4}$
z varies inversely with the cube root of t	$z = \frac{K}{\sqrt[3]{t}}$

EXAMPLE 3 The volume of a gas is inversely proportional to the pressure of the gas on its container. If a pressure of 48 pounds per square inch corresponds to a volume of 50 cubic feet, what pressure is needed to produce a volume of 100 cubic feet?

SOLUTION We can represent volume with V and pressure with P:

$$V = \frac{K}{P}$$

Using $P = 48$ and $V = 50$, we have

$$50 = \frac{K}{48}$$

$$K = 50(48)$$

$$K = 2{,}400$$

The equation that describes the relationship between P and V is

$$V = \frac{2{,}400}{P}$$

Here is a graph of this relationship.

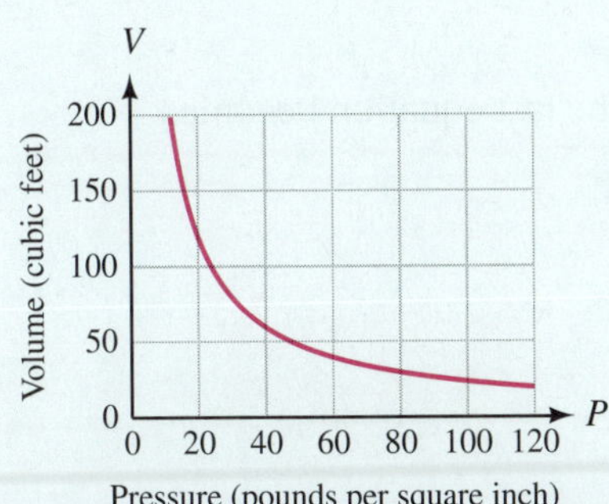

3. In Example 3, what pressure is needed to produce a volume of 150 cubic feet?

Note:
The relationship between pressure and volume as given in this example is known as Boyle's law and applies to situations such as those encountered in a piston-cylinder arrangement. It was Robert Boyle (1627–1691) who, in 1662, published the results of some of his experiments that showed, among other things, that the volume of a gas decrease as the pressure increases. This is an example of inverse variation.

Substituting $V = 100$ into our last equation, we get

$$100 = \frac{2{,}400}{P}$$

$$100P = 2{,}400$$

$$P = \frac{2{,}400}{100}$$

$$P = 24$$

A volume of 100 cubic feet is produced by a pressure of 24 pounds per square inch.

Joint Variation and Other Variation Combinations

Many times relationships among different quantities are described in terms of more than two variables. If the variable y varies directly with *two* other variables, say x and z, then we say y varies *jointly* with x and z. In addition to joint variation, there are many other combinations of direct and inverse variation involving more than two variables. The following table is a list of some variation statements and their equivalent mathematical forms:

ENGLISH PHRASE	ALGEBRAIC EQUATION
y varies jointly with x and z	$y = Kxz$
z varies jointly with r and the square of s	$z = Krs^2$
V is directly proportional to T and inversely proportional to P	$V = \frac{KT}{P}$
F varies jointly with m_1 and m_2 and inversely with the square of r	$F = \frac{Km_1m_2}{r^2}$

4. y varies jointly with x and the square of z. If y is 81 when x is 2 and z is 9, find y when x is 4 and z is 4.

EXAMPLE 4 y varies jointly with x and the square of z. When x is 5 and z is 3, y is 180. Find y when x is 2 and z is 4.

SOLUTION The general equation is given by

$$y = Kxz^2$$

Substituting $x = 5$, $z = 3$, and $y = 180$, we have

$$180 = K(5)(3)^2$$

$$180 = 45K$$

$$K = 4$$

The specific equation is

$$y = 4xz^2$$

When $x = 2$ and $z = 4$, the last equation becomes

$$y = 4(2)(4)^2$$

$$y = 128$$

Answers

3. 16 pounds per square inch
4. 32

EXAMPLE 5 In electricity, the resistance of a cable is directly proportional to its length and inversely proportional to the square of the diameter. If a 100-foot cable 0.5 inch in diameter has a resistance of 0.2 ohm, what will be the resistance of a cable made from the same material if it is 200 feet long with a diameter of 0.25 inch?

SOLUTION Let R = resistance, l = length, and d = diameter. The equation is

$$R = \frac{Kl}{d^2}$$

When $R = 0.2$, $l = 100$, and $d = 0.5$, the equation becomes

$$0.2 = \frac{K(100)}{(0.5)^2}$$

or

$$K = 0.0005$$

Using this value of K in our original equation, the result is

$$R = \frac{0.0005l}{d^2}$$

When $l = 200$ and $d = 0.25$, the equation becomes

$$R = \frac{0.0005(200)}{(0.25)^2}$$

$$R = 1.6 \text{ ohms}$$

5. Use the information in Example 5 to find the resistance of a 300-foot cable made of the same material that is 0.25 inch in diameter.

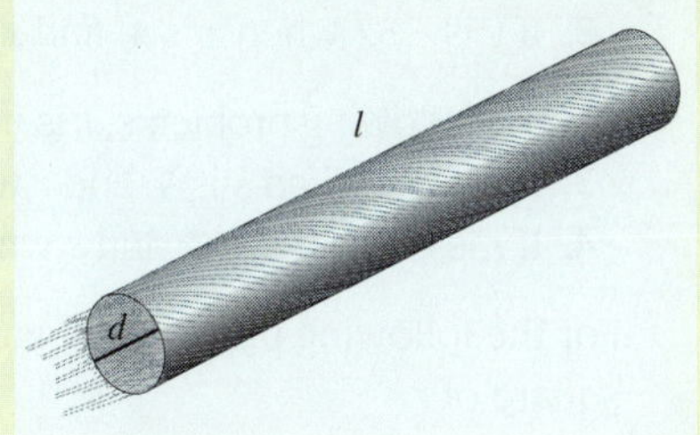

Getting Ready for Class

After reading through the preceding section, respond in your own words and in complete sentences.

A. Give an example of a direct variation statement, and then translate it into symbols.

B. Translate the equation $y = \frac{K}{x}$ into words.

C. For the inverse variation equation $y = \frac{3}{x}$, what happens to the values of y as x gets larger?

D. How are direct variation statements and linear equations in two variables related?

Answer

5. 2.4 ohms

PROBLEM SET 2.7

For the following problems, y varies directly with x.

1. If y is 10 when x is 2, find y when x is 6.
2. If y is -32 when x is 4, find x when y is -40.

For the following problems, r is inversely proportional to s.

3. If r is -3 when s is 4, find r when s is 2.
4. If r is 8 when s is 3, find s when r is 48.

For the following problems, d varies directly with the square of r.

5. If $d = 10$ when $r = 5$, find d when $r = 10$.
6. If $d = 12$ when $r = 6$, find d when $r = 9$.

For the following problems, y varies inversely with the square of x.

7. If $y = 45$ when $x = 3$, find y when x is 5.
8. If $y = 12$ when $x = 2$, find y when x is 6.

For the following problems, z varies jointly with x and the square of y.

9. If z is 54 when x and y are 3, find z when $x = 2$ and $y = 4$.
10. If z is 27 when $x = 6$ and $y = 3$, find x when $z = 50$ and $y = 4$.

For the following problems, I varies inversely with the cube of w.

11. If $I = 32$ when $w = \frac{1}{2}$, find I when $w = \frac{1}{3}$.
12. If $I = \frac{1}{25}$ when $w = 5$, find I when $w = 10$.

For the following problems, z varies jointly with y and the square of x.

13. If $z = 72$ when $x = 3$ and $y = 2$, find z when $x = 5$ and $y = 3$.
14. If $z = 240$ when $x = 4$ and $y = 5$, find z when $x = 6$ and $y = 3$.
15. If $x = 1$ when $z = 25$ and $y = 5$, find x when $z = 160$ and $y = 8$.
16. If $x = 4$ when $z = 96$ and $y = 2$, find x when $z = 108$ and $y = 1$.

For the following problems, F varies directly with m and inversely with the square of d.

17. If $F = 150$ when $m = 240$ and $d = 8$, find F when $m = 360$ and $d = 3$.
18. If $F = 72$ when $m = 50$ and $d = 5$, find F when $m = 80$ and $d = 6$.
19. If $d = 5$ when $F = 24$ and $m = 20$, find d when $F = 18.75$ and $m = 40$.
20. If $d = 4$ when $F = 75$ and $m = 20$, find d when $F = 200$ and $m = 120$.

Applying the Concepts

21. **Length of a Spring** The length a spring stretches is directly proportional to the force applied. If a force of 5 pounds stretches a spring 3 inches, how much force is necessary to stretch the same spring 10 inches?

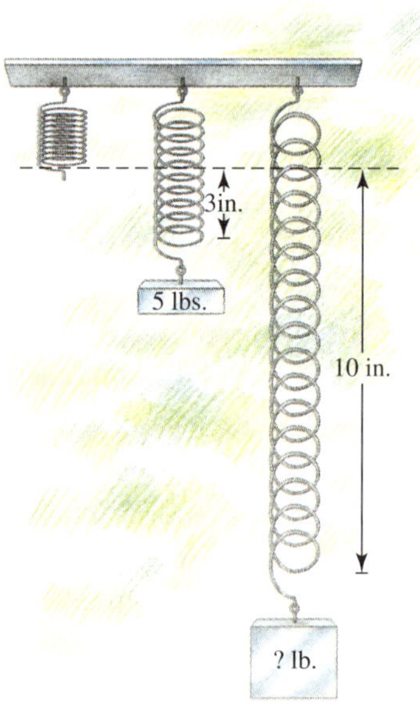

22. **Weight and Surface Area** The weight of a certain material varies directly with the surface area of that material. If 8 square feet weighs half a pound, how much will 10 square feet weigh?
23. **Pressure and Temperature** The temperature of a gas varies directly with its pressure. A temperature of 200 K produces a pressure of 50 pounds per square inch.
    (a) Find the equation that relates pressure and temperature.
    (b) Graph the equation from part (a) in the first quadrant only.
    (c) What pressure will the gas have at 280 K?
24. **Circumference and Diameter** The circumference of a wheel is directly proportional to its diameter. A wheel has a circumference of 8.5 feet and a diameter of 2.7 feet.
    (a) Find the equation that relates circumference and diameter.
    (b) Graph the equation from part (a) in the first quadrant only.
    (c) What is the circumference of a wheel that has a diameter of 11.3 feet?

25. Volume and Pressure The volume of a gas is inversely proportional to the pressure. If a pressure of 36 pounds per square inch corresponds to a volume of 25 cubic feet, what pressure is needed to produce a volume of 75 cubic feet?

26. Wave Frequency The frequency of an electromagnetic wave varies inversely with the wavelength. If a wavelength of 200 meters has a frequency of 800 kilocycles per second, what frequency will be associated with a wavelength of 500 meters?

27. *f*-Stop and Aperture Diameter The relative aperture, or *f*-stop, for a camera lens is inversely proportional to the diameter of the aperture. An *f*-stop of 2 corresponds to an aperture diameter of 40 millimeters for the lens on an automatic camera.

(a) Find the equation that relates *f*-stop and diameter.

(b) Graph the equation from part (a) in the first quadrant only.

(c) What is the *f*-stop of this camera when the aperture diameter is 10 millimeters?

28. *f*-Stop and Aperture Diameter The relative aperture, or *f*-stop, for a camera lens is inversely proportional to the diameter of the aperture. An *f*-stop of 2.8 corresponds to an aperture diameter of 75 millimeters for a certain telephoto lens.

(a) Find the equation that relates *f*-stop and diameter.

(b) Graph the equation from part (a) in the first quadrant only.

(c) What aperture diameter corresponds to an *f*-stop of 5.6?

29. Surface Area of a Cylinder The surface area of a hollow cylinder varies jointly with the height and radius of the cylinder. If a cylinder with radius 3 inches and height 5 inches has a surface area of 94 square inches, what is the surface area of a cylinder with radius 2 inches and height 8 inches?

30. Capacity of a Cylinder The capacity of a cylinder varies jointly with its height and the square of its radius. If a cylinder with a radius of 3 centimeters and a height of 6 centimeters has a capacity of 169.56 cubic centimeters, what will be the capacity of a cylinder with radius 4 centimeters and height 9 centimeters?

31. Electrical Resistance The resistance of a wire varies directly with its length and inversely with the square of its diameter. If 100 feet of wire with diameter 0.01 inch has a resistance of 10 ohms, what is the resistance of 60 feet of the same type of wire if its diameter is 0.02 inch?

32. Volume and Temperature The volume of a gas varies directly with its temperature and inversely with the pressure. If the volume of a certain gas is 30 cubic feet at a temperature of 300 K and a pressure of 20 pounds per square inch, what is the volume of the same gas at 340 K when the pressure is 30 pounds per square inch?

33. Period of a Pendulum The time it takes for a pendulum to complete one period varies directly with the square root of the length of the pendulum. A 100-centimeter pendulum takes 2.1 seconds to complete one period.

(a) Find the equation that relates period and pendulum length.

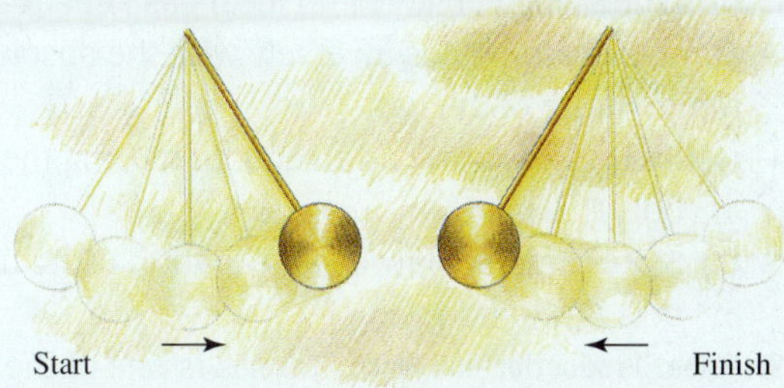

(b) Graph the equation from part (a) in quadrant I only.

(c) How long does it take to complete one period if the pendulum hangs 225 centimeters?

34. Area of a Circle The area of a circle varies directly with the square of the radius. A circle with a 5-centimeter radius has an area of 78.5 square centimeters.

(a) Find the equation that relates area and radius.

(b) Graph the equation from part (a) in quadrant I only.

(c) What is the area of a circle that has an 8-centimeter radius?

35. Music A musical tone's pitch varies inversely with its wavelength. If one tone has a pitch of 420 vibrations each second and a wavelength of 2.2 meters, find the wavelength of a tone that has a pitch of 720 vibrations each second.

36. Hooke's Law Hooke's law states that the stress (force per unit area) placed on a solid object varies directly with the strain (deformation) produced.

(a) Using the variables S_1 for stress and S_2 for strain, state this law in algebraic form.

(b) Find the constant , K, if for one type of material $S_1 = 24$ and $S_2 = 72$.

37. Gravity In Book Three of his *Principia,* Isaac Newton states that there is a single force in the universe that holds everything together, called the force of universal gravity. Newton stated that the force of universal gravity, F, is directly proportional with the product of two masses, m_1 and m_2, and inversely proportional with the square of the distance d between them. Write the equation for Newton's force of universal gravity, using the symbol G as the constant of proportionality.

38. Boyle's Law and Charles's Law Boyle's law states that for low pressures, the pressure of an ideal gas kept at a constant temperature varies inversely with the volume of the gas. Charles's law states that for low pressures, the density of an ideal gas kept at a constant pressure varies inversely with the absolute temperature of the gas.

(a) State Boyle's law as an equation using the symbols P, K, and V.

(b) State Charles's law as an equation using the symbols D, K, and T.

39. Wheat Production and Price Kansas is one of the leading states in wheat production. In Kansas, the price per bushel of wheat, P, varies inversely with wheat production, W, where W is in millions of bushels (U.S. Department of Agriculture, National Agricultural Statistics Service).

(a) In 1996, wheat production was 255 million bushels, and the average price was $4.70 per bushel. Find the proportionality constant, and write the equation relating price to production.

(b) If the wheat production in 1998 was 495 million bushels, find the average price per bushel.

(c) If the average price per bushel were $3.25, what was the wheat production that year?

40. Health Care Between 1935 and 1980, the death rate per 100,000 population from pneumonia/influenza can be estimated with inverse variation. The death rate per 100,000, R (during a given year), is inversely proportional to the square of the number of years after 1900; that is, 1940 is represented by 40 (U.S. National Center for Health Statistics, *Vital Statistics of the United States*).

(a) If the death rate per 100,000 in 1935 was 104.2, find the constant of variation and the equation that represents this relationship.

(b) Determine the death rate per 100,000 from pneumonia/influenza in 1970.

(c) In approximately what year was the death rate from pneumonia/influenza 51.1 per 100,000?

(d) The 1970 U.S. population was approximately 205,052,000. Estimate how many people died that year from pneumonia/influenza.

41. School Enrollment For the years 1990–1995, the number of high school seniors in U.S. public schools was directly proportional to the number of kindergarten children enrolled in public schools. In 1991, there were 3,686,000 kindergarten students and 2,392,000 high school seniors in public schools (U.S. National Center for Educational Statistics, *Digest of Educational Statistics,* annual).

(a) Find the proportionality constant (round to the nearest hundredth), and write the equation relating the number of high school seniors, in a given year, to the number of kindergarten children.

(b) In 1994, there were 4,047,000 kindergarten children enrolled. Estimate the number of high school seniors in public schools that year.

(c) If there are 3,200,000 high school seniors in a certain year, how many kindergarten students are enrolled that year?

42. Law of Levers Inverse variation may also be defined as occurring when the product of two variables is a constant; that is, when $x \cdot y = K$. This definition gives rise to the Law of Levers, or "seesaw principle." A seesaw is balanced when the variables on each side of the center, weight and distance from center, have the same product.

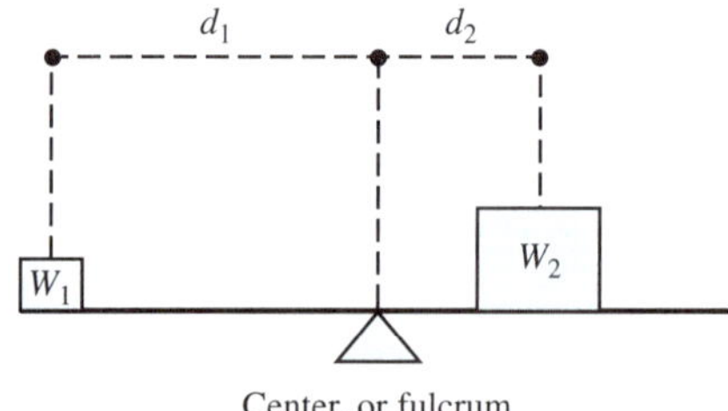

Center, or fulcrum

This implies that $W_1 \cdot d_1 = W_2 \cdot d_2$ in order to balance the seesaw in the figure. Suppose an 85-pound girl is 4 feet from the center of a seesaw. How far from the center should her 120-pound brother place himself to balance the device?

Review Problems

Factor completely.

43. $49a^2 - 144$

44. $27a^3 - 125$

45. $x^2 - x + \dfrac{1}{4}$

46. $9x^2 - 30x + 25 - y^2$
47. $4x^2(a - 5) - 9(a - 5)$
48. $35x^2 - 64x + 21$

Extending the Concepts

For the following problems, y varies inversely with the square of x.

49. What is the effect on y when x is tripled?
50. What is the effect on y when x is cut in half?

For the following problems, z varies jointly with x and the square of y.

51. What is the effect on z when x is doubled and y is tripled?
52. What is the effect on z when x is tripled and y is doubled?

2.8 Systems of Linear Equations in Two Variables

Previously we found the graph of an equation of the form $ax + by = c$ to be a straight line. Since the graph is a straight line, the equation is said to be a linear equation. Two linear equations considered together form a *linear system* of equations. For example,

$$3x - 2y = 6$$

$$2x + 4y = 20$$

is a linear system. The solution set to the system is the set of all ordered pairs that satisfy both equations. If we graph each equation on the same set of axes, we can see the solution set (Figure 1).

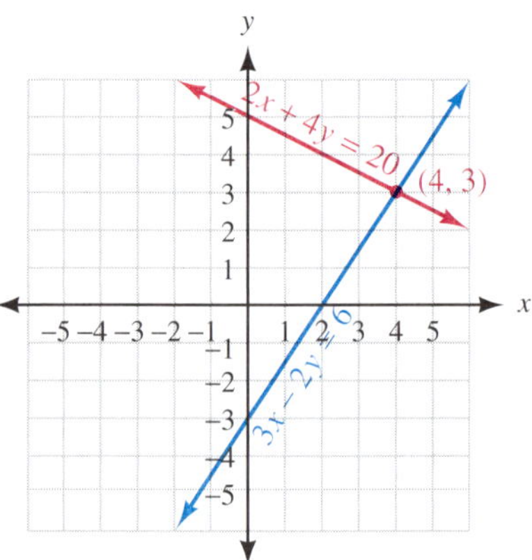

Figure 1

Note: It is important that you write solutions to systems of equations in two variables as ordered pairs, like (4, 3); that is, you should always enclose the ordered pairs in parentheses.

The point (4, 3) lies on both lines and therefore must satisfy both equations. It is obvious from the graph that it is the only point that does so. The solution set for the system is {(4, 3)}.

More generally, if $a_1x + b_1y = c_1$ and $a_2x + b_2y = c_2$ are linear equations, then the solution set for the system

$$a_1x + b_1y = c_1$$

$$a_2x + b_2y = c_2$$

can be illustrated through one of the graphs in Figure 2.

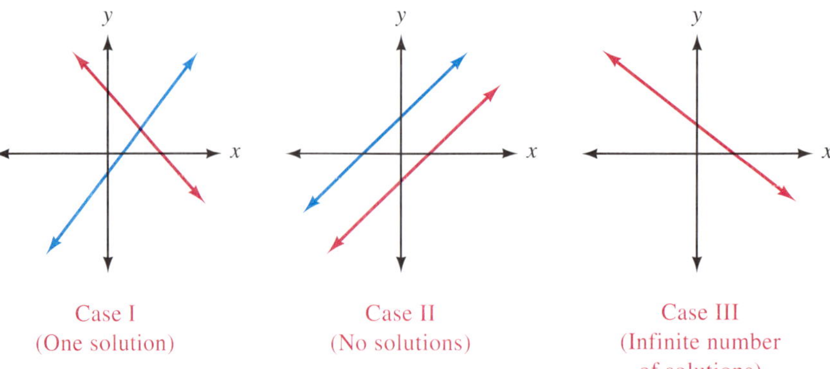

Figure 2

Case I The two lines intersect at one and only one point. The coordinates of the point give the solution to the system. This is what usually happens.

Case II The lines are parallel and therefore have no points in common. The solution set to the system is the empty set, $\varnothing$. In this case, we say the equations are *inconsistent.*

Case III The lines coincide; that is, their graphs represent the same line. The solution set consists of all ordered pairs that satisfy either equation. In this case, the equations are said to be *dependent.*

In the beginning of this section we found the solution set for the system

$$3x - 2y = 6$$
$$2x + 4y = 20$$

by graphing each equation and then reading the solution set from the graph. Solving a system of linear equations by graphing is the least accurate method. If the coordinates of the point of intersection are not integers, it can be very difficult to read the solution set from the graph. There is another method of solving a linear system that does not depend on the graph. It is called the *addition method.*

The Addition Method

EXAMPLE 1 Solve the system.

$$4x + 3y = 10$$
$$2x + y = 4$$

SOLUTION If we multiply the bottom equation by -3, the coefficients of y in the resulting equation and the top equation will be opposites:

$$4x + 3y = 10 \xrightarrow{\text{No change}} 4x + 3y = 10$$
$$2x + y = 4 \xrightarrow[\text{Multiply by } -3]{} -6x - 3y = -12$$

Adding the left and right sides of the resulting equations, we have

$$\begin{array}{r} 4x + 3y = 10 \\ -6x - 3y = -12 \\ \hline -2x \quad = -2 \end{array}$$

The result is a linear equation in one variable. We have eliminated the variable y from the equations by addition. (It is for this reason we call this method of solving a linear system the *addition method.*) Solving $-2x = -2$ for x, we have

$$x = 1$$

This is the x-coordinate of the solution to our system. To find the y-coordinate, we substitute $x = 1$ into any of the equations containing both the variables x and y. Let's try the second equation in our original system:

$$2(1) + y = 4$$
$$2 + y = 4$$
$$y = 2$$

This is the y-coordinate of the solution to our system. The ordered pair (1, 2) is the solution to the system.

Practice Problems

1. Solve the system.

$$2x - y = 7$$
$$3x + 4y = -6$$

by first multiplying the top equation by 4 and then adding the result to the bottom equation.

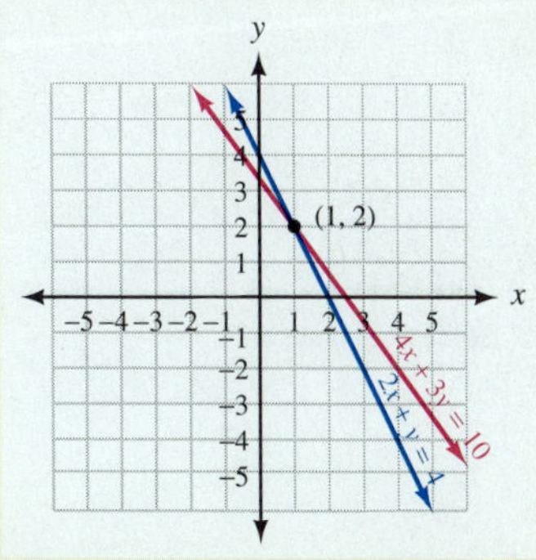

Figure 3 A visual representation of the solution to the system in Example 1

Answer

1. (2, −3)

Checking Solutions We can check our solution by substituting it into both of our equations.

Substituting $x = 1$ and $y = 2$ into $4x + 3y = 10$, we have

$$4(1) + 3(2) \stackrel{?}{=} 10$$
$$4 + 6 \stackrel{?}{=} 10$$
$$10 = 10 \quad \textbf{A true statement}$$

Substituting $x = 1$ and $y = 2$ into $2x + y = 4$, we have

$$2(1) + 2 \stackrel{?}{=} 4$$
$$2 + 2 \stackrel{?}{=} 4$$
$$4 = 4 \quad \textbf{A true statement}$$

Our solution satisfies both equations; therefore, it is a solution to our system of equations.

2. Solve the system.

$$3x - 2y = -8$$
$$-2x + 3y = 7$$

EXAMPLE 2 Solve the system.

$$3x - 5y = -2$$
$$2x - 3y = 1$$

SOLUTION We can eliminate either variable. Let's decide to eliminate the variable x. We can do so by multiplying the top equation by 2 and the bottom equation by -3, and then adding the left and right sides of the resulting equations:

$$3x - 5y = -2 \xrightarrow{\text{Multiply by 2}} 6x - 10y = -4$$
$$2x - 3y = 1 \xrightarrow[\text{Multiply by } -3]{} -6x + 9y = -3$$
$$-y = -7$$
$$y = 7$$

The y-coordinate of the solution to the system is 7. Substituting this value of y into any of the equations with both x- and y-variables gives $x = 11$. The solution to the system is (11, 7). It is the only ordered pair that satisfies both equations.

Checking Solutions Checking (11, 7) in each equation looks like this

Substituting $x = 11$ and $y = 7$ into $3x - 5y = -2$, we have

$$3(11) - 5(7) \stackrel{?}{=} -2$$
$$33 - 35 \stackrel{?}{=} -2$$
$$-2 = -2 \quad \textbf{A true statement}$$

Substituting $x = 11$ and $y = 7$ into $2x - 3y = 1$, we have

$$2(11) - 3(7) \stackrel{?}{=} 1$$
$$22 - 21 \stackrel{?}{=} 1$$
$$1 = 1 \quad \textbf{A true statement}$$

Our solution satisfies both equations; therefore, (11, 7) is a solution to our system.

3. Solve the system.

$$3x - 5y = 2$$
$$2x + 4y = 1$$

EXAMPLE 3 Solve the system.

$$2x - 3y = 4$$
$$4x + 5y = 3$$

SOLUTION We can eliminate x by multiplying the top equation by -2 and adding it to the bottom equation:

$$2x - 3y = 4 \xrightarrow{\text{Multiply by } -2} -4x + 6y = -8$$
$$4x + 5y = 3 \xrightarrow[\text{No change}]{} 4x + 5y = 3$$
$$11y = -5$$
$$y = -\frac{5}{11}$$

Answer

2. $(-2, 1)$

The y-coordinate of our solution is $-\frac{5}{11}$. If we were to substitute this value of y back into either of our original equations, we would find the arithmetic necessary to solve for x cumbersome. For this reason, it is probably best to go back to the original system and solve it a second time—for x instead of y. Here is how we do that:

$$\begin{aligned} 2x - 3y &= 4 \xrightarrow{\text{Multiply by 5}} 10x - 15y = 20 \\ 4x + 5y &= 3 \xrightarrow[\text{Multiply by 3}]{} \underline{12x + 15y = 9} \\ &\qquad\qquad\qquad 22x \qquad = 29 \\ &\qquad\qquad\qquad\quad x = \frac{29}{22} \end{aligned}$$

The solution to our system is $\left(\frac{29}{22}, -\frac{5}{11}\right)$.

The main idea in solving a system of linear equations by the addition method is to use the multiplication property of equality on one or both of the original equations, if necessary, to make the coefficients of either variable opposites. The following box shows some steps to follow when solving a system of linear equations by the addition method.

Strategy for Solving a System of Linear Equations by the Addition Method

Step 1: Decide which variable to eliminate. (In some cases, one variable will be easier to eliminate than the other. With some practice, you will notice which one it is.)

Step 2: Use the multiplication property of equality on each equation separately to make the coefficients of the variable that is to be eliminated opposites.

Step 3: Add the respective left and right sides of the system together.

Step 4: Solve for the remaining variable.

Step 5: Substitute the value of the variable from step 4 into an equation containing both variables and solve for the other variable. (Or repeat steps 2–4 to eliminate the other variable.)

Step 6: Check your solution in both equations, if necessary.

EXAMPLE 4 Solve the system.

$$\begin{aligned} 5x - 2y &= 5 \\ -10x + 4y &= 15 \end{aligned}$$

SOLUTION We can eliminate y by multiplying the first equation by 2 and adding the result to the second equation:

$$\begin{aligned} 5x - 2y &= 5 \xrightarrow{\text{Multiply by 2}} 10x - 4y = 10 \\ -10x + 4y &= 15 \xrightarrow[\text{No change}]{} \underline{-10x + 4y = 15} \\ &\qquad\qquad\qquad\qquad 0 = 25 \end{aligned}$$

The result is the false statement $0 = 25$, which indicates there is no solution to the system. If we were to graph the two lines, we would find that they are parallel. In a case like this, we say the system is *inconsistent*. Whenever both variables have been eliminated and the resulting statement is false, the solution set for the system will be the empty set, $\varnothing$.

4. Solve the system.

$$\begin{aligned} 2x + 7y &= 3 \\ 4x + 14y &= 1 \end{aligned}$$

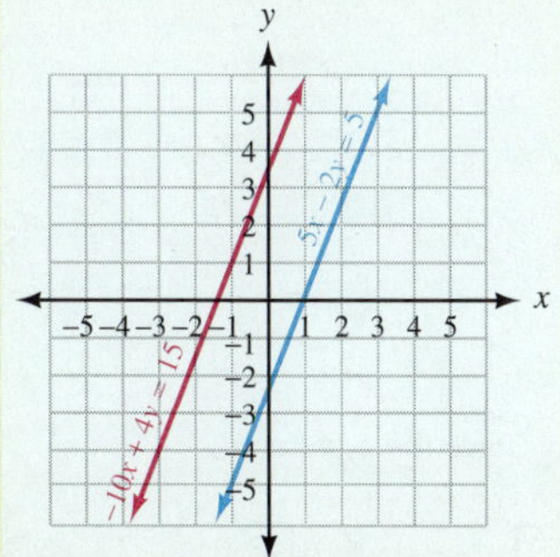

Figure 4 A visual representation of the situation in Example 4—the two lines are parallel

Answers

3. $\left(\frac{13}{22}, -\frac{1}{22}\right)$ **4.** $\varnothing$

5. Solve the system.

$$2x + 7y = 3$$
$$4x + 14y = 6$$

EXAMPLE 5 Solve the system.

$$4x + 3y = 2$$
$$8x + 6y = 4$$

SOLUTION Multiplying the top equation by −2 and adding, we can eliminate the variable x:

$$4x + 3y = 2 \xrightarrow{\text{Multiply by } -2} -8x - 6y = -4$$
$$8x + 6y = 4 \xrightarrow[\text{No change}]{} \quad 8x + 6y = 4$$
$$0 = 0$$

Both variables have been eliminated and the resulting statement $0 = 0$ is true. In this case, the lines coincide and the system is said to be *dependent*. The solution set consists of all ordered pairs that satisfy either equation. We can write the solution set as $\{(x, y) \mid 4x + 3y = 2\}$ or $\{(x, y) \mid 8x + 6y = 4\}$.

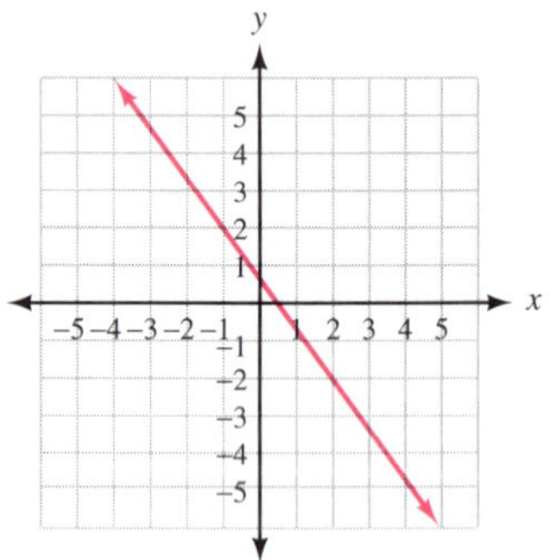

Figure 5 A visual representation of the situation in Example 5—both equations produce the same graph

Special Cases

The previous two examples illustrate the two special cases in which the graphs of the equations in the system either coincide or are parallel. In both cases the left-hand sides of the equations were multiples of each other. In the case of the dependent equations the right-hand sides were also multiples. We can generalize these observations for the system

$$a_1x + b_1y = c_1$$
$$a_2x + b_2y = c_2$$

Inconsistent System

What Happens	*Geometric Interpretation*	*Algebraic Interpretation*
Both variables are eliminated, and the resulting statement is false.	The lines are parallel, and there is no solution to the system.	$\frac{a_1}{a_2} = \frac{b_1}{b_2} \neq \frac{c_1}{c_2}$

Dependent Equations

What Happens	*Geometric Interpretation*	*Algebraic Interpretation*
Both variables are eliminated, and the resulting statement is true.	The lines coincide, and there are an infinite number of solutions to the system.	$\frac{a_1}{a_2} = \frac{b_1}{b_2} = \frac{c_1}{c_2}$

6. Solve the system.

$$\frac{1}{3}x + \frac{1}{2}y = 4$$
$$\frac{2}{3}x - \frac{1}{4}y = 3$$

EXAMPLE 6 Solve the system.

$$\frac{1}{2}x - \frac{1}{3}y = 2$$
$$\frac{1}{4}x + \frac{2}{3}y = 6$$

Answer
5. Lines coincide

SOLUTION Although we could solve this system without clearing the equations of fractions, there is probably less chance for error if we have only integer coefficients to work with. So let's begin by multiplying both sides of the top equation by 6 and both sides of the bottom equation by 12 to clear each equation of fractions:

$$\frac{1}{2}x - \frac{1}{3}y = 2 \xrightarrow{\text{Times } 6} 3x - 2y = 12$$

$$\frac{1}{4}x + \frac{2}{3}y = 6 \xrightarrow[\text{Times } 12]{} 3x + 8y = 72$$

Now we can eliminate x by multiplying the top equation by -1 and leaving the bottom equation unchanged:

$$\begin{aligned} 3x - 2y = 12 &\xrightarrow{\text{Times } -1} -3x + 2y = -12 \\ 3x + 8y = 72 &\xrightarrow[\text{No change}]{} \underline{3x + 8y = 72} \\ &\qquad\qquad\qquad 10y = 60 \\ &\qquad\qquad\qquad\quad y = 6 \end{aligned}$$

We can substitute $y = 6$ into any equation that contains both x and y. Let's use $3x - 2y = 12$.

$$\begin{aligned} 3x - 2(6) &= 12 \\ 3x - 12 &= 12 \\ 3x &= 24 \\ x &= 8 \end{aligned}$$

The solution to the system is (8, 6).

The Substitution Method

EXAMPLE 7 Solve the system.

$$\begin{aligned} 2x - 3y &= -6 \\ y &= 3x - 5 \end{aligned}$$

SOLUTION The second equation tells us y is $3x - 5$. Substituting the expression $3x - 5$ for y in the first equation, we have

$$2x - 3(3x - 5) = -6$$

The result of the substitution is the elimination of the variable y. Solving the resulting linear equation in x as usual, we have

$$\begin{aligned} 2x - 9x + 15 &= -6 \\ -7x + 15 &= -6 \\ -7x &= -21 \\ x &= 3 \end{aligned}$$

7. Solve the system

$$\begin{aligned} 4x - 2y &= -2 \\ y &= x + 3 \end{aligned}$$

by substituting the expression for y given in the second equation into the first equation.

Answer

6. (6, 4)

Putting $x = 3$ into the second equation in the original system, we have

$$\begin{aligned} y &= 3(3) - 5 \\ &= 9 - 5 \\ &= 4 \end{aligned}$$

The solution to the system is (3, 4).

Checking Solutions Checking (3, 4) in each equation looks like this

Substituting $x = 3$ and $y = 4$ into $2x - 3y = -6$, we have

$$\begin{aligned} 2(3) - 3(4) &\stackrel{?}{=} -6 \\ 6 - 12 &\stackrel{?}{=} -6 \\ -6 &= -6 \quad \textbf{A true statement} \end{aligned}$$

Substituting $x = 3$ and $y = 4$ into $y = 3x - 5$, we have

$$\begin{aligned} 4 &\stackrel{?}{=} 3(3) - 5 \\ 4 &\stackrel{?}{=} 9 - 5 \\ 4 &= 4 \quad \textbf{A true statement} \end{aligned}$$

Our solution satisfies both equations; therefore, (3, 4) is a solution to our system.

Strategy for Solving a System of Equations by the Substitution Method

Step 1: Solve either one of the equations for x or y. (This step is not necessary if one of the equations is already in the correct form, as in Example 7.)

Step 2: Substitute the expression for the variable obtained in step 1 into the other equation and solve it.

Step 3: Substitute the solution for step 2 into any equation in the system that contains both variables and solve it.

Step 4: Check your results, if necessary.

8. Solve by substitution.
$$\begin{aligned} 5x - 3y &= -4 \\ x + 2y &= 7 \end{aligned}$$

EXAMPLE 8 Solve by substitution.

$$\begin{aligned} 2x + 3y &= 5 \\ x - 2y &= 6 \end{aligned}$$

SOLUTION To use the substitution method, we must solve one of the two equations for x or y. We can solve for x in the second equation by adding $2y$ to both sides:

$$\begin{aligned} x - 2y &= 6 \\ x &= 2y + 6 \quad \textbf{Add 2y to both sides.} \end{aligned}$$

Substituting the expression $2y + 6$ for x in the first equation of our system, we have

$$\begin{aligned} 2(2y + 6) + 3y &= 5 \\ 4y + 12 + 3y &= 5 \\ 7y + 12 &= 5 \\ 7y &= -7 \\ y &= -1 \end{aligned}$$

Using $y = -1$ in either equation in the original system, we find $x = 4$. The solution is $(4, -1)$.

Note: Both the substitution method and the addition method can be used to solve any system of linear equations in two variables. Systems like the one in Example 7, however, are easier to solve using the substitution method because one of the variables is already written in terms of the other. A system like the one in Example 2 is easier to solve using the addition method because solving for one of the variables would lead to an expression involving fractions. The system in Example 8 could be solved easily by either method because solving the second equation for x is a one-step process.

Answers
7. (2, 5) 8. (1, 3)

USING TECHNOLOGY

Graphing Calculators

Solving Systems That Intersect in Exactly One Point

A graphing calculator can be used to solve a system of equations in two variables if the equations intersect in exactly one point. To solve the system shown in Example 3, we first solve each equation for y. Here is the result:

$$2x - 3y = 4 \quad \text{becomes} \quad y = \frac{4 - 2x}{-3}$$

$$4x + 5y = 3 \quad \text{becomes} \quad y = \frac{3 - 4x}{5}$$

Graphing these two functions on the calculator gives a diagram similar to the one in Figure 6.

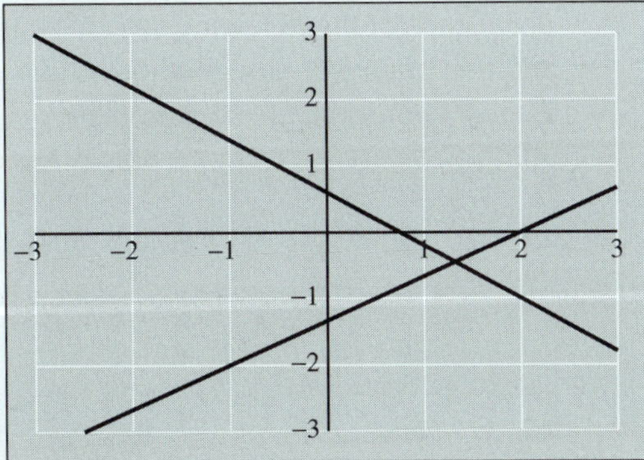

Figure 6

Using the Trace and Zoom features, we find that the two lines intersect at $x = 1.32$ and $y = -0.45$, which are the decimal equivalents (accurate to the nearest hundredth) of the fractions found in Example 3.

Getting Ready for Class

After reading through the preceding section, respond in your own words and in complete sentences.

A. Two linear equations, each with the same two variables, form a system of equations. How do we define a solution to this system? That is, what form will a solution have, and what properties does a solution possess?

B. When would substitution be more efficient than the addition method in solving two linear equations?

C. Explain what an inconsistent system of linear equations looks like graphically and what would result algebraically when attempting to solve the system.

D. When might the graphing method of solving a system of equations be more desirable than the other techniques, and when might it be less desirable?

PROBLEM SET 2.8

Solve each system by graphing both equations on the same set of axes and then reading the solution from the graph.

1. $3x - 2y = 6$
$x - y = 1$

2. $5x - 2y = 10$
$x - y = -1$

3. $y = \frac{3}{5}x - 3$
$2x - y = -4$

4. $y = \frac{1}{2}x - 2$
$2x - y = -1$

5. $y = \frac{1}{2}x$
$y = -\frac{3}{4}x + 5$

6. $y = \frac{2}{3}x$
$y = -\frac{1}{3}x + 6$

7. $3x + 3y = -2$
$y = -x + 4$

8. $2x - y = 5$
$y = 2x - 5$

Solve each of the following systems by the addition method.

9. $x + y = 5$
$3x - y = 3$

10. $x - y = 4$
$-x + 2y = -3$

11. $x + 2y = 0$
$2x - 6y = 5$

12. $x + 3y = 3$
$2x - 9y = 1$

13. $2x - 5y = 16$
$4x - 3y = 11$

14. $5x - 3y = -11$
$7x + 6y = -12$

15. $6x + 3y = -1$
$9x + 5y = 1$

16. $5x + 4y = -1$
$7x + 6y = -2$

17. $4x + 3y = 14$
$9x - 2y = 14$

18. $7x - 6y = 13$
$6x - 5y = 11$

19. $2x - 5y = 3$
$-4x + 10y = 3$

20. $3x - 2y = 1$
$-6x + 4y = -2$

21. $\frac{1}{2}x + \frac{1}{3}y = 13$
$\frac{2}{5}x + \frac{1}{4}y = 10$

22. $\frac{1}{2}x + \frac{1}{3}y = \frac{2}{3}$
$\frac{2}{3}x + \frac{2}{5}y = \frac{14}{15}$

23. $\frac{2}{3}x + \frac{2}{5}y = 4$
$\frac{1}{3}x - \frac{1}{2}y = -\frac{1}{3}$

24. $\frac{1}{2}x - \frac{1}{3}y = \frac{5}{6}$
$-\frac{2}{5}x + \frac{1}{2}y = -\frac{9}{10}$

Solve each of the following systems by the substitution method.

25. $7x - y = 24$
$x = 2y + 9$

26. $3x - y = -8$
$y = 6x + 3$

27. $6x - y = 10$
$y = -\frac{3}{4}x - 1$

28. $2x - y = 6$
$y = -\frac{4}{3}x + 1$

29. $y = 3x - 2$
$y = 4x - 4$

30. $y = 5x - 2$
$y = -2x + 5$

31. $2x - y = 5$
$4x - 2y = 10$

32. $-10x + 8y = -6$
$y = \frac{5}{4}x$

33. $\frac{1}{3}x - \frac{1}{2}y = 0$
$x = \frac{3}{2}y$

34. $\frac{2}{5}x - \frac{2}{3}y = 0$
$y = \frac{3}{5}x$

You may want to read Example 3 again before solving the systems that follow.

35. $4x - 7y = 3$
$5x + 2y = -3$

36. $3x - 4y = 7$
$6x - 3y = 5$

37. $9x - 8y = 4$
$2x + 3y = 6$

38. $4x - 7y = 10$
$-3x + 2y = -9$

39. $3x - 5y = 2$
$7x + 2y = 1$

40. $4x - 3y = -1$
$5x + 8y = 2$

Solve each of the following systems by using either the addition or substitution method. Choose the method that is most appropriate for the problem.

41. $x - 3y = 7$
$2x + y = -6$

42. $2x - y = 9$
$x + 2y = -11$

43. $y = \frac{1}{2}x + \frac{1}{3}$
$y = -\frac{1}{3}x + 2$

44. $y = \frac{3}{4}x - \frac{4}{5}$
$y = \frac{1}{2}x - \frac{1}{2}$

45. $3x - 4y = 12$
$x = \frac{2}{3}y - 4$

46. $-5x + 3y = -15$
$x = \frac{4}{5}y - 2$

47. $4x - 3y = -7$
$-8x + 6y = -11$

48. $3x - 4y = 8$
$y = \frac{3}{4}x - 2$

49. $5(2x + 3y) - 3(x - 2y) = -21$
$4(x - y) + 2(3x + y) = 34$

50. $6(x - 2y) - 3(x + y) = 72$
$5(x - 3y) - 4(3x + 2y) = 64$

51. $3(x - y) + 2 = 4(2x + y) + 35$
$4(x + 2y) - 3 = 2(2x - 3y) + 11$

52. $-3(2x - y) + 5 = 2(3x - y) - 14$
$2(3x + 2y) - 1 = 4(x + 2y) - 1$

53. $\frac{3}{4}x - \frac{1}{3}y = 1$
$y = \frac{1}{4}x$

54. $-\frac{2}{3}x + \frac{1}{2}y = -1$
$y = -\frac{1}{3}x$

55. $\frac{1}{4}x - \frac{1}{2}y = \frac{1}{3}$
$\frac{1}{3}x - \frac{1}{4}y = -\frac{2}{3}$

56. $\frac{1}{5}x - \frac{1}{10}y = -\frac{1}{5}$

$\frac{2}{3}x - \frac{1}{2}y = -\frac{1}{6}$

57. Multiply both sides of the second equation in the following system by 100, and then solve as usual.

$$x + y = 10{,}000$$
$$0.06x + 0.05y = 560$$

58. Multiply both sides of the second equation in the following system by 10, and then solve as usual.

$$x + y = 12$$
$$0.20x + 0.50y = 0.30(12)$$

59. What value of c will make the following system a dependent system (one in which the lines coincide)?

$$6x - 9y = 3$$
$$4x - 6y = c$$

60. What value of c will make the following system a dependent system?

$$5x - 7y = c$$
$$-15x + 21y = 9$$

Applying the Concepts

61. **Cost of a Phone Call** One telephone company charges 41 cents for the first minute and 32 cents for each additional minute for a certain long-distance phone call. If the number of additional minutes after the first minute is x and the cost, in cents, for the call is $C(x)$, then the function that gives the total cost, in cents, for the call is $C(x) = 32x + 41$.
   - **(a)** How much does it cost to make a 10-minute long-distance call under these conditions?
   - **(b)** If a second phone company charges 45 cents for the first minute and 30 cents for each additional minute, write the function that gives the total cost $C(x)$ of a call in terms of the number of additional minutes x.
   - **(c)** After how many additional minutes will the two companies charge an equal amount? (What is the x-coordinate of the point of intersection of the two lines?)

62. **Cost of a Taxi Ride** In a certain city, a taxi ride costs 75 cents for the first $\frac{1}{7}$ of a mile and 10 cents for every additional $\frac{1}{7}$ of a mile after the first seventh. If x is the number of additional sevenths of a mile, then the total cost $C(x)$ of a taxi ride is

$$C(x) = 10x + 75$$

   - **(a)** How much does it cost to ride a taxi for 10 miles in this city?
   - **(b)** Suppose a taxi ride in another city costs 50 cents for the first $\frac{1}{7}$ of a mile, and 15 cents for each additional $\frac{1}{7}$ of a mile. Write the function that gives the total cost $C(x)$, in cents, to ride x sevenths of a mile past the first seventh in this city.

63. **Medicine** From 1980 through 1992, outpatient surgery was linearly increasing in U.S. hospitals. For these years the percentage of all surgery performed that was outpatient surgery, $f(x)$, may be represented by the function

$$f(x) = \frac{10}{3}x + 20$$

where x is the number of years since January 1, 1980, but not beyond the year 1992. Also, from 1980 through 1992, inpatient surgery was linearly decreasing in U.S. hospitals. For these years, the percentage of all surgery performed that was inpatient surgery, $g(x)$, may be represented by the function

$$g(x) = -\frac{10}{3}x + 80$$

where x is the number of years past January 1, 1980, but not beyond the year 1992.
   - **(a)** What is the range of possible values of x in these two equations?
   - **(b)** Graph these two functions.
   - **(c)** Find the solution to this system. What year does the solution represent?

64. **Medicine** Death from heart disease has been linearly decreasing.

YEAR	DEATHS (PER 100,000 OF THE POPULATION)
1960	286
1991	148

However, cancer deaths for females have been linearly increasing.

YEAR	DEATHS (PER 100,000 OF THE POPULATION)
1970	140
1992	180

   - **(a)** Write an equation relating the number of deaths from heart disease, y, to the year x.
   - **(b)** Write an equation relating the number of cancer deaths in females, y, to the year x.
   - **(c)** In what year will the two lines in parts (a) and (b) intersect?

65. Pollution The DDT levels in fish declined in a linear manner from 1978 to 1986. In 1978 the average DDT level was 0.5 ppb (parts per billion) and in 1986 the average level was found to be only 0.18 ppb.

(a) Graph this line on a coordinate axis with the year on the horizontal axis and the DDT level on the vertical axis.

(b) Write the equation of the line found in part (a).

(c) Use the results of part (b) to find the DDT level in 1980.

(d) If the rate of DDT reduction had continued to the year 1990, what would that level have been?

66. A System With Two Solutions Given the following system:

$$y = |x|$$
$$x - 2y = -6$$

(a) Graph this system.

(b) Solve the system by inspection.

67. Marital Status The percent of men and women in the 30–34 age group who have never married is given by the following equations

Women: $P(x) = 0.55x + 6.2$

Men: $P(x) = 0.78x + 9.4$

where $P(x)$ is a percent and x is the number of years since January 1, 1970 (U.S. Bureau of the Census).

(a) For the year 1982, determine the percent of men and women, ages 30–34, who had never married.

(b) After 1970, was there a year in which the percent of never married men was equal to the percent of never married women in this age group?

68. Organ Transplants The number of procedures for heart and liver transplants have increased during the years from 1985 through 1994, according to the following functions:

Heart transplants: $N(x) = 164.2x + 719$

Liver transplants: $N(x) = 332.2x + 602$

where $N(x)$ is the number of procedures for each category and x is the number of years since January 1, 1985 (U.S. Department of Health and Human Services, Public Health Services, Division of Organ Transplantation, and United Network of Organ Sharing).

(a) Which one of the two models is increasing at a faster rate?

(b) During which year was the number of liver transplants equal to the number of heart transplants?

(c) Using the models given, predict the number of heart and liver transplants in the year 2000.

Review Problems

69. Find the slope of the line that contains $(-4, -1)$ and $(-2, 5)$.

70. A line has a slope of $\frac{2}{3}$. Find the slope of any line

(a) Parallel to it.

(b) Perpendicular to it.

71. Give the slope and y-intercept of the line $2x - 3y = 6$.

72. Give the equation of the line with slope -3 and y-intercept 5.

73. Find the equation of the line with slope $\frac{2}{3}$ that contains the point $(-6, 2)$.

74. Find the equation of the line through $(1, 3)$ and $(-1, -5)$.

75. Find the equation of the line with x-intercept 3 and y-intercept -2.

76. Find the equation of the line through $(-1, 4)$ whose graph is perpendicular to the graph of $y = 2x + 3$.

Extending the Concepts

77. The height of an object thrown in the air is given by the function $h(t) = at^2 + bt$, where t is measured in seconds and $h(t)$ is measured in feet. After 3 seconds, the object reaches a height of 24 feet; after 6 seconds, it hits the ground. Find the values of a and b.

78. The height of an object thrown in the air is given by the function $h(t) = at^2 + bt$, where t is measured in seconds and $h(t)$ is measured in feet. After 4 seconds, the object reaches a height of 40 feet; after 6 seconds, it reaches a height of 54 feet. How long will it take for the object to hit the ground?

79. The lines $y = 0$, $y = 4x$, and $y = 12 - 2x$ form a triangle in the first quadrant. Find its area. *Hint:* Draw a picture of the situation first.

80. The lines $y = 2$, $y = x$, and $y = 10 - x$ form a triangle in the first quadrant. Find its area. *Hint:* Draw a picture of the situation first.

Solve the following systems.

81. $\frac{1}{x} + \frac{2}{y} = 6$

$\frac{3}{x} - \frac{1}{y} = 4$

Hint: Let $s = \frac{1}{x}$ and $t = \frac{1}{y}$.

82. $\frac{4}{x} - \frac{3}{y} = 12$

$\frac{3}{x} + \frac{2}{y} = -6$

2.9 Systems of Linear Inequalities

Previously we graphed linear inequalities in two variables. To review, we graph the boundary line, using a solid line if the boundary is part of the solution set and a broken line if the boundary is not part of the solution set. Then we test any point that is not on the boundary line in the original inequality. A true statement tells us that the point lies in the solution set; a false statement tells us the solution set is the other region.

Figure 1 shows the graph of the inequality $x + y < 4$. Note that the boundary is not included in the solution set and is therefore drawn with a broken line. Figure 2 shows the graph of $-x + y \le 3$. Note that the boundary is drawn with a solid line because it is part of the solution set.

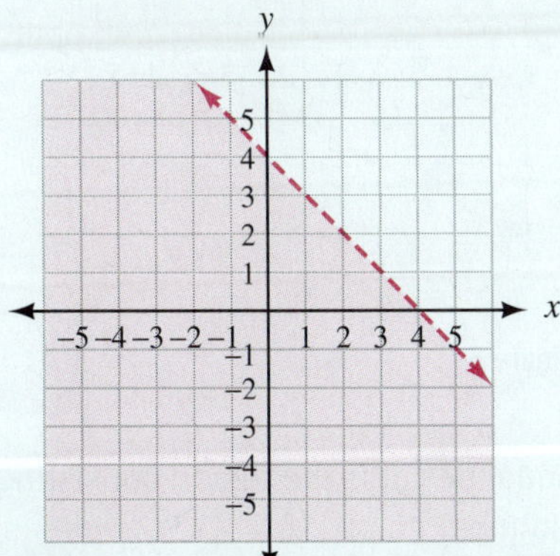

Figure 1

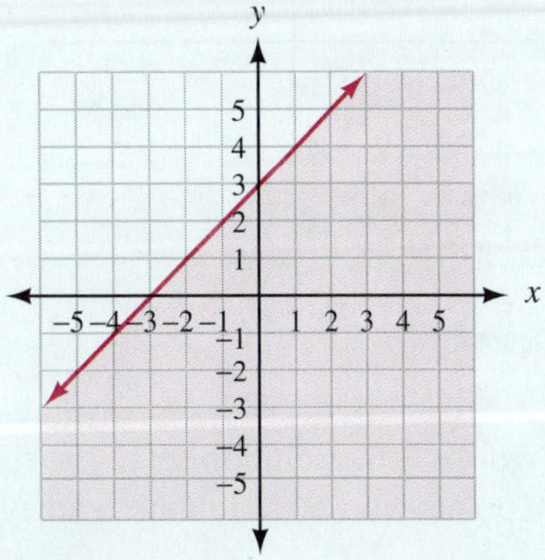

Figure 2

If we form a system of inequalities with the two inequalities, the solution set will be all the points common to both solution sets shown in the two figures above: It is the intersection of the two solution sets. Therefore, the solution set for the system of inequalities.

$$x + y < 4$$

$$-x + y \le 3$$

is all the ordered pairs that satisfy both inequalities. It is the set of points that are below the line $x + y = 4$ and also below (and including) the line $-x + y = 3$. The graph of the solution set to this system is shown in Figure 3. We have written the system in Figure 3 with the word *and* just to remind you that the solution set to a system of equations or inequalities is all the points that satisfy both equations or inequalities.

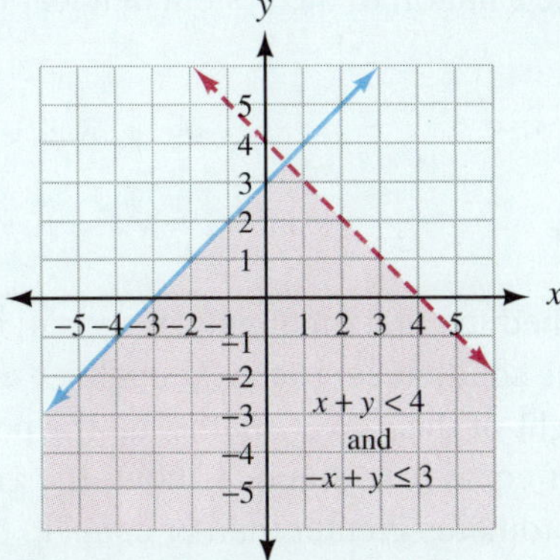

Figure 3

Practice Problems

1. Graph the solution set to the system

$$y \le \frac{1}{3}x + 1$$
$$y > \frac{1}{3}x - 3$$

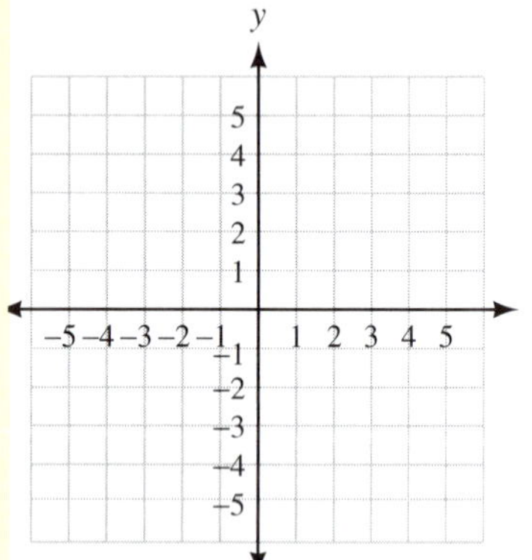

EXAMPLE 1 Graph the solution to the system of linear inequalities.

$$y < \frac{1}{2}x + 3$$

$$y \ge \frac{1}{2}x - 2$$

SOLUTION Figures 4 and 5 show the solution set for each of the inequalities separately.

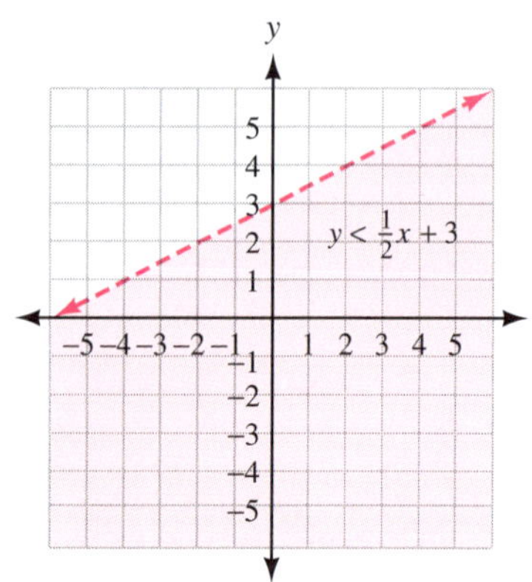

Figure 4

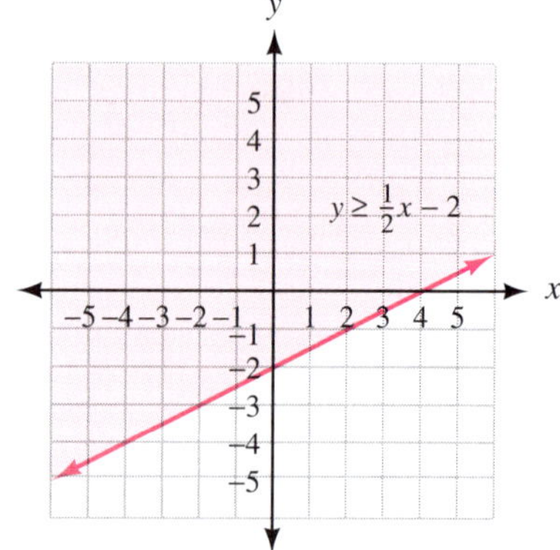

Figure 5

Figure 6 is the solution set to the system of inequalities. It is the region consisting of points whose coordinates satisfy both inequalities.

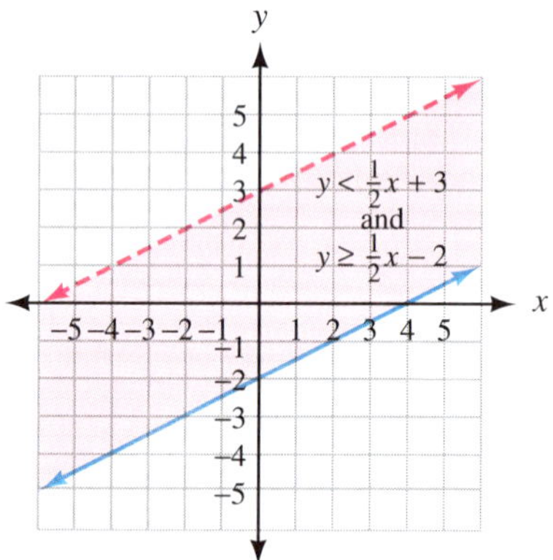

Figure 6

2. Graph the solution set to the system

$$x + y < 3$$
$$x \ge 0$$
$$y \ge 0$$

EXAMPLE 2 Graph the solution to the system of linear inequalities.

$$x + y < 4$$

$$x \ge 0$$

$$y \ge 0$$

SOLUTION We graphed the first inequality, $x + y < 4$, in Figure 1 at the beginning of this section. The solution set to the inequality $x \ge 0$, shown in Figure 7, is all the points to the right of the y-axis; that is, all the points with x-coordinates that are greater than or equal to 0. Figure 8 shows the graph of $y \ge 0$. It consists of all points with y-coordinates greater than or equal to 0; that is, all points from the x-axis up.

Answers

1.-2. See solutions section

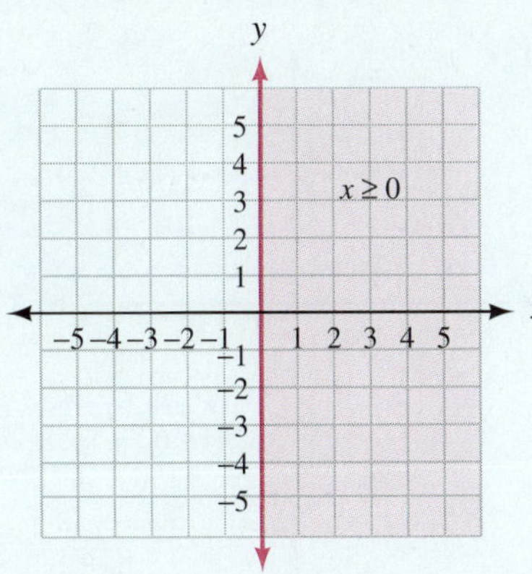

Figure 7

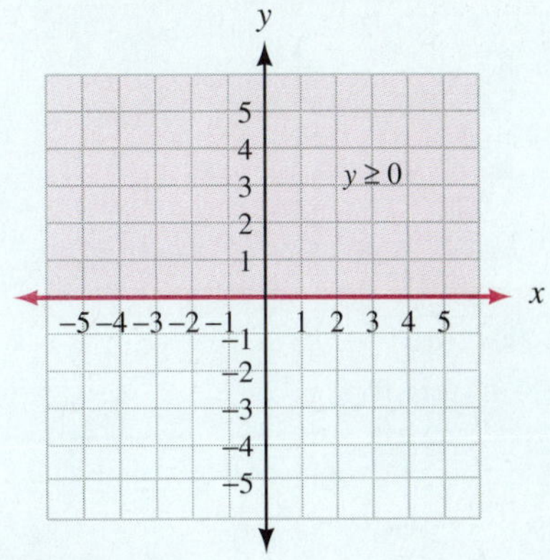

Figure 8

The regions shown in Figures 7 and 8 overlap in the first quadrant. Therefore, putting all three regions together we have the points in the first quadrant that are below the line $x + y = 4$. This region is shown in Figure 9, and it is the solution to our system of inequalities.

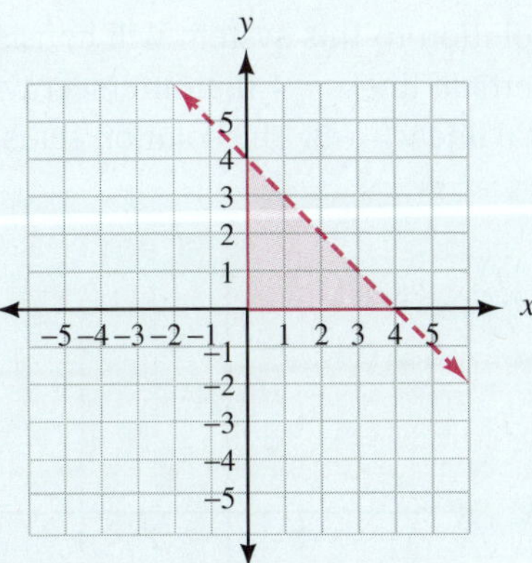

Figure 9

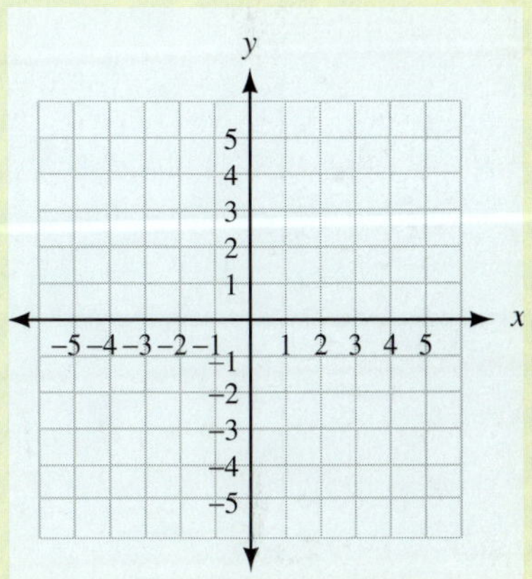

Extending the discussion in Example 2 we can name the points in each of the four quadrants using systems of inequalities.

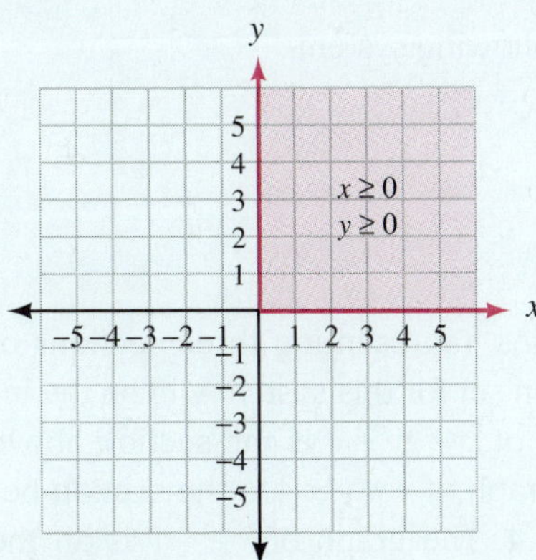

Figure 10

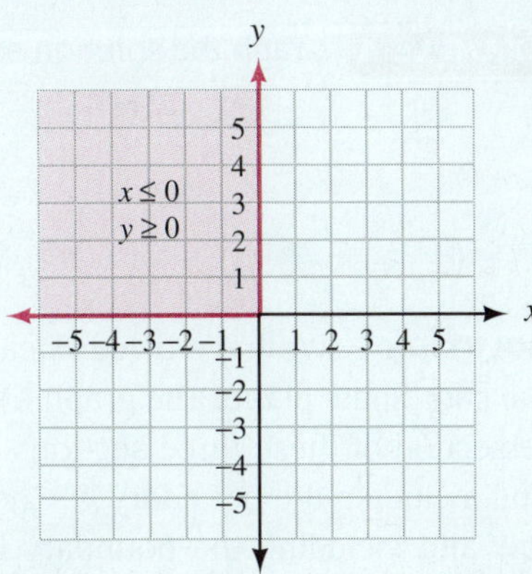

Figure 11

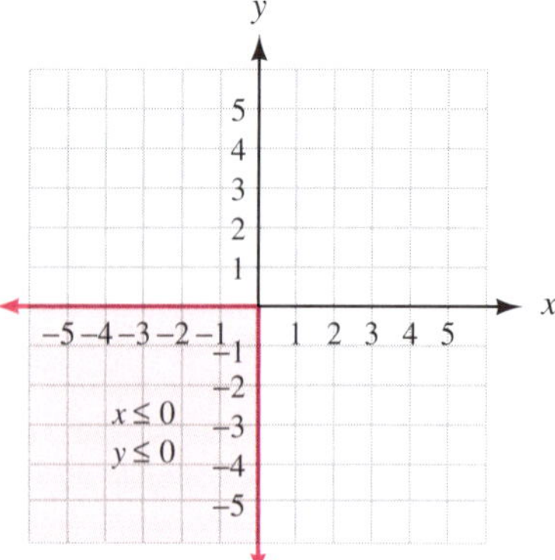

Figure 12

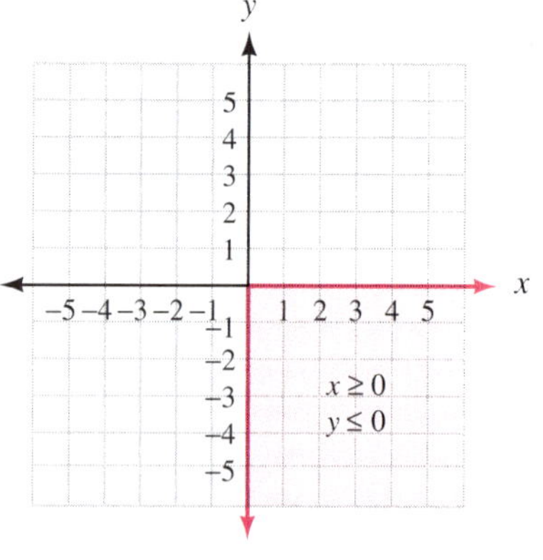

Figure 13

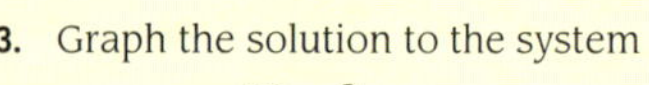

3. Graph the solution to the system

$x > -3$

$y < 4$

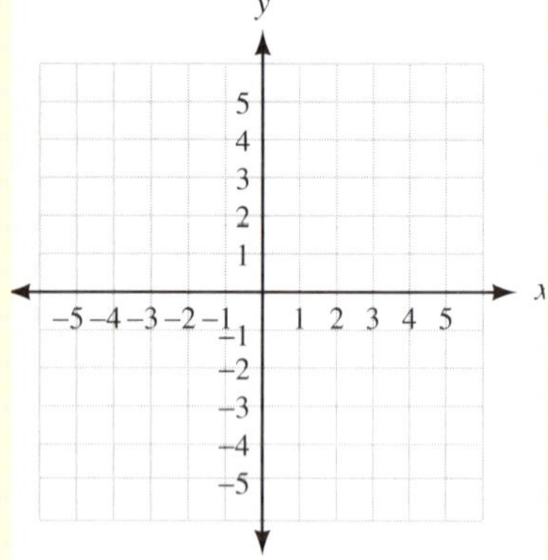

EXAMPLE 3 Graph the solution to the system of linear inequalities.

$$x \le 4$$

$$y \ge -3$$

SOLUTION The solution to this system will consist of all points to the left of and including the vertical line $x = 4$ that intersect with all points above and including the horizontal line $y = -3$. The solution set is shown in Figure 14.

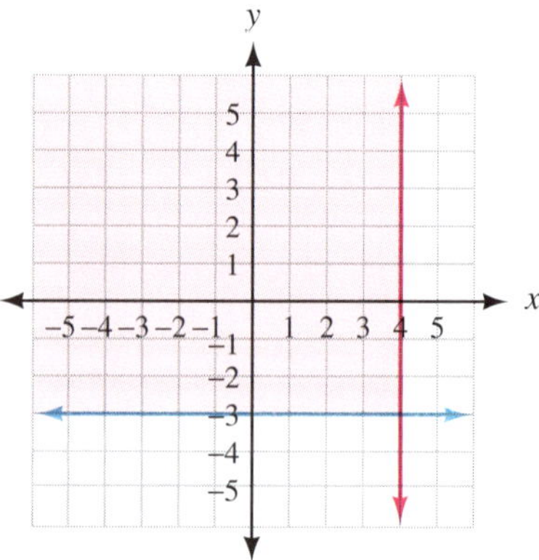

Figure 14

4. Graph the solution set for the following system.

$x - y < 5$

$x + y < 5$

$x > 1$

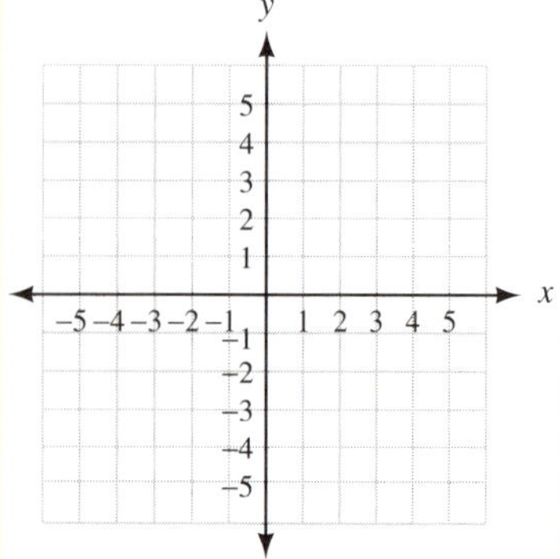

EXAMPLE 4 Graph the solution set for the following system.

$$x - 2y \le 4$$

$$x + y \le 4$$

$$x \ge -1$$

SOLUTION We have three linear inequalities, representing three sections of the coordinate plane. The graph of the solution set for this system will be the intersection of these three sections. The graph of $x - 2y \le 4$ is the section above and including the boundary $x - 2y = 4$. The graph of $x + y \le 4$ is the section below and including the boundary line $x + y = 4$. The graph of $x \ge -1$ is all the points to the right of, and including, the vertical line $x = -1$. The intersection of these three graphs is shown in Figure 15.

Answers

3.-4. See solutions section

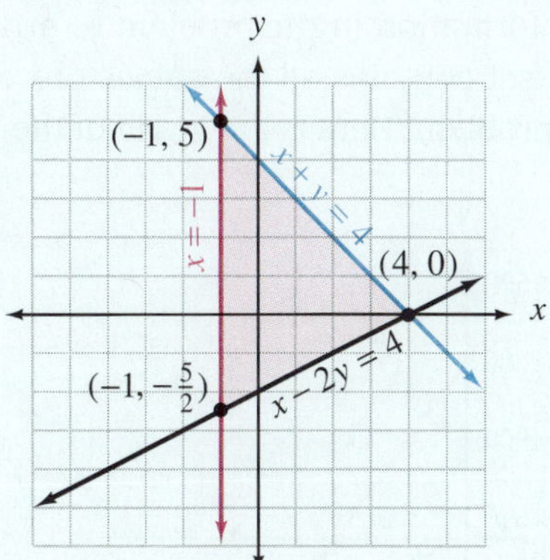

Figure 15

EXAMPLE 5 A college basketball arena plans on charging \$20 for certain seats and \$15 for others. They want to bring in more than \$18,000 from all ticket sales and have reserved at least 500 tickets at the \$15 rate. Find a system of inequalities describing all possibilities and sketch the graph. If 620 tickets are sold for \$15, at least how many tickets are sold for \$20?

5. How would the graph in Figure 16 change if they had reserved only 300 tickets at the \$15 rate?

Peter Jones/Corbis

SOLUTION Let x = the number of \$20 tickets and y = the number of \$15 tickets. We need to write a list of inequalities that describe this situation. That list will form our system of inequalities. First of all, we note that we cannot use negative numbers for either x or y. So, we have our first inequalities:

$$x \geq 0$$

$$y \geq 0$$

Next, we note that they are selling at least 500 tickets for \$15, so we can replace our second inequality with $y \geq 500$. Now our system is

$$x \geq 0$$

$$y \geq 500$$

Now the amount of money brought in by selling \$20 tickets is $20x$, and the amount of money brought in by selling \$15 tickets is $15y$. If the total income from ticket sales is to be more than \$18,000, then $20x + 15y$ must be greater than 18,000. This gives us our last inequality and completes our system.

$$20x + 15y > 18{,}000$$

$$x \geq 0$$

$$y \geq 500$$

We have used all the information in the problem to arrive at this system of inequalities. The solution set contains all the values of x and y that satisfy all the conditions given in the problem. Here is the graph of the solution set.

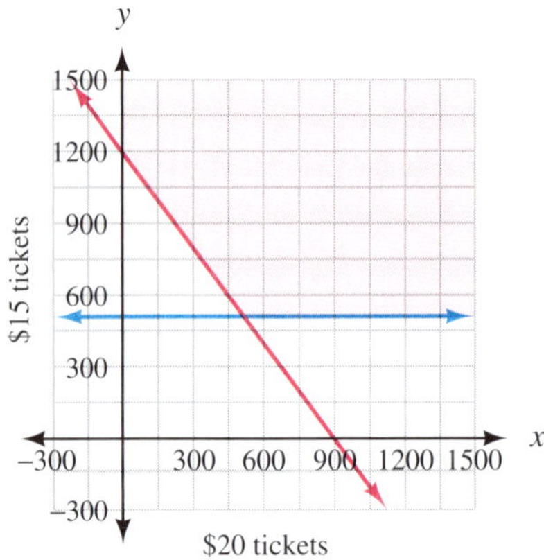

Figure 16

If 620 tickets are sold for \$15, then we substitute 620 for y in our first inequality to obtain

$20x + 15(620) > 18{,}000$	**Substitute 620 for *y***
$20x + 9{,}300 > 18{,}000$	**Multiply**
$20x > 8{,}700$	**Add −9,300 to each side**
$x > 435$	**Divide each side by 20**

If they sell 620 tickets for \$15 each, then they need to sell more than 435 tickets at \$20 each to bring in more than \$18,000.

Getting Ready for Class

After reading through the preceding section, respond in your own words and in complete sentences.

A. What is the solution set to a system of inequalities?

B. When graphing a system of linear inequalities, how do you find the equations of the boundary lines?

C. For the boundary lines of a system of linear inequalities, when do you use a dotted line rather than a solid line?

D. Once you have graphed the solution set for each inequality in a system, how do you determine the region to shade for the solution to the system of inequalities?

Answer

5. The horizontal line would move down to $y = 300$ from $y = 500$.

PROBLEM SET 2.9

Graph the solution set for each system of linear inequalities.

1. $x + y < 5$
   $2x - y > 4$
2. $x + y < 5$
   $2x - y < 4$
3. $3x + 4y > 12$
   $-3x + 2y \le 6$
4. $2x - 5y < 10$
   $2x + y > -2$
5. $y \ge x$
   $x + y > -4$
6. $y < \frac{2}{3}x$
   $x + 2y < 4$
7. $y \ge \frac{3}{4}x$
   $y \le -\frac{3}{4}x$
8. $y \le \frac{1}{2}x$
   $y \ge -2x$
9. $y < \frac{1}{3}x + 4$
   $y \ge \frac{1}{3}x - 3$
10. $y < 2x + 4$
    $y \ge 2x - 3$
11. $x \ge -3$
    $y < 2$
12. $x \le 4$
    $y \ge -2$
13. $1 \le x \le 3$
    $2 \le y \le 4$
14. $-4 \le x \le -2$
    $1 \le y \le 3$
15. $x + y \le 4$
    $x \ge 0$
    $y \ge 0$
16. $x - y \le 2$
    $x \ge 0$
    $y \le 0$
17. $x + y \le 3$
    $x - 3y \le 3$
    $x \ge -2$
18. $x - y \le 4$
    $x + 2y \le 4$
    $x \ge -1$
19. $x + y \le 2$
    $-x + y \le 2$
    $y \ge -2$
20. $x - y \le 3$
    $-x - y \le 3$
    $y \le -1$
21. $x + y < 5$
    $y > x$
    $y \ge 0$
22. $x + y < 5$
    $y > x$
    $x \ge 0$
23. $y < x$
    $x + y > -4$
    $x \le 0$
24. $y > -x$
    $-x + y < 4$
    $x \le 0$
25. $2x + 3y \le 6$
    $x \ge 0$
    $y \ge 0$
26. $x + 2y \le 10$
    $3x + y \le 12$
    $x \ge 0$
    $y \ge 0$

For each figure below, find a system of inequalities that describes the shaded region.

27.

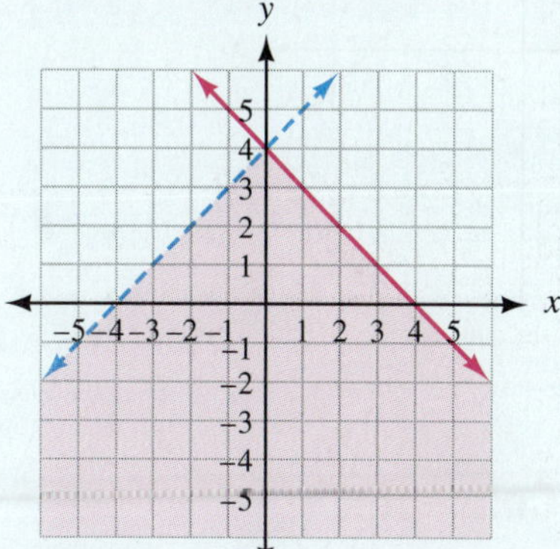

Figure 17

28.

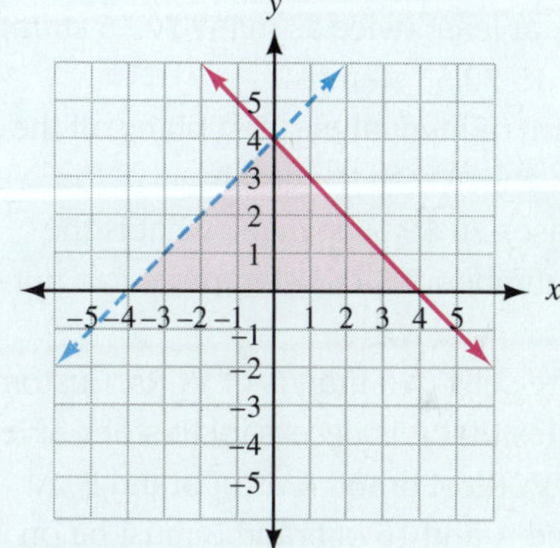

Figure 18

29.

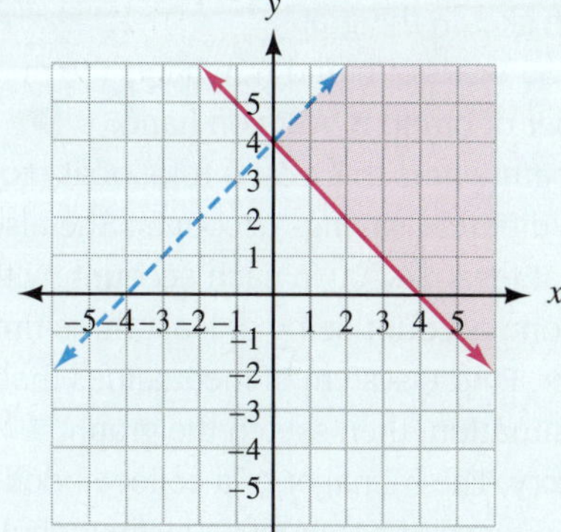

Figure 19

30.

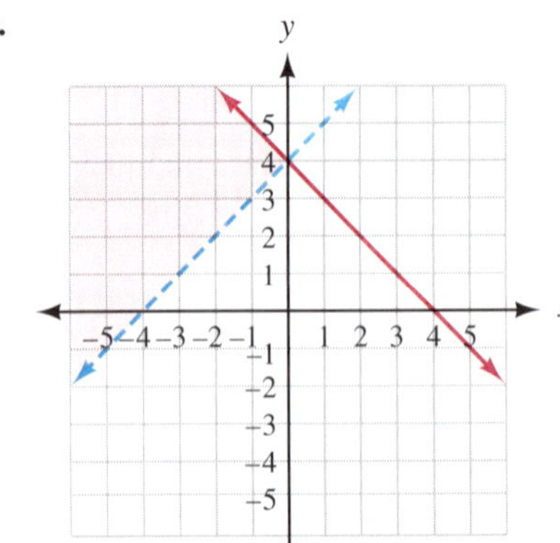

Figure 20

Applying the Concepts

31. **Office Supplies** An office worker wants to purchase some \$0.55 postage stamps and also some \$0.65 postage stamps totaling no more than \$40. It is also desired to have at least twice as many \$0.55 stamps and more than 15 \$0.55 stamps.
    (a) Find a system of inequalities describing all the possibilities and sketch the graph.
    (b) If he purchases 20 \$0.55 stamps, what is the maximum number of \$0.65 stamps he can purchase?

32. **Inventory** A store sells two brands of VCRs. Customer demand indicates that it is necessary to stock at least twice as many VCRs of brand A as of brand B. At least 30 of brand A and 15 of brand B must be on hand. In the store, there is room for not more than 100 VCRs.
    (a) Find a system of inequalities describing all possibilities, then sketch the graph.
    (b) If there are 35 VCRs of brand A, what is the maximum number of brand B VCRs on hand?

33. **Investing** Elizabeth wants to invest a total of at most \$30,000 in two different savings accounts. She also wants to have at least \$4,000 in each account, with the amount in one account being at least three times that in the other. Find a system of inequalities that describes this situation, then sketch the graph.

34. **Bookstore Inventory** The manager of a college bookstore stocks two types of notebooks, the first wholesaling for \$1.75 and the second for \$1.80. The maximum amount she can spend for notebooks is \$2,500, and an inventory of at least 450 of the \$1.75 variety and at least 650 of the \$1.80 variety is desired.
    (a) Find a system of inequalities that describes this situation, then sketch the graph.
    (b) If 525 of the \$1.75 variety are ordered, at most how many of the \$1.80 variety can be ordered?

35. **Construction** A builder specializes in the construction of car garages and tool sheds. The following timetable shows the number of hours needed to frame and roof the garages and the tool sheds. It also shows the maximum number of hours available for each kind of construction.

	HOURS TO FRAME	HOURS TO ROOF
Car Garages	10	7
Tool Sheds	4	5
Maximum hours available	80	74

    (a) Write a system of linear inequalities that describes all the conditions and sketch the graph of the system.
    (b) Can the builder construct 6 car garages and 5 tool sheds?
    (c) Can the builder construct 5 car garages and 8 tool sheds?

36. **Catering Business** A woman runs a full-service catering business. She prepares food ahead of time and she sets up the tables and provides decorations on the day of the event. The table below shows her time requirements and constraints.

	FOOD PREPARATION HOURS	SETUP HOURS
Banquets	12	9
Receptions	4	6
Maximum hours available	35	40

    (a) Write a system of linear inequalities that describes the conditions, then sketch the graph of the system.
    (b) Can she cater 2 banquets and 4 receptions?
    (c) Can she cater 2 banquets and 2 receptions?

Review Problems

Factor completely, if possible.

37. $x^2 - 16$
38. $x^4 - 16$
39. $x^2 + 16$
40. $x^2 - 64$
41. $x^2 + 64$
42. $x^3 + 64$
43. $x^2 - 16x + 64$
44. $x^2 - 16x - 64$

CHAPTER 2 SUMMARY

Examples

Linear Equations in Two Variables [2.1, 2.3]

A *linear equation in two variables* is any equation that can be put in *standard form* $ax + by = c$. The graph of every linear equation is a straight line.

1. The equation $3x + 2y = 6$ is an example of a linear equation in two variables.

Intercepts [2.1]

The *x-intercept* of an equation is the *x-coordinate* of the point where the graph crosses the x-axis. The *y-intercept* is the y-coordinate of the point where the graph crosses the y-axis. We find the y-intercept by substituting $x = 0$ into the equation and solving for y. The x-intercept is found by letting $y = 0$ and solving for x.

2. To find the x-intercept for $3x + 2y = 6$, we let $y = 0$ and get

$$3x = 6$$
$$x = 2$$

In this case the x intercept is 2, and the graph crosses the x-axis at (2, 0).

The Slope of a Line [2.2]

The *slope* of the line containing points (x_1, y_1) and (x_2, y_2) is given by

$$\text{Slope} = m = \frac{\text{Rise}}{\text{Run}} = \frac{y_2 - y_1}{x_2 - x_1}$$

Horizontal lines have 0 slope, and vertical lines have no slope.

Parallel lines have equal slopes, and perpendicular lines have slopes that are negative reciprocals.

3. The slope of the line through (6, 9) and (1, −1) is

$$m = \frac{9 - (-1)}{6 - 1} = \frac{10}{5} = 2$$

The Slope-Intercept Form of a Line [2.3]

The equation of a line with slope m and y-intercept b is given by

$$y = mx + b$$

4. The equation of the line with slope 5 and y-intercept 3 is

$$y = 5x + 3$$

The Point-Slope Form of a Line [2.3]

The equation of the line through (x_1, y_1) that has slope m can be written as

$$y - y_1 = m(x - x_1)$$

5. The equation of the line through (3, 2) with slope −4 is

$$y - 2 = -4(x - 3)$$

which can be simplified to

$$y = -4x + 14$$

Linear Inequalities in Two Variables [2.4]

An inequality of the form $ax + by < c$ is a *linear inequality in two variables.* The equation for the boundary of the solution set is given by $ax + by = c$. (This equation is found by simply replacing the inequality symbol with an equal sign.)

To graph a linear inequality, first graph the boundary, using a solid line if the boundary is included in the solution set and a broken line if the boundary is not included in the solution set. Next, choose any point not on the boundary and substitute its coordinates into the original inequality. If the resulting statement is true, the graph lies on the same side of the boundary as the test point. A false statement indicates that the solution set lies on the other side of the boundary.

6. The graph of

$$x - y \leq 3$$

is

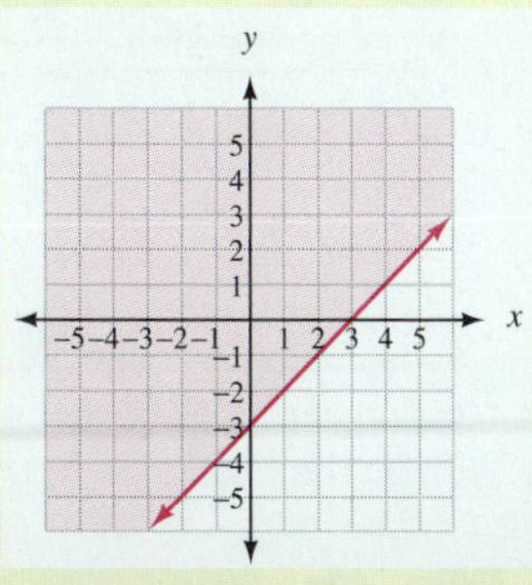

Relations and Functions [2.5]

7. The relation

$$\{(8, 1), (6, 1), (-3, 0)\}$$

is also a function because no ordered pairs have the same first coordinates. The domain is $\{8, 6, -3\}$ and the range is $\{1, 0\}$.

A *function* is a rule that pairs each element in one set, called the *domain,* with exactly one element from a second set, called the *range.*

A *relation* is any set of ordered pairs. The set of all first coordinates is called the *domain* of the relation, and the set of all second coordinates is the *range* of the relation. A function is a relation in which no two different ordered pairs have the same first coordinates.

Vertical Line Test [2.5]

8. The graph of $x = y^2$ shown in Figure 5 in Section 3.5 fails the vertical line test. It is not the graph of a function.

If a vertical line crosses the graph of a relation in more than one place, the relation cannot be a function. If no vertical line can be found that crosses the graph in more than one place, the relation must be a function.

Function Notation [2.6]

9. If $f(x) = 5x - 3$ then

$$f(0) = 5(0) - 3 = -3$$
$$f(1) = 5(1) - 3 = 2$$
$$f(-2) = 5(-2) - 3 = -13$$
$$f(a) = 5a - 3$$

The alternative notation for y is $f(x)$. It is read "f of x" and can be used instead of the variable y when working with functions. The notation y and the notation $f(x)$ are equivalent; that is, $y = f(x)$.

Variation [2.7]

10. If y varies directly with x, then

$$y = Kx$$

Then if y is 18 when x is 6,

$$18 = K \cdot 6$$

or

$$K = 3$$

So the equation can be written more specifically as

$$y = 3x$$

If we want to know what y is when x is 4, we simply substitute:

$$y = 3 \cdot 4$$
$$y = 12$$

If *y varies directly* with x (y is directly proportional to x), then

$$y = Kx$$

If *y varies inversely* with x (y is inversely proportional to x), then

$$y = \frac{K}{x}$$

If *z varies jointly* with x and y (z is directly proportional to both x and y), then

$$z = Kxy$$

In each case, K is called the *constant of variation.*

Systems of Linear Equations [2.8]

11. The solution to the system

$$x + 2y = 4$$
$$x - y = 1$$

is the ordered pair (2, 1). It is the only ordered pair that satisfies both equations.

A system of linear equations consists of two or more linear equations considered simultaneously. The solution set to a linear system in two variables is the set of ordered pairs that satisfy both equations. The solution set to a linear system in three variables consists of all the ordered triples that satisfy each equation in the system.

To Solve a System by the Addition Method [2.8]

12. We can eliminate the y-variable from the system in Example 1 by multiplying both sides of the second equation by 2 and adding the result to the first equation:

$$\begin{aligned} x + 2y = 4 &\xrightarrow{\text{No change}} x + 2y = 4 \\ x - y = 1 &\xrightarrow[\text{Times 2}]{} \underline{2x - 2y = 2} \\ & \qquad 3x \quad = 6 \\ & \qquad x \quad = 2 \end{aligned}$$

Substituting $x = 2$ into either of the original two equations gives $y = 1$. The solution is (2, 1).

Step 1: Look the system over to decide which variable will be easier to eliminate.

Step 2: Use the multiplication property of equality on each equation separately, if necessary, to ensure that the coefficients of the variable to be eliminated are opposites.

Step 3: Add the left and right sides of the system produced in step 2, and solve the resulting equation.

Step 4: Substitute the solution from step 3 back into any equation with both x- and y-variables, and solve.

Step 5: Check your solution in both equations if necessary.

To Solve a System by the Substitution Method [2.8]

Step 1: Solve either of the equations for one of the variables (this step is not necessary if one of the equations has the correct form already).

Step 2: Substitute the results of step 1 into the other equation, and solve.

Step 3: Substitute the results of step 2 into an equation with both x- and y-variables, and solve. (The equation produced in step 1 is usually a good one to use.)

Step 4: Check your solution if necessary.

13. We can apply the substitution method to the system in Example 1 by first solving the second equation for x to get

$$x = y + 1$$

Substituting this expression for x into the first equation we have

$$y + 1 + 2y = 4$$
$$3y + 1 = 4$$
$$3y = 3$$
$$y = 1$$

Using $y = 1$ in either of the original equations gives $x = 2$.

Inconsistent and Dependent Equations [2.8]

Two linear equations that have no solutions in common are said to be *inconsistent,* whereas two linear equations that have all their solutions in common are said to be *dependent.*

14. If the two lines are parallel, then the system will be inconsistent and the solution is ∅. If the two lines coincide, then the system is dependent.

Systems of Linear Inequalities [2.9]

A system of linear inequalities is two or more linear inequalities considered at the same time. To find the solution set to the system, we graph each of the inequalities on the same coordinate system. The solution set is the region that is common to all the regions graphed.

15. The solution set for the system

$$x + y < 4$$
$$-x + y \le 3$$

is shown below.

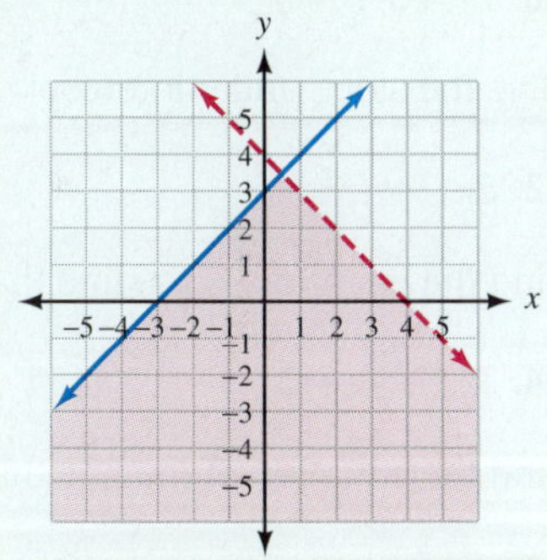

COMMON MISTAKES

1. When graphing ordered pairs, the most common mistake is to associate the first coordinate with the y-axis and the second with the x-axis. If you make this mistake you would graph (3, 1) by going up 3 and to the right 1, which is just the reverse of what you should do. Remember, the first coordinate is always associated with the horizontal axis, and the second coordinate is always associated with the vertical axis.
2. The two most common mistakes students make when first working with the formula for the slope of a line are the following:
   (a) Putting the difference of the x-coordinates over the difference of the y-coordinates.
   (b) Subtracting in one order in the numerator and then subtracting in the opposite order in the denominator.
3. When graphing linear inequalities in two variables, remember to graph the boundary with a broken line when the inequality symbol is $<$ or $>$. The only time you use a solid line for the boundary is when the inequality symbol is $\le$ or $\ge$.

CHAPTER 2 REVIEW

The problems below form a comprehensive review of the material in this chapter. They can be used to study for exams. If you would like to take a practice test on this chapter, you can use the odd-numbered problems. Give yourself an hour and work as many of the odd-numbered problems as possible. When you are finished, or when an hour has passed, check your answers with the answers in the back of the book. You can use the even-numbered problems for a second practice test.

Graph each line. [2.1]

1. $3x + 2y = 6$

2. $y = -\frac{3}{2}x + 1$

3. $x = 3$

Find the slope of the line through the following pairs of points. [8.2]

4. $(5, 2), (3, 6)$

5. $(-4, 2), (3, 2)$

Find x if the line through the two given points has the given slope. [2.2]

6. $(4, x), (1, -3); m = 2$

7. $(-4, 7), (2, x); m = -\frac{1}{3}$

8. Find the slope of any line parallel to the line through $(3, 8)$ and $(5, -2)$. [2.2]

9. The line through $(5, 3y)$ and $(2, y)$ is parallel to a line with slope 4. What is the value of y? [2.2]

Give the equation of the line with the following slope and y-intercept. [2.3]

10. $m = 3, b = 5$

11. $m = -2, b = 0$

Give the slope and y-intercept of each equation. [2.3]

12. $3x - y = 6$

13. $2x - 3y = 9$

Find the equation of the line that contains the given point and has the given slope. [2.3]

14. $(2, 4), m = 2$

15. $(-3, 1), m = -\frac{1}{3}$

Find the equation of the line that contains the given pair of points. [2.3]

16. $(2, 5), (-3, -5)$

17. $(-3, 7), (4, 7)$

18. $(-5, -1), (-3, -4)$

19. Find the equation of the line that is parallel to $2x - y = 4$ and contains the point $(2, -3)$. [2.3]

20. Find the equation of the line perpendicular to $y = -3x + 1$ that has an x-intercept of 2. [2.3]

Graph each linear inequality. [2.4]

21. $y \le 2x - 3$

22. $x \ge -1$

State the domain and range of each relation, and then indicate which relations are also functions. [2.5]

23. $\{(2, 4), (3, 3), (4, 2)\}$

24. $\{(6, 3), (-4, 3), (-2, 0)\}$

If $f = \{(2, -1), (-3, 0), (4, \frac{1}{2}), (\pi, 2)\}$ and $g = \{(2, 2), (-1, 4), (0, 0)\}$, find the following. [2.6]

25. $f(-3)$

26. $f(2) + g(2)$

Let $f(x) = 2x^2 - 4x + 1$ and $g(x) = 3x + 2$, and evaluate each of the following. [2.6]

27. $f(0)$

28. $g(a)$

29. $f[g(0)]$

30. $f[g(1)]$

For the following problems, y varies directly with x. [2.7]

31. If y is 6 when x is 2, find y when x is 8.

32. If y is -3 when x is 5, find y when x is -10.

For the following problems, y varies inversely with the square of x. [2.7]

33. If y is 9 when x is 2, find y when x is 3.

34. If y is 4 when x is 5, find y when x is 2.

Solve each application problem. [2.7]

35. Tension in a Spring The tension t in a spring varies directly with the distance d the spring is stretched. If the tension is 42 pounds when the spring is stretched 2 inches, find the tension when the spring is stretched twice as far.

36. Light Intensity The intensity of a light source varies inversely with the square of the distance from the source. Four feet from the source the intensity is 9 foot candles. What is the intensity 3 feet from the source?

Solve each system using the addition method. [2.8]

37. $x + y = 4$
$2x - y = 14$

38. $2x - 4y = 5$
$-x + 2y = 3$

39. $6x - 5y = -5$
$3x + y = 1$

40. $6x + 4y = 8$
$9x + 6y = 12$

41. $\frac{2}{3}x - \frac{1}{6}y = 0$
$\frac{4}{3}x + \frac{5}{6}y = 14$

42. $-\frac{1}{2}x + \frac{1}{3}y = -\frac{13}{6}$
$\frac{4}{5}x + \frac{3}{4}y = \frac{9}{10}$

Solve each system by the substitution method. [2.8]

43. $2x - 3y = 5$
$y = 2x - 7$

44. $x + y = 4$
$2x + 5y = 2$

45. $3x + 7y = 6$
$x = -3y + 4$

46. $5x - y = 4$
$y = 5x - 3$

Graph the solution set for each system of linear inequalities. [2.9]

47. $3x + 4y < 12$
$-3x + 2y \le 6$

48. $3x + 4y < 12$
$-3x + 2y \le 6$
$y \ge 0$

49. $3x + 4y < 12$
$x \ge 0$
$y \ge 0$

50. $x > -3$
$y > -2$

Rational Expressions

Jon Bradley/Stone/Getty Images

If you have ever put yourself on a weight loss diet, you know that you lose more weight at the beginning of the diet than you do later. If we let $W(x)$ represent a person's weight after x weeks on the diet, then the rational function

$$W(x) = \frac{80(2x + 15)}{x + 6}$$

is a mathematical model of the person's weekly progress on a diet intended to take them from 200 pounds to about 160 pounds. Rational functions are good models for quantities that fall off rapidly to begin with and then level off over time. The table shows some values for this function, while Figure 1 shows the graph of this function.

WEEKLY WEIGHT LOSS

WEEKS SINCE STARTING DIET	WEIGHT (NEAREST POUND)
0	200
4	184
8	177
12	173
16	171
20	169
24	168

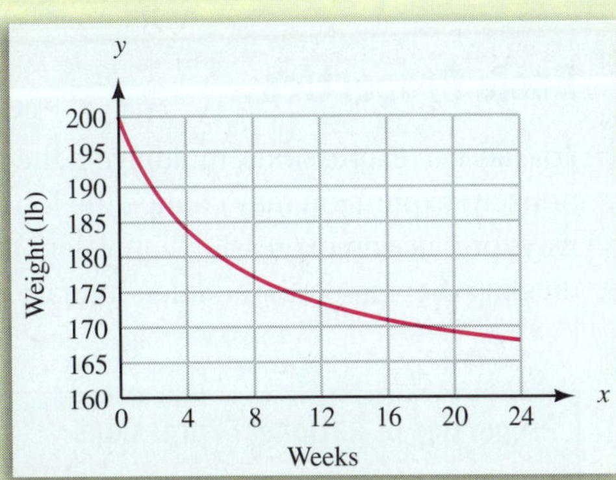

Figure 1

As you progress through this chapter, you will acquire an intuitive feel for these types of functions, and as a result, you will see why they are good models for situations such as dieting.

CHAPTER OUTLINE

3.1 Basic Properties and Reducing to Lowest Terms

This chapter is mostly concerned with simplifying a certain kind of algebraic expression. The expressions are called rational expressions because they are to algebra what rational numbers are to arithmetic. Most of the work we will do with rational expressions parallels the work you have done in previous math classes with fractions. Once we have learned to add, subtract, multiply, and divide rational expressions, we will turn our attention to equations involving rational expressions. The single most important tool needed for success in this chapter is factoring. Almost every problem encountered in this chapter involves factoring at one point or another. You may be able to understand all the theory and steps involved in solving the problems, but unless you can factor the polynomials in the problems, you will be unable to work any of them.

We will begin this section with the definition of a rational expression. We then will state the two basic properties associated with rational expressions, and go on to apply one of the properties to reduce rational expressions to lowest terms.

Recall from Chapter 1 that a *rational number* is any number that can be expressed as the ratio of two integers:

$$\text{Rational numbers} = \left\{ \frac{a}{b} \,\middle|\, a \text{ and } b \text{ are integers, } b \neq 0 \right\}$$

A rational expression is defined similarly as any expression that can be written as the ratio of two polynomials:

$$\text{Rational expressions} = \left\{ \frac{P}{Q} \,\middle|\, P \text{ and } Q \text{ are polynomials, } Q \neq 0 \right\}$$

Some examples of rational expressions are

$$\frac{2x-3}{x+5} \qquad \frac{x^2-5x-6}{x^2-1} \qquad \frac{a-b}{b-a}$$

Basic Properties

For rational expressions, multiplying the numerator and denominator by the same nonzero expression may change the form of the rational expression, but it will always produce an expression equivalent to the original one. The same is true when dividing the numerator and denominator by the same nonzero quantity.

Note: These two statements are equivalent since division is defined as multiplication by the reciprocal. We choose to state them separately for clarity.

Properties of Rational Expressions

If P, Q, and K are polynomials with $Q \neq 0$ and $K \neq 0$, then

$$\frac{P}{Q} = \frac{PK}{QK} \quad \text{and} \quad \frac{P}{Q} = \frac{P/K}{Q/K}$$

Reducing to Lowest Terms

The fraction $\frac{6}{8}$ can be written in lowest terms as $\frac{3}{4}$. The process is shown here:

$$\frac{6}{8} = \frac{3 \cdot \overset{1}{\cancel{2}}}{4 \cdot \underset{1}{\cancel{2}}} = \frac{3}{4}$$

Reducing $\frac{6}{8}$ to $\frac{3}{4}$ involves dividing the numerator and denominator by 2, the factor they have in common. Before dividing out the common factor 2, we must notice that the common factor *is* 2! (This may not be obvious since we are very familiar with the numbers 6 and 8 and therefore do not have to put much thought into finding what number divides both of them.)

We reduce rational expressions to lowest terms by first factoring the numerator and denominator and then dividing both numerator and denominator by any factors they have in common.

EXAMPLE 1 Reduce $\frac{x^2 - 9}{x - 3}$ to lowest terms.

SOLUTION Factoring, we have

$$\frac{x^2 - 9}{x - 3} = \frac{(x + 3)(x - 3)}{x - 3}$$

The numerator and denominator have the factor $x - 3$ in common. Dividing the numerator and denominator by $x - 3$, we have

$$\frac{(x + 3)\overset{1}{\cancel{(x - 3)}}}{\underset{1}{\cancel{x - 3}}} = \frac{x + 3}{1} = x + 3$$

Note For the problem in Example 1, there is an implied restriction on the variable x: it cannot be 3. If x were 3, the expression $(x^2 - 9)/(x - 3)$ would become 0/0, an expression that we cannot associate with a real number. For all problems involving rational expressions, we restrict the variable to only those values that result in a nonzero denominator. When we state the relationship

$$\frac{x^2 - 9}{x - 3} = x + 3$$

we are assuming that it is true for all values of x except $x = 3$.

Here are some other examples of reducing rational expressions to lowest terms.

EXAMPLES Reduce to lowest terms.

2. $$\frac{y^2 - 5y - 6}{y^2 - 1} = \frac{(y - 6)\cancel{(y + 1)}}{(y - 1)\cancel{(y + 1)}}$$ **Factor numerator and denominator**

$$= \frac{y - 6}{y - 1}$$ **Divide out common factor $y + 1$**

3. $$\frac{2a^3 - 16}{4a^2 - 12a + 8} = \frac{2(a^3 - 8)}{4(a^2 - 3a + 2)}$$

$$= \frac{2\cancel{(a - 2)}(a^2 + 2a + 4)}{4\cancel{(a - 2)}(a - 1)}$$ **Factor numerator and denominator**

$$= \frac{a^2 + 2a + 4}{2(a - 1)}$$ **Divide out common factor $2(a - 2)$**

Practice Problems

1. Reduce to lowest terms.
$\frac{x^2 - 9}{x + 3}$

Note: The lines are drawn through the $(x - 3)$ in the numerator and denominator indicate that we have divided through by $(x - 3)$. As the problems become more involved, these lines will help keep track of which factors have been divided out and which have not.

Reduce to lowest terms.

2. $\frac{y^2 - y - 6}{y^2 - 4}$

3. $\frac{3a^3 + 3}{6a^2 - 6a + 6}$

Answer

1. $x - 3$

4. Reduce to lowest terms.

$$\frac{x^2 + 4x + ax + 4a}{x^2 + ax + 4x + 4a}$$

Note: Beginning with Example 2 we are no longer showing the 1s that result when we divide out our common factors. If you need to show the 1s to remind you that you are *dividing*, then be sure to do so.

4. $\dfrac{x^2 - 3x + ax - 3a}{x^2 - ax - 3x + 3a} = \dfrac{x(x-3) + a(x-3)}{x(x-a) - 3(x-a)}$ **Factor numerator and denominator**

$= \dfrac{\cancel{(x-3)}(x+a)}{(x-a)\cancel{(x-3)}}$ **Divide out common factor $x - 3$**

$= \dfrac{x+a}{x-a}$

The answer to Example 4 is $(x + a)/(x - a)$. The problem cannot be reduced further. It is a fairly common mistake to attempt to divide out an x or an a in this last expression. Remember, we can divide out only the factors common to the numerator and denominator of a rational expression. For the last expression in Example 4, neither the numerator nor the denominator can be factored further; x is not a factor of the numerator or the denominator, and neither is a. The expression is in lowest terms.

The next example involves what we call a trick. The trick is to reverse the order of the terms in a difference by factoring -1 from each term. The next examples illustrate how this is done.

5. Replace a with 7 and b with 4 and then simplify.

$$\frac{a-b}{b-a}$$

EXAMPLE 5 Reduce to lowest terms: $\dfrac{a-b}{b-a}$.

SOLUTION The relationship between $a - b$ and $b - a$ is that they are opposites. We can show this fact by factoring -1 from each term in the numerator:

$\dfrac{a-b}{b-a} = \dfrac{-1(-a+b)}{b-a}$ **Factor -1 from each term in the numerator**

$= \dfrac{-1(b-a)}{b-a}$ **Reverse the order of the terms in the numerator**

$= -1$ **Divide out common factor $b - a$**

6. Reduce to lowest terms.

$$\frac{7-x}{x^2 - 49}$$

EXAMPLE 6 Reduce to lowest terms: $\dfrac{x^2 - 25}{5 - x}$.

SOLUTION We begin by factoring the numerator:

$$\frac{x^2 - 25}{5 - x} = \frac{(x-5)(x+5)}{5-x}$$

The factors $x - 5$ and $5 - x$ are similar but are not exactly the same. We can reverse the order of either by factoring -1 from it; that is: $5 - x = -1(-5 + x) = -1(x - 5)$.

$$\frac{(x-5)(x+5)}{5-x} = \frac{\cancel{(x-5)}(x+5)}{-1\cancel{(x-5)}}$$

$$= \frac{x+5}{-1}$$

$$= -(x+5)$$

Applications

Ratios You may recall from previous math classes that the ratio of a to b is the same as the fraction $\frac{a}{b}$. Here are two ratios that are used frequently in mathematics:

Answers

2. $\dfrac{y-3}{y-2}$ 3. $\dfrac{a+1}{2}$ 4. 1

5. -1 6. $\dfrac{-1}{x+7}$

1. The number π is defined as the ratio of the circumference of a circle to the diameter of a circle; that is

$$\pi = \frac{C}{d}$$

Multiplying both sides of this formula by d, we have the more common form $C = \pi d$.

Note: Since the diameter of a circle is twice the radius, the formula for circumference is sometimes written $C = 2\pi r$. Neither formula should be confused with the formula for the area of a circle, which is $A = \pi r^2$.

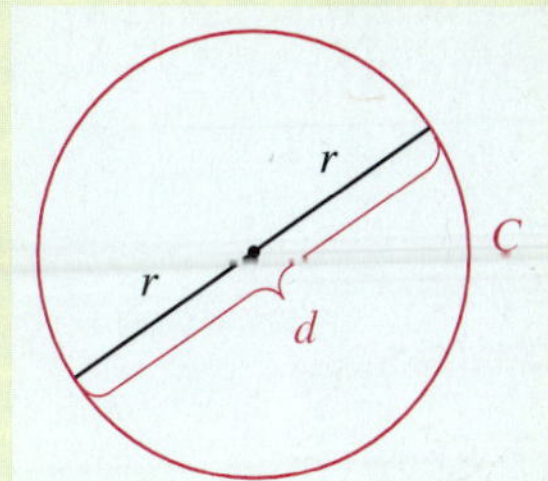

2. The *average speed* of a moving object is defined to be the ratio of distance to time. If you drive your car for 5 hours and travel a distance of 200 miles, then your average rate of speed is

$$\text{Average speed} = \frac{200 \text{ miles}}{5 \text{ hours}} = 40 \text{ miles per hour}$$

The formula we use for the relationship between average speed r, distance d, and time t is

$$r = \frac{d}{t}$$

The formula is sometimes called the *rate equation.* Multiplying both sides by t, we have an equivalent form of the rate equation, $d = rt$.

Our next example involves both the formula for the circumference of a circle and the rate equation.

EXAMPLE 7 The first Ferris wheel was designed and built by George Ferris in 1893. The diameter of the wheel was 250 feet. It had 36 carriages, equally spaced around the wheel, each of which held a maximum of 40 people. One trip around the wheel took 20 minutes. Find the average speed of a rider on the first Ferris wheel. (Use 3.14 as an approximation for π.)

7. A Ferris wheel was built in St. Louis in 1986. It is named *Colossus.* The diameter of the wheel is 165 feet. It has 40 cars, each of which holds 6 passengers. A trip around the wheel takes 40 seconds. Find the average speed of a rider on *Colossus.* (Use 3.14 as an approximation for π, and round your answer to the nearest tenth.)

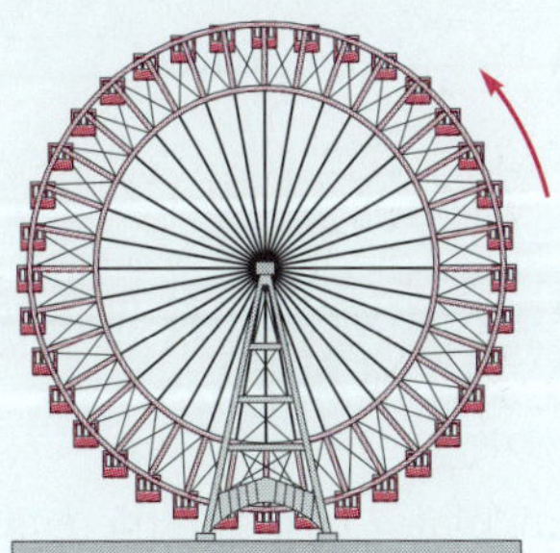

SOLUTION The distance traveled is the circumference of the wheel, which is

$$C = 250\pi = 250(3.14) = 785 \text{ feet}$$

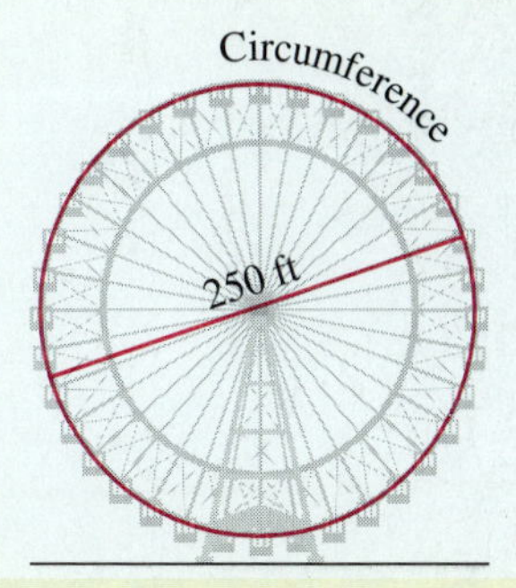

To find the average speed, we divide the distance traveled by the amount of time it took to go once around the wheel.

$$r = \frac{d}{t} = \frac{785 \text{ feet}}{20 \text{ minutes}} = 39.3 \text{ feet per minute} \quad \text{(to the nearest tenth)}$$

Later in this chapter we will convert this ratio into an equivalent ratio that gives the speed of the rider in miles per hour.

The Ferris wheel in Example 7 has a circumference of 785 feet. In the next example we graph the relationship between the average speed of a person riding this wheel and the amount of time it takes the wheel to complete one revolution.

Answer

7. 13.0 feet per second

8. Extend the table in Example 8 by finding $r(35)$. Then explain your result in words.

EXAMPLE 8 A Ferris wheel has a circumference of 785 feet. If one complete revolution of the wheel takes 10 to 30 minutes, then the relationship between the average speed of a rider on the wheel and the amount of time it takes the wheel to complete one revolution is given by the function

$$r(t) = \frac{785}{t} \qquad 10 \le t \le 30$$

where $r(t)$ is the average speed (in feet per minute) and t is the amount of time (in minutes) it takes the wheel to complete one revolution. Graph the function.

SOLUTION Since the variables r and t represent speed and time, both must be positive quantities. Therefore, the graph of this function will lie in the first quadrant only. The following table displays the values of t and $r(t)$ found from the function, along with the graph of the function (Figure 1). (Some of the numbers in the table have been rounded to the nearest tenth.)

TIME TO COMPLETE ONE REVOLUTION	SPEED (FT/MIN)
t	$r(t)$
10	78.5
15	52.3
20	39.3
25	31.4
30	26.2

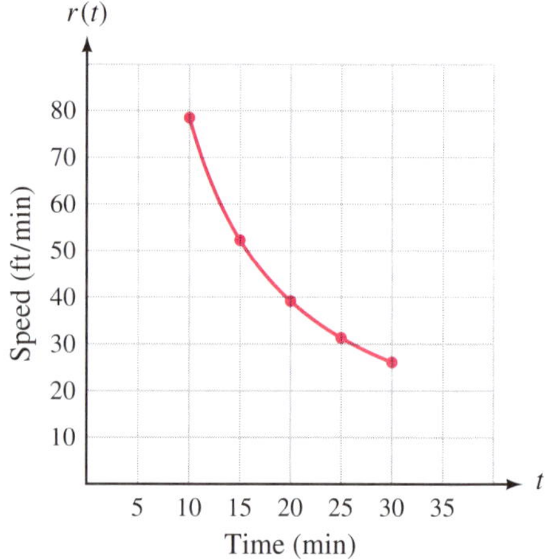

Figure 1

USING TECHNOLOGY

More About Example 8

If we use a graphing calculator to graph the equation in Example 8, it is not necessary to construct the table first. In fact, if we graph

$$Y_1 = 785/X \qquad \text{Window:} \qquad \text{X from 0 to 40, Y from 0 to 90}$$

we can use the Trace and Zoom features together to produce the numbers in the table next to Figure 1. Graph the preceding equation, and zoom in on the point with x-coordinate 20 until you are convinced that the table values for x and y are correct.

Rational Functions

The function shown in Example 8 is called a *rational function* because the right side, $785/t$, is a rational expression (the numerator, 785, is a polynomial of degree 0). We can extend our knowledge of rational expressions to functions with the following definition:

Answer

8. $r(35) = 22.4$ to the nearest tenth. If one trip around the wheel takes 35 minutes, then the rider is moving at approximately 22.4 feet per minute.

DEFINITION

A **rational function** is any function that can be written in the form

$$f(x) = \frac{P(x)}{Q(x)}$$

where $P(x)$ and $Q(x)$ are polynomials and $Q(x) \neq 0$.

EXAMPLE 9 For the rational function $f(x) = \dfrac{x-4}{x-2}$, find $f(0)$, $f(-4)$, $f(4)$, $f(-2)$, and $f(2)$.

SOLUTION To find these function values, we substitute the given value of x into the rational expression, and then simplify if possible.

$$f(0) = \frac{0-4}{0-2} = \frac{-4}{-2} = 2 \qquad f(-2) = \frac{-2-4}{-2-2} = \frac{-6}{-4} = \frac{3}{2}$$

$$f(-4) = \frac{-4-4}{-4-2} = \frac{-8}{-6} = \frac{4}{3} \qquad f(2) = \frac{2-4}{2-2} = \frac{-2}{0} \quad \text{Undefined}$$

$$f(4) = \frac{4-4}{4-2} = \frac{0}{2} = 0$$

9. If $f(x) = \dfrac{x+6}{x-3}$, find

(a) $f(0)$

(b) $f(-6)$

(c) $f(6)$

(d) $f(3)$

Because the rational function in Example 9 is not defined when x is 2, the domain of that function does not include 2. We have more to say about the domain of a rational function next.

The Domain of a Rational Function

In Example 8 the domain of the rational function is specified as $10 \le t \le 30$, and the function is defined for all values of t in that domain. If the domain of a rational function is not specified, it is assumed to be all real numbers for which the function is defined; that is, the domain of the rational function

$$f(x) = \frac{P(x)}{Q(x)}$$

is all x for which $Q(x)$ is nonzero. For example:

The domain for $r(t) = \dfrac{785}{t}$, $10 \le t \le 30$, is $\{t \mid 10 \le t \le 30\}$.

The domain for $f(x) = \dfrac{x-4}{x-2}$ is $\{x \mid x \neq 2\}$.

The domain for $g(x) = \dfrac{x^2+5}{x+1}$ is $\{x \mid x \neq -1\}$.

The domain for $h(x) = \dfrac{x}{x^2-9}$ is $\{x \mid x \neq -3, x \neq 3\}$.

Notice that, for these functions, $f(2)$, $g(-1)$, $h(-3)$, and $h(3)$ are all undefined, and that is why the domains are written as shown.

Answer

9. (a) -2 (b) 0 (c) 4 (d) undefined

10. Graph the equation $y = \frac{x^2 - 1}{x + 1}$ and explain how this graph differs from the graph of $y = x - 1$.

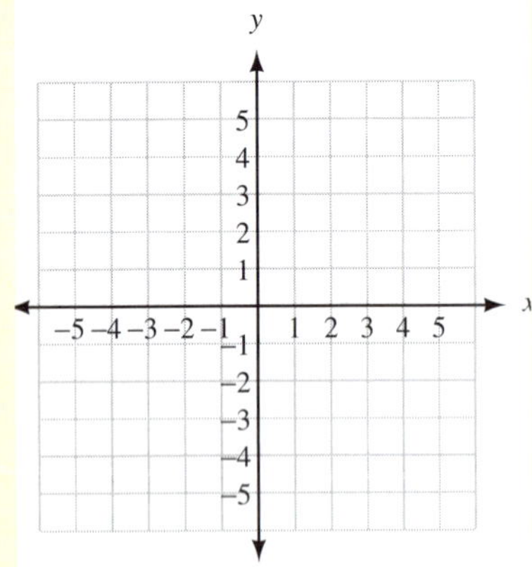

Answer

10. See Solutions Section.

EXAMPLE 10 Graph the equation $y = \frac{x^2 - 9}{x - 3}$. How is this graph different from the graph of $y = x + 3$?

SOLUTION We know from the discussion in Example 1 that

$$y = \frac{x^2 - 9}{x - 3} = \frac{(x + 3)(x - 3)}{x - 3} = x + 3$$

This relationship is true for all x except $x = 3$ because the rational expressions with $x - 3$ in the denominator are undefined when x is 3. However, for all other values of x, the expressions

$$\frac{x^2 - 9}{x - 3} \quad \text{and} \quad x + 3$$

are equal. Therefore, the graphs of

$$y = \frac{x^2 - 9}{x - 3} \quad \text{and} \quad y = x + 3$$

will be the same except when x is 3. In the first equation, there is no value of y to correspond to $x = 3$. In the second equation, $y = x + 3$, so y is 6 when x is 3.

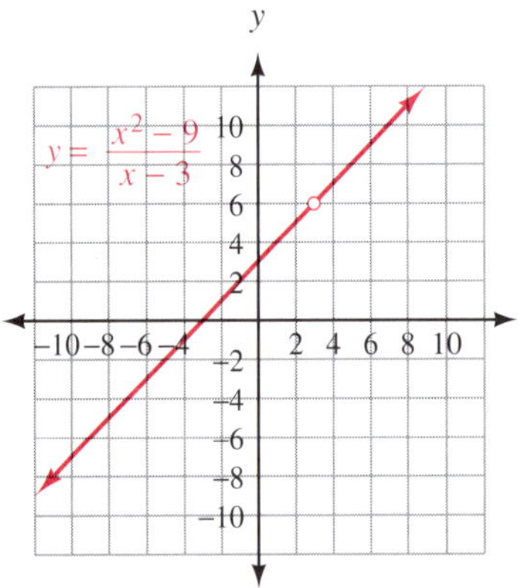

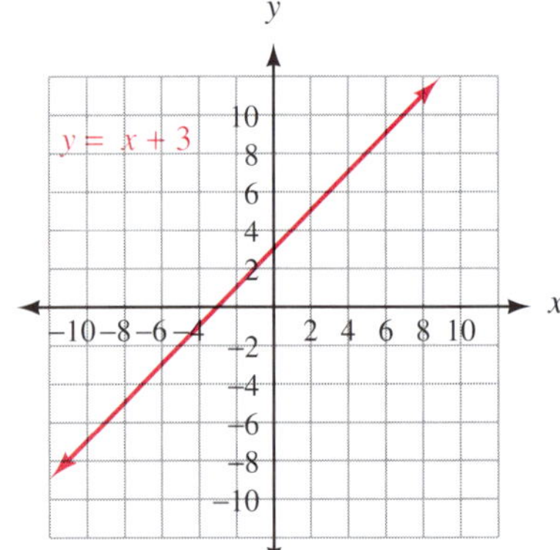

Figure 2 Graphs of $y = \frac{x^2 - 9}{x - 3}$ (left) and $y = x + 3$ (right)

Now you can see the difference in the graphs of the two equations. To show that there is no y value for $x = 3$ in the graph on the left in Figure 2, we draw an open circle at that point on the line.

Notice that the two graphs shown in Figure 2 are both graphs of functions. Suppose we use function notation to designate them as follows:

$$f(x) = \frac{x^2 - 9}{x - 3} \quad \text{and} \quad g(x) = x + 3$$

The two functions, f and g, are equivalent except when $x = 3$ because $f(3)$ is undefined, while $g(3) = 6$. The domain of the function f is all real numbers except $x = 3$, while the domain for g is all real numbers, with no restrictions.

Getting Ready for Class

After reading through the preceding section, respond in your own words and in complete sentences.

A. What is a rational expression?

B. Explain how to determine if a rational expression is in "lowest terms."

C. When is a rational expression undefined?

D. Explain the process we use to reduce a rational expression or a fraction to lowest terms.

PROBLEM SET 3.1

1. If $g(x) = \dfrac{x+3}{x-1}$, find $g(0)$, $g(-3)$, $g(3)$, $g(-1)$, and $g(1)$, if possible.

2. If $g(x) = \dfrac{x-2}{x-1}$, find $g(0)$, $g(-2)$, $g(2)$, $g(-1)$, and $g(1)$, if possible.

3. If $h(t) = \dfrac{t-3}{t+1}$, find $h(0)$, $h(-3)$, $h(3)$, $h(-1)$, and $h(1)$, if possible.

4. If $h(t) = \dfrac{t-2}{t+1}$, find $h(0)$, $h(-2)$, $h(2)$, $h(-1)$, and $h(1)$, if possible.

State the domain for each rational function.

5. $f(x) = \dfrac{x-3}{x-1}$

6. $f(x) = \dfrac{x+4}{x-2}$

7. $g(x) = \dfrac{x^2-4}{x-2}$

8. $g(x) = \dfrac{x^2-9}{x-3}$

9. $h(t) = \dfrac{t-4}{t^2-16}$

10. $h(t) = \dfrac{t-5}{t^2-25}$

Reduce each rational expression to lowest terms.

11. $\dfrac{x^2-16}{6x+24}$

12. $\dfrac{12x-9y}{3x^2+3xy}$

13. $\dfrac{a^4-81}{a-3}$

14. $\dfrac{a^2-4a-12}{a^2+8a+12}$

15. $\dfrac{20y^2-45}{10y^2-5y-15}$

16. $\dfrac{20x^2-93x+34}{4x^2-9x-34}$

17. $\dfrac{12y-2xy-2x^2y}{6y-4xy-2x^2y}$

18. $\dfrac{250a+100ax+10ax^2}{50a-2ax^2}$

19. $\dfrac{(x-3)^2(x+2)}{(x+2)^2(x-3)}$

20. $\dfrac{(x-4)^3(x+3)}{(x+3)^2(x-4)}$

21. $\dfrac{x^3+1}{x^2-1}$

22. $\dfrac{x^3-1}{x^2-1}$

23. $\dfrac{4am-4an}{3n-3m}$

24. $\dfrac{ad-ad^2}{d-1}$

25. $\dfrac{ab-a+b-1}{ab+a+b+1}$

26. $\dfrac{6cd-4c-9d+6}{6d^3-13d+6}$

27. $\dfrac{21x^2-23x+6}{21x^2+x-10}$

28. $\dfrac{36x^2-11x-12}{20x^2-39x+18}$

29. $\dfrac{8x^2-6x-9}{8x^2-18x+9}$

30. $\dfrac{42x^2+23x-10}{14x^2+45x-14}$

31. $\dfrac{4x^2+29x+45}{8x^2-10x-63}$

32. $\dfrac{30x^2-61x+30}{60x^2+22x-60}$

33. $\dfrac{a^3+b^3}{a^2-b^2}$

34. $\dfrac{a^2-b^2}{a^3-b^3}$

35. $\dfrac{8x^4-8x}{4x^4+4x^3+4x^2}$

36. $\dfrac{6x^2+7xy-3y^2}{6x^2+xy-y^2}$

37. $\dfrac{ax+2x+3a+6}{ay+2y-4a-8}$

38. $\dfrac{x^2-3ax-2x+6a}{x^2-3ax+2x-6a}$

39. $\dfrac{x^3+3x^2-4x-12}{x^2+x-6}$

40. $\dfrac{x^3+5x^2-4x-20}{x^2+7x+10}$

41. $\dfrac{3x^3+21x^2+36x}{3x^4+12x^3-27x^2-108x}$

42. $\dfrac{2x^4+14x^3+20x^2}{2x^5+4x^4-50x^3-100x^2}$

43. $\dfrac{4x^4-25}{6x^3-4x^2+15x-10}$

44. $\dfrac{16x^4-49}{8x^3-12x^2+14x-21}$

Refer to Examples 5 and 6 in this section, and reduce the following to lowest terms.

45. $\dfrac{x-4}{4-x}$

46. $\dfrac{6-x}{x-6}$

47. $\dfrac{y^2-36}{6-y}$

48. $\dfrac{1-y}{y^2-1}$

49. $\dfrac{1-9a^2}{9a^2-6a+1}$

50. $\dfrac{1-a^2}{a^2-2a+1}$

51. $\dfrac{y^2-2y+1}{3-2y-y^2}$

52. $\dfrac{5+4r-r^2}{r^2+2r+1}$

53. $\dfrac{m^2-n^2}{n^2-m^2}$

54. $\dfrac{d^3-c^3}{c^3-d^3}$

55. $\dfrac{z^2-10z+25}{20+z-z^2}$

56. $\dfrac{12r^2-25rt+12t^2}{18t^2-32r^2}$

Reduce each rational expression to lowest terms; then subtract.

57. $\dfrac{28x^2-41x+15}{7x-5} - \dfrac{12x^2-41x+24}{3x-8}$

58. $\dfrac{42x^2+47x-55}{7x-5} - \dfrac{18x^2-15x-88}{3x-8}$

59. $\dfrac{x^3-8}{x-2} - \dfrac{x^3+8}{x+2}$

60. $\dfrac{x^4-16}{x+2} - \dfrac{x^4-16}{x-2}$

Let $f(x) = \dfrac{x^2 - 4}{x - 2}$ and $g(x) = x + 2$, and evaluate the following expressions, if possible.

61. $f(0)$ and $g(0)$
62. $f(1)$ and $g(1)$
63. $f(2)$ and $g(2)$
64. $f(3)$ and $g(3)$

Let $f(x) = \dfrac{x^2 - 1}{x - 1}$ and $g(x) = x + 1$, and evaluate the following expressions, if possible.

65. $f(0)$ and $g(0)$
66. $f(1)$ and $g(1)$
67. $f(2)$ and $g(2)$
68. $f(3)$ and $g(3)$

69. Graph the equation $y = \dfrac{x^2 - 4}{x - 2}$. Then explain how this graph is different from the graph of $y = x + 2$.

70. Graph the equation $y = \dfrac{x^2 - 1}{x - 1}$. Then explain how this graph is different from the graph of $y = x + 1$.

Applying the Concepts

71. Diet The following rational function is the one we mentioned in the introduction to this chapter. The quantity $W(x)$ is the weight (in pounds) of the person after x weeks of dieting. Use the function to fill in the table. Then compare your results with the graph in the chapter introduction.

$$W(x) = \frac{80(2x + 15)}{x + 6}$$

WEEKS x	WEIGHT (lb) $W(x)$
0	
1	
4	
12	
24	

72. Drag Racing The following rational function gives the speed $V(x)$, in miles per hour, of a dragster at each second x during a quarter-mile race. Use the function to fill in the table.

$$V(x) = \frac{340x}{x + 3}$$

TIME (sec) x	SPEED (mi/hr) $V(x)$
0	
1	
2	
3	
4	
5	
6	

For Problems 73–76, round all answers to the nearest tenth.

73. Average Speed A jogger covers 3.5 miles in 29.75 minutes. Find the average speed of the jogger in miles per minute.

74. Average Speed A bullet fired from a gun travels a distance of 4,750 feet in 3.2 seconds. Find the average speed of the bullet in feet per second.

75. Fuel Consumption A pickup truck travels 175.8 miles on 16.3 gallons of gas. Give the average rate of fuel consumption of the truck in miles per gallon.

76. Fuel Consumption A luxury car travels 200 miles on 16.5 gallons of gas. Give the average fuel consumption of the car in miles per gallon.

For Problems 77 and 78, use 3.14 as an approximation for π. Round answers to the nearest tenth.

77. Average Speed A person riding a Ferris wheel with a diameter of 65 feet travels once around the wheel in 30 seconds. What is the average speed of the rider in feet per second?

78. Average Speed A person riding a Ferris wheel with a diameter of 102 feet travels once around the wheel in 3.5 minutes. What is the average speed of the rider in feet per minute?

The abbreviation "rpm" stands for revolutions per minute. If a point on a circle rotates at 300 rpm, then it rotates through one complete revolution 300 times every minute. The length of time it takes to rotate once around the circle is $\frac{1}{300}$ minute. Use 3.14 as an approximation for π.

79. Average Speed A $3\frac{1}{2}$-inch diskette, when placed in the disk drive of a computer, rotates at 300 rpm (1 revolution takes $\frac{1}{300}$ minute). Find the average speed of a point 2 inches from the center of the diskette. Then find the average speed of a point 1.5 inches from the center of the diskette.

80. Average Speed A 5-inch fixed disk in a computer rotates at 3,600 rpm. Find the average speed of a point 2 inches from the center of the disk. Then find the average speed of a point 1.5 inches from the center.

81. Average Speed The Ferris wheel in Problem 77 has a circumference of 204 feet (to the nearest foot). If a ride on the wheel takes from 20 to 50 seconds, then the relationship between the average speed of a rider and the amount of time it takes to complete one revolution is given by the function

$$r(t) = \frac{204}{t} \qquad 20 \le t \le 50$$

where $r(t)$ is in feet per second and t is in seconds.

(a) State the domain for this function.

(b) Graph the function.

82. Average Speed The Ferris wheel in Problem 78 has a circumference of 320 feet (to the nearest foot). If a ride on the wheel takes from 3 to 5 minutes, then the relationship between the average speed of a rider and the amount of time it takes to complete one revolution is given by the function

$$r(t) = \frac{320}{t} \quad 3 \le t \le 5$$

where $r(t)$ is in feet per minute and t is in minutes.

(a) State the domain for this function.

(b) Graph the function.

83. Intensity of Light The relationship between the intensity of light that falls on a surface from a 100-watt light bulb and the distance from that surface is given by the rational function

$$I(d) = \frac{120}{d^2} \qquad \text{for} \qquad 1 \le d \le 6$$

where $I(d)$ is the intensity of light (in lumens per square foot) and d is the distance (in feet) from the light bulb to the surface.

(a) State the domain for this function.

(b) Graph the function.

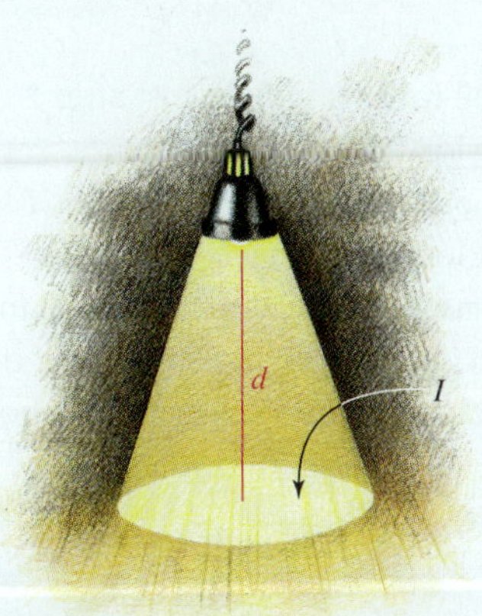

84. Average Speed If it takes Maria t minutes to run a mile, then her average speed $s(t)$ is given by the rational function

$$s(t) = \frac{60}{t} \qquad \text{for} \qquad 6 \le t \le 12$$

where $s(t)$ is in miles per hour and t is in minutes.

(a) State the domain for this function.

(b) Graph the function.

85. Dieting The following is a graph of a person's weight loss (in pounds) after x weeks of dieting:

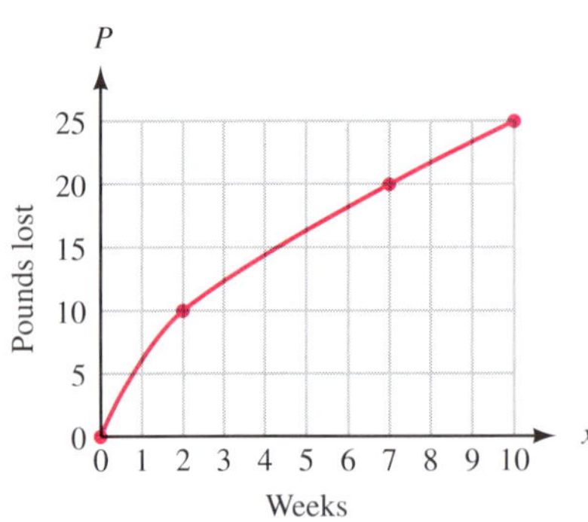

(a) Find the ratio of weight lost after 2 weeks to weight lost after 10 weeks.

(b) Find the ratio of weight lost after 7 weeks to weight lost after 10 weeks.

86. Traveling Time Suppose you travel 220 miles on the freeway to get to a relative's house. If the speed limit has been changed from 55 miles per hour to 70 miles per hour, how much time will you save with the new speed limit?

The **traffic capacity** of a road can be approximated with the formula

$$N = \frac{1{,}760V}{l}$$

where N is the traffic capacity in vehicles per hour, V is the speed of the vehicles in miles per hour, and l is the distance, in yards, from the front of one vehicle to the front of the next vehicle for vehicles in the same lane. Use this formula to solve the following problems.

87. Traffic Capacity Suppose $V = 70$ miles per hour and $l = 40$ yards. Under these conditions, what is the traffic capacity of a four-lane highway?

88. Traffic Capacity Suppose $V = 75$ miles per hour and $l = 42.5$ yards. What is the traffic capacity of a four-lane highway under these conditions?

Review Problems

Solve each system by the addition method.

89. $4x + 3y = 10$
$2x + y = 4$

90. $3x - 5y = -2$
$2x - 3y = 1$

91. $4x + 5y = 5$
$\frac{6}{5}x + y = 2$

92. $4x + 2y = -2$
$\frac{1}{2}x + y = 0$

Solve each system by the substitution method.

93. $x + y = 3$
$y = x + 3$

94. $x + y = 6$
$y = x - 4$

95. $2x - 3y = -6$
$y = 3x - 5$

96. $7x - y = 24$
$x = 2y + 9$

Extending the Concepts

Reduce to lowest terms.

97. $\dfrac{(x + y)^3 - 8z^3}{x + y - 2z}$

98. $\dfrac{(x + y)^3 + 8z^3}{x + y + 2z}$

99. $\dfrac{a^3 + (b - c)^3}{a + b - c}$

100. $\dfrac{a^3 - (b - c)^3}{a - b + c}$

101. $\dfrac{x^2 - (y + 1)^2}{x + y + 1}$

102. $\dfrac{x^2 - (y - 1)^2}{x + y - 1}$

103. $\dfrac{(x^2 - 1)(x + 1)}{(x^2 - 2x + 1)^2}$

104. $\dfrac{x^4 - y^4}{(x^2 + 2xy + y^2)(x^2 + y^2)}$

105. The graphs of two rational functions are given in Figures 3 and 4. Use the graphs to find the following.

(a) $f(2)$
(b) $f(-1)$
(c) $f(0)$
(d) $g(3)$
(e) $g(6)$
(f) $g(-1)$
(g) $f(g(6))$
(h) $g(f(-2))$

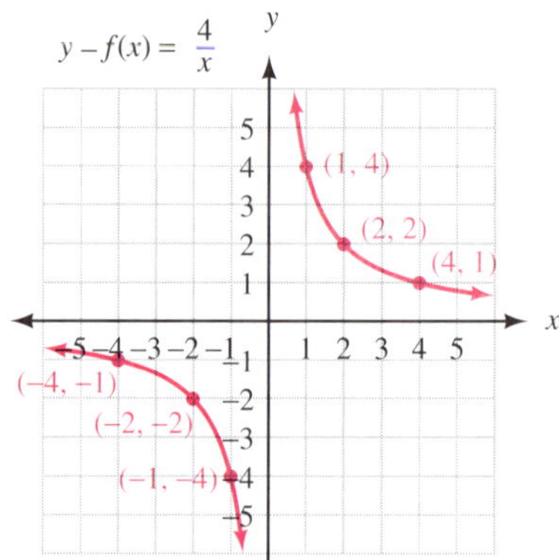

Figure 3

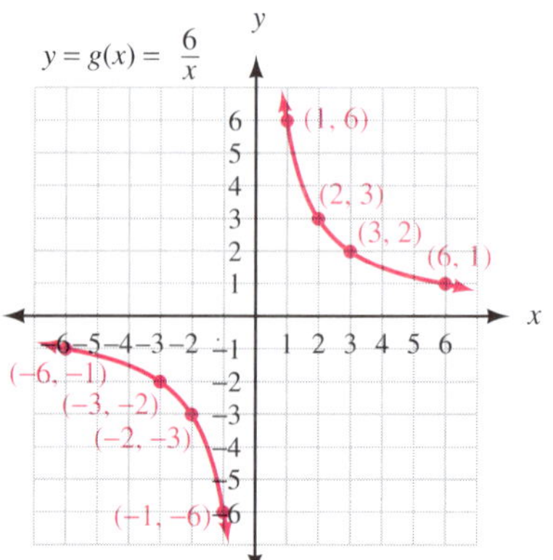

Figure 4

3.2 Division of Polynomials

First Bank of San Luis Obispo charges \$2.00 per month and \$0.15 per check for a regular checking account. So, if you write x checks in one month, the total monthly cost of the checking account will be $C(x) = 2.00 + 0.15x$. From this formula, we see that the more checks we write in a month, the more we pay for the account. But it is also true that the more checks we write in a month, the lower the cost per check. To find the cost per check, we use the average cost function. To find the average cost function, we divide the total cost by the number of checks written.

$$\text{Average Cost} = \overline{C}(x) = \frac{C(x)}{x} = \frac{2.00 + 0.15x}{x}$$

This last expression gives us the average cost per check for each of the x checks written. To work with this last expression, we need to know something about division with polynomials, and that is what we will cover in this section.

We begin this section by considering division of a polynomial by a monomial. This is the simplest kind of polynomial division. The rest of the section is devoted to division of a polynomial by a polynomial. This kind of division is similar to long division with whole numbers.

Dividing a Polynomial by a Monomial

To divide a polynomial by a monomial, we use the definition of division and apply the distributive property. The following example illustrates the procedure.

EXAMPLE 1 Divide $\dfrac{10x^5 - 15x^4 + 20x^3}{5x^2}$.

SOLUTION

$$= (10x^5 - 15x^4 + 20x^3) \cdot \frac{1}{5x^2}$$ **Dividing by $5x^2$ is the same as multiplying by $\frac{1}{5x^2}$**

$$= 10x^5 \cdot \frac{1}{5x^2} - 15x^4 \cdot \frac{1}{5x^2} + 20x^3 \cdot \frac{1}{5x^2}$$ **Distributive property**

$$= \frac{10x^5}{5x^2} - \frac{15x^4}{5x^2} + \frac{20x^3}{5x^2}$$ **Multiplying by $\frac{1}{5x^2}$ is the same as dividing by $5x^2$**

$$= 2x^3 - 3x^2 + 4x$$ **Divide coefficients, subtract exponents**

Notice that division of a polynomial by a monomial is accomplished by dividing each term of the polynomial by the monomial. The first two steps are usually not shown in a problem like this. They are part of Example 1 to justify distributing $5x^2$ under all three terms of the polynomial $10x^5 - 15x^4 + 20x^3$.

Here are some more examples of this kind of division.

Practice Problems

1. Divide $\dfrac{12x^4 - 18x^3 + 24x^2}{6x}$.

Answer

1. $2x^3 - 3x^2 + 4x$

Divide.

2. $\dfrac{27x^4y^7 - 81x^5y^3}{-9x^3y^2}$

3. $\dfrac{12a^5 + 8a^4 + 16a^3 + 4a^2}{8a^4}$

EXAMPLES Divide. Write all results with positive exponents.

2. $$\frac{8x^3y^5 - 16x^2y^2 + 4x^4y^3}{-2x^2y} = \frac{8x^3y^5}{-2x^2y} + \frac{-16x^2y^2}{-2x^2y} + \frac{4x^4y^3}{-2x^2y}$$
$$= -4xy^4 + 8y - 2x^2y^2$$

3. $$\frac{10a^4b^2 + 8ab^3 - 12a^3b + 6ab}{4a^2b^2} = \frac{10a^4b^2}{4a^2b^2} + \frac{8ab^3}{4a^2b^2} - \frac{12a^3b}{4a^2b^2} + \frac{6ab}{4a^2b^2}$$
$$= \frac{5a^2}{2} + \frac{2b}{a} - \frac{3a}{b} + \frac{3}{2ab}$$

Notice in Example 3 that the result is not a polynomial because of the last three terms. If we were to write each as a product, some of the variables would have negative exponents. For example, the second term would be

$$\frac{2b}{a} = 2a^{-1}b$$

The divisor in each of the previous examples was a monomial. We now want to turn our attention to division of polynomials in which the divisor has two or more terms.

Dividing a Polynomial by a Polynomial

4. Divide $\dfrac{2x^2 - 5xy + 3y^2}{x - y}$.

EXAMPLE 4 Divide $\dfrac{x^2 - 6xy - 7y^2}{x + y}$.

SOLUTION In this case, we can factor the numerator and perform our division by simply dividing out common factors, just like we did in the previous section:

$$\frac{x^2 - 6xy - 7y^2}{x + y} = \frac{\cancel{(x + y)}(x - 7y)}{\cancel{x + y}}$$
$$= x - 7y$$

Difference Quotients The diagram in Figure 1 is an important diagram from calculus. Although it may look complicated, the point of it is simple: The slope of the line passing through the points P and Q is given by the formula

$$\text{Slope of line through } PQ = m = \frac{f(x) - f(a)}{x - a}$$

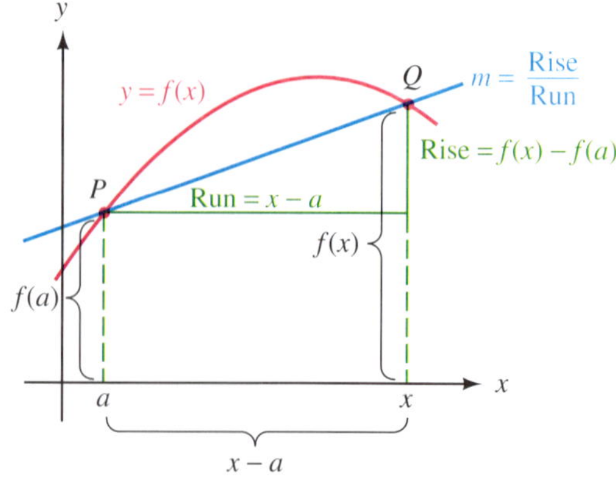

Figure 1

The expression $\dfrac{f(x) - f(a)}{x - a}$ is called a difference quotient. When $f(x)$ is a polynomial, the difference quotient will be a rational expression.

Answers
2. $-3xy^5 + 9x^2y$
3. $\frac{3a}{2} + 1 + \frac{2}{a} + \frac{1}{2a^2}$
4. $2x - 3y$

EXAMPLE 5 If $f(x) = 3x - 5$, find $\frac{f(x) - f(a)}{x - a}$.

SOLUTION

$$\frac{f(x) - f(a)}{x - a} = \frac{(3x - 5) - (3a - 5)}{x - a}$$

$$= \frac{3x - 5 - 3a + 5}{x - a}$$

$$= \frac{3x - 3a}{x - a}$$

$$= \frac{3(x - a)}{x - a}$$

$$= 3$$

5. If $f(x) = 2x + 3$, find $\frac{f(x) - f(a)}{x - a}$.

EXAMPLE 6 If $f(x) = x^2 - 4$, find $\frac{f(x) - f(a)}{x - a}$ and simplify.

SOLUTION Because $f(x) = x^2 - 4$ and $f(a) = a^2 - 4$, we have

$$\frac{f(x) - f(a)}{x - a} = \frac{(x^2 - 4) - (a^2 - 4)}{x - a}$$

$$= \frac{x^2 - 4 - a^2 + 4}{x - a}$$

$$= \frac{x^2 - a^2}{x - a}$$

$$= \frac{(x + a)\cancel{(x - a)}}{\cancel{x - a}} \quad \text{Factor and divide out common factor}$$

$$= x + a$$

6. If $f(x) = x^2 - 1$, find $\frac{f(x) - f(a)}{x - a}$.

Long Division

For the type of division shown in Examples 4 through 6, the denominator must be a factor of the numerator. When the denominator is not a factor of the numerator, or in the case where we can't factor the numerator, the method used in Examples 4 through 6 won't work. We need to develop a new method for these cases. Since this new method is very similar to long division with whole numbers, we will review it here.

EXAMPLE 7 Divide $25\overline{)4,628}$.

SOLUTION

```
      1     ← Estimate 25 into 46
25)4,628
   2 5      ← Multiply 1 × 25 = 25
   2 1      ← Subtract 46 − 25 = 21
```

```
      1
25)4,628
   2 5↓
   2 12     ← Bring down the 2
```

7. Divide $35\overline{)7,546}$.

Note: You may realize when looking over this example that you don't have a very good idea why you proceed as you do with the steps in long division. What you do know is the process always works. We are going to approach the explanation for division of two polynomials with this in mind; that is, we won't always be sure why the steps we use are important, only that they always produce the correct result.

Answers
5. 2 6. $x + a$

These are the four basic steps in long division: estimate, multiply, subtract, and bring down the next term. To complete the problem, we simply perform the same four steps:

$$\begin{array}{r l} 18 & \leftarrow \textbf{8 is the estimate} \\ 25\overline{)4{,}628} & \\ 2\,5 & \\ \hline 2\,12 & \\ 2\,00\downarrow & \leftarrow \textbf{Multiply to get 200} \\ \hline 128 & \leftarrow \textbf{Subtract to get 12, then bring down the 8} \end{array}$$

One more time:

$$\begin{array}{r l} 185 & \leftarrow \textbf{5 is the estimate} \\ 25\overline{)4{,}628} & \\ 2\,5 & \\ \hline 2\,12 & \\ 2\,00\downarrow & \\ \hline 128 & \\ 125 & \leftarrow \textbf{Multiply to get 125} \\ \hline 3 & \leftarrow \textbf{Subtract to get 3} \end{array}$$

Since 3 is less than 25 and we have no more terms to bring down, we have our answer:

$$\frac{4{,}628}{25} = 185 + \frac{3}{25}$$

To check our answer, we multiply 185 by 25 and then add 3 to the result:

$$25(185) + 3 = 4{,}625 + 3 = 4{,}628$$

Long division with polynomials is very similar to long division with whole numbers. Both use the same four basic steps: estimate, multiply, subtract, and bring down the next term. We use long division with polynomials when the denominator has two or more terms and is not a factor of the numerator. Here is an example.

8. Divide $\frac{3x^2 - 8x - 1}{x - 3}$.

EXAMPLE 8 Divide $\frac{2x^2 - 7x + 9}{x - 2}$.

SOLUTION

$$\begin{array}{r l} 2x & \leftarrow \textbf{Estimate } 2x^2 \div x = 2x \\ x - 2\overline{)\;2x^2 - 7x + 9} & \\ \not{+}\,2x^2 \not{-}\,4x & \leftarrow \textbf{Multiply } 2x(x-2) = 2x^2 - 4x \\ \hline -3x & \leftarrow \textbf{Subtract } (2x^2 - 7x) - (2x^2 - 4x) = -3x \end{array}$$

$$\begin{array}{r l} 2x & \\ x - 2\overline{)\;2x^2 - 7x + 9} & \\ \not{+}\,2x^2 \not{-}\,4x & \\ \hline -3x + 9 & \leftarrow \textbf{Bring down the 9} \end{array}$$

Notice we change the signs on $2x^2 - 4x$ and add in the subtraction step. Subtracting a polynomial is equivalent to adding its opposite.

Answer

7. $215 + \frac{3}{5}$

We repeat the four steps.

$$\begin{array}{r l}
2x - 3 & \leftarrow \textbf{-3 is the estimate: } -3x \div x = -3 \\
x - 2\overline{\big)\ 2x^2 - 7x + 9} & \\
\not{+}\,2x^2 \not{-}\,4x & \\
\hline
-3x + 9 & \\
\not{-}\,3x \not{+}\,6 & \leftarrow \textbf{Multiply } -3(x-2) = -3x + 6 \\
\hline
3 & \leftarrow \textbf{Subtract } (-3x + 9) - (-3x + 6) = 3
\end{array}$$

Since we have no other term to bring down, we have our answer:

$$\frac{2x^2 - 7x + 9}{x - 2} = 2x - 3 + \frac{3}{x - 2}$$

To check, we multiply $(2x - 3)(x - 2)$ to get $2x^2 - 7x + 6$; then, adding the remainder 3 to this result, we have $2x^2 - 7x + 9$.

In setting up a division problem involving two polynomials, we must remember two things: (1) both polynomials should be in decreasing powers of the variable, and (2) neither should skip any powers from the highest power down to the constant term. If there are any missing terms, they can be filled in using a coefficient of 0.

EXAMPLE 9 Divide $2x - 4\overline{)4x^3 - 6x - 11}$.

9. Divide $x - 2\overline{)3x^3 + 3x + 1}$.

SOLUTION Since the trinomial is missing a term in x^2, we can fill it in with $0x^2$:

$$4x^3 - 6x - 11 = 4x^3 + 0x^2 - 6x - 11$$

Adding $0x^2$ does not change our original problem.

$$\begin{array}{r}
2x^2 + 4x + 5 \\
2x - 4\overline{\big)\ 4x^3 + 0x^2 - 6x - 11} \\
\not{+}\,4x^3 \not{-}\,8x^2 \qquad\qquad \\
\hline
+8x^2 - 6x \qquad \\
\not{+}\,8x^2 \not{-}\,16x \qquad \\
\hline
+10x - 11 \\
\not{+}\,10x \not{-}\,20 \\
\hline
+9
\end{array}$$

***Notice:* Adding the $0x^2$ term gives us a column in which to write $-8x^2$**

$$\frac{4x^3 - 6x - 11}{2x - 4} = 2x^2 + 4x + 5 + \frac{9}{2x - 4}$$

To check this result, we multiply $2x - 4$ and $2x^2 + 4x + 5$:

$$\begin{array}{r}
2x^2 + 4x + 5 \\
2x - 4 \\
\hline
4x^3 + 8x^2 + 10x \qquad \\
-8x^2 - 16x - 20 \\
\hline
4x^3 \qquad\qquad - 6x - 20
\end{array}$$

Adding 9 (the remainder) to this result gives us the polynomial $4x^3 - 6x - 11$. Our answer checks.

Answers

8. $3x + 1 + \frac{2}{x - 3}$
9. $3x^2 + 6x + 15 + \frac{31}{x - 2}$

10. Divide $\frac{2x^2 - 5xy + 3y^2}{x - y}$.

For our last example in this section, let's do Example 4 again, but this time use long division.

EXAMPLE 10 Divide $\frac{x^2 - 6xy - 7y^2}{x + y}$.

SOLUTION

$$\begin{array}{r} x \quad - 7y \\ x + y \overline{) \; x^2 - 6xy - 7y^2} \\ \underline{\not+ x^2 \not+ \; xy \quad\quad} \\ - 7xy - 7y^2 \\ \underline{\not- 7xy \not- 7y^2} \\ 0 \end{array}$$

In this case, the remainder is 0 and we have

$$\frac{x^2 - 6xy - 7y^2}{x + y} = x - 7y$$

which is easy to check since

$$(x + y)(x - 7y) = x^2 - 6xy - 7y^2$$

Getting Ready for Class

After reading through the preceding section, respond in your own words and in complete sentences.

A. What are the four steps used in long division with polynomials?
B. What does it mean to have a remainder of 0?
C. When must long division be performed, and when can factoring be used to divide polynomials?
D. What property of real numbers is the key to dividing a polynomial by a monomial?

Answer
10. $2x - 3y$

PROBLEM SET 3.2

Find the following quotients.

1. $\dfrac{4x^3 - 8x^2 + 6x}{2x}$
2. $\dfrac{6x^3 + 12x^2 - 9x}{3x}$
3. $\dfrac{10x^4 + 15x^3 - 20x^2}{-5x^2}$
4. $\dfrac{12x^5 - 18x^4 - 6x^3}{6x^3}$
5. $\dfrac{8y^5 + 10y^3 - 6y}{4y^3}$
6. $\dfrac{6y^4 - 3y^3 + 18y^2}{9y^2}$
7. $\dfrac{5x^3 - 8x^2 - 6x}{-2x^2}$
8. $\dfrac{-9x^5 + 10x^3 - 12x}{-6x^4}$
9. $\dfrac{28a^3b^5 + 42a^4b^3}{7a^2b^2}$
10. $\dfrac{a^2b + ab^2}{ab}$
11. $\dfrac{10x^3y^2 - 20x^2y^3 - 30x^3y^3}{-10x^2y}$
12. $\dfrac{9x^4y^4 + 18x^3y^4 - 27x^2y^4}{-9xy^3}$

Divide by factoring numerators and then dividing out common factors.

13. $\dfrac{x^2 - x - 6}{x - 3}$
14. $\dfrac{x^2 - x - 6}{x + 2}$
15. $\dfrac{2a^2 - 3a - 9}{2a + 3}$
16. $\dfrac{2a^2 + 3a - 9}{2a - 3}$
17. $\dfrac{5x^2 - 14xy - 24y^2}{x - 4y}$
18. $\dfrac{5x^2 - 26xy - 24y^2}{5x + 4y}$
19. $\dfrac{x^3 - y^3}{x - y}$
20. $\dfrac{x^3 + 8}{x + 2}$
21. $\dfrac{y^4 - 16}{y - 2}$
22. $\dfrac{y^4 - 81}{y - 3}$
23. $\dfrac{x^3 + 2x^2 - 25x - 50}{x - 5}$
24. $\dfrac{x^3 + 2x^2 - 25x - 50}{x + 5}$
25. $\dfrac{4x^3 + 12x^2 - 9x - 27}{x + 3}$
26. $\dfrac{9x^3 + 18x^2 - 4x - 8}{x + 2}$

Divide using the long division method.

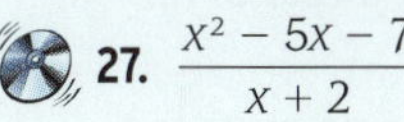

27. $\dfrac{x^2 - 5x - 7}{x + 2}$
28. $\dfrac{x^2 + 4x - 8}{x - 3}$
29. $\dfrac{6x^2 + 7x - 18}{3x - 4}$
30. $\dfrac{8x^2 - 26x - 9}{2x - 7}$
31. $\dfrac{2x^3 - 3x^2 - 4x + 5}{x + 1}$
32. $\dfrac{3x^3 - 5x^2 + 2x - 1}{x - 2}$
33. $\dfrac{2y^3 - 9y^2 - 17y + 39}{2y - 3}$
34. $\dfrac{3y^3 - 19y^2 + 17y + 4}{3y - 4}$
35. $\dfrac{2x^3 - 9x^2 + 11x - 6}{2x^2 - 3x + 2}$
36. $\dfrac{6x^3 + 7x^2 - x + 3}{3x^2 - x + 1}$
37. $\dfrac{6y^3 - 8y + 5}{2y - 4}$
38. $\dfrac{9y^3 - 6y^2 + 8}{3y - 3}$
39. $\dfrac{a^4 - 2a + 5}{a - 2}$
40. $\dfrac{a^4 + a^3 - 1}{a + 2}$
41. $\dfrac{y^4 - 16}{y - 2}$
42. $\dfrac{y^4 - 81}{y - 3}$
43. $\dfrac{x^4 + x^3 - 3x^2 - x + 2}{x^2 + 3x + 2}$
44. $\dfrac{2x^4 + x^3 + 4x - 3}{2x^2 - x + 3}$

For the functions below, evaluate each difference quotient.

(a) $\dfrac{f(x + h) - f(x)}{h}$ **(b)** $\dfrac{f(x) - f(a)}{x - a}$

45. $f(x) = 4x$
46. $f(x) = -3x$
47. $f(x) = 5x + 3$
48. $f(x) = 6x - 5$
49. $f(x) = x^2$
50. $f(x) = 3x^2$
51. $f(x) = x^2 + 1$
52. $f(x) = x^2 - 3$
53. $f(x) = x^2 - 3x + 4$
54. $f(x) = x^2 + 4x - 7$

55. $f(x) = 2x^2 + 3x - 4$

56. $f(x) = 5x^2 + 3x - 2$

57. Factor $x^3 + 6x^2 + 11x + 6$ completely if one of its factors is $x + 3$.

58. Factor $x^3 + 10x^2 + 29x + 20$ completely if one of its factors is $x + 4$.

59. Factor $x^3 + 5x^2 - 2x - 24$ completely if one of its factors is $x + 3$.

60. Factor $x^3 + 3x^2 - 10x - 24$ completely if one of its factors is $x + 2$.

61. Problems 21 and 41 are the same problem. Are the two answers you obtained equivalent?

62. Problems 22 and 42 are the same problem. Are the two answers you obtained equivalent?

63. Find $P(-2)$ if $P(x) = x^2 - 5x - 7$. Compare it with the remainder in Problem 27.

64. Find $P(3)$ if $P(x) = x^2 + 4x - 8$. Compare it with the remainder in Problem 28.

Applying the Concepts

65. The Factor Theorem The factor theorem of algebra states that if $x - a$ is a factor of a polynomial, $P(x)$, then $P(a) = 0$. Verify the following.

(a) That $x - 2$ is a factor of $P(x) = x^3 - 3x^2 + 5x - 6$, and that $P(2) = 0$

(b) That $x - 5$ is a factor of $P(x) = x^4 - 5x^3 - x^2 + 6x - 5$, and that $P(5) = 0$

66. The Remainder Theorem The remainder theorem of algebra states that if a polynomial $P(x)$, is divided by $x - a$, then the remainder is $P(a)$. Verify the remainder theorem by showing that when $P(x) = x^2 - x + 3$ is divided by $x - 2$ the remainder is 5, and that $P(2) = 5$.

67. Checking Account First Bank of San Luis Obispo charges \$2.00 per month and \$0.15 per check for a regular checking account. As we mentioned in the introduction to this section, the total monthly cost of this account is $C(x) = 2.00 + 0.15x$. To find the average cost of each of the x checks, we divide the total cost by the number of checks written. That is,

$$\overline{C}(x) = \frac{C(x)}{x}$$

(a) Use the total cost function to fill in the following table.

x	1	5	10	15	20
$C(x)$					

(b) Find the formula for the average cost function, $\overline{C}(x)$.

(c) Use the average cost function to fill in the following table.

x	1	5	10	15	20
$\overline{C}(x)$					

(d) What happens to the average cost as more checks are written?

(e) Assume that you write at least 1 check a month, but never more than 20 checks per month, and graph both $y = C(x)$ and $y = \overline{C}(x)$ on the same set of axes.

(f) Give the domain and range of each of the functions you graphed in part (e).

68. Average Cost A company that manufactures computer diskettes uses the function $C(x) = 200 + 2x$ to represent the daily cost of producing x diskettes.

(a) Find the average cost function, $\overline{C}(x)$.

(b) Use the average cost function to fill in the following table: (Round to the nearest dollar.)

x	1	5	10	15	20
$\overline{C}(x)$					

(c) What happens to the average cost as more items are produced?

(d) Graph the function $y = \overline{C}(x)$ for $x > 0$.

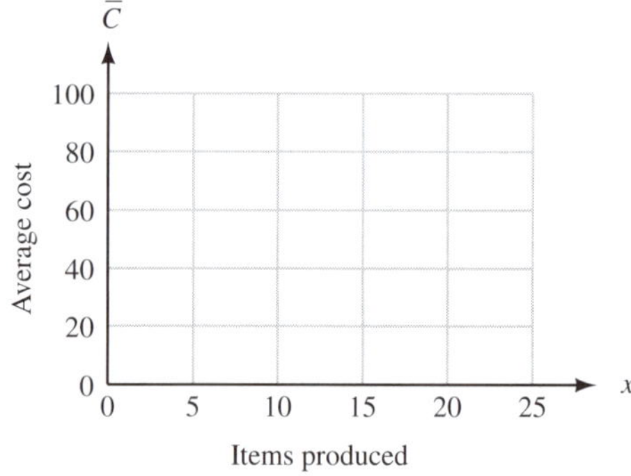

69. Average Cost For long distance service, a particular phone company charges a monthly fee of \$4.95 plus \$0.07 per minute of calling time used. The relationship between the number of minutes of calling time used, m, and the amount of the monthly phone bill $T(m)$ is given by the function $T(m) = 4.95 + 0.07m$.

(a) Find the total cost when 100, 400, and 500 minutes of calling time is used in 1 month.

(b) Find a formula for the average cost per minute function $\overline{T}(m)$.

(c) Find the average cost per minute of calling time used when 100, 400, and 500 minutes are used in 1 month.

70. Average Cost A company manufactures electric pencil sharpeners. Each month they have fixed costs of \$40,000 and variable costs of \$8.50 per sharpener. Therefore, the total monthly cost to manufacture x sharpeners is given by the function $C(x) = 40{,}000 + 8.5x$.

(a) Find the total cost to manufacture 1,000, 5,000, and 10,000 sharpeners a month.

(b) Write an expression for the average cost per sharpener function $\overline{C}(x)$.

(c) Find the average cost per sharpener to manufacture 1,000, 5,000, and 10,000 sharpeners per month.

Review Problems

Divide.

71. $\frac{3}{5} \div \frac{2}{7}$

72. $\frac{2}{7} \div \frac{3}{5}$

73. $\frac{3}{4} \div \frac{6}{11}$

74. $\frac{6}{8} \div \frac{3}{5}$

75. $\frac{4}{9} \div 8$

76. $\frac{3}{7} \div 6$

77. $8 \div \frac{1}{4}$

78. $12 \div \frac{2}{3}$

Extending the Concepts

Divide.

79. $\frac{4x^5 - x^4 - 20x^3 + 8x^2 - 15}{x^2 - 5}$

80. $\frac{4x^5 + 2x^4 + x^3 - 20x^2 - 10x - 5}{x^3 - 5}$

81. $\frac{0.5x^3 - 0.3x^2 + 0.22x + 0.06}{x + 0.2}$

82. $\frac{0.6x^3 - 1.1x^2 - 0.1x + 0.6}{0.3x + 0.2}$

83. $\frac{3x^2 + x - 10}{2x + 4}$

84. $\frac{2x^2 - x - 10}{3x + 6}$

85. $\frac{2x^2 + \frac{1}{3}x + \frac{5}{3}}{3x - 1}$

86. $\frac{x^2 + \frac{3}{5}x + \frac{8}{5}}{5x - 2}$

3.3 Multiplication and Division of Rational Expressions

If you have ever taken a home videotape to be duplicated, you know the amount you pay for the duplication service depends on the number of copies you have made: The more copies you have made, the lower the charge per copy. The following demand function gives the price (in dollars) per tape $p(x)$ a company charges for making x copies of a 30-minute videotape. As you can see, it is a rational function.

$$p(x) = \frac{2(x + 60)}{x + 5}$$

The graph in Figure 1 shows this function from $x = 0$ to $x = 100$. As you can see, the more copies that are made, the lower the price per copy.

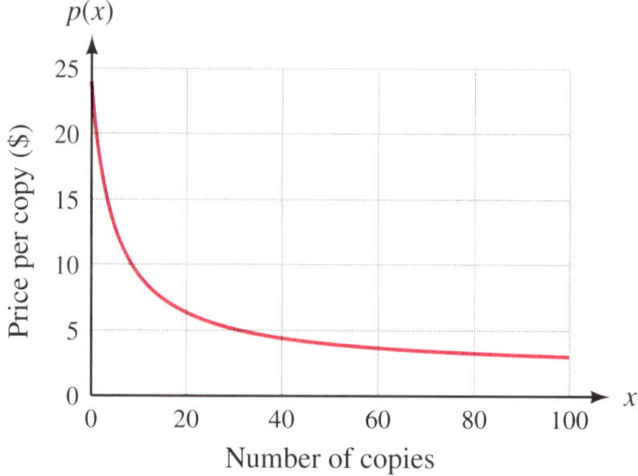

Figure 1

If we were interested in finding the revenue function for this situation, we would multiply the number of copies made x by the price per copy $p(x)$. This involves multiplication with a rational expression, which is one of the topics we cover in this section.

In the previous section we found the process of reducing rational expressions to lowest terms to be the same process used in reducing fractions to lowest terms. The similarity also holds for the process of multiplication or division of rational expressions.

Multiplication with fractions is the simplest of the four basic operations. To multiply two fractions we simply multiply numerators and multiply denominators; that is, if a, b, c, and d are real numbers, with $b \neq 0$ and $d \neq 0$, then

$$\frac{a}{b} \cdot \frac{c}{d} = \frac{ac}{bd}$$

Practice Problems

1. Multiply $\frac{3}{4} \cdot \frac{12}{27}$.

EXAMPLE 1 Multiply $\frac{6}{7} \cdot \frac{14}{18}$.

SOLUTION

$$\frac{6}{7} \cdot \frac{14}{18} = \frac{6(14)}{7(18)} \quad \text{Multiply numerators and denominators}$$

$$= \frac{\cancel{2} \cdot \cancel{3}(2 \cdot \cancel{7})}{\cancel{7}(\cancel{2} \cdot \cancel{3} \cdot 3)} \quad \text{Factor}$$

$$= \frac{2}{3} \quad \text{Divide out common factors}$$

Answer

1. $\frac{1}{3}$

Our next example is similar to some of the problems we worked in an earlier chapter. We multiply fractions whose numerators and denominators are monomials by multiplying numerators and multiplying denominators and then reducing to lowest terms. Here is how it looks.

EXAMPLE 2 Multiply $\frac{8x^3}{27y^8} \cdot \frac{9y^3}{12x^2}$.

SOLUTION We multiply numerators and denominators without actually carrying out the multiplication:

$$\frac{8x^3}{27y^8} \cdot \frac{9y^3}{12x^2} = \frac{8 \cdot 9x^3y^3}{27 \cdot 12x^2y^8}$$ **Multiply numerators** **Multiply denominators**

$$= \frac{\cancel{4} \cdot 2 \cdot \cancel{9}x^3y^3}{\cancel{9} \cdot 3 \cdot \cancel{4} \cdot 3x^2y^8}$$ **Factor coefficients**

$$= \frac{2x}{9y^5}$$ **Divide out common factors**

The product of two rational expressions is the product of their numerators over the product of their denominators.

Once again, we should mention that the little slashes we have drawn through the factors are used to denote the factors we have divided out of the numerator and denominator.

EXAMPLE 3 Multiply $\frac{x-3}{x^2-4} \cdot \frac{x+2}{x^2-6x+9}$.

SOLUTION We begin by multiplying numerators and denominators. We then factor all polynomials and divide out factors common to the numerator and denominator:

$$\frac{x-3}{x^2-4} \cdot \frac{x+2}{x^2-6x+9} = \frac{(x-3)(x+2)}{(x^2-4)(x^2-6x+9)}$$ **Multiply**

$$= \frac{\cancel{(x-3)}\cancel{(x+2)}}{\cancel{(x+2)}(x-2)\cancel{(x-3)}(x-3)}$$ **Factor**

$$= \frac{1}{(x-2)(x-3)}$$ **Divide out common factors**

The first two steps can be combined to save time. We can perform the multiplication and factoring steps together.

EXAMPLE 4 Multiply $\frac{2y^2-4y}{2y^2-2} \cdot \frac{y^2-2y-3}{y^2-5y+6}$.

SOLUTION

$$\frac{2y^2-4y}{2y^2-2} \cdot \frac{y^2-2y-3}{y^2-5y+6} = \frac{\cancel{2}y\cancel{(y-2)}\cancel{(y-3)}\cancel{(y+1)}}{\cancel{2}\cancel{(y+1)}(y-1)\cancel{(y-3)}\cancel{(y-2)}}$$

$$= \frac{y}{y-1}$$

Notice in both of the preceding examples that we did not actually multiply the polynomials as we did in the chapter on exponents and polynomials. It would be senseless to do that since we would then have to factor each of the resulting products to reduce them to lowest terms.

The quotient of two rational expressions is the product of the first and the reciprocal of the second; that is, we find the quotient of two rational expressions

2. Multiply $\frac{6x^4}{4y^9} \cdot \frac{12y^5}{3x^2}$.

Note: Notice how we factor the coefficients just enough so that we can see the factors they have in common. If you want to show this step without showing the factoring, it would look like this:

$$\frac{\overset{2}{\cancel{8}} \cdot \overset{1}{\cancel{9}}x^3y^3}{\underset{3}{\cancel{27}} \cdot \underset{3}{\cancel{12}}x^2y^8}$$

3. Multiply $\frac{x+5}{x^2-25} \cdot \frac{x-5}{x^2-10x+25}$.

4. Multiply.

$$\frac{3y^2-3y}{3y-12} \cdot \frac{y^2-2y-8}{y^2+3y+2}$$

Answers

2. $\frac{6x^2}{y^4}$ 3. $\frac{1}{(x-5)^2}$ 4. $\frac{y(y-1)}{y+1}$

the same way we find the quotient of two fractions. Here is an example that reviews division with fractions.

5. Divide $\frac{5}{9} \div \frac{10}{27}$.

EXAMPLE 5 Divide $\frac{6}{8} \div \frac{3}{5}$.

SOLUTION

$$\frac{6}{8} \div \frac{3}{5} = \frac{6}{8} \cdot \frac{5}{3}$$ **Write division in terms of multiplication**

$$= \frac{6(5)}{8(3)}$$ **Multiply numerators and denominators**

$$= \frac{\not{2} \cdot \not{3}(5)}{\not{2} \cdot 2 \cdot 2(\not{3})}$$ **Factor**

$$= \frac{5}{4}$$ **Divide out common factors**

To divide one rational expression by another, we use the definition of division to multiply by the reciprocal of the expression that follows the division symbol.

6. Divide $\frac{9x^4}{4y^3} \div \frac{3x^2}{8y^5}$.

EXAMPLE 6 Divide $\frac{8x^3}{5y^2} \div \frac{4x^2}{10y^6}$.

SOLUTION First we rewrite the problem in terms of multiplication. Then we multiply.

$$\frac{8x^3}{5y^2} \div \frac{4x^2}{10y^6} = \frac{8x^3}{5y^2} \cdot \frac{10y^6}{4x^2}$$

$$= \frac{\overset{2}{\not{8}} \cdot \overset{2}{\not{10}}x^3y^6}{\not{4} \cdot \not{5}x^2y^2}$$

$$= 4xy^4$$

7. Divide.
$$\frac{xy^2 - y^3}{x^2 - y^2} \div \frac{x^3 + y^3}{x^2 + 2xy + y^2}$$

EXAMPLE 7 Divide $\frac{x^2 - y^2}{x^2 - 2xy + y^2} \div \frac{x^3 + y^3}{x^3 - x^2y}$.

SOLUTION We begin by writing the problem as the product of the first and the reciprocal of the second and then proceed as in the previous two examples:

$$\frac{x^2 - y^2}{x^2 - 2xy + y^2} \div \frac{x^3 + y^3}{x^3 - x^2y}$$ **Multiply by the reciprocal of the divisor**

$$= \frac{x^2 - y^2}{x^2 - 2xy + y^2} \cdot \frac{x^3 - x^2y}{x^3 + y^3}$$

$$= \frac{\cancel{(x - y)}\cancel{(x + y)}(x^2)\cancel{(x - y)}}{\cancel{(x - y)}\cancel{(x - y)}\cancel{(x + y)}(x^2 - xy + y^2)}$$ **Factor and multiply**

$$= \frac{x^2}{x^2 - xy + y^2}$$ **Divide out common factors**

Here are some more examples of multiplication and division with rational expressions.

8. Perform the indicated operations.
$$\frac{a^2 + 3a - 4}{a - 4} \cdot \frac{a + 3}{a^2 - 4a + 3} \div \frac{a + 1}{a^2 - 2a - 3}$$

EXAMPLE 8 Perform the indicated operations.

$$\frac{a^2 - 8a + 15}{a + 4} \cdot \frac{a + 2}{a^2 - 5a + 6} \div \frac{a^2 - 3a - 10}{a^2 + 2a - 8}$$

Answers

5. $\frac{3}{2}$ 6. $6x^2y^2$ 7. $\frac{y^2}{x^2 - xy + y^2}$

SOLUTION First we rewrite the division as multiplication by the reciprocal. Then we proceed as usual.

$$\frac{a^2 - 8a + 15}{a + 4} \cdot \frac{a + 2}{a^2 - 5a + 6} \div \frac{a^2 - 3a - 10}{a^2 + 2a - 8}$$

$$= \frac{(a^2 - 8a + 15)(a + 2)(a^2 + 2a - 8)}{(a + 4)(a^2 - 5a + 6)(a^2 - 3a - 10)}$$ **Change division to multiplication by the reciprocal**

$$= \frac{(a - 5)(a - 3)(a + 2)(a + 4)(a - 2)}{(a + 4)(a - 3)(a - 2)(a - 5)(a + 2)}$$ **Factor**

$$= 1$$ **Divide out common factors**

Our next example involves factoring by grouping. As you may have noticed, working the problems in this chapter gives you a very detailed review of factoring.

EXAMPLE 9 Multiply $\frac{xa + xb + ya + yb}{xa - xb - ya + yb} \cdot \frac{xa + xb - ya - yb}{xa - xb + ya - yb}$

SOLUTION We will factor each polynomial by grouping, which takes two steps.

$$\frac{xa + xb + ya + yb}{xa - xb - ya + yb} \cdot \frac{xa + xb - ya - yb}{xa - xb + ya - yb}$$

$$= \frac{x(a + b) + y(a + b)}{x(a - b) - y(a - b)} \cdot \frac{x(a + b) - y(a + b)}{x(a - b) + y(a - b)}$$

$$= \frac{(a + b)(x + y)(a + b)(x - y)}{(a - b)(x - y)(a - b)(x + y)}$$ **Factor by grouping**

$$= \frac{(a + b)^2}{(a - b)^2}$$

EXAMPLE 10 Multiply $(4x^2 - 36) \cdot \frac{12}{4x + 12}$.

SOLUTION We can think of $4x^2 - 36$ as having a denominator of 1. Thinking of it in this way allows us to proceed as we did in the previous examples.

$$(4x^2 - 36) \cdot \frac{12}{4x + 12}$$

$$= \frac{4x^2 - 36}{1} \cdot \frac{12}{4x + 12}$$ **Write $4x^2 - 36$ with denominator 1**

$$= \frac{4(x - 3)(x + 3)12}{4(x + 3)}$$ **Factor**

$$= 12(x - 3)$$ **Divide out common factors**

Getting Ready for Class

After reading through the preceding section, respond in your own words and in complete sentences.

A. Summarize the steps used to multiply fractions.
B. What is the first step in multiplying two rational expressions?
C. Why is factoring important when multiplying and dividing rational expressions?
D. How is division with rational expressions different than multiplication of rational expressions?

9. Multiply.
$$\frac{xa + xb - ya - yb}{xa + 2x + ya + 2y} \cdot \frac{xa + 2x + ya + 2y}{xa + xb + ya + yb}$$

10. Multiply.
$$(5x^2 - 45) \cdot \frac{3}{5x - 15}$$

Answers

8. $\frac{(a + 4)(a + 3)}{a - 4}$ **9.** $\frac{x - y}{x + y}$

10. $3(x + 3)$

PROBLEM SET 3.3

Perform the indicated operations.

1. $\frac{2}{9} \cdot \frac{3}{4}$

2. $\frac{5}{6} \cdot \frac{7}{8}$

3. $\frac{3}{4} \div \frac{1}{3}$

4. $\frac{3}{8} \div \frac{5}{4}$

5. $\frac{3}{7} \cdot \frac{14}{24} \div \frac{1}{2}$

6. $\frac{6}{5} \cdot \frac{10}{36} \div \frac{3}{4}$

7. $\frac{10x^2}{5y^2} \cdot \frac{15y^3}{2x^4}$

8. $\frac{8x^3}{7y^4} \cdot \frac{14y^6}{16x^2}$

9. $\frac{11a^2b}{5ab^2} \div \frac{22a^3b^2}{10ab^4}$

10. $\frac{8ab^3}{9a^2b} \div \frac{16a^2b^2}{18ab^3}$

11. $\frac{6x^2}{5y^3} \cdot \frac{11z^2}{2x^2} \div \frac{33z^5}{10y^8}$

12. $\frac{4x^3}{7y^2} \cdot \frac{6z^5}{5x^6} \div \frac{24z^2}{35x^6}$

Perform the indicated operations. Be sure to write all answers in lowest terms.

13. $\frac{x^2 - 9}{x^2 - 4} \cdot \frac{x - 2}{x - 3}$

14. $\frac{x^2 - 16}{x^2 - 25} \cdot \frac{x - 5}{x - 4}$

15. $\frac{y^2 - 1}{y + 2} \cdot \frac{y^2 + 5y + 6}{y^2 + 2y - 3}$

16. $\frac{y - 1}{y^2 - y - 6} \cdot \frac{y^2 + 5y + 6}{y^2 - 1}$

17. $\frac{3x - 12}{x^2 - 4} \cdot \frac{x^2 + 6x + 8}{x - 4}$

18. $\frac{x^2 + 5x + 1}{4x - 4} \cdot \frac{x - 1}{x^2 + 5x + 1}$

19. $\frac{xy}{xy + 1} \div \frac{x}{y}$

20. $\frac{y}{x} \div \frac{xy}{xy - 1}$

21. $\frac{1}{x^2 - 9} \div \frac{1}{x^2 + 9}$

22. $\frac{1}{x^2 - 9} \div \frac{1}{(x - 3)^2}$

23. $\frac{y - 3}{y^2 - 6y + 9} \cdot \frac{y - 3}{4}$

24. $\frac{y - 3}{y^2 - 6y + 9} \div \frac{y - 3}{4}$

25. $\frac{5x + 2y}{25x^2 - 5xy - 6y^2} \cdot \frac{20x^2 - 7xy - 3y^2}{4x + y}$

26. $\frac{7x + 3y}{42x^2 - 17xy - 15y^2} \cdot \frac{12x^2 - 4xy - 5y^2}{2x + y}$

27. $\frac{a^2 - 5a + 6}{a^2 - 2a - 3} \div \frac{a - 5}{a^2 + 3a + 2}$

28. $\frac{a^2 + 7a + 12}{a - 5} \div \frac{a^2 + 9a + 18}{a^2 - 7a + 10}$

29. $\frac{4t^2 - 1}{6t^2 + t - 2} \div \frac{8t^3 + 1}{27t^3 + 8}$

30. $\frac{9t^2 - 1}{6t^2 + 7t - 3} \div \frac{27t^3 + 1}{8t^3 + 27}$

31. $\frac{2x^2 - 5x - 12}{4x^2 + 8x + 3} \div \frac{x^2 - 16}{2x^2 + 7x + 3}$

32. $\frac{x^2 - 2x + 1}{3x^2 + 7x - 20} \div \frac{x^2 + 3x - 4}{3x^2 - 2x - 5}$

33. $\frac{2a^2 - 21ab - 36b^2}{a^2 - 11ab - 12b^2} \div \frac{10a + 15b}{a^2 - b^2}$

34. $\frac{3a^2 + 7ab - 20b^2}{a^2 + 5ab + 4b^2} \div \frac{3a^2 - 17ab + 20b^2}{3a - 12b}$

35. $\frac{6c^2 - c - 15}{9c^2 - 25} \cdot \frac{15c^2 + 22c - 5}{6c^2 + 5c - 6}$

36. $\frac{m^2 + 4m - 21}{m^2 - 12m + 27} \cdot \frac{m^2 - 7m + 12}{m^2 + 3m - 28}$

37. $\frac{6a^2b + 2ab^2 - 20b^3}{4a^2b - 16b^3} \cdot \frac{10a^2 - 22ab + 4b^2}{27a^3 - 125b^3}$

38. $\frac{12a^2b - 3ab^2 - 42b^3}{9a^2 - 36b^2} \cdot \frac{6a^2 - 15ab + 6b^2}{8a^3b - b^4}$

39. $\frac{360x^3 - 490x}{36x^2 + 84x + 49} \cdot \frac{30x^2 + 83x + 56}{150x^3 + 65x^2 - 280x}$

40. $\frac{490x^2 - 640}{49x^2 - 112x + 64} \cdot \frac{28x^2 - 95x + 72}{56x^3 - 62x^2 - 144x}$

41. $\frac{x^5 - x^2}{5x^2 - 5x} \cdot \frac{10x^4 - 10x^2}{2x^4 + 2x^3 + 2x^2}$

42. $\frac{2x^4 - 16x}{3x^6 - 48x^2} \cdot \frac{6x^5 + 24x^3}{4x^4 + 8x^3 + 16x^2}$

43. $\frac{a^2 - 16b^2}{a^2 - 8ab + 16b^2} \cdot \frac{a^2 - 9ab + 20b^2}{a^2 - 7ab + 12b^2} \div \frac{a^2 - 25b^2}{a^2 - 6ab + 9b^2}$

44. $\frac{a^2 - 6ab + 9b^2}{a^2 - 4b^2} \cdot \frac{a^2 - 5ab + 6b^2}{(a - 3b)^2} \div \frac{a^2 - 9b^2}{a^2 - ab - 6b^2}$

45. $\frac{2y^2 - 7y - 15}{42y^2 - 29y - 5} \cdot \frac{12y^2 - 16y + 5}{7y^2 - 36y + 5} \div \frac{4y^2 - 9}{49y^2 - 1}$

46. $\frac{8y^2 + 18y - 5}{21y^2 - 16y + 3} \cdot \frac{35y^2 - 22y + 3}{6y^2 + 17y + 5} \div \frac{16y^2 - 1}{9y^2 - 1}$

47. $\frac{xy - 2x + 3y - 6}{xy + 2x - 4y - 8} \cdot \frac{xy + x - 4y - 4}{xy - x + 3y - 3}$

48. $\frac{ax + bx + 2a + 2b}{ax - 3a + bx - 3b} \cdot \frac{ax - bx - 3a + 3b}{ax - bx - 2a + 2b}$

49. $\frac{xy^2 - y^2 + 4xy - 4y}{xy - 3y + 4x - 12} \div \frac{xy^3 + 2xy^2 + y^3 + 2y^2}{xy^2 - 3y^2 + 2xy - 6y}$

50. $\frac{4xb - 8b + 12x - 24}{xb^2 + 3b^2 + 3xb + 9b} \div \frac{4xb - 8b - 8x + 16}{xb^2 + 3b^2 - 2xb - 6b}$

51. $\frac{2x^3 + 10x^2 - 8x - 40}{x^3 + 4x^2 - 9x - 36} \cdot \frac{x^2 + x - 12}{2x^2 + 14x + 20}$

52. $\dfrac{x^3 + 2x^2 - 9x - 18}{x^4 + 3x^3 - 4x^2 - 12x} \cdot \dfrac{x^3 + 5x^2 + 6x}{x^2 - x - 6}$

53. $\dfrac{w^3 - w^2x}{wy - w} \div \left(\dfrac{w - x}{y - 1}\right)^2$

54. $\dfrac{a^3 - a^2b}{ac - a} \div \left(\dfrac{a - b}{c - 1}\right)^2$

55. $\dfrac{mx + my + 2x + 2y}{6x^2 - 5xy - 4y^2} \div \dfrac{2mx - 4x + my - 2y}{3mx - 6x - 4my + 8y}$

56. $\dfrac{ax - 2a + 2xy - 4y}{ax + 2a - 2xy - 4y} \div \dfrac{ax + 2a + 2xy + 4y}{ax - 2a - 2xy + 4y}$

57. $\dfrac{1 - 4d^2}{(d - c)^2} \cdot \dfrac{d^2 - c^2}{1 + 2d}$

58. $\dfrac{k^2 - 1}{k^3 + 1} \div \dfrac{k^3 - 1}{k - 1} \cdot (k^2 + k + 1)$

59. $\dfrac{r^2 - s^2}{r^2 + rs + s^2} \cdot \dfrac{r^3 - s^3}{r^2 + s^2} \div \dfrac{r^4 - s^4}{r^2 - s^2}$

60. $\dfrac{r^3 - s^3}{s - r} \cdot \dfrac{(r + s)^2}{r^2 + s^2}$

Use the method shown in Example 10 to find the following products.

61. $(3x - 6) \cdot \dfrac{x}{x - 2}$

62. $(4x + 8) \cdot \dfrac{x}{x + 2}$

63. $(x^2 - 25) \cdot \dfrac{2}{x - 5}$

64. $(x^2 - 49) \cdot \dfrac{5}{x + 7}$

65. $(x^2 - 3x + 2) \cdot \dfrac{3}{3x - 3}$

66. $(x^2 - 3x + 2) \cdot \dfrac{-1}{x - 2}$

67. $(y - 3)(y - 4)(y + 3) \cdot \dfrac{-1}{y^2 - 9}$

68. $(y + 1)(y + 4)(y - 1) \cdot \dfrac{3}{y^2 - 1}$

69. $a(a + 5)(a - 5) \cdot \dfrac{a + 1}{a^2 + 5a}$

70. $a(a + 3)(a - 3) \cdot \dfrac{a - 1}{a^2 - 3a}$

Applying the Concepts

At the beginning of this section, we introduced the demand equation shown here. Use it to work Problems 71–74.

$$p(x) = \frac{2(x + 60)}{x + 5}$$

71. **Demand Equation** Use the demand equation to fill in the table. Then compare your results with the graph shown in Figure 1 of this section.

NUMBER OF COPIES x	PRICE PER COPY (\$) $p(x)$
1	
10	
20	
50	
100	

72. **Demand Equation** To find the revenue for selling 50 copies of a tape, we multiply the price per tape by 50. Find the revenue for selling 50 tapes.

73. **Revenue** Find the revenue for selling 100 tapes.

74. **Revenue** Find the revenue equation $R(x)$.

75. **Thinking Critically** Replace each question mark with either multiplication or division to make a true statement.

(a) $\dfrac{a^2b}{a^3} \; ? \; \dfrac{ab^2}{b^3} \; ? \; \dfrac{a^4}{b} = \dfrac{b}{a^4}$

(b) $\dfrac{a^2b}{a^3} \; ? \; \dfrac{ab^2}{b^3} \; ? \; \dfrac{a^4}{b} = \dfrac{a^4}{b}$

(c) $\dfrac{a^2b}{a^3} \; ? \; \dfrac{ab^2}{b^3} \; ? \; \dfrac{a^4}{b} = a^2b$

76. **Area** The following box has a square top. The front face of the box has an area of $A = x^3 - 2x^2 - 2x - 3$. The height of the box is $h = x^2 + x + 1$. Find a formula for the area of the top square in terms of x.

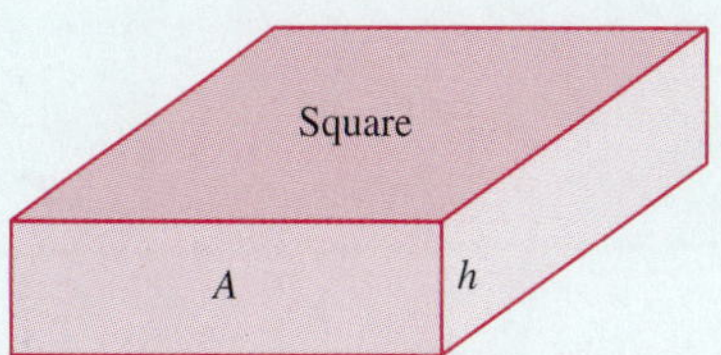

Surface Area of a Cylinder The surface area of the cylinder in the figure is defined as the area of its two circular bases and the lateral, or side, area. The surface area may be found by the formula

$$A = 2\pi r^2 + 2\pi rh$$

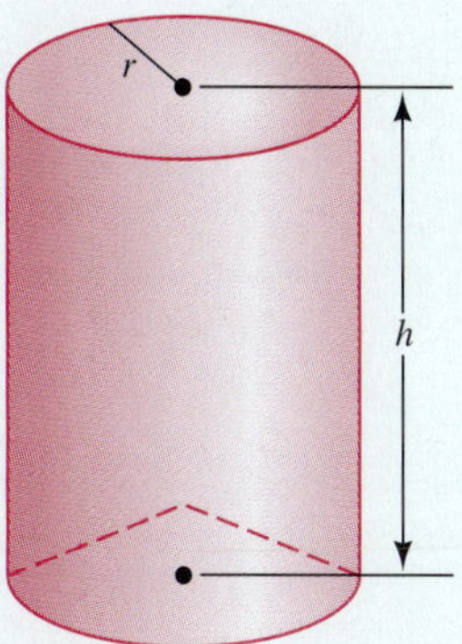

77. If the surface area is 6π, find the value of $(r^2 + rh)$.

78. If the surface area is 6π, and $h = 2$, find r.

Review Problems

If $f(x) = 3x^2 - 4x + 2$, find:

79. $f(0)$

80. $f(-1)$

81. $f(a)$

82. $f(a + 1)$

Extending the Concepts

Divide.

83. $\dfrac{x^6 + y^6}{x^4 + 4x^2y^2 + 3y^4} \div \dfrac{x^4 + 3x^2y^2 + 2y^4}{x^4 + 5x^2y^2 + 6y^4}$

84. $\dfrac{x^2 + 9xy + 8y^2}{x^2 + 7xy - 8y^2} \div \dfrac{x^2 - y^2}{x^2 + 5xy - 6y^2}$

85. $\dfrac{a^2(2a + b) + 6a(2a + b) + 5(2a + b)}{3a^2(2a + b) - 2a(2a + b) + (2a + b)} \div \dfrac{a + 1}{a - 1}$

86. $\dfrac{2x^2(x - 3z) - 5x(x - 3z) + 2(x - 3z)}{4x^2(x - 3z) - 11x(x - 3z) + 6(x - 3z)} \div \dfrac{4x - 3}{4x + 1}$

87. $\dfrac{a^3 - a^2b}{ac - a} \div \left(\dfrac{a - b}{c - 1}\right)^2$

88. $\dfrac{p^3 + q^3}{q - p} \div \dfrac{(p + q)^2}{p^2 - q^2}$

3.4 Addition and Subtraction of Rational Expressions

This section is concerned with addition and subtraction of rational expressions. In the first part of this section we will look at addition of expressions that have the same denominator. In the second part of this section we will look at addition of expressions that have different denominators.

Addition and Subtraction with the Same Denominator

To add two expressions that have the same denominator, we simply add numerators and put the sum over the common denominator. Since the process we use to add and subtract rational expressions is the same process used to add and subtract fractions, we will begin with an example involving fractions.

EXAMPLE 1 Add $\frac{4}{9} + \frac{2}{9}$.

SOLUTION We add fractions with the same denominator by using the distributive property. Here is a detailed look at the steps involved.

$$\frac{4}{9} + \frac{2}{9} = 4\left(\frac{1}{9}\right) + 2\left(\frac{1}{9}\right)$$

$$= (4 + 2)\left(\frac{1}{9}\right) \quad \textbf{Distributive property}$$

$$= 6\left(\frac{1}{9}\right)$$

$$= \frac{6}{9}$$

$$= \frac{2}{3} \quad \textbf{Divide numerator and denominator by common factor 3}$$

Note that the important thing about the fractions in this example is that they each have a denominator of 9. If they did not have the same denominator, we could not have written them as two terms with a factor of $\frac{1}{9}$ in common. Without the $\frac{1}{9}$ common to each term, we couldn't apply the distributive property. And without the distributive property, we would not have been able to add the two fractions.

In the examples that follow, we will not show all the steps we showed in Example 1. The steps are shown in Example 1 so that you will see why both fractions must have the same denominator before we can add them. In practice we simply add numerators and place the result over the common denominator.

We add and subtract rational expressions with the same denominator by combining numerators and writing the result over the common denominator. Then we reduce the result to lowest terms, if possible. Example 2 shows this process in detail. If you see the similarities between operations on rational numbers and operations on rational expressions, this chapter will look like an extension of rational numbers rather than a completely new set of topics.

EXAMPLE 2 Add $\frac{x}{x^2 - 1} + \frac{1}{x^2 - 1}$.

SOLUTION Since the denominators are the same, we simply add numerators:

$$\frac{x}{x^2 - 1} + \frac{1}{x^2 - 1} = \frac{x + 1}{x^2 - 1} \quad \textbf{Add numerators}$$

$$= \frac{\cancel{x + 1}}{(x - 1)\cancel{(x + 1)}} \quad \textbf{Factor denominator}$$

$$= \frac{1}{x - 1} \quad \textbf{Divide out common factor } x + 1$$

Practice Problems

1. Add $\frac{3}{8} + \frac{1}{8}$.

2. Add $\frac{x}{x^2 - 9} + \frac{3}{x^2 - 9}$.

Answers

1. $\frac{1}{2}$ 2. $\frac{1}{x - 3}$

Our next example involves subtraction of rational expressions. Pay careful attention to what happens to the signs of the terms in the numerator of the second expression when we subtract it from the first expression.

3. Subtract $\frac{2x-7}{x-2} - \frac{x-5}{x-2}$.

EXAMPLE 3 Subtract $\frac{2x-5}{x-2} - \frac{x-3}{x-2}$.

SOLUTION Since each expression has the same denominator, we simply subtract the numerator in the second expression from the numerator in the first expression and write the difference over the common denominator $x - 2$. We must be careful, however, that we subtract both terms in the second numerator. To ensure that we do, we will enclose that numerator in parentheses.

$$\frac{2x-5}{x-2} - \frac{x-3}{x-2} = \frac{2x-5-(x-3)}{x-2} \quad \text{Subtract numerators}$$

$$= \frac{2x-5-x+3}{x-2} \quad \text{Remove parentheses}$$

$$= \frac{x-2}{x-2} \quad \text{Combine similar terms in the numerator}$$

$$= 1 \quad \text{Reduce (or divide)}$$

Note the $+3$ in the numerator of the second step. It is a very common mistake to write that as -3 by forgetting to subtract both terms in the numerator of the second expression. Whenever the expression we are subtracting has two or more terms in its numerator, we have to watch for this mistake.

Note: If you got an answer of

$$\frac{x-12}{x-2}$$

for Practice Problem 3, then you made the mistake talked about in Example 3. Look at the first two lines in the solution to Example 3 again and see if you can find your mistake.

Next we consider addition and subtraction of fractions and rational expressions that have different denominators.

Addition and Subtraction with Different Denominators

Before we look at an example of addition of fractions with different denominators, we need to review the definition for the least common denominator.

DEFINITION

The **least common denominator,** abbreviated LCD, for a set of denominators is the smallest expression that is divisible by each of the denominators.

The first step in combining two fractions is to find the LCD. Once we have the common denominator, we rewrite each fraction as an equivalent fraction with the common denominator. After that, we simply add or subtract as we did in our first three examples.

Example 4 is a review of the step-by-step procedure used to add two fractions with different denominators.

4. Add $\frac{3}{10} + \frac{11}{42}$.

EXAMPLE 4 Add $\frac{3}{14} + \frac{7}{30}$.

SOLUTION

Step 1: Find the LCD.
To do this, we first factor both denominators into prime factors.

Factor 14: $14 = 2 \cdot 7$

Factor 30: $30 = 2 \cdot 3 \cdot 5$

Answer
3. 1

Since the LCD must be divisible by 14, it must have factors of $2 \cdot 7$. It must also be divisible by 30 and, therefore, have factors of $2 \cdot 3 \cdot 5$. We do not need to repeat the 2 that appears in both the factors of 14 and those of 30. Therefore,

$$\text{LCD} = 2 \cdot 3 \cdot 5 \cdot 7 = 210$$

Step 2: Change to equivalent fractions.

Since we want each fraction to have a denominator of 210 and at the same time keep its original value, we multiply each by 1 in the appropriate form.

Change $\frac{3}{14}$ to a fraction with denominator 210:

$$\frac{3}{14} \cdot \frac{\mathbf{15}}{\mathbf{15}} = \frac{45}{210}$$

Change $\frac{7}{30}$ to a fraction with denominator 210:

$$\frac{7}{30} \cdot \frac{\mathbf{7}}{\mathbf{7}} = \frac{49}{210}$$

Note: When we multiply $\frac{3}{14}$ by $\frac{15}{15}$ we obtain a fraction with the same value as $\frac{3}{14}$ (because we multiplied by 1) but with the common denominator 210.

Step 3: Add numerators of equivalent fractions found in step 2:

$$\frac{45}{210} + \frac{49}{210} = \frac{94}{210}$$

Step 4: Reduce to lowest terms if necessary:

$$\frac{94}{210} = \frac{47}{105}$$

The main idea in adding fractions is to write each fraction again with the LCD for a denominator. In doing so, we must be sure not to change the value of either of the original fractions.

EXAMPLE 5 Add $\dfrac{-2}{x^2 - 2x - 3} + \dfrac{3}{x^2 - 9}$.

5. Add $\dfrac{-3}{x^2 - 2x - 8} + \dfrac{4}{x^2 - 16}$.

SOLUTION ***Step 1:*** Factor each denominator and build the LCD from the factors:

$$\left.\begin{aligned} x^2 - 2x - 3 &= (x - 3)(x + 1) \\ x^2 - 9 &= (x - 3)(x + 3) \end{aligned}\right\} \text{LCD} = (x - 3)(x + 3)(x + 1)$$

Step 2: Change each rational expression to an equivalent expression that has the LCD for a denominator:

$$\frac{-2}{x^2 - 2x - 3} = \frac{-2}{(x - 3)(x + 1)} \cdot \frac{\mathbf{(x + 3)}}{\mathbf{(x + 3)}} = \frac{-2x - 6}{(x - 3)(x + 3)(x + 1)}$$

$$\frac{3}{x^2 - 9} = \frac{3}{(x - 3)(x + 3)} \cdot \frac{\mathbf{(x + 1)}}{\mathbf{(x + 1)}} = \frac{3x + 3}{(x - 3)(x + 3)(x + 1)}$$

Step 3: Add numerators of the rational expressions found in step 2:

$$\frac{-2x - 6}{(x - 3)(x + 3)(x + 1)} + \frac{3x + 3}{(x - 3)(x + 3)(x + 1)} = \frac{x - 3}{(x - 3)(x + 3)(x + 1)}$$

Step 4: Reduce to lowest terms by dividing out the common factor $x - 3$:

$$= \frac{1}{(x + 3)(x + 1)}$$

Answers

4. $\frac{59}{105}$ 5. $\dfrac{1}{(x + 4)(x + 2)}$

6. Add $\frac{x-4}{2x-6}+\frac{3}{x^2-9}$.

EXAMPLE 6 Subtract $\frac{x+4}{2x+10}-\frac{5}{x^2-25}$.

SOLUTION We begin by factoring each denominator:

$$\frac{x+4}{2x+10}-\frac{5}{x^2-25}=\frac{x+4}{2(x+5)}-\frac{5}{(x+5)(x-5)}$$

The LCD is $2(x+5)(x-5)$. Completing the problem we have

$$=\frac{x+4}{2(x+5)}\cdot\frac{\mathbf{(x-5)}}{\mathbf{(x-5)}}-\frac{5}{(x+5)(x-5)}\cdot\frac{\mathbf{2}}{\mathbf{2}}$$

$$=\frac{x^2-x-20}{2(x+5)(x-5)}-\frac{10}{2(x+5)(x-5)}$$

$$=\frac{x^2-x-30}{2(x+5)(x-5)}$$

To see if this expression will reduce, we factor the numerator into $(x-6)(x+5)$.

$$=\frac{(x-6)\cancel{(x+5)}}{2\cancel{(x+5)}(x-5)}$$

$$=\frac{x-6}{2(x-5)}$$

7. Subtract.
$\frac{2x-4}{x^2+5x+4}-\frac{x-4}{x^2+6x+8}$

EXAMPLE 7 Subtract $\frac{2x-2}{x^2+4x+3}-\frac{x-1}{x^2+5x+6}$.

SOLUTION We factor each denominator and build the LCD from those factors:

$$\frac{2x-2}{x^2+4x+3}-\frac{x-1}{x^2+5x+6}$$

$$=\frac{2x-2}{(x+3)(x+1)}-\frac{x-1}{(x+3)(x+2)}$$

$$=\frac{2x-2}{(x+3)(x+1)}\cdot\frac{\mathbf{(x+2)}}{\mathbf{(x+2)}}-\frac{x-1}{(x+3)(x+2)}\cdot\frac{\mathbf{(x+1)}}{\mathbf{(x+1)}}$$ **The LCD is $(x+1)(x+2)(x+3)$**

$$=\frac{2x^2+2x-4}{(x+1)(x+2)(x+3)}-\frac{x^2-1}{(x+1)(x+2)(x+3)}$$ **Multiply out each numerator**

$$=\frac{(2x^2+2x-4)-(x^2-1)}{(x+1)(x+2)(x+3)}$$

$$=\frac{x^2+2x-3}{(x+1)(x+2)(x+3)}$$ **Subtract numerators**

$$=\frac{\cancel{(x+3)}(x-1)}{(x+1)(x+2)\cancel{(x+3)}}$$ **Factor numerator to see if we can reduce**

$$=\frac{x-1}{(x+1)(x+2)}$$ **Reduce**

Answers

6. $\frac{x+2}{2(x+3)}$ 7. $\frac{x-1}{(x+1)(x+2)}$

EXAMPLE 8 Add $\frac{x^2}{x-7} + \frac{6x+7}{7-x}$.

SOLUTION In the first section of this chapter we were able to reverse the terms in a factor such as $7 - x$ by factoring -1 from each term. In a problem like this, the same result can be obtained by multiplying the numerator and denominator by -1:

$$\frac{x^2}{x-7} + \frac{6x+7}{7-x} \cdot \frac{\mathbf{-1}}{\mathbf{-1}} = \frac{x^2}{x-7} + \frac{-6x-7}{x-7}$$

$$= \frac{x^2-6x-7}{x-7} \qquad \textbf{Add numerators}$$

$$= \frac{\cancel{(x-7)}(x+1)}{\cancel{(x-7)}} \qquad \textbf{Factor numerator}$$

$$= x + 1 \qquad \textbf{Divide out } x - 7$$

For our next example we will look at a problem in which we combine a whole number and a rational expression.

EXAMPLE 9 Subtract $2 - \frac{9}{3x+1}$.

SOLUTION To subtract these two expressions, we think of 2 as a rational expression with a denominator of 1.

$$2 - \frac{9}{3x+1} = \frac{2}{1} - \frac{9}{3x+1}$$

The LCD is $3x + 1$. Multiplying the numerator and denominator of the first expression by $3x + 1$ gives us a rational expression equivalent to 2 but with a denominator of $3x + 1$.

$$\frac{2}{1} \cdot \frac{\mathbf{(3x+1)}}{\mathbf{(3x+1)}} - \frac{9}{3x+1} = \frac{6x+2-9}{3x+1}$$

$$= \frac{6x-7}{3x+1}$$

The numerator and denominator of this last expression do not have any factors in common other than 1, so the expression is in lowest terms.

EXAMPLE 10 Write an expression for the sum of a number and twice its reciprocal. Then, simplify that expression.

SOLUTION If x is the number, then its reciprocal is $\frac{1}{x}$. Twice its reciprocal is $\frac{2}{x}$. The sum of the number and twice its reciprocal is

$$x + \frac{2}{x}$$

To combine these two expressions, we think of the first term x as a rational expression with a denominator of 1. The least common denominator is x:

$$x + \frac{2}{x} = \frac{x}{1} + \frac{2}{x}$$

$$= \frac{x}{1} \cdot \frac{\mathbf{x}}{\mathbf{x}} + \frac{2}{x}$$

$$= \frac{x^2+2}{x}$$

8. Add.

$$\frac{x^2}{x-4} + \frac{x+12}{4-x}$$

9. Add.

$$2 + \frac{25}{5x-1}$$

10. One number is three times another. Write an expression for the sum of the reciprocals of the two numbers. Then simplify that expression.

Answers

8. $x + 3$ **9.** $\frac{10x+23}{5x-1}$ **10.** $\frac{4}{3x}$

Getting Ready for Class

After reading through the preceding section, respond in your own words and in complete sentences.

A. Briefly describe how you would add two rational expressions that have the same denominator.

B. Why is factoring important in finding a least common denominator?

C. What is the last step in adding or subtracting two rational expressions?

D. Explain how you would change the fraction $\frac{5}{x-3}$ to an equivalent fraction with denominator $x^2 - 9$.

PROBLEM SET 3.4

Combine the following fractions.

1. $\frac{3}{4} + \frac{1}{2}$

2. $\frac{5}{6} + \frac{1}{3}$

3. $\frac{2}{5} - \frac{1}{15}$

4. $\frac{5}{8} - \frac{1}{4}$

5. $\frac{5}{6} + \frac{7}{8}$

6. $\frac{3}{4} + \frac{2}{3}$

7. $\frac{9}{48} - \frac{3}{54}$

8. $\frac{6}{28} - \frac{5}{42}$

9. $\frac{3}{4} - \frac{1}{8} + \frac{2}{3}$

10. $\frac{1}{3} - \frac{5}{6} + \frac{5}{12}$

Combine the following rational expressions. Reduce all answers to lowest terms.

11. $\frac{x}{x+3} + \frac{3}{x+3}$

12. $\frac{5x}{5x+2} + \frac{2}{5x+2}$

13. $\frac{4}{y-4} - \frac{y}{y-4}$

14. $\frac{8}{y+8} + \frac{y}{y+8}$

15. $\frac{x}{x^2-y^2} - \frac{y}{x^2-y^2}$

16. $\frac{x}{x^2-y^2} + \frac{y}{x^2-y^2}$

17. $\frac{2x-3}{x-2} - \frac{x-1}{x-2}$

18. $\frac{2x-4}{x+2} - \frac{x-6}{x+2}$

19. $\frac{1}{a} + \frac{2}{a^2} - \frac{3}{a^3}$

20. $\frac{3}{a} + \frac{2}{a^2} - \frac{1}{a^3}$

21. $\frac{7x-2}{2x+1} - \frac{5x-3}{2x+1}$

22. $\frac{7x-1}{3x+2} - \frac{4x-3}{3x+2}$

Combine the following rational expressions. Reduce all answers to lowest terms.

23. $\frac{2}{t^2} - \frac{3}{2t}$

24. $\frac{5}{3t} - \frac{4}{t^2}$

25. $\frac{3x+1}{2x-6} - \frac{x+2}{x-3}$

26. $\frac{x+1}{x-2} - \frac{4x+7}{5x-10}$

27. $\frac{6x+5}{5x-25} - \frac{x+2}{x-5}$

28. $\frac{4x+2}{3x+12} - \frac{x-2}{x+4}$

29. $\frac{x+1}{2x-2} - \frac{2}{x^2-1}$

30. $\frac{x+7}{2x+12} + \frac{6}{x^2-36}$

31. $\frac{1}{a-b} - \frac{3ab}{a^3-b^3}$

32. $\frac{1}{a+b} + \frac{3ab}{a^3+b^3}$

33. $\frac{1}{2y-3} - \frac{18y}{8y^3-27}$

34. $\frac{1}{3y-2} - \frac{18y}{27y^3-8}$

35. $\frac{x}{x^2-5x+6} - \frac{3}{3-x}$

36. $\frac{x}{x^2+4x+4} - \frac{2}{2+x}$

37. $\frac{2}{4t-5} + \frac{9}{8t^2-38t+35}$

38. $\frac{3}{2t-5} + \frac{21}{8t^2-14t-15}$

39. $\frac{1}{a^2-5a+6} + \frac{3}{a^2-a-2}$

40. $\frac{-3}{a^2+a-2} + \frac{5}{a^2-a-6}$

41. $\frac{1}{8x^3-1} - \frac{1}{4x^2-1}$

42. $\frac{1}{27x^3-1} - \frac{1}{9x^2-1}$

43. $\frac{4}{4x^2-9} - \frac{6}{8x^2-6x-9}$

44. $\frac{9}{9x^2+6x-8} - \frac{6}{9x^2-4}$

45. $\frac{4a}{a^2+6a+5} - \frac{3a}{a^2+5a+4}$

46. $\frac{3a}{a^2+7a+10} - \frac{2a}{a^2+6a+8}$

47. $\frac{2x-1}{x^2+x-6} - \frac{x+2}{x^2+5x+6}$

48. $\frac{4x+1}{x^2+5x+4} - \frac{x+3}{x^2+4x+3}$

49. $\frac{2x-8}{3x^2+8x+4} + \frac{x+3}{3x^2+5x+2}$

50. $\frac{5x+3}{2x^2+5x+3} - \frac{3x+9}{2x^2+7x+6}$

51. $\frac{2}{x^2+5x+6} - \frac{4}{x^2+4x+3} + \frac{3}{x^2+3x+2}$

52. $\frac{-5}{x^2+3x-4} + \frac{5}{x^2+2x-3} + \frac{1}{x^2+7x+12}$

53. $\frac{2x+8}{x^2+5x+6} - \frac{x+5}{x^2+4x+3} - \frac{x-1}{x^2+3x+2}$

54. $\frac{2x+11}{x^2+9x+20} - \frac{x+1}{x^2+7x+12} - \frac{x+6}{x^2+8x+15}$

55. $2 + \frac{3}{2x+1}$

56. $3 - \frac{2}{2x+3}$

57. $5 + \frac{2}{4-t}$

58. $7 + \frac{3}{5-t}$

59. $x - \frac{4}{2x+3}$

60. $x - \frac{5}{3x+4} + 1$

61. $\dfrac{x}{x+2} + \dfrac{1}{2x+4} - \dfrac{3}{x^2+2x}$

62. $\dfrac{x}{x+3} + \dfrac{7}{3x+9} - \dfrac{2}{x^2+3x}$

63. $\dfrac{1}{x} + \dfrac{x}{2x+4} - \dfrac{2}{x^2+2x}$

64. $\dfrac{1}{x} + \dfrac{x}{3x+9} - \dfrac{3}{x^2+3x}$

Applying the Concepts

65. Optometry The formula

$$P = \frac{1}{a} + \frac{1}{b}$$

is used by optometrists to help determine how strong to make the lenses for a pair of eyeglasses. If a is 10 and b is 0.2, find the corresponding value of P.

66. Optometry Show that the formula in Problem 65 can be written

$$P = \frac{a+b}{ab}$$

Then let $a = 10$ and $b = 0.2$ in this new form of the formula to find P.

67. Comparing Expressions Show that the expressions $(x+y)^{-1}$ and $x^{-1} + y^{-1}$ are not equal when $x = 3$ and $y = 4$.

68. Comparing Expressions Show that the expressions $(x+y)^{-1}$ and $x^{-1} + y^{-1}$ are not equal. (Begin by writing each with positive exponents only.)

69. Elliptical Orbits Consider two objects, A and B, that move in the same direction along an elliptical path at constant but different velocities.

It can be shown that the time, T, it takes for the two objects to meet can be found from the formula

$$\frac{1}{T} = \frac{1}{t_A} - \frac{1}{t_B}$$

where t_A = time required for object A to orbit, and t_B = time required for object B to orbit.

(a) If $t_A = 24$ months and $t_B = 30$ months, when will these two objects meet?

(b) If $t_A = t_B$ what can one conclude?

70. Average Velocity If a car travels at a constant velocity, v_1, for 10 miles and then at a constant, but different velocity, v_2, for the next 10 miles, it can be shown that the car's average velocity, v_{avg}, over these 20 miles satisfies the equation

$$\frac{2}{v_{avg}} = \frac{1}{v_1} + \frac{1}{v_2}$$

Find the average velocity of a car that travels a constant 45 miles per hour for 10 miles and then increases to a constant 60 miles per hour for the next 10 miles.

71. Number Problem Write an expression for the sum of a number and 4 times its reciprocal. Then, simplify that expression.

72. Number Problem Write an expression for the sum of a number and 3 times its reciprocal. Then, simplify that expression.

73. Number Problem Write an expression for the sum of the reciprocals of two consecutive integers. Then, simplify that expression.

74. Number Problem Write an expression for the sum of the reciprocals of two consecutive even integers. Then, simplify that expression.

Review Problems

For each relation that follows, state the domain and the range, and indicate which are also functions.

75. $\{(1, 2), (3, 4), (4, 2)\}$

76. $\{(0, 0), (1, 1), (0, 1)\}$

77. $\{(3, 1), (2, 3), (1, 2)\}$

78. $\{(-1, 1), (2, -2), (-3, -3)\}$

Extending the Concepts

Simplify.

79. $\left(1 - \dfrac{1}{x}\right)\left(1 - \dfrac{1}{x+1}\right)\left(1 - \dfrac{1}{x+2}\right)\left(1 - \dfrac{1}{x+3}\right)$

80. $\left(1 + \dfrac{1}{x}\right)\left(1 + \dfrac{1}{x+1}\right)\left(1 + \dfrac{1}{x+2}\right)\left(1 + \dfrac{1}{x+3}\right)$

81. $\left(\dfrac{a^2 - b^2}{u^2 - v^2}\right)\left(\dfrac{av - au}{b - a}\right) + \left(\dfrac{a^2 - av}{u + v}\right)\left(\dfrac{1}{a}\right)$

82. $\left(\dfrac{6r^2}{r^2 - 1}\right)\left(\dfrac{r+1}{3}\right) - \dfrac{2r^2}{r-1}$

83. $\dfrac{18x - 19}{4x^2 + 27x - 7} - \dfrac{12x - 41}{3x^2 + 17x - 28}$

84. $\dfrac{42 - 22y}{3y^2 - 13y - 10} - \dfrac{21 - 13y}{2y^2 - 9y - 5}$

85. $\left(\dfrac{1}{y^2 - 1} \div \dfrac{1}{y^2 + 1}\right)\left(\dfrac{y^3 + 1}{y^4 - 1}\right) + \dfrac{1}{(y+1)^2(y-1)}$

86. $\left(\dfrac{a^3 - 64}{a^2 - 16} \div \dfrac{a^2 - 4a + 16}{a^2 - 4} \div \dfrac{a^2 + 4a + 16}{a^3 + 64}\right) + 4 - a^2$

3.5 Complex Fractions

The quotient of two fractions or two rational expressions is called a *complex fraction.* This section is concerned with the simplification of complex fractions.

EXAMPLE 1 Simplify $\dfrac{\frac{3}{4}}{\frac{5}{8}}$.

SOLUTION There are generally two methods that can be used to simplify complex fractions.

Method 1 We can multiply the numerator and denominator of the complex fraction by the LCD for both of the fractions, which in this case is 8.

$$\frac{\frac{3}{4}}{\frac{5}{8}} = \frac{\frac{3}{4}\cdot \mathbf{8}}{\frac{5}{8}\cdot \mathbf{8}} = \frac{6}{5}$$

Method 2 Instead of dividing by $\frac{5}{8}$ we can multiply by $\frac{8}{5}$.

$$\frac{\frac{3}{4}}{\frac{5}{8}} = \frac{3}{4}\cdot\frac{8}{5} = \frac{24}{20} = \frac{6}{5}$$

Here are some examples of complex fractions involving rational expressions. Most can be solved using either of the two methods shown in Example 1.

EXAMPLE 2 Simplify $\dfrac{\frac{1}{x}+\frac{1}{y}}{\frac{1}{x}-\frac{1}{y}}$.

SOLUTION This problem is most easily solved using Method 1. We begin by multiplying both the numerator and denominator by the quantity xy, which is the LCD for all the fractions:

$$\frac{\frac{1}{x}+\frac{1}{y}}{\frac{1}{x}-\frac{1}{y}} = \frac{\left(\frac{1}{x}+\frac{1}{y}\right)\cdot \mathbf{xy}}{\left(\frac{1}{x}-\frac{1}{y}\right)\cdot \mathbf{xy}}$$

$$= \frac{\frac{1}{x}(xy)+\frac{1}{y}(xy)}{\frac{1}{x}(xy)-\frac{1}{y}(xy)}$$

Apply the distributive property to distribute xy over both terms in the numerator and denominator

$$= \frac{y+x}{y-x}$$

EXAMPLE 3 Simplify $\dfrac{\frac{x-2}{x^2-9}}{\frac{x^2-4}{x+3}}$.

SOLUTION Applying method 2, we have

$$\frac{\frac{x-2}{x^2-9}}{\frac{x^2-4}{x+3}} = \frac{x-2}{x^2-9}\cdot\frac{x+3}{x^2-4}$$

$$= \frac{\cancel{(x-2)}\cancel{(x+3)}}{\cancel{(x+3)}(x-3)(x+2)\cancel{(x-2)}}$$

$$= \frac{1}{(x-3)(x+2)}$$

Practice Problems

1. Simplify $\dfrac{\frac{2}{3}}{\frac{5}{6}}$. (Try both methods.)

Note: You should become proficient at both methods. Each method can be useful in simplifying complex fractions.

2. Simplify $\dfrac{\frac{1}{x}-\frac{1}{3}}{\frac{1}{x}+\frac{1}{3}}$.

3. Simplify $\dfrac{\frac{x+5}{x^2-16}}{\frac{x^2-25}{x-4}}$.

Note: We could have used method 1 just as easily: we would have multiplied the numerator and denominator of the complex fraction by $(x+3)(x-3)$.

Answers

1. $\frac{4}{5}$ 2. $\dfrac{3-x}{3+x}$

3. $\dfrac{1}{(x+4)(x-5)}$

4. Simplify $\dfrac{1 - \dfrac{9}{x^2}}{1 - \dfrac{1}{x} - \dfrac{6}{x^2}}$.

5. Simplify $2 + \dfrac{5}{x - \frac{1}{5}}$.

EXAMPLE 4 Simplify $\dfrac{1 - \dfrac{4}{x^2}}{1 - \dfrac{1}{x} - \dfrac{6}{x^2}}$.

SOLUTION The simplest way to simplify this complex fraction is to multiply the numerator and denominator by the LCD, x^2:

$$\frac{1 - \dfrac{4}{x^2}}{1 - \dfrac{1}{x} - \dfrac{6}{x^2}} = \frac{\mathbf{x^2}\left(1 - \dfrac{4}{x^2}\right)}{\mathbf{x^2}\left(1 - \dfrac{1}{x} - \dfrac{6}{x^2}\right)} \quad \textbf{Multiply numerator and denominator by } \boldsymbol{x^2}$$

$$= \frac{x^2 \cdot 1 - x^2 \cdot \dfrac{4}{x^2}}{x^2 \cdot 1 - x^2 \cdot \dfrac{1}{x} - x^2 \cdot \dfrac{6}{x^2}} \quad \textbf{Distributive property}$$

$$= \frac{x^2 - 4}{x^2 - x - 6} \quad \textbf{Simplify}$$

$$= \frac{(x - 2)\cancel{(x + 2)}}{(x - 3)\cancel{(x + 2)}} \quad \textbf{Factor}$$

$$= \frac{x - 2}{x - 3} \quad \textbf{Reduce}$$

EXAMPLE 5 Simplify $2 - \dfrac{3}{x + \frac{1}{3}}$.

SOLUTION First we simplify the expression that follows the subtraction sign.

$$2 - \frac{3}{x + \dfrac{1}{3}} = 2 - \frac{\mathbf{3} \cdot 3}{\mathbf{3}\left(x + \dfrac{1}{3}\right)} = 2 - \frac{9}{3x + 1}$$

Now we subtract by rewriting the first term, 2, with the LCD, $3x + 1$.

$$2 - \frac{9}{3x + 1} = \frac{2}{1} \cdot \frac{\mathbf{3x + 1}}{\mathbf{3x + 1}} - \frac{9}{3x + 1}$$

$$= \frac{6x + 2 - 9}{3x + 1} = \frac{6x - 7}{3x + 1}$$

Getting Ready for Class

After reading through the preceding section, respond in your own words and in complete sentences.

A. What is a complex fraction?
B. Explain how a least common denominator can be used to simplify a complex fraction.
C. Explain how some complex fractions can be converted to division problems. When is it more efficient to convert a complex fraction to a division problem of rational expressions?
D. Which method of simplifying complex fractions do you prefer? Why?

Answers

4. $\dfrac{x + 3}{x + 2}$ 5. $\dfrac{10x + 23}{5x - 1}$

PROBLEM SET 3.5

Simplify each of the following as much as possible.

1. $\dfrac{\frac{3}{4}}{\frac{2}{3}}$

2. $\dfrac{\frac{5}{9}}{\frac{7}{12}}$

3. $\dfrac{\frac{1}{3}-\frac{1}{4}}{\frac{1}{2}+\frac{1}{8}}$

4. $\dfrac{\frac{1}{6}-\frac{1}{3}}{\frac{1}{4}-\frac{1}{8}}$

5. $\dfrac{3+\frac{2}{5}}{1-\frac{3}{7}}$

6. $\dfrac{2+\frac{5}{6}}{1-\frac{7}{8}}$

7. $\dfrac{\frac{1}{x}}{1+\frac{1}{x}}$

8. $\dfrac{1-\frac{1}{x}}{\frac{1}{x}}$

9. $\dfrac{1+\frac{1}{a}}{1-\frac{1}{a}}$

10. $\dfrac{1-\frac{2}{a}}{1-\frac{3}{a}}$

11. $\dfrac{\frac{1}{x}-\frac{1}{y}}{\frac{1}{x}+\frac{1}{y}}$

12. $\dfrac{\frac{1}{x}+\frac{2}{y}}{\frac{2}{x}+\frac{1}{y}}$

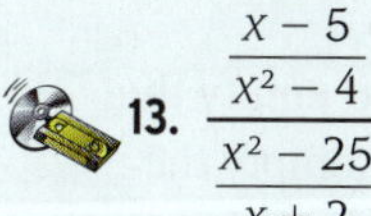

13. $\dfrac{\frac{x-5}{x^2-4}}{\frac{x^2-25}{x+2}}$

14. $\dfrac{\frac{3x+1}{x^2-49}}{\frac{9x^2-1}{x-7}}$

15. $\dfrac{\frac{4a}{2a^3+2}}{\frac{8a}{4a+4}}$

16. $\dfrac{\frac{2a}{3a^3-3}}{\frac{4a}{6a-6}}$

17. $\dfrac{1-\frac{9}{x^2}}{1-\frac{1}{x}-\frac{6}{x^2}}$

18. $\dfrac{4-\frac{1}{x^2}}{4+\frac{4}{x}+\frac{1}{x^2}}$

19. $\dfrac{2+\frac{5}{a}-\frac{3}{a^2}}{2-\frac{5}{a}+\frac{2}{a^2}}$

20. $\dfrac{3+\frac{5}{a}-\frac{2}{a^2}}{3-\frac{10}{a}+\frac{3}{a^2}}$

21. $\dfrac{27-\frac{8}{x^3}}{3+\frac{1}{x}-\frac{2}{x^2}}$

22. $\dfrac{64+\frac{1}{x^3}}{4-\frac{11}{x}-\frac{3}{x^2}}$

23. $\dfrac{1+\frac{2}{x}+\frac{4}{x^2}+\frac{8}{x^3}}{1-\frac{16}{x^4}}$

24. $\dfrac{27+\frac{9}{x}+\frac{3}{x^2}+\frac{1}{x^3}}{81-\frac{1}{x^4}}$

25. $\dfrac{2+\frac{3}{x}-\frac{18}{x^2}-\frac{27}{x^3}}{2+\frac{9}{x}+\frac{9}{x^2}}$

26. $\dfrac{3+\frac{5}{x}-\frac{12}{x^2}-\frac{20}{x^3}}{3+\frac{11}{x}+\frac{10}{x^2}}$

27. $\dfrac{1+\frac{1}{x+3}}{1-\frac{1}{x+3}}$

28. $\dfrac{1+\frac{1}{x-2}}{1-\frac{1}{x-2}}$

29. $\dfrac{1-\frac{1}{a+1}}{1+\frac{1}{a-1}}$

30. $\dfrac{\frac{1}{a-1}+1}{\frac{1}{a+1}-1}$

31. $\dfrac{\frac{1}{x+3}+\frac{1}{x-3}}{\frac{1}{x+3}-\frac{1}{x-3}}$

32. $\dfrac{\frac{1}{x+a}+\frac{1}{x-a}}{\frac{1}{x+a}-\frac{1}{x-a}}$

33. $\dfrac{\frac{y+1}{y-1}+\frac{y-1}{y+1}}{\frac{y+1}{y-1}-\frac{y-1}{y+1}}$

34. $\dfrac{\frac{y-1}{y+1}-\frac{y+1}{y-1}}{\frac{y-1}{y+1}+\frac{y+1}{y-1}}$

35. $1-\dfrac{x}{1-\frac{1}{x}}$

36. $x-\dfrac{1}{x-\frac{1}{2}}$

37. $\dfrac{1-\dfrac{1}{x+\frac{1}{2}}}{1+\dfrac{1}{x+\frac{1}{2}}}$

38. $\dfrac{2+\dfrac{1}{x-\frac{1}{3}}}{2-\dfrac{1}{x-\frac{1}{3}}}$

39. $\dfrac{\frac{1}{x+h}-\frac{1}{x}}{h}$

40. $\dfrac{\frac{1}{(x+h)^2}-\frac{1}{x^2}}{h}$

41. $\dfrac{\frac{3}{ab}+\frac{4}{bc}-\frac{2}{ac}}{\frac{5}{abc}}$

42. $\dfrac{\frac{x}{yz}+\frac{y}{xz}+\frac{z}{xy}}{\frac{1}{x^2y^2}-\frac{1}{x^2z^2}+\frac{1}{y^2z^2}}$

43. $\dfrac{\frac{t^2-2t-8}{t^2+7t+6}}{\frac{t^2-t-6}{t^2+2t+1}}$

44. $\dfrac{\frac{y^2-5y-14}{y^2+3y-10}}{\frac{y^2-8y+7}{y^2+6y+5}}$

45. $\dfrac{5 + \dfrac{4}{b-1}}{\dfrac{7}{b+5} - \dfrac{3}{b-1}}$

46. $\dfrac{\dfrac{6}{x+5} - 7}{\dfrac{8}{x+5} - \dfrac{9}{x+3}}$

47. $\dfrac{\dfrac{3}{x^2 - x - 6}}{\dfrac{2}{x+2} - \dfrac{4}{x-3}}$

48. $\dfrac{\dfrac{9}{a-7} + \dfrac{8}{2a+3}}{\dfrac{10}{2a^2 - 11a - 21}}$

49. $\dfrac{\dfrac{1}{m-4} + \dfrac{1}{m-5}}{\dfrac{1}{m^2 - 9m + 20}}$

50. $\dfrac{\dfrac{1}{k^2 - 7k + 12}}{\dfrac{1}{k-3} + \dfrac{1}{k-4}}$

Applying the Concepts

51. Difference Quotient For each rational function below, find the difference quotient

$$\frac{f(x) - f(a)}{x - a}$$

(a) $f(x) = \dfrac{4}{x}$

(b) $f(x) = \dfrac{1}{x+1}$

(c) $f(x) = \dfrac{1}{x^2}$

52. Difference Quotient For each rational function below, find the difference quotient

$$\frac{f(x+h) - f(x)}{h}$$

(a) $f(x) = \dfrac{4}{x}$

(b) $f(x) = \dfrac{1}{x+1}$

(c) $f(x) = \dfrac{1}{x^2}$

53. Optics The formula $f = \frac{ab}{a+b}$ is used in optics to find the focal length of a lens. Show that the formula $f = (a^{-1} + b^{-1})^{-1}$ is equivalent to the preceding formula by rewriting it without the negative exponents and then simplifying the result.

54. Average Speed Professor Vaughen drives to his office at Miami Dade College at an average speed of 40 miles per hour. On the way home from work he averages 60 miles per hour. Assume the distance to work and the distance home from work are the same, and find Professor Vaughen's average speed for the total trip to and from work. (His average speed will be the total distance divided by total time.)

55. Doppler Effect The change in the pitch of a sound (such as a train whistle) as an object passes is called the Doppler effect, named after C. J. Doppler (1803–1853). A person will *hear* a sound with a frequency, h, according to the formula

$$h = \frac{f}{1 + \dfrac{v}{s}}$$

where f is the actual frequency of the sound being produced, s is the speed of sound (about 740 miles per hour), and v is the velocity of the moving object.

(a) Examine this fraction, and then explain why h and f approach the same value as v becomes smaller and smaller.

(b) Solve this formula for v.

56. Mean of Two Numbers The mean of two numbers is a point on the number line midway between them.

(a) Given two numbers, a and b, determine a formula for the mean of a and b.

(b) Given two numbers, a and b, determine a formula for the mean of their reciprocals.

(c) If $a = 4$ and $b = 6$, find the mean of their reciprocals.

57. Work Problem A water storage tank has two drains. It can be shown that the time it takes to empty the tank if both drains are open is given by the formula

$$\frac{1}{\dfrac{1}{a} + \dfrac{1}{b}}$$

where a = time it takes for the first drain to empty the tank, and b = time for the second drain to empty the tank.

(a) Simplify this complex fraction.

(b) Find the amount of time needed to empty the tank using both drains if, used alone, the first drain empties the tank in 4 hours and the second drain can empty the tank in 3 hours.

58. Investing If an amount of money, P, is invested for 1 year at 6% interest, compounded n times per year, then the amount of money in the account at the end of 1 year may be found from the formula

$$A = P\left(1 + \frac{0.06}{n}\right)^n$$

Determine A if $n = 10$ and if $n = 11$; if $n = 50$ and if $n = 51$; if $n = 100$ and if $n = 101$. (Round your answers to six decimal places to the right of the decimal point.)

Review Problems

Solve each equation.

59. $10 - 2(x + 3) = x + 1$

60. $15 - 3(x - 1) = x - 2$

61. $x^2 - x - 12 = 0$

62. $3x^2 + x - 10 = 0$

63. $(x + 1)(x - 6) = -12$

64. $(x + 1)(x - 4) = -6$

Extending the Concepts

Simplify each expression.

65. $\dfrac{\left(\frac{1}{3}\right) - \left(\frac{1}{3}\right)^2}{1 - \frac{1}{3}}$

66. $\dfrac{\left(\frac{1}{2}\right) - \left(\frac{1}{2}\right)^2}{1 - \frac{1}{2}}$

67. $\dfrac{\left(\frac{1}{9}\right) - \frac{1}{9}\left(\frac{1}{3}\right)^4}{1 - \frac{1}{3}}$

68. $\dfrac{\left(\frac{1}{6}\right) - \frac{1}{6}\left(\frac{1}{2}\right)^4}{1 - \frac{1}{2}}$

69. $\dfrac{1 + \dfrac{1}{1 - \frac{a}{b}}}{1 - \dfrac{3}{1 - \frac{a}{b}}}$

70. $\dfrac{1 - \dfrac{1}{\frac{a}{b} + 2}}{1 + \dfrac{3}{\frac{a}{2b} + 1}}$

71. $\dfrac{a^{-1} + b^{-1}}{(ab)^{-1}}$

72. $\dfrac{(r^{-1} - s^{-1})^{-1}}{(rs)^{-2}}$

73. $\dfrac{(q^{-2} - t^{-2})^{-1}}{(t^{-1} - q^{-1})^{-1}}$

74. $\dfrac{(q^{-2} + t^{-2})^{-1}}{(t^{-1} + q^{-1})^{-1}}$

3.6 Equations Involving Rational Expressions

The first step in solving an equation that contains one or more rational expressions is to find the LCD for all denominators in the equation. We then multiply both sides of the equation by the LCD to clear the equation of all fractions—that is, after we have multiplied through by the LCD, each term in the resulting equation will have a denominator of 1.

Practice Problems

1. Solve $\frac{x}{3} + 1 = \frac{1}{2}$.

EXAMPLE 1 Solve $\frac{x}{2} - 3 = \frac{2}{3}$.

SOLUTION The LCD for 2 and 3 is 6. Multiplying both sides by 6, we have

$$\mathbf{6}\left(\frac{x}{2} - 3\right) = \mathbf{6}\left(\frac{2}{3}\right)$$

$$6\left(\frac{x}{2}\right) - 6(3) = 6\left(\frac{2}{3}\right)$$

$$3x - 18 = 4$$

$$3x = 22$$

$$x = \frac{22}{3}$$

Multiplying both sides of an equation by the LCD clears the equation of fractions because the LCD has the property that all the denominators divide it evenly.

2. Solve $\frac{2}{a+5} = \frac{1}{3}$.

EXAMPLE 2 Solve $\frac{6}{a-4} = \frac{3}{8}$.

SOLUTION The LCD for $a - 4$ and 8 is $8(a - 4)$. Multiplying both sides by this quantity yields

$$\mathbf{8(a - 4)} \cdot \frac{6}{a-4} = \mathbf{8(a - 4)} \cdot \frac{3}{8}$$

$$48 = (a - 4) \cdot 3$$

$$48 = 3a - 12$$

$$60 = 3a$$

$$20 = a$$

The solution set is {20}, which checks in the original equation.

When we multiply both sides of an equation by an expression containing the variable, we must be sure to check our solutions. The multiplication property of equality does not allow multiplication by 0. If the expression we multiply by contains the variable, then it has the possibility of being 0. In the last example we multiplied both sides by $8(a - 4)$. This gives a restriction $a \neq 4$ for any solution we come up with.

Answers

1. $-\frac{3}{2}$ 2. 1

EXAMPLE 3 Solve $\frac{x}{x-2} + \frac{2}{3} = \frac{2}{x-2}$.

SOLUTION The LCD is $3(x-2)$. We are assuming $x \neq 2$ when we multiply both sides of the equation by $3(x-2)$:

$$\mathbf{3(x-2)} \cdot \left[\frac{x}{x-2} + \frac{2}{3}\right] = \mathbf{3(x-2)} \cdot \frac{2}{x-2}$$

$$3x + (x-2) \cdot 2 = 3 \cdot 2$$

$$3x + 2x - 4 = 6$$

$$5x - 4 = 6$$

$$5x = 10$$

$$x = 2$$

The only possible solution is $x = 2$. Checking this value back in the original equation gives

$$\frac{2}{2-2} + \frac{2}{3} \stackrel{?}{=} \frac{2}{2-2}$$

$$\frac{2}{0} + \frac{2}{3} \stackrel{?}{=} \frac{2}{0}$$

The first and last terms are undefined. The proposed solution, $x = 2$, does not check in the original equation. The solution set is the empty set. There is no solution to the original equation.

When the proposed solution to an equation is not actually a solution, it is called an *extraneous* solution. In the last example, $x = 2$ is an extraneous solution.

EXAMPLE 4 Solve $\frac{5}{x^2 - 3x + 2} - \frac{1}{x-2} = \frac{1}{3x-3}$.

SOLUTION Writing the equation again with the denominators in factored form, we have

$$\frac{5}{(x-2)(x-1)} - \frac{1}{x-2} = \frac{1}{3(x-1)}$$

The LCD is $3(x-2)(x-1)$. Multiplying through by the LCD, we have

$$\mathbf{3(x-2)(x-1)} \cdot \frac{5}{(x-2)(x-1)} - \mathbf{3(x-2)(x-1)} \cdot \frac{1}{(x-2)}$$

$$= \mathbf{3(x-2)(x-1)} \cdot \frac{1}{3(x-1)}$$

$$3 \cdot 5 - 3(x-1) \cdot 1 = (x-2) \cdot 1$$

$$15 - 3x + 3 = x - 2$$

$$-3x + 18 = x - 2$$

$$-4x + 18 = -2$$

$$-4x = -20$$

$$x = 5$$

3. Solve $\frac{x}{x+1} - \frac{1}{2} = \frac{-1}{x+1}$.

Note: In the process of solving the equation, we multiplied both sides by $3(x-2)$, solved for x, and got $x = 2$ for our solution. But when x is 2, the quantity $3(x-2) = 3(2-2) = 3(0) = 0$, which means we multiplied both sides of our equation by 0, which is not allowed under the multiplication property of equality.

4. Solve.
$\frac{x}{x^2 - 9} - \frac{1}{x+3} = \frac{1}{4x-12}$

Note: We can check the proposed solution in any of the equations obtained before multiplying through by the LCD. We cannot check the proposed solution in an equation obtained *after* multiplying through by the LCD since, if we have multiplied by 0, the resulting equations will not be equivalent to the original one.

Answers

3. No solution **4.** 9

Checking the proposed solution $x = 5$ in the original equation yields a true statement. Try it and see.

5. Solve $1 - \frac{2}{x} = \frac{8}{x^2}$.

EXAMPLE 5 Solve $3 + \frac{1}{x} = \frac{10}{x^2}$.

SOLUTION To clear the equation of denominators, we multiply both sides by x^2:

$$\mathbf{x^2}\left(3 + \frac{1}{x}\right) = \mathbf{x^2}\left(\frac{10}{x^2}\right)$$

$$3(x^2) + \left(\frac{1}{x}\right)(x^2) = \left(\frac{10}{x^2}\right)(x^2)$$

$$3x^2 + x = 10$$

Rewrite in standard form, and solve:

$$3x^2 + x - 10 = 0$$

$$(3x - 5)(x + 2) = 0$$

$$3x - 5 = 0 \quad \text{or} \quad x + 2 = 0$$

$$x = \frac{5}{3} \quad \text{or} \quad x = -2$$

The solution set is $\{-2, \frac{5}{3}\}$. Both solutions check in the original equation. Remember: We have to check *all solutions* any time we multiply both sides of the equation by an expression that contains the variable, just to be sure we haven't multiplied by 0.

6. Solve $\frac{y + 1}{3(y + 4)} = \frac{8}{(y + 4)(y - 4)}$.

EXAMPLE 6 Solve $\frac{y - 4}{y^2 - 5y} = \frac{2}{y^2 - 25}$.

SOLUTION Factoring each denominator, we find the LCD is $y(y - 5)(y + 5)$. Multiplying each side of the equation by the LCD clears the equation of denominators and leads us to our possible solutions:

$$\mathbf{y(y - 5)(y + 5)} \cdot \frac{y - 4}{y(y - 5)} = \frac{2}{(y - 5)(y + 5)} \cdot \mathbf{y(y - 5)(y + 5)}$$

$$(y + 5)(y - 4) = 2y$$

$$y^2 + y - 20 = 2y \quad \textbf{Multiply out the left side}$$

$$y^2 - y - 20 = 0 \quad \textbf{Add } -2y \textbf{ to each side}$$

$$(y - 5)(y + 4) = 0$$

$$y - 5 = 0 \quad \text{or} \quad y + 4 = 0$$

$$y = 5 \quad \text{or} \quad y = -4$$

The two possible solutions are 5 and -4. If we substitute -4 for y in the original equation, we find that it leads to a true statement. It is, therefore, a solution. However, on the other hand, if we substitute 5 for y in the original equation, we find that both sides of the equation are undefined. The only solution to our original equation is $y = -4$. The other possible solution $y = 5$ is extraneous.

Answers

5. $-2, 4$ 6. 7

EXAMPLE 7 Solve for y: $x = \dfrac{y-4}{y-2}$.

SOLUTION To solve for y, we first multiply each side by $y - 2$ to obtain

$$x(y-2) = y - 4$$

$$xy - 2x = y - 4 \quad \text{Distributive property}$$

$$xy - y = 2x - 4 \quad \text{Collect all terms containing } y \text{ on the left side}$$

$$y(x-1) = 2x - 4 \quad \text{Factor } y \text{ from each term on the left side}$$

$$y = \frac{2x-4}{x-1} \quad \text{Divide each side by } x - 1$$

7. Solve for y: $x = \dfrac{y+2}{y-1}$.

EXAMPLE 8 Solve the formula $\frac{1}{x} = \frac{1}{b} + \frac{1}{a}$ for x.

SOLUTION We begin by multiplying both sides by the least common denominator xab. As you can see from our previous examples, multiplying both sides of an equation by the LCD is equivalent to multiplying each term of both sides by the LCD:

$$\mathbf{xab} \cdot \frac{1}{x} = \frac{1}{b} \cdot \mathbf{xab} + \frac{1}{a} \cdot \mathbf{xab}$$

$$ab = xa + xb$$

$$ab = (a+b)x \quad \text{Factor } x \text{ from the right side}$$

$$\frac{ab}{a+b} = x$$

We know we are finished because the variable we were solving for is alone on one side of the equation and does not appear on the other side.

8. Solve the formula $\dfrac{1}{a} = \dfrac{1}{x} + \dfrac{1}{b}$ for x.

Getting Ready for Class

After reading through the preceding section, respond in your own words and in complete sentences.

A. Explain how a least common denominator can be used to simplify an equation.

B. What is an extraneous solution?

C. Is it possible for an equation containing rational expressions to have no solutions?

D. What is the last step in solving an equation that contains rational expressions?

Answers

7. $y = \dfrac{x+2}{x-1}$ 8. $x = \dfrac{ab}{b-a}$

PROBLEM SET 3.6

Solve each of the following equations.

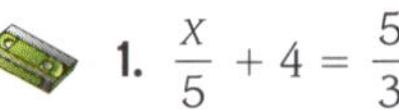

1. $\frac{x}{5} + 4 = \frac{5}{3}$
2. $\frac{x}{5} = \frac{x}{2} - 9$
3. $\frac{a}{3} + 2 = \frac{4}{5}$
4. $\frac{a}{4} + \frac{1}{2} = \frac{2}{3}$
5. $\frac{y}{2} + \frac{y}{4} + \frac{y}{6} = 3$
6. $\frac{y}{3} - \frac{y}{6} + \frac{y}{2} = 1$
7. $\frac{5}{2x} = \frac{1}{x} + \frac{3}{4}$
8. $\frac{1}{2a} = \frac{2}{a} - \frac{3}{8}$
9. $\frac{1}{x} = \frac{1}{3} - \frac{2}{3x}$
10. $\frac{5}{2x} = \frac{2}{x} - \frac{1}{12}$
11. $\frac{2x}{x-3} + 2 = \frac{2}{x-3}$
12. $\frac{2}{x+5} = \frac{2}{5} - \frac{x}{x+5}$
13. $1 - \frac{1}{x} = \frac{12}{x^2}$
14. $2 + \frac{5}{x} = \frac{3}{x^2}$
15. $y - \frac{4}{3y} = -\frac{1}{3}$
16. $\frac{y}{2} - \frac{4}{y} = -\frac{7}{2}$
17. $\frac{x+2}{x+1} = \frac{1}{x+1} + 2$
18. $\frac{x+6}{x+3} = \frac{3}{x+3} + 2$
19. $\frac{3}{a-2} = \frac{2}{a-3}$
20. $\frac{5}{a+1} = \frac{4}{a+2}$
21. $6 - \frac{5}{x^2} = \frac{7}{x}$
22. $10 - \frac{3}{x^2} = -\frac{1}{x}$
23. $\frac{1}{x-1} - \frac{1}{x+1} = \frac{3x}{x^2-1}$
24. $\frac{5}{x-1} + \frac{2}{x-1} = \frac{4}{x+1}$
25. $\frac{2}{x-3} + \frac{x}{x^2-9} = \frac{4}{x+3}$
26. $\frac{2}{x+5} + \frac{3}{x+4} = \frac{2x}{x^2+9x+20}$
27. $\frac{3}{2} - \frac{1}{x-4} = \frac{-2}{2x-8}$
28. $\frac{2}{x} - \frac{1}{x+1} = \frac{-2}{5x+5}$
29. $\frac{t-4}{t^2-3t} = \frac{-2}{t^2-9}$
30. $\frac{t+3}{t^2-2t} = \frac{10}{t^2-4}$
31. $\frac{3}{y-4} - \frac{2}{y+1} = \frac{5}{y^2-3y-4}$
32. $\frac{1}{y+2} - \frac{2}{y-3} = \frac{2y}{y^2-y-6}$
33. $\frac{2}{1+a} = \frac{3}{1-a} + \frac{5}{a}$
34. $\frac{1}{a+3} - \frac{a}{a^2-9} = \frac{2}{3-a}$
35. $\frac{3}{2x-6} - \frac{x+1}{4x-12} = 4$
36. $\frac{2x-3}{5x+10} + \frac{3x-2}{4x+8} = 1$
37. $\frac{y+2}{y^2-y} - \frac{6}{y^2-1} = 0$
38. $\frac{y+3}{y^2-y} - \frac{8}{y^2-1} = 0$
39. $\frac{4}{2x-6} - \frac{12}{4x+12} = \frac{12}{x^2-9}$
40. $\frac{1}{x+2} + \frac{1}{x-2} = \frac{4}{x^2-4}$
41. $\frac{2}{y^2-7y+12} - \frac{1}{y^2-9} = \frac{4}{y^2-y-12}$
42. $\frac{1}{y^2+5y+4} + \frac{3}{y^2-1} = \frac{-1}{y^2+3y-4}$
43. Solve the equation $6x^{-1} + 4 = 7$ by multiplying both sides by x. (*Remember:* $x^{-1} \cdot x = x^{-1} \cdot x^1 = x^0 = 1$.)
44. Solve the equation $3x^{-1} - 5 = 2x^{-1} - 3$ by multiplying both sides by x.
45. Solve the equation $1 + 5x^{-2} = 6x^{-1}$ by multiplying both sides by x^2.
46. Solve the equation $1 + 3x^{-2} = 4x^{-1}$ by multiplying both sides by x^2.
47. Solve the formula $\frac{1}{x} = \frac{1}{b} - \frac{1}{a}$ for x.
48. Solve $\frac{1}{x} = \frac{1}{a} - \frac{1}{b}$ for x.
49. Solve for R in the formula $\frac{1}{R} = \frac{1}{R_1} + \frac{1}{R_2}$.
50. Solve for R in the formula

$$\frac{1}{R} = \frac{1}{R_1} + \frac{1}{R_2} + \frac{1}{R_3}$$

Solve for y.

51. $x = \frac{y-3}{y-1}$
52. $x = \frac{y-2}{y-3}$
53. $x = \frac{2y+1}{3y+1}$
54. $x = \frac{3y+2}{5y+1}$

Graph each function. Show the vertical asymptote.

55. $f(x) = \frac{1}{x - 3}$

56. $f(x) = \frac{1}{x + 3}$

57. $f(x) = \frac{4}{x + 2}$

58. $f(x) = \frac{4}{x - 2}$

59. $g(x) = \frac{2}{x - 4}$

60. $g(x) = \frac{2}{x + 4}$

61. $g(x) = \frac{6}{x + 1}$

62. $g(x) = \frac{6}{x - 1}$

Let $f(x) = \frac{1}{x - 3}$ and $g(x) = \frac{1}{x + 3}$, and evaluate the following.

63. $f(0)$ and $f(6)$

64. $g(0)$ and $g(-6)$

65. $f(1)$ and $f(5)$

66. $g(1)$ and $g(-7)$

67. Give the domain for the function f as defined above.

68. Give the domain for the function g as defined above.

Let $f(x) = \frac{4}{x + 2}$ and $g(x) = \frac{4}{x - 2}$, and evaluate the following.

69. $f(0)$ and $f(-4)$

70. $g(0)$ and $g(4)$

71. $f(2)$ and $f(-6)$

72. $g(1)$ and $g(-5)$

73. Give the domain for the function f as defined above.

74. Give the domain for the function g as defined above.

Applying the Concepts

75. Geometry From plane geometry and the principle of similar triangles, the relationship between y_1, y_2, and h shown in Figure 1 can be expressed as

$$\frac{1}{h} = \frac{1}{y_1} + \frac{1}{y_2}$$

Two poles are 12 feet high and 8 feet high. If a cable is attached to the top of each one and stretched to the bottom of the other, what is the height above the ground at which the two wires will meet?

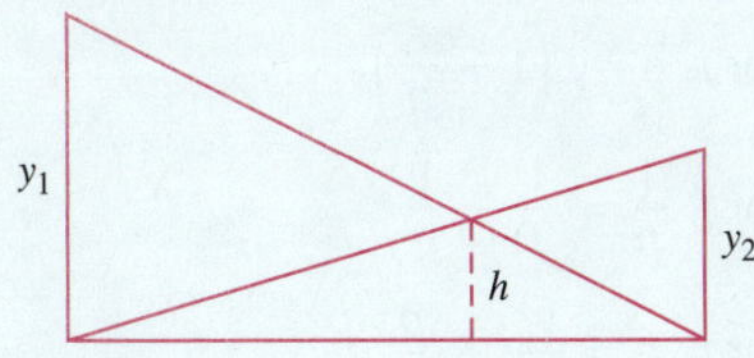

Figure 1

76. Geometry We can use the diagram in Figure 2 to develop the formula

$$\frac{1}{h} = \frac{1}{y_1} + \frac{1}{y_2}$$

by taking the following steps:

Step 1: By similar triangles, $\frac{y}{h} = \frac{x + y}{?}$.

Step 2: Solve the proportion from step 1 for h.

Step 3: Solve the proportion from step 2 for $\frac{1}{h}$.

Step 4: Write the right side of the proportion from step 3 as two fractions.

Step 5: Use the fact that $y \cdot y_1 = x \cdot y_2$, and substitute in the proportion from step 4 to obtain the desired result.

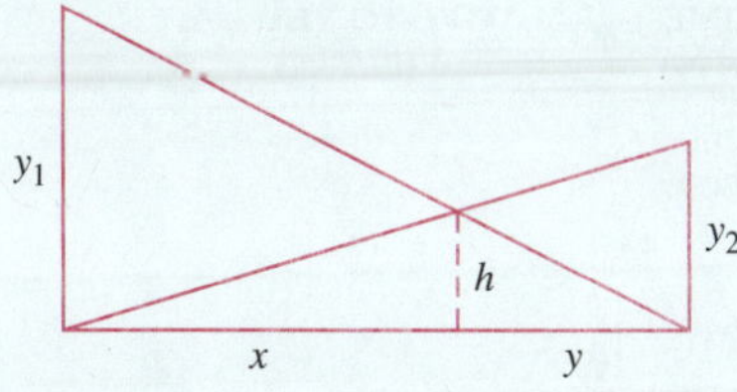

Figure 2

77. An Identity An identity is an equation that is true for any value of the variable for which the expression is defined. Verify the following expression is an identity by simplifying the left side of the expression.

$$\frac{2}{x - y} - \frac{1}{y - x} = \frac{3}{x - y}$$

78. Batting Average In baseball, a player's batting average is the ratio of the number of his hits to the number of his official at-bats. There is a dispute about who had the highest batting average in 1910. Although many sources list Ty Cobb of the Detroit Tigers as having the highest batting average of .385, some baseball researchers claim that his average was actually .382, and that he was second that year to Napoleon Lajoie of the Cleveland Indians, who had an average of .383. Assuming that both Cobb and Lajoie each had about 550 official at-bats in 1910, and that the researchers are correct and Cobb hit "only" .382 that year, how many more hits than Cobb did Lajoie have?

79. Kayak Race In a kayak race, the participants must paddle a kayak 450 meters down a river and then return 450 meters up the river to the starting point (Figure 3). Susan has correctly deduced that the total time t (in seconds) depends on the speed c (in meters per second) of the water according to the following expression:

$$t = \frac{450}{v+c} + \frac{450}{v-c}$$

where v is the speed of the kayak relative to the water (the speed of the kayak in still water).

(a) Fill in the following table.

TIME t(sec)	SPEED OF KAYAK RELATIVE TO THE WATER v(m/sec)	CURRENT OF THE RIVER c(m/sec)
240		1
300		2
	4	3
	3	1
540	3	
	3	3

(b) If the kayak race were conducted in the still waters of a lake, do you think that the total time of a given participant would be greater than, equal to, or smaller than the time in the river? Justify your answer.

(c) Suppose Peter can paddle his kayak at 4.1 meters per second and that the speed of the current is 4.1 meters per second. What will happen when Peter makes the turn and tries to come back up the river? How does this situation show up in the equation for total time?

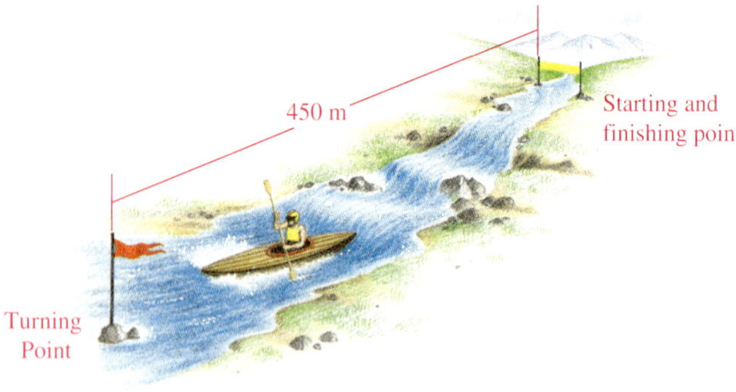

Figure 3

80. Harmonic Mean A number, h, is the harmonic mean of two numbers, n_1 and n_2, if $1/h$ is the mean (average) of $1/n_1$ and $1/n_2$.

(a) Write an equation relating the harmonic mean, h, to two numbers, n_1 and n_2. Then solve for h.

(b) Find the harmonic mean of 3 and 5.

Review Problems

These problems are taken from the book *Algebra for the Practical Man,* written by J. E. Thompson and published by D. Van Nostrand Company in 1931.

81. A man spent \$112.80 for 108 geese and ducks, each goose costing 14 dimes and each duck 6 dimes. How many of each did he buy?

82. If 15 pounds of tea and 10 pounds of coffee together cost \$15.50, while 25 pounds of tea and 13 pounds of coffee at the same prices cost \$24.55, find the price per pound of each.

83. A number of oranges at the rate of 3 for \$0.10 and apples at \$0.15 a dozen cost, together, \$6.80. Five times as many oranges and $\frac{1}{4}$ as many apples at the same rates would have cost \$25.45. How many of each were bought?

84. An estate is divided among three persons, A, B, and C. A's share is 3 times that of B, and B's share is twice that of C. If A receives \$9,000 more than C, how much does each receive?

Extending the Concepts

Solve each equation.

85. $\frac{12}{x} + \frac{8}{x^2} - \frac{75}{x^3} - \frac{50}{x^4} = 0$

86. $\frac{45}{x} + \frac{18}{x^2} - \frac{80}{x^3} - \frac{32}{x^4} = 0$

87. $\frac{1}{x^3} - \frac{1}{3x^2} - \frac{1}{4x} + \frac{1}{12} = 0$

88. $\frac{1}{x^3} - \frac{1}{2x^2} - \frac{1}{9x} + \frac{1}{18} = 0$

89. Solve for x. $\frac{2}{x} + \frac{4}{x+a} = \frac{-6}{a-x}$

90. Solve for x. $\frac{1}{b-x} - \frac{1}{x} = \frac{-2}{b+x}$

91. Solve for v. $\frac{s-vt}{t^2} = -16$

92. Solve for r. $A = P\left(1 + \frac{r}{n}\right)$

93. Solve for f. $\frac{1}{p} = \frac{1}{f} + \frac{1}{g}$

94. Solve for p. $h = \frac{v^2}{2g} + \frac{p}{c}$

3.7 Applications

We begin this section with some application problems, the solutions to which involve equations that contain rational expressions. As you will see, the solutions to the examples show only the essential steps from our Blueprint for Problem Solving. Recall that step 1 was done mentally; we read the problem and mentally list the items that are known and the items that are unknown. This is an essential part of problem solving. Now that you have had experience with application problems, however, you are doing step 1 automatically.

Also in this section we will look at a method of solving conversion problems that is called *unit analysis.* With unit analysis, we can convert expressions with units of feet per minute to equivalent expressions in miles per hour. This method of converting between different units of measure is used often in chemistry, physics, and engineering classes.

EXAMPLE 1 One number is twice another. The sum of their reciprocals is 2. Find the numbers.

SOLUTION Let x = the smaller number. The larger number is $2x$. Their reciprocals are $\frac{1}{x}$ and $\frac{1}{2x}$. The equation that describes the situation is

$$\frac{1}{x} + \frac{1}{2x} = 2$$

Multiplying both sides by the LCD $2x$, we have

$$\mathbf{2x} \cdot \frac{1}{x} + \mathbf{2x} \cdot \frac{1}{2x} = \mathbf{2x}(2)$$

$$2 + 1 = 4x$$

$$3 = 4x$$

$$x = \frac{3}{4}$$

The smaller number is $\frac{3}{4}$. The larger is $2\left(\frac{3}{4}\right) = \frac{6}{4} = \frac{3}{2}$. Adding their reciprocals, we have

$$\frac{4}{3} + \frac{2}{3} = \frac{6}{3} = 2$$

The sum of the reciprocals of $\frac{3}{4}$ and $\frac{3}{2}$ is 2.

Practice Problems

1. One number is 3 times another. The sum of their reciprocals is $\frac{4}{3}$. Find the numbers.

EXAMPLE 2 The speed of a boat in still water is 20 miles per hour. It takes the same amount of time for the boat to travel 3 miles downstream (with the current) as it does to travel 2 miles upstream (against the current). Find the speed of the current.

SOLUTION The following table will be helpful in finding the equation necessary to solve this problem.

2. A boat can travel at 15 miles per hour in still water. If it takes the same amount of time for the boat to travel 2 miles downstream as it does to travel 1 mile upstream, find the speed of the current.

If we let x = the speed of the current, the speed

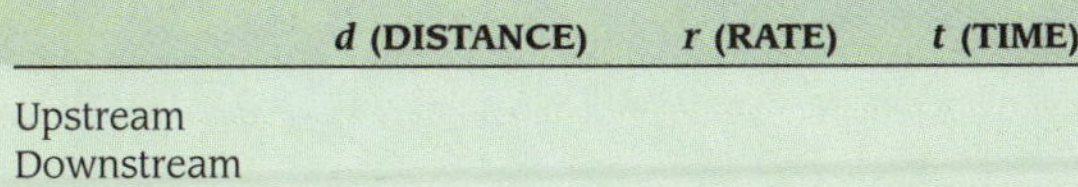

	d (DISTANCE)	*r* (RATE)	*t* (TIME)
Upstream			
Downstream			

Answer

1. 1, 3

(rate) of the boat upstream is $(20 - x)$ since it is traveling against the current. The rate downstream is $(20 + x)$ since the boat is then traveling with the current. The distance traveled upstream is 2 miles, and the distance traveled downstream is 3 miles. Putting the information given here into the table, we have

	d	*r*	*t*
Upstream	2	$20 - x$	
Downstream	3	$20 + x$	

To fill in the last two spaces in the table we must use the relationship $d = r \cdot t$. Since we know the spaces to be filled in are in the time column, we solve the equation $d = r \cdot t$ for t and get

$$t = \frac{d}{r}$$

The completed table then is

	d	*r*	*t*
Upstream	2	$20 - x$	$\frac{2}{20 - x}$
Downstream	3	$20 + x$	$\frac{3}{20 + x}$

Reading the problem again, we find that the time moving upstream is equal to the time moving downstream, or

$$\frac{2}{20 - x} = \frac{3}{20 + x}$$

Multiplying both sides by the LCD $(20 - x)(20 + x)$ gives

$$(20 + x) \cdot 2 = 3(20 - x)$$

$$40 + 2x = 60 - 3x$$

$$5x = 20$$

$$x = 4$$

The speed of the current is 4 miles per hour.

3. The current of a river is 2 miles per hour. It takes a motorboat a total of 3 hours to travel 8 miles upstream and return 8 miles downstream. What is the speed of the boat in still water?

EXAMPLE 3 The current of a river is 3 miles per hour. It takes a motorboat a total of 3 hours to travel 12 miles upstream and return 12 miles downstream. What is the speed of the boat in still water?

SOLUTION This time we let x = the speed of the boat in still water. Then, we fill in as much of the table as possible using the information given in the problem. For instance, since we let x = the speed of the boat in still water, the rate upstream (against the current) must be $x - 3$. The rate downstream (with the current) is $x + 3$.

	d	*r*	*t*
Upstream	12	$x - 3$	
Downstream	12	$x + 3$	

Answer
2. 5 mph

The last two boxes can be filled in using the relationship

$t = \frac{d}{r}$

	d	r	t
Upstream	12	$x - 3$	$\frac{12}{x-3}$
Downstream	12	$x + 3$	$\frac{12}{x+3}$

The total time for the trip up and back is 3 hours:

$$\text{Time upstream} + \text{Time downstream} = \text{Total time}$$

$$\frac{12}{x-3} + \frac{12}{x+3} = 3$$

Multiplying both sides by $(x - 3)(x + 3)$, we have

$$12(x + 3) + 12(x - 3) = 3(x^2 - 9)$$

$$12x + 36 + 12x - 36 = 3x^2 - 27$$

$$3x^2 - 24x - 27 = 0$$

$$x^2 - 8x - 9 = 0 \qquad \textbf{Divide both sides by 3}$$

$$(x - 9)(x + 1) = 0$$

$$x = 9 \quad \text{or} \quad x = -1$$

The speed of the motorboat in still water is 9 miles per hour. (We don't use $x = -1$ because the speed of the motorboat cannot be a negative number.)

EXAMPLE 4 An inlet pipe can fill a pool in 10 hours, and the drain can empty it in 12 hours. If the pool is empty and both the inlet pipe and drain are open, how long will it take to fill the pool?

4. The hot-water faucet can fill a sink in 3 minutes. The drain will empty the sink in 4 minutes. If the hot-water faucet is on and the drain is open, how long will it take to fill the sink?

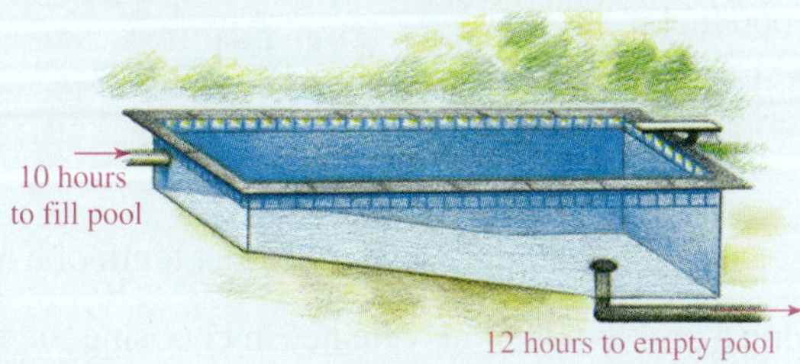

SOLUTION It is helpful to think in terms of how much work is done by each pipe in 1 hour.

Let x = the time it takes to fill the pool with both pipes open.

If the inlet pipe can fill the pool in 10 hours, then in 1 hour it is $\frac{1}{10}$ full. If the outlet pipe empties the pool in 12 hours, then in 1 hour it is $\frac{1}{12}$ empty. If the pool can be filled in x hours with both the inlet pipe and the drain open, then in 1 hour it is $\frac{1}{x}$ full when both pipes are open.

Here is the equation:

In 1 hour

$$\begin{bmatrix}\text{Amount filled by} \\ \text{inlet pipe}\end{bmatrix} - \begin{bmatrix}\text{Amount emptied by} \\ \text{the drain}\end{bmatrix} = \begin{bmatrix}\text{Fraction of pool filled} \\ \text{with both pipes open}\end{bmatrix}$$

$$\frac{1}{10} - \frac{1}{12} = \frac{1}{x}$$

Answer

3. 6 mph

Multiplying through by $60x$, we have

$$\mathbf{60x} \cdot \frac{1}{10} - \mathbf{60x} \cdot \frac{1}{12} = \mathbf{60x} \cdot \frac{1}{x}$$

$$6x - 5x = 60$$

$$x = 60$$

It takes 60 hours to fill the pool if both the inlet pipe and the drain are open.

Unit Analysis

In the 1950s the United States had a spy plane, the U-2, that could fly at an altitude of 65,000 feet. Do you know how many miles are in 65,000 feet?

We can solve problems like this by using a method called *unit analysis.* With unit analysis, we analyze the units we are given and the units for which we are asked, and then multiply by the appropriate *conversion factor.* Since 1 mile is 5,280 feet, the conversion factor we use is

$$\frac{1 \text{ mile}}{5{,}280 \text{ feet}}$$

which is the number 1. Multiplying 65,000 feet by this conversion factor we have the following:

$$65{,}000 \text{ feet} = \frac{65{,}000 \text{ feet}}{1} \cdot \frac{1 \text{ mile}}{5{,}280 \text{ feet}}$$

We treat the units common to the numerator and denominator in the same way we treat factors common to the numerator and denominator: We divide out common units, just as we divide out common factors. In the preceding expression we have feet common to the numerator and denominator. Dividing them out leaves us with miles only. Here is the complete problem.

$$65{,}000 \text{ feet} = \frac{65{,}000 \text{ \sout{feet}}}{1} \cdot \frac{1 \text{ mile}}{5{,}280 \text{ \sout{feet}}}$$

$$= \frac{65{,}000}{5{,}280} \text{ mile}$$

$$= 12.3 \text{ miles, to the nearest tenth of a mile}$$

The key to solving a problem like this one lies in choosing the appropriate conversion factor. The fact that 1 mile = 5,280 feet yields two conversion factors, each of which is equal to the number 1. They are

$$\frac{1 \text{ mile}}{5{,}280 \text{ feet}} \quad \text{and} \quad \frac{5{,}280 \text{ feet}}{1 \text{ mile}}$$

The conversion factor we choose depends on the units we are given and the units with which we want to end up. Multiplying any expression by either of the two conversion factors leaves the value of the original expression unchanged because each of the conversion factors is simply the number 1.

Answer
4. 12 minutes

Table 1 lists the units of volume in the U.S. system and their conversion factors.

Table 1

UNITS OF VOLUME IN THE U.S. SYSTEM

THE RELATIONSHIP BETWEEN	IS	TO CONVERT FROM ONE TO THE OTHER, MULTIPLY BY
cubic inches (in³) and cubic feet (ft³)	$1 \text{ ft}^3 = 1{,}728 \text{ in}^3$	$\frac{1{,}728 \text{ in}^3}{1 \text{ ft}^3}$ or $\frac{1 \text{ ft}^3}{1{,}728 \text{ in}^3}$
cubic feet and cubic yards (yd³)	$1 \text{ yd}^3 = 27 \text{ ft}^3$	$\frac{27 \text{ ft}^3}{1 \text{ yd}^3}$ or $\frac{1 \text{ yd}^3}{27 \text{ ft}^3}$
fluid ounces (fl oz) and pints (pt)	1 pt = 16 fl oz	$\frac{16 \text{ fl oz}}{1 \text{ pt}}$ or $\frac{1 \text{ pt}}{16 \text{ fl oz}}$
pints and quarts (qt)	1 qt = 2 pt	$\frac{2 \text{ pt}}{1 \text{ qt}}$ or $\frac{1 \text{ qt}}{2 \text{ pt}}$
quarts and gallons (gal)	1 gal = 4 qt	$\frac{4 \text{ qt}}{1 \text{ gal}}$ or $\frac{1 \text{ gal}}{4 \text{ qt}}$

EXAMPLE 5 What is the capacity (volume) in pints of a 1-gallon container of milk?

SOLUTION We change from gallons to quarts and then quarts to pints by multiplying by the appropriate conversion factors as given in Table 1.

$$1 \text{ gal} = 1 \cancel{\text{gal}} \times \frac{4 \cancel{\text{qt}}}{1 \cancel{\text{gal}}} \times \frac{2 \text{ pt}}{1 \cancel{\text{qt}}}$$
$$= 1 \times 4 \times 2 \text{ pt}$$
$$= 8 \text{ pt}$$

A 1-gallon container has the same capacity as 8 1-pint containers.

5. How many pints are in a 5-gallon pail?

EXAMPLE 6 A dairy herd produces 1,800 quarts of milk each day. How many gallons is this equivalent to?

SOLUTION Converting 1,800 quarts to gallons, we have

$$1{,}800 \text{ qt} = 1{,}800 \cancel{\text{qt}} \times \frac{1 \text{ gal}}{4 \cancel{\text{qt}}}$$
$$= \frac{1{,}800}{4} \text{gal}$$
$$= 450 \text{ gal}$$

We see that 1,800 quarts is equivalent to 450 gallons.

6. A dairy herd produces 2,000 quarts of milk each day. How many 10-gallon containers will this milk fill?

Answers

5. 40 pt 6. 50 containers

7. In Practice Problem 7 of the first section of this chapter we found a rider on the Ferris wheel called *Colossus* was traveling at approximately 13.0 feet per second. Give the speed in miles per hour. Round to the nearest tenth.

EXAMPLE 7 In the first section of this chapter we found a rider on the first Ferris wheel was traveling at approximately 39.3 feet per minute. Convert 39.3 feet per minute to miles per hour.

SOLUTION We know that 5,280 feet = 1 mile and 60 minutes = 1 hour. Therefore, we have the following conversion factors, each of which is equal to 1.

$$\frac{5{,}280 \text{ feet}}{1 \text{ mile}} \qquad \frac{1 \text{ mile}}{5{,}280 \text{ feet}} \qquad \frac{60 \text{ minutes}}{1 \text{ hour}} \qquad \frac{1 \text{ hour}}{60 \text{ minutes}}$$

The conversion factors we choose to multiply by are the ones that will allow us to divide out the units we are converting from and leave us with the units we are converting to. Specifically, we want to get rid of feet and be left with miles. Likewise, we want to get rid of minutes and be left with hours. Here is the conversion process that will accomplish these goals:

$$\begin{aligned} 39.3 \text{ feet per minute} &= \frac{39.3 \text{ feet}}{1 \text{ minute}} \cdot \frac{1 \text{ mile}}{5{,}280 \text{ feet}} \cdot \frac{60 \text{ minutes}}{1 \text{ hour}} \\ &= \frac{39.3 \cdot 60 \text{ miles}}{5{,}280 \text{ hours}} \\ &= 0.45 \text{ miles per hour, to the nearest hundredth} \end{aligned}$$

8. Assume that the mistake in the advertisement is that feet per second should read feet per minute. Is 1,100 feet per minute a reasonable speed for a chair lift?

EXAMPLE 8 In 1993, a ski resort in Vermont advertised their new high-speed chair lift as "the world's fastest chair lift, with a speed of 1,100 feet per second." Show why the speed cannot be correct.

SOLUTION To solve this problem, we can convert feet per second into miles per hour, a unit of measure we are more familiar with on an intuitive level.

$$\begin{aligned} 1{,}100 \text{ feet per second} &= \frac{1{,}100 \text{ feet}}{1 \text{ second}} \cdot \frac{1 \text{ mile}}{5{,}280 \text{ feet}} \cdot \frac{60 \text{ seconds}}{1 \text{ minute}} \cdot \frac{60 \text{ minutes}}{1 \text{ hour}} \\ &= \frac{1{,}100 \cdot 60 \cdot 60 \text{ miles}}{5{,}280 \text{ hours}} \\ &= 750 \text{ miles per hour} \end{aligned}$$

Obviously, there is a mistake in the advertisement.

Getting Ready for Class

After reading through the preceding section, respond in your own words and in complete sentences.

A. Briefly list the steps in the Blueprint for Problem Solving that you have used previously to solve application problems.

B. Write an application problem for which the solution depends on solving the equation $\frac{1}{2} + \frac{1}{3} = \frac{1}{x}$.

C. What is a conversion factor in unit analysis?

D. What conversion factors would you use to convert feet per second to miles per hour?

Answers

7. 8.9 mph

8. 12.5 mph is a reasonable speed for a chair lift.

PROBLEM SET 3.7

Solve each of the following word problems. Be sure to show the equation in each case.

Number Problems

1. One number is 3 times another. The sum of their reciprocals is $\frac{20}{3}$. Find the numbers.
2. One number is 3 times another. The sum of their reciprocals is $\frac{4}{9}$. Find the numbers.
3. The sum of a number and its reciprocals is $\frac{10}{3}$. Find the number.
4. The sum of a number and twice its reciprocal is $\frac{27}{5}$. Find the number.
5. The sum of the reciprocals of two consecutive integers is $\frac{7}{12}$. Find the two integers.
6. Find two consecutive even integers, the sum of whose reciprocals is $\frac{3}{4}$.
7. If a certain number is added to the numerator and denominator of $\frac{7}{9}$, the result is $\frac{5}{6}$. Find the number.
8. Find the number you would add to both the numerator and denominator of $\frac{8}{11}$ so that the result would be $\frac{6}{7}$.

Rate Problems

9. The speed of a boat in still water is 5 miles per hour. If the boat travels 3 miles downstream in the same amount of time it takes to travel 1.5 miles upstream, what is the speed of the current?
10. A boat, which moves at 18 miles per hour in still water, travels 14 miles downstream in the same amount of time it takes to travel 10 miles upstream. Find the speed of the current.
11. The current of a river is 2 miles per hour. A boat travels to a point 8 miles upstream and back again in 3 hours. What is the speed of the boat in still water?
12. A motorboat travels at 4 miles per hour in still water. It goes 12 miles upstream and 12 miles back again in a total of 8 hours. Find the speed of the current of the river.
13. Train A has a speed 15 miles per hour greater than that of train B. If train A travels 150 miles in the same time train B travels 120 miles, what are the speeds of the two trains?
14. A train travels 30 miles per hour faster than a car. If the train covers 120 miles in the same time the car covers 80 miles, what is the speed of each of them?
15. A small airplane flies 810 miles from Los Angeles to Portland, OR, with an average speed of 270 miles per hour. An hour and a half after the plane leaves, a Boeing 747 leaves Los Angeles for Portland. Both planes arrive in Portland at the same time. What was the average speed of the 747?
16. Lou leaves for a cross-country excursion on a bicycle traveling at 20 miles per hour. His friends are driving the trip and will meet him at several rest stops along the way. The first stop is scheduled 30 miles from the original starting point. If the people driving leave 15 minutes after Lou from the same place, how fast will they have to drive to reach the first rest stop at the same time as Lou?
17. A tour bus leaves Sacramento every Friday evening at 5:00 P.M. for a 270-mile trip to Las Vegas. This week, however, the bus leaves at 5:30 P.M. To arrive in Las Vegas on time, the driver drives 6 miles per hour faster than usual. What is the bus's usual speed?
18. A bakery delivery truck leaves the bakery at 5:00 A.M each morning on its 140-mile route. One day the driver gets a late start and does not leave the bakery until 5:30 A.M. To finish her route on time the driver drives 5 miles per hour faster than usual. At what speed does she usually drive?

Work Problems

19. A water tank can be filled by an inlet pipe in 8 hours. It takes twice that long for the outlet pipe to empty the tank. How long will it take to fill the tank if both pipes are open?

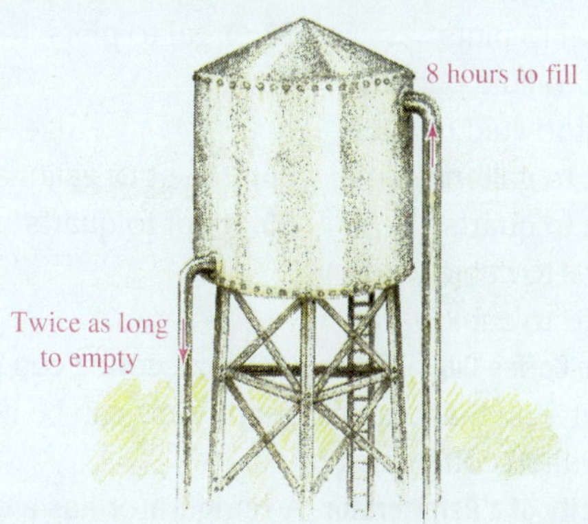

20. A sink can be filled from the faucet in 5 minutes. It takes only 3 minutes to empty the sink when the drain is open. If the sink is full and both the faucet and the drain are open, how long will it take to empty the sink?

21. It takes 10 hours to fill a pool with the inlet pipe. It can be emptied in 15 hours with the outlet pipe. If the pool is half full to begin with, how long will it take to fill it from there if both pipes are open?

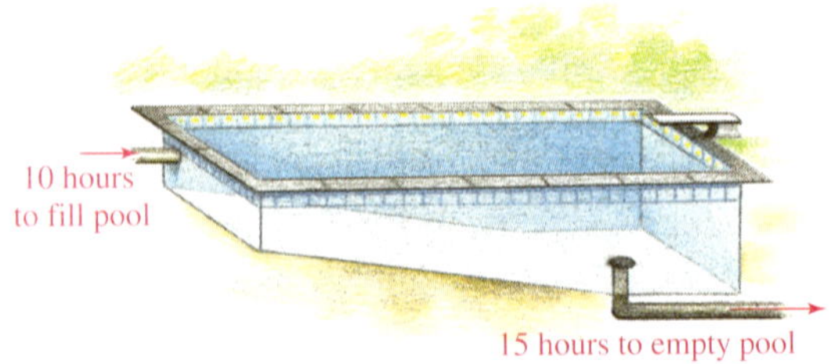

22. A sink is one-quarter full when both the faucet and the drain are opened. The faucet alone can fill the sink in 6 minutes, while it takes 8 minutes to empty it with the drain. How long will it take to fill the remaining three quarters of the sink?
23. A sink has two faucets: one for hot water and one for cold water. The sink can be filled by a cold water faucet in 3.5 minutes. If both faucets are open, the sink is filled in 2.1 minutes. How long does it take to fill the sink with just the hot water faucet open?
24. A water tank is being filled by two inlet pipes. Pipe A can fill the tank in $4\frac{1}{2}$ hours, but both pipes together can fill the tank in 2 hours. How long does it take to fill the tank using only pipe B?

Unit Analysis Problems

Make the following conversions using the conversion factors given in Table 1.

25. 5 yd^3 to cubic feet
26. 3.8 yd^3 to cubic feet
27. 3 pt to fluid ounces
28. 8 pt to fluid ounces
29. 2 gal to quarts
30. 12 gal to quarts
31. 2.5 gal to pints
32. 7 gal to pints
33. 15 qt to fluid ounces
34. 5.9 qt to fluid ounces
35. 64 pt to gallons
36. 256 pt to gallons
37. 12 pt to quarts
38. 18 pt to quarts
39. 243 ft^3 to cubic yards
40. 864 ft^3 to cubic yards
41. **Filling Coffee Cups** If a regular-size coffee cup holds about $\frac{1}{2}$ pint, about how many cups can be filled from a 1-gallon coffee maker?
42. **Capacity of a Refrigerator** A refrigerator has a capacity of 20 cubic feet. What is the capacity of the refrigerator in cubic inches?
43. **Volume of a Tank** The gasoline tank on a car holds 18 gallons of gas. What is the volume of the tank in quarts?
44. **Filling Glasses** How many 8-fluid-ounce glasses of water will it take to fill a 3-gallon aquarium?

Give your answers to the following problems to the nearest tenth.

45. The South Coast Shopping Mall in Costa Mesa, California, covers an area of 2,224,750 square feet. If 1 acre = 43,560 square feet, how many acres does the South Coast Shopping Mall cover?
46. The relationship between liters and cubic inches, both of which are measures of volume, is 0.0164 liters = 1 cubic inch. If a Ford Mustang has a motor with a displacement of 4.9 liters, what is the displacement in cubic inches?
47. The Forest chair lift at the Northstar ski resort in Lake Tahoe is 5,750 feet long. If a ride on this chair lift takes 11 minutes, what is the average speed of the lift in miles per hour?
48. The Bear Paw chair lift at the Northstar ski resort in Lake Tahoe is 790 feet long. If a ride on this chair lift takes 2.2 minutes, what is the average speed of the lift in miles per hour?
49. A sprinter runs 100 meters in 10.8 seconds. What is the sprinter's average speed in miles per hour? (1 meter = 3.28 feet)
50. A runner covers 400 meters in 49.8 seconds. What is the average speed of the runner in miles per hour?
51. A person riding a Ferris wheel with a diameter of 65 feet travels once around the wheel in 30 seconds. What is the average speed of the rider in miles per hour?
52. A person riding a Ferris wheel with a diameter of 102 feet travels once around the wheel in 3.5 minutes. What is the average speed of the rider in miles per hour?
53. A $3\frac{1}{2}$-inch diskette, when placed in the disk drive of a computer, rotates at 300 rpm (meaning one revolution takes 1/300 minute). Find the average speed of a point 1.5 inches from the center of the diskette in miles per hour?
54. A 5-inch fixed disk in a computer rotates at 3,600 rpm. Find the average speed of a point 2 inches from the center of the disk in miles per hour.
55. **Unit Analysis** An ad similar to the one shown in Figure 1 appeared recently in a national magazine. The ad indicates that 3,241,440 minutes is 6 years. Is this statement correct?

Courtesy of Timex Corp

Figure 1

56. **Unit Analysis and Thomas Jefferson** In addition to being our third president, Thomas Jefferson was also this nation's first Secretary of State. One of his official duties in 1791 was to summarize the units presently in use in the new United States. The units of capacity that he listed in his official report were as follows:

Measures of Capacity

Let the bushel be divided into 10 pottles;
Each pottle into 10 demi-pints;
Each demi-pint into 10 metres, which will be of a cubic inch each.

(a) Change 8 bushels into demi-pints.
(b) Change 76 metres into pottles.
(c) What should the number of bushels be multiplied by to convert bushels into metres?

Miscellaneous Problems

57. **Rhind Papyrus** Nearly 4,000 years ago, Egyptians worked mathematical exercises involving reciprocals. The *Rhind Papyrus* contains a wealth of such problems, and one of them is as follows:

> A quantity and its two thirds are added together, one third of this is added, then one-third of the sum is taken, and the result is 10.

Write an equation and solve this exercise.

58. **Photograph** For clear photographs, a camera must be properly focused. Professional photographers use a mathematical relationship relating the distance from the camera lens to the object being photographed, a; the distance from the lens to the film, b; and the focal length of the lens, f. These quantities, a, b, and f, are related by the equation

$$\frac{1}{a} + \frac{1}{b} = \frac{1}{f}$$

A camera has a focal length of 3 inches. If the lens is 5 inches from the film, how far should the lens be placed from the object being photographed for the camera to be perfectly focused?

The Periodic Table If you take a chemistry class, you will work with the Periodic Table of Elements. Figure 2 shows three of the elements listed in the periodic table. As you can see, the bottom number in each figure is the molecular weight of the element. In chemistry, a *mole* is the amount of a substance that will give the weight in grams equal to the molecular weight. For example, 1 mole of lead is 207.2 grams.

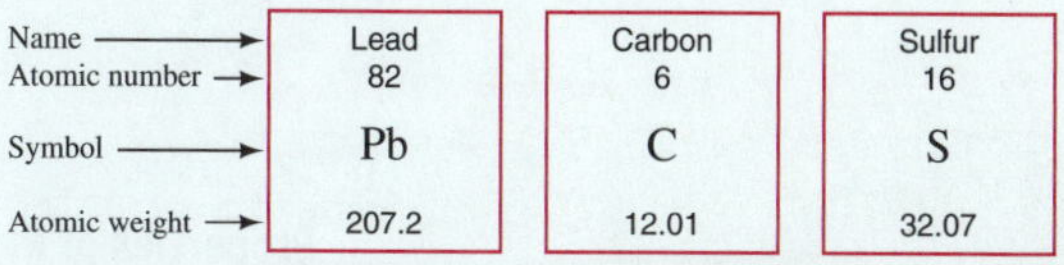

Name →	Lead	Carbon	Sulfur
Atomic number →	82	6	16
Symbol →	Pb	C	S
Atomic weight →	207.2	12.01	32.07

Figure 2

59. **Chemistry** For the element carbon, 1 mole = 12.01 grams.
(a) To the nearest gram, how many grams of carbon are in 2.5 moles of carbon?
(b) How many moles of carbon are in 39 grams of carbon? Round to the nearest hundredth.

60. **Chemistry** For the element of sulfur, 1 mole = 32.07 grams.
(a) How many grams of sulfur are in 3 moles of sulfur?
(b) How many moles of sulfur are found in 80.2 grams of sulfur?

Review Problems

Perform the indicated operations.

61. $\dfrac{2a+10}{a^3} \cdot \dfrac{a^2}{3a+15}$

62. $\dfrac{4a+8}{a^2-a-6} \div \dfrac{a^2+7a+12}{a^2-9}$

63. $(x^2-9)\left(\dfrac{x+2}{x+3}\right)$

64. $\dfrac{1}{x+4} + \dfrac{8}{x^2-16}$

65. $\dfrac{2x-7}{x-2} - \dfrac{x-5}{x-2}$

66. $2 + \dfrac{25}{5x-1}$

Simplify each expression.

67. $\dfrac{\frac{1}{x} - \frac{1}{3}}{\frac{1}{x} + \frac{1}{3}}$

68. $\dfrac{1 - \frac{9}{x^2}}{1 - \frac{1}{x} - \frac{6}{x^2}}$

Solve each equation.

69. $\dfrac{x}{x-3} + \dfrac{3}{2} = \dfrac{3}{x-3}$

70. $1 - \dfrac{3}{x} = \dfrac{-2}{x^2}$

CHAPTER 3 SUMMARY

Examples

1. $\frac{3}{4}$ is a rational number. $\frac{x-3}{x^2-9}$ is a rational expression.

Rational Numbers and Expressions [3.1]

A *rational number* is any number that can be expressed as the ratio of two integers:

$$\text{Rational numbers} = \left\{ \frac{a}{b} \,\middle|\, a \text{ and } b \text{ are integers, } b \neq 0 \right\}$$

A *rational expression* is any quantity that can be expressed as the ratio of two polynomials:

$$\text{Rational expressions} = \left\{ \frac{P}{Q} \,\middle|\, P \text{ and } Q \text{ are polynomials, } Q \neq 0 \right\}$$

Properties of Rational Expressions [3.1]

If P, Q, and K are polynomials with $Q \neq 0$ and $K \neq 0$, then

$$\frac{P}{Q} = \frac{PK}{QK} \quad \text{and} \quad \frac{P}{Q} = \frac{P/K}{Q/K}$$

which is to say that multiplying or dividing the numerator and denominator of a rational expression by the same nonzero quantity always produces an equivalent rational expression.

2. $$\frac{x-3}{x^2-9} = \frac{\cancel{x-3}}{(\cancel{x-3})(x+3)} = \frac{1}{x+3}$$

Reducing to Lowest Terms [3.1]

To reduce a rational expression to lowest terms, we first factor the numerator and denominator and then divide the numerator and denominator by any factors they have in common.

3. $$\frac{15x^3 - 20x^2 + 10x}{5x} = 3x^2 - 4x + 2$$

Dividing a Polynomial by a Monomial [3.2]

To divide a polynomial by a monomial, divide each term of the polynomial by the monomial.

4. $$\begin{array}{r} x \; - 2 \\ x - 3 \overline{) \; x^2 - 5x + 8} \\ \underline{\mp x^2 \pm 3x} \\ -2x + 8 \\ \underline{\pm 2x \mp 6} \\ 2 \end{array}$$

Long Division with Polynomials [3.2]

If division with polynomials cannot be accomplished by dividing out factors common to the numerator and denominator, then we use a process similar to long division with whole numbers. The steps in the process are estimate, multiply, subtract, and bring down the next term.

5. $$\frac{x+1}{x^2-4} \cdot \frac{x+2}{3x+3} = \frac{(\cancel{x+1})(\cancel{x+2})}{(x-2)(\cancel{x+2})(3)(\cancel{x+1})} = \frac{1}{3(x-2)}$$

Multiplication [3.3]

To multiply two rational numbers or rational expressions, multiply numerators and multiply denominators. In symbols,

$$\frac{P}{Q} \cdot \frac{R}{S} = \frac{PR}{QS} \qquad (Q \neq 0 \text{ and } S \neq 0)$$

In practice, we don't really multiply, but rather, we factor and then divide out common factors.

6. $$\frac{x^2-y^2}{x^3+y^3} \div \frac{x-y}{x^2-xy+y^2} = \frac{x^2-y^2}{x^3+y^3} \cdot \frac{x^2-xy+y^2}{x-y} = \frac{(\cancel{x+y})(\cancel{x-y})(x^2-xy+y^2)}{(\cancel{x+y})(x^2-xy+y^2)(\cancel{x-y})} = 1$$

Division [3.3]

To divide one rational expression by another, we use the definition of division to rewrite our division problem as an equivalent multiplication problem. Instead of dividing by a rational expression we multiply by its reciprocal. In symbols,

$$\frac{P}{Q} \div \frac{R}{S} = \frac{P}{Q} \cdot \frac{S}{R} = \frac{PS}{QR} \qquad (Q \neq 0, S \neq 0, R \neq 0)$$

Least Common Denominator [3.4]

The *least common denominator,* LCD, for a set of denominators is the smallest quantity divisible by each of the denominators.

7. The LCD for $\frac{2}{x-3}$ and $\frac{3}{5}$ is $5(x-3)$.

Addition and Subtraction [3.4]

If P, Q, and R represent polynomials, $R \neq 0$, then

$$\frac{P}{R} + \frac{Q}{R} = \frac{P+Q}{R} \quad \text{and} \quad \frac{P}{R} - \frac{Q}{R} = \frac{P-Q}{R}$$

When adding or subtracting rational expressions with different denominators, we must find the LCD for all denominators and change each rational expression to an equivalent expression that has the LCD.

8. $\frac{2}{x-3} + \frac{3}{5}$

$$= \frac{2}{x-3} \cdot \frac{\mathbf{5}}{\mathbf{5}} + \frac{3}{5} \cdot \frac{\mathbf{x-3}}{\mathbf{x-3}}$$

$$= \frac{3x+1}{5(x-3)}$$

Complex Fractions [3.5]

A rational expression that contains, in its numerator or denominator, other rational expressions is called a *complex fraction.* One method of simplifying a complex fraction is to multiply the numerator and denominator by the LCD for all denominators.

9. $$\frac{\frac{1}{x}+\frac{1}{y}}{\frac{1}{x}-\frac{1}{y}} = \frac{\mathbf{xy}\left(\frac{1}{x}+\frac{1}{y}\right)}{\mathbf{xy}\left(\frac{1}{x}-\frac{1}{y}\right)} = \frac{y+x}{y-x}$$

Equations Involving Rational Expressions [3.6]

To solve an equation involving rational expressions, we first find the LCD for all denominators appearing on either side of the equation. We then multiply both sides by the LCD to clear the equation of all fractions and solve as usual.

10. Solve $\frac{x}{2} + 3 = \frac{1}{3}$.

$$\mathbf{6}\left(\tfrac{x}{2}\right) + \mathbf{6} \cdot 3 = \mathbf{6} \cdot \tfrac{1}{3}$$

$$3x + 18 = 2$$

$$x = -\tfrac{16}{3}$$

COMMON MISTAKES

1. Attempting to divide the numerator and denominator of a rational expression by a quantity that is not a factor of both. Like this:

$$\frac{\cancel{x^2} - \overset{3}{\cancel{9x}} + \overset{2}{\cancel{20}}}{\cancel{x^2} - \underset{1}{\cancel{3x}} - \underset{1}{\cancel{10}}} \quad \textbf{Mistake}$$

This makes no sense at all. The numerator and denominator must be factored completely before any factors they have in common can be recognized:

$$\frac{x^2 - 9x + 20}{x^2 - 3x - 10} = \frac{\cancel{(x-5)}(x-4)}{\cancel{(x-5)}(x+2)} = \frac{x-4}{x+2}$$

2. Forgetting to check solutions to equations involving rational expressions. When we multiply both sides of an equation by a quantity containing the variable, we must be sure to check for extraneous solutions.

CHAPTER 3 REVIEW

The problems below form a comprehensive review of the material in this chapter. They can be used to study for exams. If you would like to take a practice test on this chapter, you can use the odd-numbered problems. Give yourself an hour and work as many of the odd-numbered problems as possible. When you are finished, or when an hour has passed, check your answers with the answers in the back of the book. You can use the even-numbered problems for a second practice test.

Reduce to lowest terms. [3.1]

1. $\dfrac{125x^4yz^3}{35x^2y^4z^3}$ **2.** $\dfrac{a^3-ab^2}{4a+4b}$ **3.** $\dfrac{x^2-25}{x^2+10x+25}$ **4.** $\dfrac{ax+x-5a-5}{ax-x-5a+5}$

Divide. If the denominator is a factor of the numerator, you may want to factor the numerator and divide out the common factor. [3.2]

5. $\dfrac{12x^3+8x^2+16x}{4x^2}$ **6.** $\dfrac{27a^2b^3-15a^3b^2+21a^4b^4}{-3a^2b^2}$

7. $\dfrac{x^{6n}-x^{5n}}{x^{3n}}$ **8.** $\dfrac{x^2-x-6}{x-3}$ **9.** $\dfrac{5x^2-14xy-24y^2}{x-4y}$ **10.** $\dfrac{y^4-16}{y-2}$

11. $\dfrac{8x^2-26x-9}{2x-7}$ **12.** $\dfrac{2y^3-9y^2-17y+39}{2y-3}$

Multiply and divide as indicated. [3.3]

13. $\dfrac{3}{4}\cdot\dfrac{12}{15}\div\dfrac{1}{3}$ **14.** $\dfrac{15x^2y}{8xy^2}\div\dfrac{10xy}{4x}$

15. $\dfrac{x^3-1}{x^4-1}\cdot\dfrac{x^2-1}{x^2+x+1}$ **16.** $\dfrac{a^2+5a+6}{a+1}\cdot\dfrac{a+5}{a^2+2a-3}\div\dfrac{a^2+7a+10}{a^2-1}$

17. $\dfrac{ax+bx+2a+2b}{ax-3a+bx-3b}\div\dfrac{ax-bx-2a+2b}{ax-bx-3a+3b}$ **18.** $(4x^2-9)\cdot\dfrac{x+3}{2x+3}$

Add and subtract as indicated. [3.4]

19. $\dfrac{3}{5}-\dfrac{1}{10}+\dfrac{8}{15}$ **20.** $\dfrac{5}{x-5}-\dfrac{x}{x-5}$

21. $\dfrac{1}{x}+\dfrac{1}{x^2}+\dfrac{1}{x^3}$ **22.** $\dfrac{8}{y^2-16}-\dfrac{7}{y^2-y-12}$

23. $\dfrac{x-2}{x^2+5x+4}-\dfrac{x-4}{2x^2+12x+16}$ **24.** $3+\dfrac{4}{5x-2}$

Simplify each complex fraction. [3.5]

25. $\dfrac{1+\dfrac{2}{3}}{1-\dfrac{2}{3}}$ **26.** $\dfrac{\dfrac{4a}{2a^3+2}}{\dfrac{8a}{4a+4}}$ **27.** $1+\dfrac{1}{x+\dfrac{1}{x}}$ **28.** $\dfrac{1-\dfrac{9}{x^2}}{1-\dfrac{1}{x}-\dfrac{6}{x^2}}$

Solve each equation. [3.6]

29. $\dfrac{3}{x-1}=\dfrac{3}{5}$ **30.** $\dfrac{x+1}{3}+\dfrac{x-3}{4}=\dfrac{1}{6}$ **31.** $\dfrac{5}{y+1}=\dfrac{4}{y+2}$ **32.** $\dfrac{x+6}{x+3}-2=\dfrac{3}{x+3}$

33. $\dfrac{4}{x^2-x-12}+\dfrac{1}{x^2-9}=\dfrac{2}{x^2-7x+12}$ **34.** $\dfrac{a+4}{a^2+5a}=\dfrac{-2}{a^2-25}$

35. **Distance, Rate, and Time** A car makes a 120-mile trip 10 miles per hour faster than a truck. The truck takes 2 hours longer to make the trip. What are the speeds of the car and the truck? [3.7]

36. **Average Speed** A jogger covers 3.5 miles in 28 minutes. Find the average speed of the jogger in miles per hour. [3.7]

37. **Unit Analysis** The speed of sound is 1,088 feet per second. Convert the speed of sound to miles per hour. Round your answer to the nearest whole number. [3.7]

38. **Filling a Pool** An inlet pipe can fill a pool in 10 hours, and the drain can empty it in 15 hours. If the pool is half full and both the inlet pipe and the drain are left open, how long will it take to fill the pool the rest of the way? [3.7]

39. **Unit Analysis** The top of Mount Whitney, the highest point in California, is 14,494 feet above sea level. Give this height in miles to the nearest tenth of a mile. [3.7]

40. **Unit Analysis** A bullet fired from a gun travels a distance of 4,750 feet in 3.2 seconds. Find the average speed of the bullet in miles per hour. Round to the nearest whole number. [3.7]

Rational Exponents and Roots

Chip Parter/Stone/Getty Images

Ecology and conservation are topics that interest most college students. If our rivers and oceans are to be preserved for future generations, we need to work to eliminate pollution from our waters. If a river is flowing at 1 meter per second and a pollutant is entering the river at a constant rate, the shape of the pollution plume can often be modeled by the simple equation

$$y = \sqrt{x}$$

The following table and graph were produced from the equation.

WIDTH OF A POLLUTANT PLUME	
DISTANCE FROM SOURCE (METERS) x	WIDTH OF PLUME (METERS) y
0	0
1	1
4	2
9	3
16	4

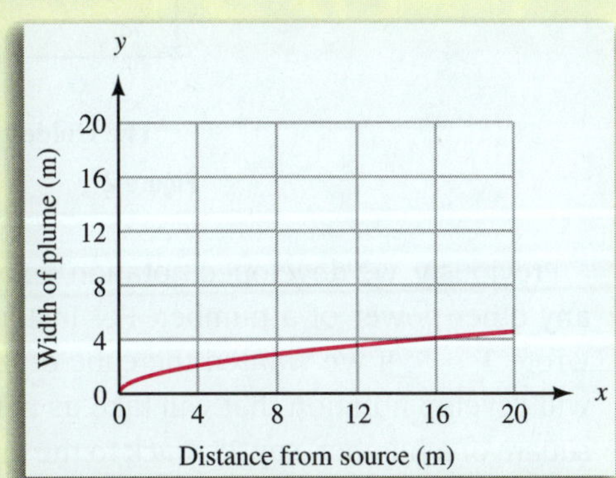

Figure 1

To visualize how Figure 1 models the pollutant plume, imagine that the river is flowing from left to right, parallel to the x-axis, with the x-axis as one of its banks. The pollutant is entering the river from the bank at (0, 0).

By modeling pollution with mathematics, we can use our knowledge of mathematics to help control and eliminate pollution.

CHAPTER OUTLINE

4.1 Rational Exponents

Figure 1 shows a square in which each of the four sides is 1 inch long. To find the square of the length of the diagonal c, we apply the Pythagorean theorem:

$$c^2 = 1^2 + 1^2$$
$$c^2 = 2$$

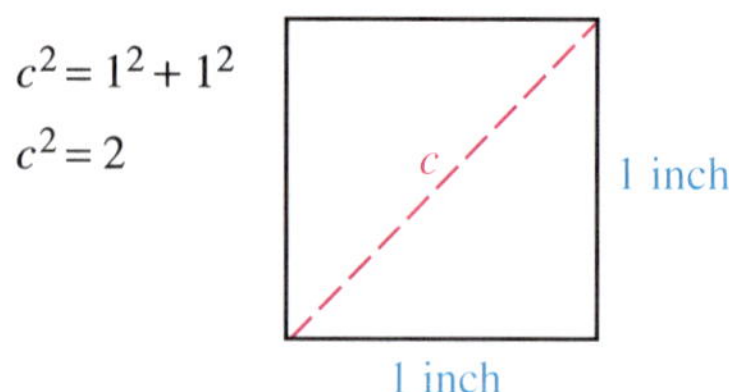

Figure 1

Because we know that c is positive and that its square is 2, we call c the *positive square root* of 2, and we write $c = \sqrt{2}$. Associating numbers, such as $\sqrt{2}$, with the diagonal of a square or rectangle allows us to analyze some interesting items from geometry. One particularly interesting geometric object that we will study in this section is shown in Figure 2. It is constructed from a right triangle, and the length of the diagonal is found from the Pythagorean theorem. We will come back to this figure at the end of this section.

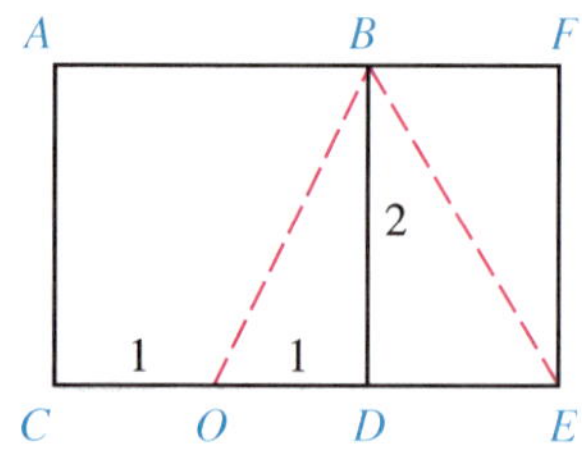

The Golden Rectangle

Figure 2

Previously, we developed notation (exponents) to give us the square, cube, or any other power of a number. For instance, if we wanted the square of 3, we wrote $3^2 = 9$. If we wanted the cube of 3, we wrote $3^3 = 27$. In this section, we will develop notation that will take us in the reverse direction—that is, from the square of a number, say 25, back to the original number, 5.

Note: It is a common mistake to assume that an expression like $\sqrt{25}$ indicates both square roots, 5 and -5. The expression $\sqrt{25}$ indicates only the positive square root of 25, which is 5. If we want the negative square root, we must use a negative sign: $-\sqrt{25} = -5$.

DEFINITION

If x is a nonnegative real number, then the expression $\sqrt{x}$ is called the **positive square root** of x and is such that

$$(\sqrt{x})^2 = x$$

In words: $\sqrt{x}$ is the positive number we square to get x.

The negative square root of x, $-\sqrt{x}$, is defined in a similar manner.

Practice Problems

1. Give the two square roots of 36.

EXAMPLE 1 The positive square root of 64 is 8 because 8 is the positive number with the property $8^2 = 64$. The negative square root of 64 is -8 since -8 is the negative number whose square is 64. We can summarize both of these facts by saying

$$\sqrt{64} = 8 \quad \text{and} \quad -\sqrt{64} = -8$$

Answer

1. 6, -6

The higher roots, cube roots, fourth roots, and so on are defined by definitions similar to that of square roots.

DEFINITION

If x is a real number and n is a positive integer, then

Positive square root of x, $\sqrt{x}$, is such that $(\sqrt{x})^2 = x$ $\quad x \geq 0$

Cube root of x, $\sqrt[3]{x}$, is such that $(\sqrt[3]{x})^3 = x$

Positive fourth root of x, $\sqrt[4]{x}$, is such that $(\sqrt[4]{x})^4 = x$ $\quad x \geq 0$

Fifth root of x, $\sqrt[5]{x}$, is such that $(\sqrt[5]{x})^5 = x$

$\vdots$

The n**th root of** x, $\sqrt[n]{x}$, is such that $(\sqrt[n]{x})^n = x$ $\quad x \geq 0$ if n is even

Note: We have restricted the even roots in this definition to nonnegative numbers. Even roots of negative numbers exist but are not represented by real numbers; that is, $\sqrt{-4}$ is not a real number since there is no real number whose square is -4.

The following is a table of the most common roots used in this book. Any of the roots that are unfamiliar should be memorized.

SQUARE ROOTS		CUBE ROOTS	FOURTH ROOTS
$\sqrt{0} = 0$	$\sqrt{49} = 7$	$\sqrt[3]{0} = 0$	$\sqrt[4]{0} = 0$
$\sqrt{1} = 1$	$\sqrt{64} = 8$	$\sqrt[3]{1} = 1$	$\sqrt[4]{1} = 1$
$\sqrt{4} = 2$	$\sqrt{81} = 9$	$\sqrt[3]{8} = 2$	$\sqrt[4]{16} = 2$
$\sqrt{9} = 3$	$\sqrt{100} = 10$	$\sqrt[3]{27} = 3$	$\sqrt[4]{81} = 3$
$\sqrt{16} = 4$	$\sqrt{121} = 11$	$\sqrt[3]{64} = 4$	
$\sqrt{25} = 5$	$\sqrt{144} = 12$	$\sqrt[3]{125} = 5$	
$\sqrt{36} = 6$	$\sqrt{169} = 13$		

Notation An expression like $\sqrt[3]{8}$ that involves a root is called a *radical expression.* In the expression $\sqrt[3]{8}$, the ^{3} is called the *index,* the $\sqrt{\ }$ is the *radical sign,* and 8 is called the *radicand.* The index of a radical must be a positive integer greater than 1. If no index is written, it is assumed to be 2.

Roots and Negative Numbers

When dealing with negative numbers and radicals, the only restriction concerns negative numbers under even roots. We can have negative signs in front of radicals and negative numbers under odd roots and still obtain real numbers. Here are some examples to help clarify this. In the last section of this chapter we will see how to deal with even roots of negative numbers.

EXAMPLES Simplify each expression, if possible.

2. $\sqrt[3]{-8} = -2$ because $(-2)^3 = -8$.
3. $\sqrt{-4}$ is not a real number since there is no real number whose square is -4.
4. $-\sqrt{25} = -5$ is the negative square root of 25.
5. $\sqrt[5]{-32} = -2$ because $(-2)^5 = -32$.
6. $\sqrt[4]{-81}$ is not a real number since there is no real number we can raise to the fourth power and obtain -81.

Simplify, if possible.

2. $\sqrt[3]{-64}$
3. $\sqrt{-25}$
4. $-\sqrt{4}$
5. $\sqrt[5]{-1}$
6. $\sqrt[4]{-16}$

Variables Under a Radical

From the preceding examples it is clear that we must be careful that we do not try to take an even root of a negative number. For this reason, we will assume that all variables appearing under a radical sign represent nonnegative numbers.

Answers

2. -4 3. Not a real number
4. -2 5. -1
6. Not a real number

Simplify.

7. $\sqrt{81a^4b^8}$

8. $\sqrt[3]{8x^3y^9}$

9. $\sqrt[4]{81a^4b^8}$

EXAMPLES Assume all variables represent nonnegative numbers and simplify each expression as much as possible.

7. $\sqrt{25a^4b^6} = 5a^2b^3$ because $(5a^2b^3)^2 = 25a^4b^6$.

8. $\sqrt[3]{x^6y^{12}} = x^2y^4$ because $(x^2y^4)^3 = x^6y^{12}$.

9. $\sqrt[4]{81r^8s^{20}} = 3r^2s^5$ because $(3r^2s^5)^4 = 81r^8s^{20}$.

Rational Numbers as Exponents

Next we develop a second kind of notation involving exponents that will allow us to designate square roots, cube roots, and so on in another way.

Consider the equation $x = 8^{1/3}$. Although we have not encountered fractional exponents before, let's assume that all the properties of exponents hold in this case. Cubing both sides of the equation, we have

$$x^3 = (8^{1/3})^3$$

$$x^3 = 8^{(1/3)(3)}$$

$$x^3 = 8^1$$

$$x^3 = 8$$

The last line tells us that x is the number whose cube is 8. It must be true, then, that x is the cube root of 8, $x = \sqrt[3]{8}$. Since we started with $x = 8^{1/3}$, it follows that

$$8^{1/3} = \sqrt[3]{8}$$

It seems reasonable, then, to define fractional exponents as indicating roots. Here is the formal definition.

DEFINITION

If x is a real number and n is a positive integer greater than 1, then

$$x^{1/n} = \sqrt[n]{x} \qquad (x \geq 0 \text{ when } n \text{ is even})$$

In words: The quantity $x^{1/n}$ is the nth root of x.

With this definition we have a way of representing roots with exponents. Here are some examples.

Simplify.

10. $9^{1/2}$

11. $27^{1/3}$

12. $-49^{1/2}$

13. $(-49)^{1/2}$

14. $\left(\frac{16}{25}\right)^{1/2}$

EXAMPLES Write each expression as a root and then simplify, if possible.

10. $8^{1/3} = \sqrt[3]{8} = 2$

11. $36^{1/2} = \sqrt{36} = 6$

12. $-25^{1/2} = -\sqrt{25} = -5$

13. $(-25)^{1/2} = \sqrt{-25}$, which is not a real number

14. $\left(\frac{4}{9}\right)^{1/2} = \sqrt{\frac{4}{9}} = \frac{2}{3}$

The properties of exponents developed in a previous chapter were applied to integer exponents only. We will now extend these properties to include rational exponents also. We do so without proof.

Answers

7. $9a^2b^4$ 8. $2xy^3$ 9. $3ab^2$

10. 3 11. 3 12. -7

13. Not a real number 14. $\frac{4}{5}$

Properties of Exponents

If a and b are real numbers and r and s are rational numbers, and a and b are nonnegative whenever r and s indicate even roots, then

1. $a^r \cdot a^s = a^{r+s}$
2. $(a^r)^s = a^{rs}$
3. $(ab)^r = a^r b^r$
4. $a^{-r} = \dfrac{1}{a^r}$ $(a \neq 0)$
5. $\left(\dfrac{a}{b}\right)^r = \dfrac{a^r}{b^r}$ $(b \neq 0)$
6. $\dfrac{a^r}{a^s} = a^{r-s}$ $(a \neq 0)$

There are times when rational exponents can simplify our work with radicals. Here are Examples 8 and 9 again, but this time we will work them using rational exponents.

EXAMPLES Write each radical with a rational exponent and then simplify.

15. $\sqrt[3]{x^6y^{12}} = (x^6y^{12})^{1/3}$
$= (x^6)^{1/3}(y^{12})^{1/3}$
$= x^2y^4$

16. $\sqrt[4]{81r^8s^{20}} = (81r^8s^{20})^{1/4}$
$= 81^{1/4}(r^8)^{1/4}(s^{20})^{1/4}$
$= 3r^2s^5$

Simplify.

15. $\sqrt[3]{8x^3y^9}$

16. $\sqrt[4]{81a^4b^8}$

So far, the numerators of all the rational exponents we have encountered have been 1. The next theorem extends the work we can do with rational exponents to rational exponents with numerators other than 1.

We can extend our properties of exponents with the following theorem.

The Rational Exponent Theorem

If a is a nonnegative real number, m is an integer, and n is a positive integer, then

$$a^{m/n} = (a^{1/n})^m = (a^m)^{1/n}$$

Proof We can prove this theorem using the properties of exponents. Since $\frac{m}{n} = m\left(\frac{1}{n}\right)$ we have

$$a^{m/n} = a^{m(1/n)} = (a^m)^{1/n} \qquad a^{m/n} = a^{(1/n)(m)} = (a^{1/n})^m$$

Here are some examples that illustrate how we use this theorem.

Note: On a scientific calculator, Example 17 would look like this:

8 [y^x] [(] 2 [÷] 3 [)] [=]

EXAMPLES Simplify as much as possible.

17. $8^{2/3} = (8^{1/3})^2$	**Rational exponent theorem**
$= 2^2$	**Definition of fractional exponents**
$= 4$	**The square of 2 is 4**
18. $25^{3/2} = (25^{1/2})^3$	**Rational exponent theorem**
$= 5^3$	**Definition of fractional exponents**
$= 125$	**The cube of 5 is 125**

Simplify.

17. $9^{3/2}$

18. $16^{3/4}$

Answers

15. $2xy^3$ **16.** $3ab^2$ **17.** 27 **18.** 8

19. $8^{-2/3}$

20. $\left(\frac{16}{81}\right)^{-3/4}$

19. $9^{-3/2} = (9^{1/2})^{-3}$ **Rational exponent theorem**

$= 3^{-3}$ **Definition of fractional exponents**

$= \frac{1}{3^3}$ **Property 4 for exponents**

$= \frac{1}{27}$ **The cube of 3 is 27**

20. $\left(\frac{27}{8}\right)^{-4/3} = \left[\left(\frac{27}{8}\right)^{1/3}\right]^{-4}$ **Rational exponent theorem**

$= \left(\frac{3}{2}\right)^{-4}$ **Definition of fractional exponents**

$= \left(\frac{2}{3}\right)^{4}$ **Property 4 for exponents**

$= \frac{16}{81}$ **The fourth power of $\frac{2}{3}$ is $\frac{16}{81}$**

Simplify.

21. $x^{1/2} \cdot x^{1/4}$

22. $(y^{3/5})^{5/6}$

23. $\frac{z^{3/4}}{z^{2/3}}$

24. $\frac{(x^{1/3}y^{-3})^6}{x^4y^{10}}$

EXAMPLES Assume all variables represent positive quantities and simplify as much as possible.

21. $x^{1/3} \cdot x^{5/6} = x^{1/3+5/6}$ **Property 1**

$= x^{2/6+5/6}$ **LCD is 6**

$= x^{7/6}$ **Add fractions**

22. $(y^{2/3})^{3/4} = y^{(2/3)(3/4)}$ **Property 2**

$= y^{1/2}$ **Multiply fractions: $\frac{2}{3} \cdot \frac{3}{4} = \frac{6}{12} = \frac{1}{2}$**

23. $\frac{z^{1/3}}{z^{1/4}} = z^{1/3-1/4}$ **Property 6**

$= z^{4/12-3/12}$ **LCD is 12**

$= z^{1/12}$ **Subtract fractions**

24. $\frac{(x^{-3}y^{1/2})^4}{x^{10}y^{3/2}} = \frac{(x^{-3})^4(y^{1/2})^4}{x^{10}y^{3/2}}$ **Property 3**

$= \frac{x^{-12}y^2}{x^{10}y^{3/2}}$ **Property 2**

$= x^{-22}y^{1/2}$ **Property 6**

$= \frac{y^{1/2}}{x^{22}}$ **Property 4**

Facts From Geometry: The Pythagorean Theorem (Again) and the Golden Rectangle

Now that we have had some experience working with square roots, we can rewrite the Pythagorean theorem using a square root. If triangle ABC is a right triangle with $C = 90°$, then the length of the longest side is the *positive square root* of the sum of the squares of the other two sides (see Figure 3).

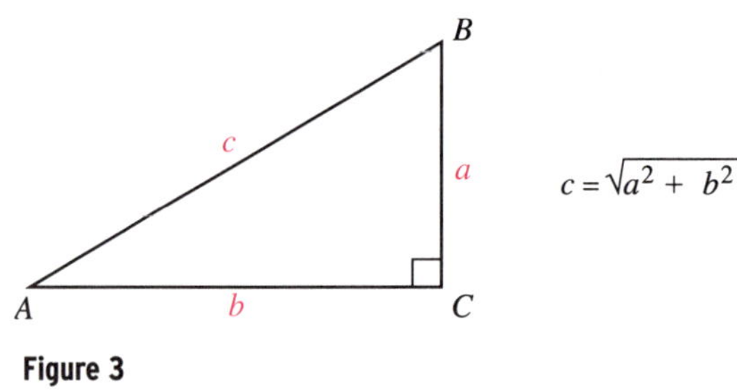

Figure 3

Answers

19. $\frac{1}{4}$ 20. $\frac{27}{8}$ 21. $x^{3/4}$ 22. $y^{1/2}$ 23. $z^{1/12}$ 24. $\frac{1}{x^2y^{28}}$

Constructing a Golden Rectangle from a Square of Side 2

Step 1: Draw a square with a side of length two. Connect the midpoint of side *CD* to corner *B*. (Note that we have labeled the midpoint of segment *CD* with the letter *O*.)

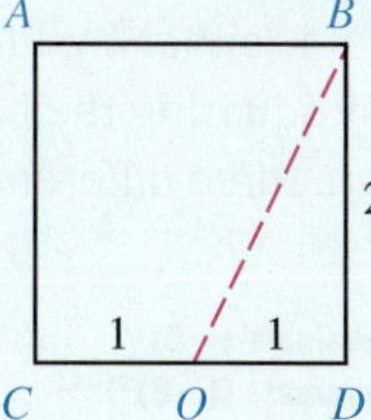

Step 2: Drop the diagonal from step 1 down so it aligns with side *CD*.

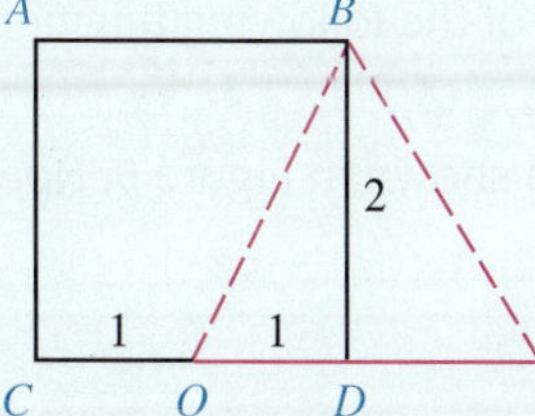

Step 3: Form rectangle *ACEF*. This is a golden rectangle.

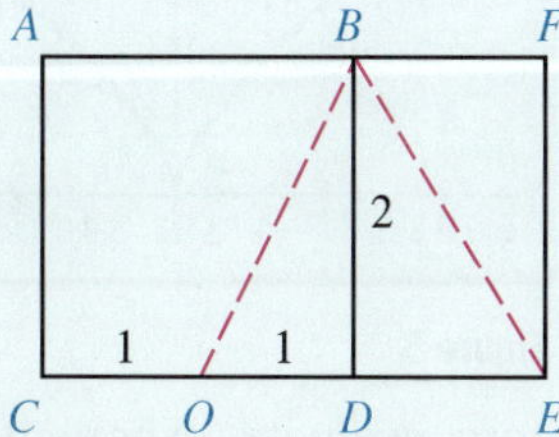

Note: In the introduction to this section we mentioned the golden rectangle. Its origins can be traced back more than 2,000 years to the Greek civilization that produced Pythagoras, Socrates, Plato, Aristotle, and Euclid. The most important mathematical work to come from that Greek civilization was Euclid's *Elements,* an elegantly written summary of all that was known about geometry at that time in history. Euclid's *Elements,* according to Howard Eves, an authority on the history of mathematics, exercised a greater influence on scientific thinking than any other work. Here is how we construct a golden rectangle from a square of side 2, using the same method that Euclid used in his *Elements.*

All golden rectangles are constructed from squares. Every golden rectangle, no matter how large or small it is, will have the same shape. To associate a number with the shape of the golden rectangle, we use the ratio of its length to its width. This ratio is called the *golden ratio.* To calculate the golden ratio, we must first find the length of the diagonal we used to construct the golden rectangle. Figure 4 shows the golden rectangle that we constructed from a square of side 2. The length of the diagonal *OB* is found by applying the Pythagorean theorem to triangle *OBD*.

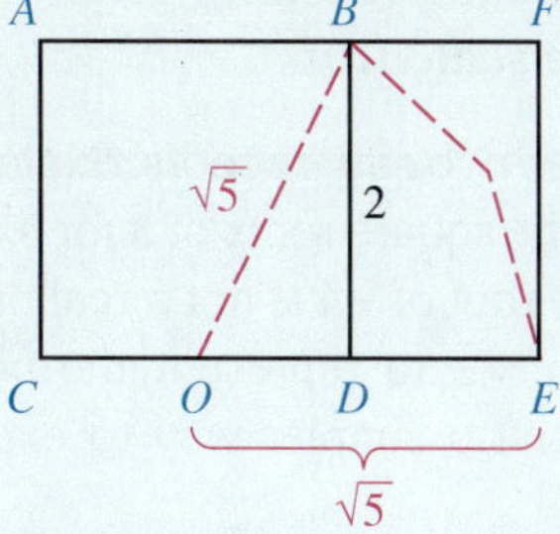

Figure 4

The length of segment *OE* is equal to the length of diagonal *OB*; both are $\sqrt{5}$. Since the distance from *C* to *O* is 1, the length *CE* of the golden rectangle is $1 + \sqrt{5}$. Now we can find the golden ratio:

$$\text{Golden ratio} = \frac{\text{length}}{\text{width}} = \frac{CE}{EF} = \frac{1 + \sqrt{5}}{2}$$

USING TECHNOLOGY

Graphing Calculators—A Word of Caution

Some graphing calculators give surprising results when evaluating expressions such as $(-8)^{2/3}$. As you know from reading this section, the expression $(-8)^{2/3}$ simplifies to 4, either by taking the cube root first and then squaring the result, or by squaring the base first and then taking the cube root of the result. Here are three different ways to evaluate this expression on your calculator:

1. (−8)^(2/3) **To evaluate $(-8)^{2/3}$**
2. ((−8)^2)^(1/3) **To evaluate $((-8)^2)^{1/3}$**
3. ((−8)^(1/3))^2 **To evaluate $((-8)^{1/3})^2$**

Note any difference in the results.

Next, graph each of the following functions, one at a time.

1. $Y_1 = X^{2/3}$ **2.** $Y_2 = (X^2)^{1/3}$ **3.** $Y_3 = (X^{1/3})^2$

The correct graph is shown in Figure 5. Note which of your graphs match the correct graph.

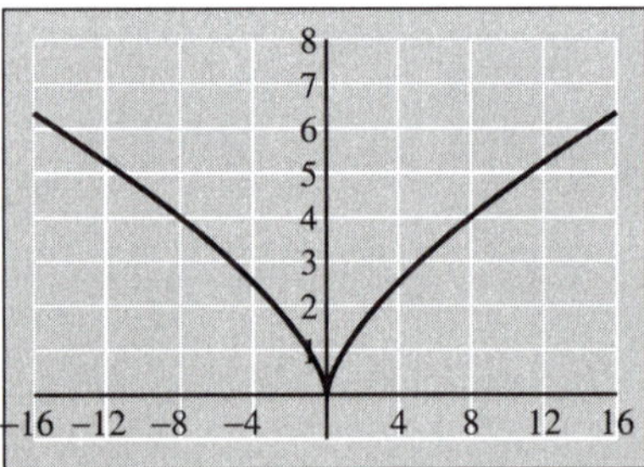

Figure 5

Different calculators evaluate exponential expressions in different ways. You should use the method (or methods) that gave you the correct graph.

Getting Ready for Class

After reading through the preceding section, respond in your own words and in complete sentences.

A. Every real number has two square roots. Explain the notation we use to tell them apart. Use the square roots of 3 for examples.

B. Explain why a square root of −4 is not a real number.

C. We use the notation $\sqrt{2}$ to represent the positive square root of 2. Explain why there isn't a simpler way to express the positive square root of 2.

D. For the expression $a^{m/n}$, explain the significance of the numerator m and the significance of the denominator n in the exponent.

PROBLEM SET 4.1

Use the definition of rational exponents to write each of the following with the appropriate root. Then simplify.

1. $36^{1/2}$
2. $49^{1/2}$
3. $-9^{1/2}$
4. $-16^{1/2}$
5. $8^{1/3}$
6. $-8^{1/3}$
7. $(-8)^{1/3}$
8. $-27^{1/3}$
9. $32^{1/5}$
10. $81^{1/4}$
11. $144^{1/2}$
12. $4^{1/2}$
13. $(-144)^{1/2}$
14. $(-4)^{1/2}$
15. $-49^{1/2}$
16. $(-49)^{1/2}$
17. $(-27)^{1/3}$
18. $-81^{1/4}$
19. $16^{1/4}$
20. $-16^{1/4}$
21. $(-16)^{1/4}$
22. $-(-16)^{1/4}$
23. $(0.04)^{1/2}$
24. $(0.81)^{1/2}$
25. $(0.008)^{1/3}$
26. $(0.125)^{1/3}$
27. $\left(\frac{81}{25}\right)^{1/2}$
28. $\left(\frac{9}{16}\right)^{1/2}$
29. $\left(\frac{64}{125}\right)^{1/3}$
30. $\left(\frac{8}{27}\right)^{1/3}$

Simplify each expression. Assume all variables represent nonnegative numbers.

31. $\sqrt{36a^8}$
32. $\sqrt{49a^{10}}$
33. $\sqrt[3]{27a^{12}}$
34. $\sqrt[3]{8a^{15}}$
35. $\sqrt[3]{x^3y^6}$
36. $\sqrt[3]{x^6y^3}$
37. $\sqrt[5]{32x^{10}y^5}$
38. $\sqrt[5]{32x^5y^{10}}$
39. $\sqrt[4]{16a^{12}b^{20}}$
40. $\sqrt[4]{81a^{24}b^8}$

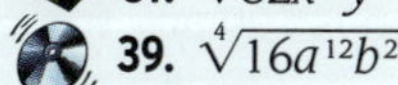

Use the rational exponent theorem to simplify each of the following as much as possible.

41. $27^{2/3}$
42. $8^{4/3}$
43. $25^{3/2}$
44. $9^{3/2}$
45. $16^{3/4}$
46. $81^{3/4}$

Simplify each expression. Remember, negative exponents give reciprocals.

47. $27^{-1/3}$
48. $9^{-1/2}$

49. $81^{-3/4}$
50. $4^{-3/2}$
51. $\left(\frac{25}{36}\right)^{-1/2}$
52. $\left(\frac{16}{49}\right)^{-1/2}$
53. $\left(\frac{81}{16}\right)^{-3/4}$
54. $\left(\frac{27}{8}\right)^{-2/3}$
55. $16^{1/2} + 27^{1/3}$
56. $25^{1/2} + 100^{1/2}$
57. $8^{-2/3} + 4^{-1/2}$
58. $49^{-1/2} + 25^{-1/2}$

Use the properties of exponents to simplify each of the following as much as possible. Assume all bases are positive.

59. $x^{3/5} \cdot x^{1/5}$
60. $x^{3/4} \cdot x^{5/4}$
61. $(a^{3/4})^{4/3}$
62. $(a^{2/3})^{3/4}$
63. $\frac{x^{1/5}}{x^{3/5}}$
64. $\frac{x^{2/7}}{x^{5/7}}$
65. $\frac{x^{5/6}}{x^{2/3}}$
66. $\frac{x^{7/8}}{x^{8/7}}$
67. $(x^{3/5}y^{5/6}z^{1/3})^{3/5}$
68. $(x^{3/4}y^{1/8}z^{5/6})^{4/5}$
69. $\frac{a^{3/4}b^2}{a^{7/8}b^{1/4}}$
70. $\frac{a^{1/3}b^4}{a^{3/5}b^{1/3}}$
71. $\frac{(y^{2/3})^{3/4}}{(y^{1/3})^{3/5}}$
72. $\frac{(y^{5/4})^{2/5}}{(y^{1/4})^{4/3}}$
73. $\left(\frac{a^{-1/4}}{b^{1/2}}\right)^8$
74. $\left(\frac{a^{-1/5}}{b^{1/3}}\right)^{15}$
75. $\frac{(r^{-2}s^{1/3})^6}{r^8s^{3/2}}$
76. $\frac{(r^{-5}s^{1/2})^4}{r^{12}s^{5/2}}$
77. $\frac{(25a^6b^4)^{1/2}}{(8a^{-9}b^3)^{-1/3}}$
78. $\frac{(27a^3b^6)^{1/3}}{(81a^8b^{-4})^{1/4}}$
79. Show that the expression $(a^{1/2} + b^{1/2})^2$ is not equal to $a + b$ by replacing a with 9 and b with 4 in both expressions and then simplifying each.
80. Show that the statement $(a^2 + b^2)^{1/2} = a + b$ is not, in general, true by replacing a with 3 and b with 4 and then simplifying both sides.
81. You may have noticed, if you have been using a calculator to find roots, that you can find the fourth root of a number by pressing the square root button twice. Written in symbols, this fact looks like this:

$$\sqrt{\sqrt{a}} = \sqrt[4]{a} \qquad (a \geq 0)$$

Show that this statement is true by rewriting each side with exponents instead of radical notation and then simplifying the left side.

82. Show that the statement is true by rewriting each side with exponents instead of radical notation and then simplifying the left side.

$$\sqrt[3]{\sqrt{a}} = \sqrt[6]{a} \qquad (a \geq 0)$$

Applying the Concepts

83. **Maximum Speed** The maximum speed (v) that an automobile can travel around a curve of radius r without skidding is given by the equation

$$v = \left(\frac{5r}{2}\right)^{1/2}$$

where v is in miles per hour and r is measured in feet. What is the maximum speed a car can travel around a curve with a radius of 250 feet without skidding?

84. Relativity The equation

$$L = \left(1 - \frac{v^2}{c^2}\right)^{1/2}$$

gives the relativistic length of a 1-foot ruler traveling with velocity v. Find L if

$$\frac{v}{c} = \frac{3}{5}$$

85. Golden Ratio The golden ratio is the ratio of the length to the width in any golden rectangle. The exact value of this number is $\frac{1+\sqrt{5}}{2}$. Use a calculator to find a decimal approximation to this number and round it to the nearest thousandth.

86. Golden Ratio The reciprocal of the golden ratio is $\frac{2}{1+\sqrt{5}}$. Find a decimal approximation to this number that is accurate to the nearest thousandth.

87. Sequences Find the next term in the following sequence. Then explain how this sequence is related to the Fibonacci sequence,

$$\frac{3}{2}, \frac{5}{3}, \frac{8}{5}, \ldots$$

88. Sequences Write the first 10 terms in the sequence shown in Problem 87. Then find a decimal approximation to each of the 10 terms, rounding each to the nearest thousandth.

89. Chemistry Figure 6 shows part of a model of a magnesium oxide (MgO) crystal. Each corner of the square is at the center of one oxygen ion (O^{2-}), and the center of the middle ion is at the center of the square. The radius for each oxygen ion is 126 picometers (pm), and the radius for each magnesium ion (Mg^{2+}) is 86 picometers.

(a) Find the length of the side of the square. Write your answer in picometers.

(b) Find the length of the diagonal of the square. Write your answer in picometers.

(c) If 1 meter is 10^{12} picometers, give the length of the diagonal of the square in meters.

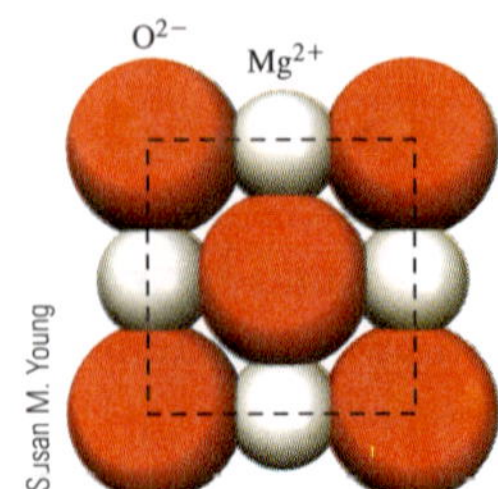

Figure 6

90. Chemistry Figure 7 shows part of a model of crystallized aluminum. Each corner of the square is at the center of one aluminum atom, and the center of the middle atom is at the center of the square. The radius for each aluminum atom is 143 picometers (pm).

(a) Find the length of the side of the square. Write your answer in picometers.

(b) If 1 meter is 10^{12} picometers, give the length of the side of the square in meters.

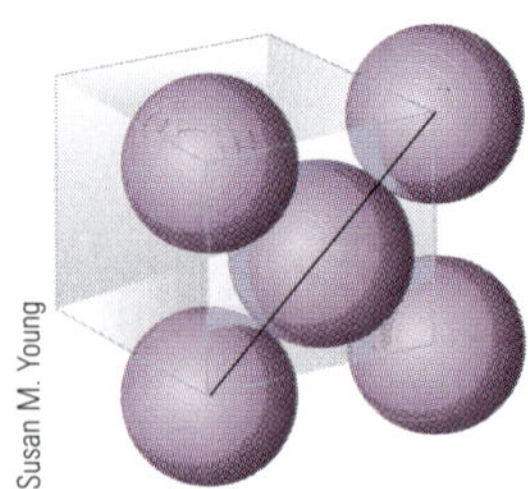

Figure 7

91. Geometry The length of each side of the cube shown in Figure 8 is 1 inch.

(a) Find the length of the diagonal CH.

(b) Find the length of the diagonal CF.

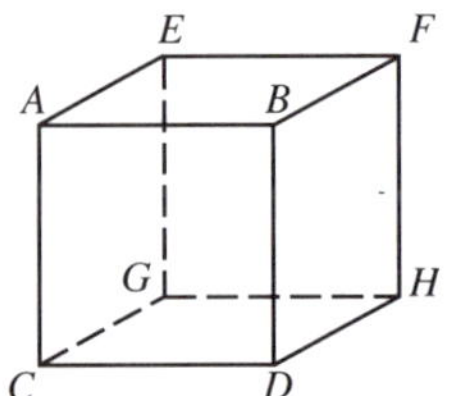

Figure 8

92. Chemistry Figure 9 shows part of a model of a crystal containing three atoms. The endpoints of the diagonal of the cube are at the centers of the two outside atoms, and the center of the cube is at the center of the middle atom. The radius of each atom is 100 picometers.

(a) Find the length of the diagonal of the cube.

(b) Find the length of the side of the cube.

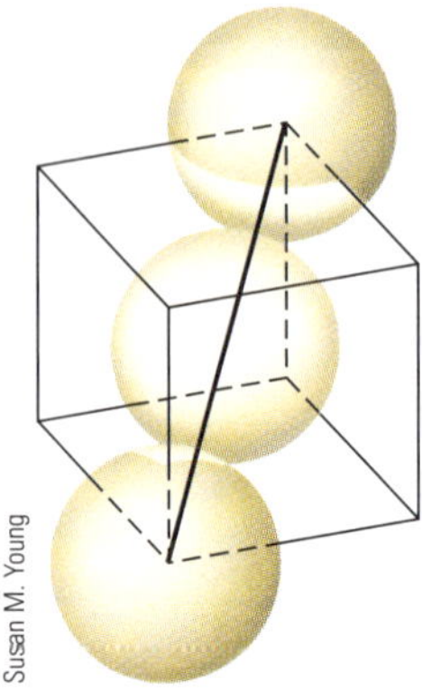

Figure 9

93. Comparing Graphs Identify the graph with the correct equation.

(a) $y = x$

(b) $y = x^2$

(c) $y = x^{2/3}$

(d) What are the two points of intersection of all three graphs?

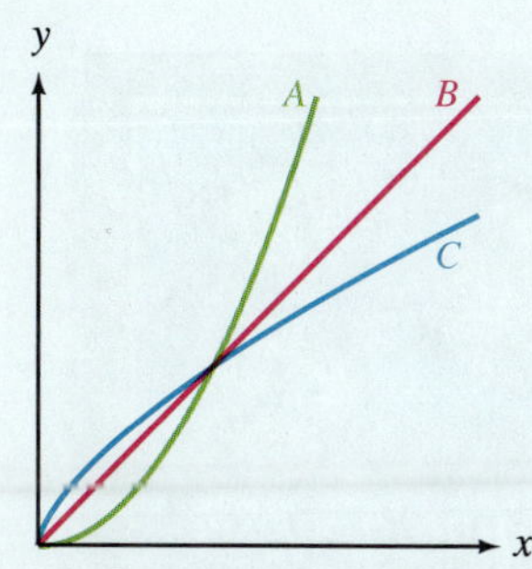

94. Falling Objects The time t in seconds it takes an object to fall d feet is given by the equation

$$t = \frac{1}{4}\sqrt{d}$$

(a) The Sears Tower in Chicago is 1,450 feet tall. How long would it take a penny to fall to the ground from the top of the Sears Tower?

(b) An object took 30 seconds to fall to the ground. From what distance must it have been dropped?

Review Problems

Multiply.

95. $x^2(x^4 - x)$

96. $5x^2(2x^3 - x)$

97. $(x - 3)(x + 5)$

98. $(x - 2)(x + 2)$

99. $(x^2 - 5)^2$

100. $(x^2 + 5)^2$

101. $(x - 3)(x^2 + 3x + 9)$

102. $(x + 3)(x^2 - 3x + 9)$

Extending the Concepts

Graph the equation $y = x^{3/4}$, and then use the Trace and Zoom feature to approximate each of the following to the nearest tenth.

103. $2^{3/4}$

104. $3^{3/4}$

105. $10^{3/4}$

106. $16^{3/4}$

Graph $y = x^{3/4}$ and $y = x^{4/3}$ on the same coordinate system, using a window with X from -4 to 4 and Y from -1 to 4. Use the graphs to answer the following questions.

107. Where do the two graphs intersect?

108. For what values of x is $x^{3/4} \geq x^{4/3}$?

Carbon-14 dating is used extensively in science to find the age of fossils. If at one time a fossilized substance contains an amount A of carbon-14, then t years later the amount of carbon-14 it contains is given by the expression

$$A \cdot 2^{-t/5{,}600}$$

Use this expression and a calculator to solve the following problems.

109. A fossilized substance contains 3 micrograms of carbon-14. How much carbon-14 will be left at the end of

(a) 5,000 years?

(b) 10,000 years?

(c) 56,000 years?

(d) 112,000 years?

110. A fossilized substance contains 5 micrograms of carbon-14. How much carbon-14 will be left at the end of

(a) 500 years?

(b) 5,000 years?

(c) 56,000 years?

(d) 112,000 years?

4.2 More Expressions Involving Rational Exponents

Suppose you purchased 10 silver proof coin sets in 1997 for \$21 each, for a total investment of \$210. Three years later, in 2000, you find that each set is worth \$30, which means that your 10 sets have a total value of \$300.

You can calculate the annual rate of return on this investment using a formula that involves rational exponents. The annual rate of return will tell you at what interest rate you would have to invest your original \$210 for it to be worth \$300 three years later. As you will see at the end of this section, the annual rate of return on this investment is 12.6%, which is a good return on your money.

In this section we will look at multiplication, division, factoring, and simplification of some expressions that resemble polynomials but contain rational exponents. The problems in this section will be of particular interest to you if you are planning to take either an engineering calculus class or a business calculus class. As was the case in the previous section, we will assume all variables represent nonnegative real numbers. That way, we will not have to worry about the possibility of introducing undefined terms—even roots of negative numbers—into any of our examples. Let's begin this section with a look at multiplication of expressions containing rational exponents.

Practice Problems

1. Multiply $x^{1/3}(x^{2/3} - x^{1/3})$.

EXAMPLE 1 Multiply $x^{2/3}(x^{4/3} - x^{1/3})$.

SOLUTION Applying the distributive property and then simplifying the resulting terms, we have:

$$\begin{aligned} x^{2/3}(x^{4/3} - x^{1/3}) &= x^{2/3}x^{4/3} - x^{2/3}x^{1/3} && \textbf{Distributive property} \\ &= x^{6/3} - x^{3/3} && \textbf{Add exponents} \\ &= x^2 - x && \textbf{Simplify} \end{aligned}$$

2. Multiply $(x^{3/5} + 2)(x^{3/5} - 7)$.

EXAMPLE 2 Multiply $(x^{2/3} - 3)(x^{2/3} + 5)$.

SOLUTION Applying the FOIL method, we multiply as if we were multiplying two binomials:

$$\begin{aligned} (x^{2/3} - 3)(x^{2/3} + 5) &= x^{2/3}x^{2/3} + 5x^{2/3} - 3x^{2/3} - 15 \\ &= x^{4/3} + 2x^{2/3} - 15 \end{aligned}$$

Answers

1. $x - x^{2/3}$ 2. $x^{6/5} - 5x^{3/5} - 14$

EXAMPLE 3 Multiply $(3a^{1/3} - 2b^{1/3})(4a^{1/3} - b^{1/3})$.

SOLUTION Again, we use the FOIL method to multiply:

$$(3a^{1/3} - 2b^{1/3})(4a^{1/3} - b^{1/3})$$
$$= 3a^{1/3}4a^{1/3} - 3a^{1/3}b^{1/3} - 2b^{1/3}4a^{1/3} + 2b^{1/3}b^{1/3}$$
$$= 12a^{2/3} - 11a^{1/3}b^{1/3} + 2b^{2/3}$$

EXAMPLE 4 Expand $(t^{1/2} - 5)^2$.

SOLUTION We can use the definition of exponents and the FOIL method:

$$(t^{1/2} - 5)^2 = (t^{1/2} - 5)(t^{1/2} - 5)$$
$$= t^{1/2}t^{1/2} - 5t^{1/2} - 5t^{1/2} + 25$$
$$= t - 10t^{1/2} + 25$$

We can obtain the same result by using the formula for the square of a binomial, $(a - b)^2 = a^2 - 2ab + b^2$.

$$(t^{1/2} - 5)^2 = (t^{1/2})^2 - 2t^{1/2} \cdot 5 + 5^2$$
$$= t - 10t^{1/2} + 25$$

EXAMPLE 5 Multiply $(x^{3/2} - 2^{3/2})(x^{3/2} + 2^{3/2})$.

SOLUTION This product has the form $(a - b)(a + b)$, which will result in the difference of two squares, $a^2 - b^2$:

$$(x^{3/2} - 2^{3/2})(x^{3/2} + 2^{3/2}) = (x^{3/2})^2 - (2^{3/2})^2$$
$$= x^3 - 2^3$$
$$= x^3 - 8$$

EXAMPLE 6 Multiply $(a^{1/3} - b^{1/3})(a^{2/3} + a^{1/3}b^{1/3} + b^{2/3})$.

SOLUTION We can find this product by multiplying in columns:

$$\begin{array}{rrrr}
a^{2/3} & + a^{1/3}b^{1/3} & + b^{2/3} & \\
 & a^{1/3} & - b^{1/3} & \\
\hline
a & + a^{2/3}b^{1/3} & + a^{1/3}b^{2/3} & \\
 & - a^{2/3}b^{1/3} & - a^{1/3}b^{2/3} & - b \\
\hline
a & & & - b
\end{array}$$

The product is $a - b$.

Our next example involves division with expressions that contain rational exponents. As you will see, this kind of division is very similar to division of a polynomial by a monomial.

EXAMPLE 7 Divide $\dfrac{15x^{2/3}y^{1/3} - 20x^{4/3}y^{2/3}}{5x^{1/3}y^{1/3}}$.

SOLUTION We can approach this problem in the same way we approached division by a monomial. We simply divide each term in the numerator by the term in the denominator:

$$\frac{15x^{2/3}y^{1/3} - 20x^{4/3}y^{2/3}}{5x^{1/3}y^{1/3}} = \frac{15x^{2/3}y^{1/3}}{5x^{1/3}y^{1/3}} - \frac{20x^{4/3}y^{2/3}}{5x^{1/3}y^{1/3}}$$
$$= 3x^{1/3} - 4xy^{1/3}$$

3. Multiply.
$(5a^{1/2} - 4b^{1/2})(3a^{1/2} - b^{1/2})$

4. Expand $(t^{1/3} + 3)^2$.

5. Multiply.
$(x^{5/2} - 3^{1/2})(x^{5/2} + 3^{1/2})$

6. Multiply.
$(a^{1/3} + b^{1/3})(a^{2/3} - a^{1/3}b^{1/3} + b^{2/3})$

7. Divide $\dfrac{36x^{3/4}y^{1/4} - 18x^{5/4}y^{1/4}}{6x^{1/4}y^{1/4}}$.

Answers

3. $15a - 17a^{1/2}b^{1/2} + 4b$
4. $t^{2/3} + 6t^{1/3} + 9$
5. $x^5 - 3$ 6. $a + b$
7. $6x^{1/2} - 3x$

The next three examples involve factoring. In the first example, we are told what to factor from each term of an expression.

8. Factor $2(x-3)^{1/2}$ from $8(x-3)^{3/2} - 6(x-3)^{1/2}$.

EXAMPLE 8 Factor $3(x-2)^{1/3}$ from $12(x-2)^{4/3} - 9(x-2)^{1/3}$, and then simplify, if possible.

SOLUTION This solution is similar to factoring out the greatest common factor:

$$12(x-2)^{4/3} - 9(x-2)^{1/3} = 3(x-2)^{1/3}[4(x-2)-3]$$
$$= 3(x-2)^{1/3}(4x-11)$$

Although an expression containing rational exponents is not a polynomial—remember, a polynomial must have exponents that are whole numbers—we are going to treat the expressions that follow as if they were polynomials.

9. Factor $x^{2/3} - 4x^{1/3} - 21$.

EXAMPLE 9 Factor $x^{2/3} - 3x^{1/3} - 10$ as if it were a trinomial.

SOLUTION We can think of $x^{2/3} - 3x^{1/3} - 10$ as if it is a trinomial in which the variable is $x^{1/3}$. To see this, replace $x^{1/3}$ with y to get

$$y^2 - 3y - 10$$

Since this trinomial in y factors as $(y-5)(y+2)$, we can factor our original expression similarly:

$$x^{2/3} - 3x^{1/3} - 10 = (x^{1/3}-5)(x^{1/3}+2)$$

Remember, with factoring, we can always multiply our factors to check that we have factored correctly.

10. Factor $6x^{2/3} + 19x^{1/3} + 10$.

EXAMPLE 10 Factor $6x^{2/5} + 11x^{1/5} - 10$ as if it were a trinomial.

SOLUTION We can think of the expression in question as a trinomial in $x^{1/5}$.

$$6x^{2/5} + 11x^{1/5} - 10 = (3x^{1/5}-2)(2x^{1/5}+5)$$

In our next example, we combine two expressions by applying the methods we used to add and subtract fractions or rational expressions.

11. Subtract.

$$(x^2-3)^{1/2} - \frac{x^2}{(x^2-3)^{1/2}}$$

EXAMPLE 11 Subtract $(x^2+4)^{1/2} - \dfrac{x^2}{(x^2+4)^{1/2}}$.

SOLUTION To combine these two expressions, we need to find a least common denominator, change to equivalent fractions, and subtract numerators. The least common denominator is $(x^2+4)^{1/2}$.

$$(x^2+4)^{1/2} - \frac{x^2}{(x^2+4)^{1/2}} = \frac{(x^2+4)^{1/2}}{1} \cdot \frac{\mathbf{(x^2+4)^{1/2}}}{\mathbf{(x^2+4)^{1/2}}} - \frac{x^2}{(x^2+4)^{1/2}}$$
$$= \frac{x^2+4-x^2}{(x^2+4)^{1/2}}$$
$$= \frac{4}{(x^2+4)^{1/2}}$$

Answers

8. $2(x-3)^{1/2}(4x-15)$
9. $(x^{1/3}-7)(x^{1/3}+3)$
10. $(3x^{1/3}+2)(2x^{1/3}+5)$
11. $\dfrac{-3}{(x^2-3)^{1/2}}$

EXAMPLE 12 If you purchase an investment for P dollars and t years later it is worth A dollars, then the annual rate of return r on that investment is given by the formula

$$r = \left(\frac{A}{P}\right)^{1/t} - 1$$

Find the annual rate of return on a coin collection that was purchased for \$210 and sold 3 years later for \$300.

SOLUTION Using $A = 300$, $P = 210$, and $t = 3$ in the formula, we have

$$r = \left(\frac{300}{210}\right)^{1/3} - 1$$

The easiest way to simplify this expression is with a calculator.

[(] 300 [÷] 210 [)] [^] [(] 1 [÷] 3 [)] [−] 1 [=]

Allowing three decimal places, the result is 0.126. The annual return on the coin collection is approximately 12.6%. To do as well with a savings account, we would have to invest the original \$210 in an account that paid 12.6%, compounded annually.

12. Find the annual rate of return on a coin collection that was purchased for \$600 and sold 3 years later for \$800.

Getting Ready for Class

After reading through the preceding section, respond in your own words and in complete sentences.

A. When multiplying expressions with fractional exponents, when do we add the fractional exponents?

B. Is it possible to multiply two expressions with fractional exponents and end up with an expression containing only integer exponents? Support your answer with examples.

C. Write an application modeled by the equation $r = \left(\frac{1{,}000}{600}\right)^{1/8} - 1$.

D. When can you use the FOIL method with expressions that contain rational exponents?

Answer

12. 10.1%

PROBLEM SET 4.2

Multiply. (Assume all variables in this problem set represent nonnegative real numbers.)

1. $x^{2/3}(x^{1/3} + x^{4/3})$
2. $x^{2/5}(x^{3/5} - x^{8/5})$

3. $a^{1/2}(a^{3/2} - a^{1/2})$
4. $a^{1/4}(a^{3/4} + a^{7/4})$
5. $2x^{1/3}(3x^{8/3} - 4x^{5/3} + 5x^{2/3})$
6. $5x^{1/2}(4x^{5/2} + 3x^{3/2} + 2x^{1/2})$
7. $4x^{1/2}y^{3/5}(3x^{3/2}y^{-3/5} - 9x^{-1/2}y^{7/5})$
8. $3x^{4/5}y^{1/3}(4x^{6/5}y^{-1/3} - 12x^{-4/5}y^{5/3})$
9. $(x^{2/3} - 4)(x^{2/3} + 2)$
10. $(x^{2/3} - 5)(x^{2/3} + 2)$
11. $(a^{1/2} - 3)(a^{1/2} - 7)$
12. $(a^{1/2} - 6)(a^{1/2} - 2)$
13. $(4y^{1/3} - 3)(5y^{1/3} + 2)$
14. $(5y^{1/3} - 2)(4y^{1/3} + 3)$
15. $(5x^{2/3} + 3y^{1/2})(2x^{2/3} + 3y^{1/2})$
16. $(4x^{2/3} - 2y^{1/2})(5x^{2/3} - 3y^{1/2})$
17. $(t^{1/2} + 5)^2$
18. $(t^{1/2} - 3)^2$

19. $(x^{3/2} + 4)^2$
20. $(x^{3/2} - 6)^2$
21. $(a^{1/2} - b^{1/2})^2$
22. $(a^{1/2} + b^{1/2})^2$
23. $(2x^{1/2} - 3y^{1/2})^2$
24. $(5x^{1/2} + 4y^{1/2})^2$
25. $(a^{1/2} - 3^{1/2})(a^{1/2} + 3^{1/2})$
26. $(a^{1/2} - 5^{1/2})(a^{1/2} + 5^{1/2})$
27. $(x^{3/2} + y^{3/2})(x^{3/2} - y^{3/2})$
28. $(x^{5/2} + y^{5/2})(x^{5/2} - y^{5/2})$
29. $(t^{1/2} - 2^{3/2})(t^{1/2} + 2^{3/2})$
30. $(t^{1/2} - 5^{3/2})(t^{1/2} + 5^{3/2})$
31. $(2x^{3/2} + 3^{1/2})(2x^{3/2} - 3^{1/2})$
32. $(3x^{1/2} + 2^{3/2})(3x^{1/2} - 2^{3/2})$
33. $(x^{1/3} + y^{1/3})(x^{2/3} - x^{1/3}y^{1/3} + y^{2/3})$
34. $(x^{1/3} - y^{1/3})(x^{2/3} + x^{1/3}y^{1/3} + y^{2/3})$
35. $(a^{1/3} - 2)(a^{2/3} + 2a^{1/3} + 4)$
36. $(a^{1/3} + 3)(a^{2/3} - 3a^{1/3} + 9)$
37. $(2x^{1/3} + 1)(4x^{2/3} - 2x^{1/3} + 1)$
38. $(3x^{1/3} - 1)(9x^{2/3} + 3x^{1/3} + 1)$
39. $(t^{1/4} - 1)(t^{1/4} + 1)(t^{1/2} + 1)$
40. $(t^{1/4} - 2)(t^{1/4} + 2)(t^{1/2} + 4)$

Divide.

41. $\dfrac{18x^{3/4} + 27x^{1/4}}{9x^{1/4}}$
42. $\dfrac{25x^{1/4} + 30x^{3/4}}{5x^{1/4}}$
43. $\dfrac{12x^{2/3}y^{1/3} - 16x^{1/3}y^{2/3}}{4x^{1/3}y^{1/3}}$
44. $\dfrac{12x^{4/3}y^{1/3} - 18x^{1/3}y^{4/3}}{6x^{1/3}y^{1/3}}$
45. $\dfrac{21a^{7/5}b^{3/5} - 14a^{2/5}b^{8/5}}{7a^{2/5}b^{3/5}}$
46. $\dfrac{24a^{9/5}b^{3/5} - 16a^{4/5}b^{8/5}}{8a^{4/5}b^{3/5}}$
47. **(a)** Factor $3a^3$ from $12a^4 - 9a^3$.
    (b) Factor $3(y + 2)^2$ from $12(y + 2)^3 - 9(y + 2)^2$.
    (c) Factor $3(x - 2)^{1/2}$ from $12(x - 2)^{3/2} - 9(x - 2)^{1/2}$.
48. **(a)** Factor $4t^4$ from $4t^5 + 8t^4$.
    (b) Factor $4(a + 2)^3$ from $4(a + 2)^4 + 8(a + 2)^3$.
    (c) Factor $4(x + 1)^{1/3}$ from $4(x + 1)^{4/3} + 8(x + 1)^{1/3}$.
49. **(a)** Factor $5y^7$ from $5y^8 - 15y^7$.
    (b) Factor $5(n - 4)^5$ from $5(n - 4)^6 - 15(n - 4)^5$.
    (c) Factor $5(x - 3)^{7/5}$ from $5(x - 3)^{12/5} - 15(x - 3)^{7/5}$.
50. **(a)** Factor $6b^8$ from $6b^9 - 12b^8$.
    (b) Factor $6(y - 1)^7$ from $6(y - 1)^8 - 12(y - 1)^7$.
    (c) Factor $6(x + 3)^{8/7}$ from $6(x + 3)^{15/7} - 12(x + 3)^{8/7}$.
51. **(a)** Factor $9ab^3 + 6b^2$.
    (b) Factor $9m(n + 4)^2 + 6(n + 4)$.
    (c) Factor $9x(x + 1)^{3/2} + 6(x + 1)^{1/2}$.
52. **(a)** Factor $4m^2n^2 + 8mn^3$.
    (b) Factor $4a^2(b - 1) + 8a(b - 1)^2$.
    (c) Factor $4x^2(x + 1)^{1/2} + 8x(x + 1)^{3/2}$.

Factor each of the following as if it were a trinomial.

53. $x^{2/3} - 5x^{1/3} + 6$
54. $x^{2/3} - x^{1/3} - 6$
55. $a^{2/5} - 2a^{1/5} - 8$
56. $a^{2/5} + 2a^{1/5} - 8$
57. $2y^{2/3} - 5y^{1/3} - 3$
58. $3y^{2/3} + 5y^{1/3} - 2$
59. $9t^{2/5} - 25$
60. $16t^{2/5} - 49$
61. $4x^{2/7} + 20x^{1/7} + 25$
62. $25x^{2/7} - 20x^{1/7} + 4$

Simplify each of the following to a single fraction.

63. $\dfrac{3}{x^{1/2}} + x^{1/2}$
64. $\dfrac{2}{x^{1/2}} - x^{1/2}$
65. $x^{2/3} + \dfrac{5}{x^{1/3}}$
66. $x^{3/4} - \dfrac{7}{x^{1/4}}$
67. $\dfrac{3x^2}{(x^3 + 1)^{1/2}} + (x^3 + 1)^{1/2}$
68. $\dfrac{x^3}{(x^2 - 1)^{1/2}} + 2x(x^2 - 1)^{1/2}$

69. $\dfrac{x^2}{(x^2+4)^{1/2}} - (x^2+4)^{1/2}$

70. $\dfrac{x^5}{(x^2-2)^{1/2}} + 4x^3(x^2-2)^{1/2}$

Use a calculator to find approximations to each of the following. Round your answers for Problems 75 and 76 to three places past the decimal point.

71. $16^{0.25}$

72. $81^{0.25}$

73. $9^{1.5}$

74. $32^{0.4}$

75. $\left(\frac{1}{2}\right)^{1/5}$

76. $\left(\frac{1}{2}\right)^{1/10}$

Applying the Concepts

77. **Investing** A coin collection is purchased as an investment for $500 and sold 4 years later for $900. Find the annual rate of return on the investment.

78. **Investing** An investor buys stock in a company for $800. Five years later, the same stock is worth $1,600. Find the annual rate of return on the stocks.

79. **Investing** Find the annual rate of return on a home that is purchased for $60,000 and is sold 5 years later for $80,000.

80. **Investing** Find the annual rate of return on a home that is purchased for $75,000 and is sold 10 years later for $150,000.

81. **Falling Object** If an object is dropped, the time, in seconds, that it takes to fall d feet is given by the equation $t = \left(\frac{d}{16}\right)^{1/2}$. How long does it take for a ball that is dropped off a 200-foot cliff to hit the bottom?

82. **Stopping Distance** Police can measure the skid marks left by a car and calculate the speed that the car was traveling before it slammed on the brakes. If it takes d feet to stop, then the initial speed of the car, in miles per hour, is given approximately by $v = \left(\frac{d}{k}\right)^{1/2}$, where k is a constant that depends on the particular car. According to *Consumer Reports,* the 1998 Mercury Mountaineer takes 147 feet to stop at 60 mph on dry pavement. This allows us to calculate the value of k for this sport utility vehicle as $k \approx 0.0408$. Suppose it took a 1998 Mercury Mountaineer 240 feet to stop on dry pavement. How fast was it initially going?

83. **Kepler's Law** Kepler's third law of planetary motion says the period T of the orbit of a planet or satellite is related to the radius r of its orbit by the equation $T = kr^{3/2}$, where k is a constant that depends on the planetary system. The earth is about 93 million miles from the sun and orbits the sun in 1 year. Therefore, if we measure r in millions of miles and T in years, we can calculate the value of the constant k as $k \approx 0.001115$. Mars is about 141.4 million miles from the sun. How long does it take for Mars to complete one orbit?

84. **Population Growth** If a population of P_0 grows to a size of P in t years, the annual percentage growth rate is given by the equation $r = \left(\frac{P}{P_0}\right)^{1/t} - 1$. In 1970, the population of Algeria was 13.9 million. In 1980, the population had grown to 18.8 million people. What is the annual growth rate of the population? Express your answer as a percentage rounded to two decimal places.

Review Problems

Reduce to lowest terms.

85. $\dfrac{x^2-9}{x^4-81}$

86. $\dfrac{6-a-a^2}{3-2a-a^2}$

Divide.

87. $\dfrac{15x^2y - 20x^4y^2}{5xy}$

88. $\dfrac{12x^3y^2 - 24x^2y^3}{6xy}$

Divide using long division.

89. $\dfrac{10x^2+7x-12}{2x+3}$

90. $\dfrac{6x^2-x-35}{2x-5}$

91. $\dfrac{x^3-125}{x-5}$

92. $\dfrac{x^3+64}{x+4}$

Extending the Concepts

93. Choose values for x and y to show the following: $(x^{1/2}+y^{1/2})^2 \neq x+y$.

94. Choose values for x and y to show the following: $(x^2+y^2)^{1/2} \neq x+y$.

Write the following with positive exponents and simplify.

95. $\dfrac{x+y}{x^{-1}+y^{-1}}$

96. $\dfrac{x^2-y^2}{x^{-1}-y^{-1}}$

Write the following with rational exponents and simplify.

97. $\sqrt[5]{x^3} \cdot \sqrt[4]{x}$

98. $\sqrt[6]{x^5} \cdot \sqrt[3]{x^2}$

99. $\sqrt[4]{\sqrt[3]{m}}$

100. $\sqrt[3]{\sqrt{k}}$

4.3 Simplified Form for Radicals

Earlier in this chapter, we showed how the Pythagorean theorem can be used to construct a golden rectangle. In a similar manner, the Pythagorean theorem can be used to construct the attractive spiral shown here.

The Spiral of Roots

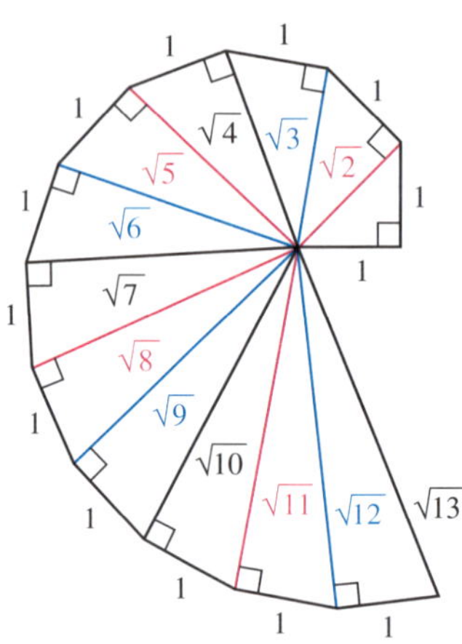

This spiral is called the Spiral of Roots because each of the diagonals is the positive square root of one of the positive integers. At the end of this section, we will use the Pythagorean theorem and some of the material in this section to construct this spiral.

In this section we will use radical notation instead of rational exponents. We will begin by stating two properties of radicals. Following this, we will give a definition for simplified form for radical expressions. The examples in this section show how we use the properties of radicals to write radical expressions in simplified form.

There are two properties of radicals. For these two properties, we will assume a and b are nonnegative real numbers whenever n is an even number.

Property 1 for Radicals

$$\sqrt[n]{ab} = \sqrt[n]{a}\sqrt[n]{b}$$

In words: The nth root of a product is the product of the nth roots.

Proof of Property 1

$$\sqrt[n]{ab} = (ab)^{1/n} \quad \textbf{Definition of fractional exponents}$$

$$= a^{1/n}b^{1/n} \quad \textbf{Exponents distribute over products}$$

$$= \sqrt[n]{a}\sqrt[n]{b} \quad \textbf{Definition of fractional exponents}$$

Note: There is no property for radicals that says the nth root of a sum is the sum of the nth roots; that is,

$$\sqrt[n]{a+b} \neq \sqrt[n]{a} + \sqrt[n]{b}$$

Property 2 for Radicals

$$\sqrt[n]{\frac{a}{b}} = \frac{\sqrt[n]{a}}{\sqrt[n]{b}} \qquad (b \neq 0)$$

In words: The nth root of a quotient is the quotient of the nth roots.

The proof of property 2 is similar to the proof of property 1.

The two properties of radicals allow us to change the form of and simplify radical expressions without changing their value.

Simplified Form for Radical Expressions

A radical expression is in *simplified form* if

1. None of the factors of the radicand (the quantity under the radical sign) can be written as powers greater than or equal to the index—that is, no perfect squares can be factors of the quantity under a square root sign, no perfect cubes can be factors of what is under a cube root sign, and so forth;
2. There are no fractions under the radical sign; and
3. There are no radicals in the denominator.

Note: Writing a radical expression in simplified form does not always result in a simpler-looking expression. Simplified form for radicals is a way of writing radicals so they are easiest to work with.

Satisfying the first condition for simplified form actually amounts to taking as much out from under the radical sign as possible. The following examples illustrate the first condition for simplified form.

EXAMPLE 1 Write $\sqrt{50}$ in simplified form.

SOLUTION The largest perfect square that divides 50 is 25. We write 50 as $25 \cdot 2$ and apply property 1 for radicals:

$$\sqrt{50} = \sqrt{25 \cdot 2} \qquad \mathbf{50 = 25 \cdot 2}$$

$$= \sqrt{25}\,\sqrt{2} \qquad \textbf{Property 1}$$

$$= 5\sqrt{2} \qquad \mathbf{\sqrt{25} = 5}$$

We have taken as much as possible out from under the radical sign—in this case, factoring 25 from 50 and then writing $\sqrt{25}$ as 5.

EXAMPLE 2 Write in simplified form: $\sqrt{48x^4y^3}$, where $x, y \geq 0$.

SOLUTION The largest perfect square that is a factor of the radicand is $16x^4y^2$. Applying property 1 again, we have

$$\sqrt{48x^4y^3} = \sqrt{16x^4y^2 \cdot 3y}$$

$$= \sqrt{16x^4y^2}\,\sqrt{3y}$$

$$= 4x^2y\sqrt{3y}$$

EXAMPLE 3 Write $\sqrt[3]{40a^5b^4}$ in simplified form.

SOLUTION We now want to factor the largest perfect cube from the radicand. We write $40a^5b^4$ as $8a^3b^3 \cdot 5a^2b$ and proceed as we did in Examples 1 and 2.

$$\sqrt[3]{40a^5b^4} = \sqrt[3]{8a^3b^3 \cdot 5a^2b}$$

$$= \sqrt[3]{8a^3b^3}\,\sqrt[3]{5a^2b}$$

$$= 2ab\sqrt[3]{5a^2b}$$

Here are some further examples concerning the first condition for simplified form.

Practice Problems

1. Write $\sqrt{18}$ in simplified form.
2. Write $\sqrt{50x^2y^3}$ in simplified form. Assume $x, y \geq 0$.
3. Write $\sqrt[3]{54a^4b^3}$ in simplified form.

Answers

1. $3\sqrt{2}$ 2. $5xy\sqrt{2y}$
3. $3ab\sqrt[3]{2a}$

Write each expression in simplified form.

4. $\sqrt{75x^5y^8}$

5. $\sqrt[4]{48a^8b^5c^4}$

EXAMPLES Write each expression in simplified form.

4. $\sqrt{12x^7y^6} = \sqrt{4x^6y^6 \cdot 3x}$

$= \sqrt{4x^6y^6}\sqrt{3x}$

$= 2x^3y^3\sqrt{3x}$

5. $\sqrt[3]{54a^6b^2c^4} = \sqrt[3]{27a^6c^3 \cdot 2b^2c}$

$= \sqrt[3]{27a^6c^3}\sqrt[3]{2b^2c}$

$= 3a^2c\sqrt[3]{2b^2c}$

The second property of radicals is used to simplify a radical that contains a fraction.

6. Simplify $\sqrt{\frac{5}{9}}$.

EXAMPLE 6 Simplify $\sqrt{\frac{3}{4}}$.

SOLUTION Applying property 2 for radicals, we have

$$\sqrt{\frac{3}{4}} = \frac{\sqrt{3}}{\sqrt{4}} \qquad \textbf{Property 2}$$

$$= \frac{\sqrt{3}}{2} \qquad \sqrt{4} = 2$$

The last expression is in simplified form because it satisfies all three conditions for simplified form.

7. Write $\sqrt{\frac{2}{3}}$ in simplified form.

EXAMPLE 7 Write $\sqrt{\frac{5}{6}}$ in simplified form.

SOLUTION Proceeding as in Example 6, we have

$$\sqrt{\frac{5}{6}} = \frac{\sqrt{5}}{\sqrt{6}}$$

The resulting expression satisfies the second condition for simplified form since neither radical contains a fraction. It does, however, violate condition 3 since it has a radical in the denominator. Getting rid of the radical in the denominator is called *rationalizing the denominator* and is accomplished, in this case, by multiplying the numerator and denominator by $\sqrt{6}$:

Note: The idea behind rationalizing the denominator is to produce a perfect square under the square root sign in the denominator. This is accomplished by multiplying both the numerator and denominator by the appropriate radical.

$$\frac{\sqrt{5}}{\sqrt{6}} = \frac{\sqrt{5}}{\sqrt{6}} \cdot \frac{\sqrt{\mathbf{6}}}{\sqrt{\mathbf{6}}}$$

$$= \frac{\sqrt{30}}{\sqrt{6^2}}$$

$$= \frac{\sqrt{30}}{6}$$

8. Rationalize the denominator.

$$\frac{5}{\sqrt{2}}$$

EXAMPLES Rationalize the denominator.

8. $\frac{4}{\sqrt{3}} = \frac{4}{\sqrt{3}} \cdot \frac{\sqrt{\mathbf{3}}}{\sqrt{\mathbf{3}}}$

$= \frac{4\sqrt{3}}{\sqrt{3^2}}$

$= \frac{4\sqrt{3}}{3}$

Answers

4. $5x^2y^4\sqrt{3x}$ 5. $2a^2bc\sqrt[4]{3b}$

6. $\frac{\sqrt{5}}{3}$ 7. $\frac{\sqrt{6}}{3}$

9. $\frac{2\sqrt{3x}}{\sqrt{5y}} = \frac{2\sqrt{3x}}{\sqrt{5y}} \cdot \frac{\sqrt{5y}}{\sqrt{5y}}$

$= \frac{2\sqrt{15xy}}{\sqrt{(5y)^2}}$

$= \frac{2\sqrt{15xy}}{5y}$

When the denominator involves a cube root, we must multiply by a radical that will produce a perfect cube under the cube root sign in the denominator, as our next example illustrates.

EXAMPLE 10 Rationalize the denominator in $\frac{7}{\sqrt[3]{4}}$.

SOLUTION Since $4 = 2^2$, we can multiply both numerator and denominator by $\sqrt[3]{2}$ and obtain $\sqrt[3]{2^3}$ in the denominator.

$$\frac{7}{\sqrt[3]{4}} = \frac{7}{\sqrt[3]{2^2}}$$

$$= \frac{7}{\sqrt[3]{2^2}} \cdot \frac{\sqrt[3]{2}}{\sqrt[3]{2}}$$

$$= \frac{7\sqrt[3]{2}}{\sqrt[3]{2^3}}$$

$$= \frac{7\sqrt[3]{2}}{2}$$

EXAMPLE 11 Simplify $\sqrt{\frac{12x^5y^3}{5z}}$.

SOLUTION We use property 2 to write the numerator and denominator as two separate radicals:

$$\sqrt{\frac{12x^5y^3}{5z}} = \frac{\sqrt{12x^5y^3}}{\sqrt{5z}}$$

Simplifying the numerator, we have

$$\frac{\sqrt{12x^5y^3}}{\sqrt{5z}} = \frac{\sqrt{4x^4y^2}\sqrt{3xy}}{\sqrt{5z}}$$

$$= \frac{2x^2y\sqrt{3xy}}{\sqrt{5z}}$$

To rationalize the denominator, we multiply the numerator and denominator by $\sqrt{5z}$:

$$\frac{2x^2y\sqrt{3xy}}{\sqrt{5z}} \cdot \frac{\sqrt{5z}}{\sqrt{5z}} = \frac{2x^2y\sqrt{15xyz}}{\sqrt{(5z)^2}}$$

$$= \frac{2x^2y\sqrt{15xyz}}{5z}$$

The Square Root of a Perfect Square

So far in this chapter we have assumed that all our variables are nonnegative when they appear under a square root symbol. There are times, however, when this is not the case.

9. Rationalize the denominator.

$\frac{3\sqrt{5x}}{\sqrt{2y}}$

10. Rationalize the denominator.

$\frac{5}{\sqrt[3]{9}}$

11. Simplify $\sqrt{\frac{48x^3y^4}{7z}}$.

Answers

8. $\frac{5\sqrt{2}}{2}$ 9. $\frac{3\sqrt{10xy}}{2y}$

10. $\frac{5\sqrt[3]{3}}{3}$ 11. $\frac{4xy^2\sqrt{21xz}}{7z}$

Consider the following two statements:

$$\sqrt{3^2} = \sqrt{9} = 3 \quad \text{and} \quad \sqrt{(-3)^2} = \sqrt{9} = 3$$

Whether we operate on 3 or −3, the result is the same: Both expressions simplify to 3. The other operation we have worked with in the past that produces the same result is absolute value; that is,

$$|3| = 3 \quad \text{and} \quad |-3| = 3$$

This leads us to the next property of radicals.

Property 3 for Radicals

If a is a real number, then $\sqrt{a^2} = |a|$.

The result of this discussion and property 3 is simply this:

If we know a is positive, then $\sqrt{a^2} = a$.

If we know a is negative, then $\sqrt{a^2} = |a|$.

If we don't know if a is positive or negative, then $\sqrt{a^2} = |a|$.

Simplify each expression. Do not assume the variables represent nonnegative numbers.

12. $\sqrt{16x^2}$
13. $\sqrt{25x^3}$
14. $\sqrt{x^2 + 10x + 25}$
15. $\sqrt{2x^3 + 7x^2}$

EXAMPLES Simplify each expression. Do *not* assume the variables represent positive numbers.

12. $\sqrt{9x^2} = 3|x|$
13. $\sqrt{x^3} = |x|\sqrt{x}$
14. $\sqrt{x^2 - 6x + 9} = \sqrt{(x-3)^2} = |x - 3|$
15. $\sqrt{x^3 - 5x^2} = \sqrt{x^2(x-5)} = |x|\sqrt{x-5}$

As you can see, we must use absolute value symbols when we take a square root of a perfect square, unless we know the base of the perfect square is a positive number. The same idea holds for higher even roots, but not for odd roots. With odd roots, no absolute value symbols are necessary.

Simplify each expression.

16. $\sqrt[3]{(-3)^3}$
17. $\sqrt[3]{(-1)^3}$

EXAMPLES Simplify each expression.

16. $\sqrt[3]{(-2)^3} = \sqrt[3]{-8} = -2$
17. $\sqrt[3]{(-5)^3} = \sqrt[3]{-125} = -5$

We can extend this discussion to all roots as follows:

Extending Property 3 for Radicals

If a is a real number, then

$$\sqrt[n]{a^n} = |a| \quad \text{if} \quad n \text{ is even}$$

$$\sqrt[n]{a^n} = a \quad \text{if} \quad n \text{ is odd}$$

Answers

12. $4|x|$ 13. $5|x|\sqrt{x}$
14. $|x+5|$ 15. $|x|\sqrt{2x+7}$
16. -3 17. -1

The Spiral of Roots

To visualize the square roots of the positive integers, we can construct the spiral of roots that we mentioned in the introduction to this section. To begin, we draw two line segments, each of length 1, at right angles to each other. Then we use the Pythagorean theorem to find the length of the diagonal. Figure 1 illustrates this procedure.

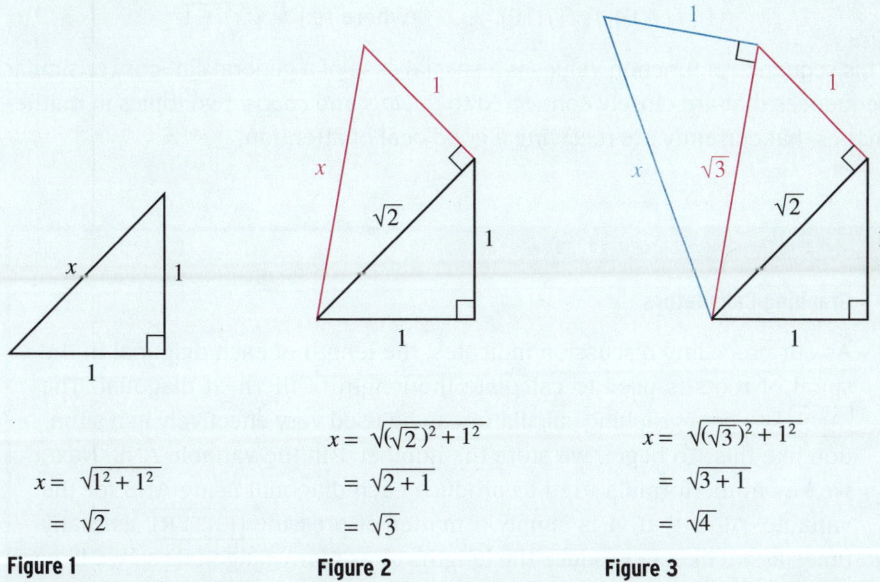

$$x = \sqrt{1^2 + 1^2} = \sqrt{2}$$

Figure 1

$$x = \sqrt{(\sqrt{2})^2 + 1^2} = \sqrt{2 + 1} = \sqrt{3}$$

Figure 2

$$x = \sqrt{(\sqrt{3})^2 + 1^2} = \sqrt{3 + 1} = \sqrt{4}$$

Figure 3

Next, we construct a second triangle by connecting a line segment of length 1 to the end of the first diagonal so that the angle formed is a right angle. We find the length of the second diagonal using the Pythagorean theorem. Figure 2 illustrates this procedure. Continuing to draw new triangles by connecting line segments of length 1 to the end of each new diagonal, so that the angle formed is a right angle, the spiral of roots begins to appear (Figure 3).

The Spiral of Roots and Function Notation

Looking over the diagrams and calculations in the preceding discussion, we see that each diagonal in the spiral of roots is found by using the length of the previous diagonal.

First diagonal: $\sqrt{1^2 + 1^2} = \sqrt{2}$

Second diagonal: $\sqrt{(\sqrt{2})^2 + 1^2} = \sqrt{3}$

Third diagonal: $\sqrt{(\sqrt{3})^2 + 1^2} = \sqrt{4}$

Fourth diagonal: $\sqrt{(\sqrt{4})^2 + 1^2} = \sqrt{5}$

A process like this one, in which the answer to one calculation is used to find the answer to the next calculation, is called a *recursive* process. In this particular case, we can use function notation to model the process. If we let x represent the length of any diagonal, then the length of the next diagonal is given by

$$f(x) = \sqrt{x^2 + 1}$$

To begin the process of finding the diagonals, we let $x = 1$:

$$f(1) = \sqrt{1^2 + 1} = \sqrt{2}$$

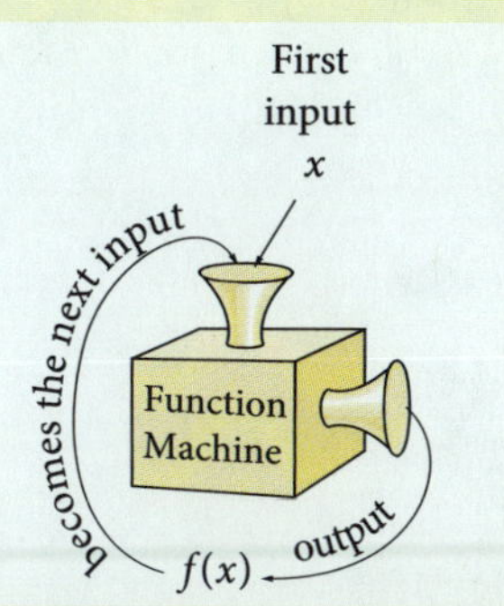

To find the next diagonal, we substitute $\sqrt{2}$ for x to obtain

$$f[f(1)] = f(\sqrt{2}) = \sqrt{(\sqrt{2})^2 + 1} = \sqrt{3}$$

$$f(f[f(1)]) = f(\sqrt{3}) = \sqrt{(\sqrt{3})^2 + 1} = \sqrt{4}$$

We can describe this process of finding the diagonals of the spiral of roots concisely this way:

$$f(1), f[f(1)], f(f[f(1)]), \ldots \qquad \text{where } f(x) = \sqrt{x^2 + 1}$$

This sequence of function values is a special case of a general category of similar sequences that are closely connected to *fractals* and *chaos,* two topics in mathematics that currently are receiving a good deal of attention.

USING TECHNOLOGY

Graphing Calculators

As our preceding discussion indicates, the length of each diagonal in the spiral of roots is used to calculate the length of the next diagonal. The [ANS] key on a graphing calculator can be used very effectively in a situation like this. To begin, we store the number 1 in the variable ANS. Next, we key in the formula used to produce each diagonal using ANS for the variable. After that, it is simply a matter of pressing [ENTER], as many times as we like, to produce the lengths of as many diagonals as we like. Here is a summary of what we do:

Enter This	*Display Shows*
1 [ENTER]	1.000
$\sqrt{\ }$ (ANS^2 + 1) [ENTER]	1.414
[ENTER]	1.732
[ENTER]	2.000
[ENTER]	2.236

If you continue to press the [ENTER] key, you will produce decimal approximations for as many of the diagonals in the spiral of roots as you like.

Getting Ready for Class

After reading through the preceding section, respond in your own words and in complete sentences.

A. Explain why this statement is false: "The square root of a sum is the sum of the square roots."

B. What is simplified form for an expression that contains a square root?

C. Why is it not necessarily true that $\sqrt{a^2} = a$?

D. What does it mean to rationalize the denominator in an expression?

PROBLEM SET 4.3

Use Property 1 for radicals to write each of the following expressions in simplified form. (Assume all variables are nonnegative through Problem 70.)

1. $\sqrt{8}$
2. $\sqrt{32}$
3. $\sqrt{98}$
4. $\sqrt{75}$
5. $\sqrt{288}$
6. $\sqrt{128}$
7. $\sqrt{80}$
8. $\sqrt{200}$
9. $\sqrt{48}$
10. $\sqrt{27}$
11. $\sqrt{675}$
12. $\sqrt{972}$
13. $\sqrt[3]{54}$
14. $\sqrt[3]{24}$
15. $\sqrt[3]{128}$
16. $\sqrt[3]{162}$
17. $\sqrt[3]{432}$
18. $\sqrt[3]{1,536}$
19. $\sqrt[5]{64}$
20. $\sqrt[4]{48}$
21. $\sqrt{18x^3}$
22. $\sqrt{27x^5}$
23. $\sqrt[4]{32y^7}$
24. $\sqrt[5]{32y^7}$
25. $\sqrt[3]{40x^4y^7}$
26. $\sqrt[3]{128x^6y^2}$
27. $\sqrt{48a^2b^3c^4}$
28. $\sqrt{72a^4b^3c^2}$
29. $\sqrt[3]{48a^2b^3c^4}$
30. $\sqrt[3]{72a^4b^3c^2}$
31. $\sqrt[5]{64x^8y^{12}}$
32. $\sqrt[4]{32x^9y^{10}}$
33. $\sqrt[5]{243x^7y^{10}z^5}$
34. $\sqrt[5]{64x^8y^4z^{11}}$

Substitute the given numbers into the expression $\sqrt{b^2 - 4ac}$, and then simplify.

35. $a = 2, b = -6, c = 3$
36. $a = 6, b = 7, c = -5$
37. $a = 1, b = 2, c = 6$
38. $a = 2, b = 5, c = 3$
39. $a = \frac{1}{2}, b = -\frac{1}{2}, c = -\frac{5}{4}$
40. $a = \frac{7}{4}, b = -\frac{3}{4}, c = -2$

Rationalize the denominator in each of the following expressions.

41. $\frac{2}{\sqrt{3}}$
42. $\frac{3}{\sqrt{2}}$
43. $\frac{5}{\sqrt{6}}$
44. $\frac{7}{\sqrt{5}}$
45. $\sqrt{\frac{1}{2}}$
46. $\sqrt{\frac{1}{3}}$
47. $\sqrt{\frac{1}{5}}$
48. $\sqrt{\frac{1}{6}}$
49. $\frac{4}{\sqrt[3]{2}}$
50. $\frac{5}{\sqrt[3]{3}}$
51. $\frac{2}{\sqrt[3]{9}}$
52. $\frac{3}{\sqrt[3]{4}}$
53. $\sqrt[4]{\frac{3}{2x^2}}$
54. $\sqrt[4]{\frac{5}{3x^2}}$
55. $\sqrt[4]{\frac{8}{y}}$
56. $\sqrt[4]{\frac{27}{y}}$
57. $\sqrt[3]{\frac{4x}{3y}}$
58. $\sqrt[3]{\frac{7x}{6y}}$
59. $\sqrt[3]{\frac{2x}{9y}}$
60. $\sqrt[3]{\frac{5x}{4y}}$
61. $\sqrt[4]{\frac{1}{8x^3}}$
62. $\sqrt[4]{\frac{8}{9x^3}}$

Write each of the following in simplified form.

63. $\sqrt{\frac{27x^3}{5y}}$
64. $\sqrt{\frac{12x^5}{7y}}$
65. $\sqrt{\frac{75x^3y^2}{2z}}$
66. $\sqrt{\frac{50x^2y^3}{3z}}$
67. $\sqrt{\frac{16a^4b^3}{9c}}$
68. $\sqrt[3]{\frac{54a^5b^4}{25c^2}}$
69. $\sqrt[3]{\frac{8x^3y^6}{9z}}$
70. $\sqrt[3]{\frac{27x^6y^3}{2z^2}}$

Simplify each expression. Do *not* assume the variables represent positive numbers.

71. $\sqrt{25x^2}$
72. $\sqrt{49x^2}$
73. $\sqrt{27x^3y^2}$
74. $\sqrt{40x^3y^2}$
75. $\sqrt{x^2 - 10x + 25}$
76. $\sqrt{x^2 - 16x + 64}$
77. $\sqrt{4x^2 + 12x + 9}$
78. $\sqrt{16x^2 + 40x + 25}$
79. $\sqrt{4a^4 + 16a^3 + 16a^2}$
80. $\sqrt{9a^4 + 18a^3 + 9a^2}$
81. $\sqrt{4x^3 - 8x^2}$
82. $\sqrt{18x^3 - 9x^2}$
83. Show that the statement $\sqrt{a + b} = \sqrt{a} + \sqrt{b}$ is not true by replacing a with 9 and b with 16 and simplifying both sides.
84. Find a pair of values for a and b that will make the statement $\sqrt{a + b} = \sqrt{a} + \sqrt{b}$ true.

Applying the Concepts

85. **Diagonal Distance** The distance d between opposite corners of a rectangular room with length l and width w is given by

$$d = \sqrt{l^2 + w^2}$$

How far is it between opposite corners of a living room that measures 10 by 15 feet?

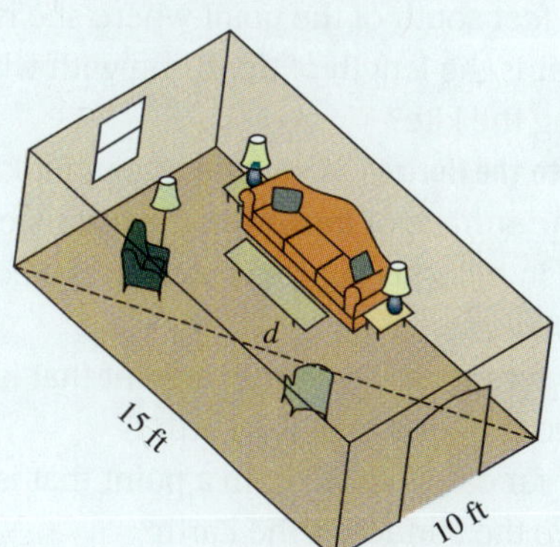

86. **Radius of a Sphere** The radius r of a sphere with volume V can be found by using the formula

$$r = \sqrt[3]{\frac{3V}{4\pi}}$$

Find the radius of a sphere with volume 9 cubic feet. Write your answer in simplified form. (Use $\frac{22}{7}$ for π.)

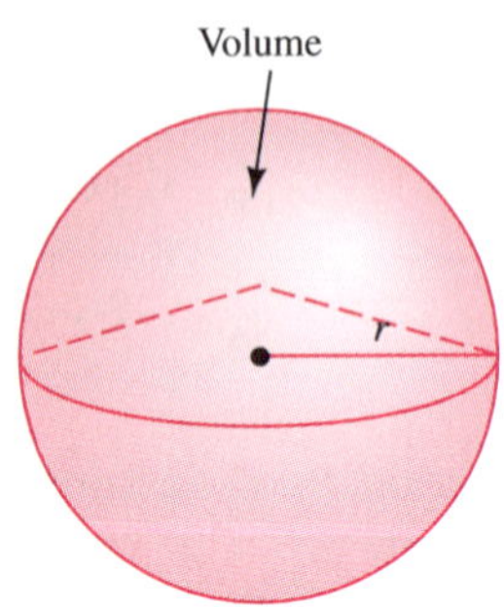

87. **Diagonal of a Box** The length of the diagonal of a rectangular box with length l, width w, and height h is given by $d = \sqrt{l^2 + w^2 + h^2}$.

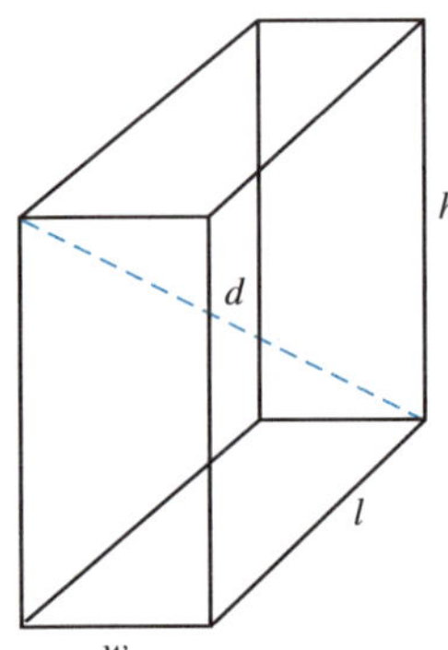

(a) Find the length of the diagonal of a rectangular box that is 3 feet wide, 4 feet long, and 12 feet high.

(b) Find the length of the diagonal of a rectangular box that is 2 feet wide, 4 feet high, and 6 feet long.

88. **Flying a Kite** Mary is flying a kite. Her kite is flying 250 feet off the ground, above a point that is 100 feet east and 300 feet south of the point where she is standing. What is the length of the string with which Mary is holding the kite?

89. **Distance to the Horizon** If you are at a point k miles above the surface of the Earth, the distance you can see, in miles, is approximated by the equation $d = \sqrt{8000k + k^2}$.

(a) How far can you see from a point that is 1 mile above the surface of the Earth?

(b) How far can you see from a point that is 2 miles above the surface of the Earth?

(c) How far can you see from a point that is 3 miles above the surface of the Earth?

90. **Investing** If you invest P dollars and you want the investment to grow to A dollars in t years, the interest rate that must be earned if interest is compounded annually is given by the formula $r = \sqrt[t]{\frac{A}{P}} - 1$. If you invest \$4,000 and want to have \$7,000 in 8 years, what interest rate must be earned?

91. **Spiral of Roots** Construct your own spiral of roots by using a ruler. Draw the first triangle by using two 1-inch lines. The first diagonal will have a length of $\sqrt{2}$ inches. Each new triangle will be formed by drawing a 1-inch line segment at the end of the previous diagonal so the angle formed is 90°.

92. **Spiral of Roots** Construct a spiral of roots by using line segments of length 2 inches. The length of the first diagonal will be $2\sqrt{2}$ inches. The length of the second diagonal will be $2\sqrt{3}$ inches.

93. **Spiral of Roots** If $f(x) = \sqrt{x^2 + 1}$, find the first six terms in the following sequence. Use your results to predict the value of the 10th term and the 100th term.

$$f(1), f[f(1)], f(f[f(1)]), \ldots$$

94. **Spiral of Roots** If $f(x) = \sqrt{x^2 + 4}$, find the first six terms in the following sequence. Use your results to predict the value of the 10th term and the 100th term. (The numbers in this sequence are the lengths of the diagonals of the spiral you drew in Problem 92.)

$$f(2), f[f(2)], f(f[f(2)]), \ldots$$

Review Problems

Perform the indicated operations.

95. $\frac{8xy^3}{9x^2y} \div \frac{16x^2y^2}{18xy^3}$

96. $\frac{25x^2}{5y^4} \cdot \frac{30y^3}{2x^5}$

97. $\frac{12a^2 - 4a - 5}{2a + 1} \cdot \frac{7a + 3}{42a^2 - 17a - 15}$

98. $\frac{20a^2 - 7a - 3}{4a + 1} \cdot \frac{25a^2 - 5a - 6}{5a + 2}$

99. $\frac{8x^3 + 27}{27x^3 + 1} \div \frac{6x^2 + 7x - 3}{9x^2 - 1}$

100. $\frac{27x^3 + 8}{8x^3 + 1} \div \frac{6x^2 + x - 2}{4x^2 - 1}$

Extending the Concepts

Factor each radicand into the product of prime factors. Then simplify each radical.

101. $\sqrt[3]{8{,}640}$

102. $\sqrt{8{,}640}$

103. $\sqrt[3]{10{,}584}$

104. $\sqrt{10{,}584}$

Assume a is a positive number, and rationalize each denominator.

105. $\frac{1}{\sqrt[10]{a^3}}$ **106.** $\frac{1}{\sqrt[12]{a^7}}$

107. $\frac{1}{\sqrt[20]{a^{11}}}$ **108.** $\frac{1}{\sqrt[15]{a^{13}}}$

109. Show that the two expressions $\sqrt{x^2+1}$ and $x+1$ are not, in general, equal to each other by graphing $y=\sqrt{x^2+1}$ and $y=x+1$ in the same viewing window.

110. Show that the two expressions $\sqrt{x^2+9}$ and $x+3$ are not, in general, equal to each other by graphing $y=\sqrt{x^2+9}$ and $y=x+3$ in the same viewing window.

111. Approximately how far apart are the graphs in Problem 109 when $x=2$?

112. Approximately how far apart are the graphs in Problem 110 when $x=2$?

113. For what value of x are the expressions $\sqrt{x^2+1}$ and $x+1$ equal?

114. For what value of x are the expressions $\sqrt{x^2+9}$ and $x+3$ equal?

115. Heron's Formula Heron's formula for the area of a triangle is

$$A=\sqrt{s(s-a)(s-b)(s-c)}$$

in which a, b, and c are the lengths of the sides of the triangle and s is one-half the perimeter of the triangle.

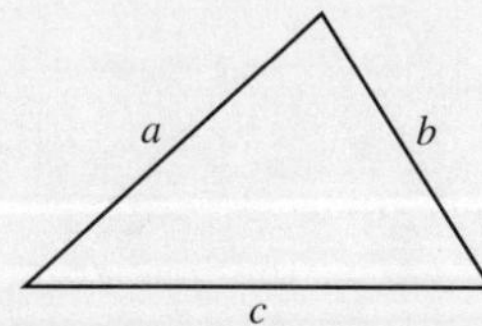

(a) Write a formula to find s in terms of a, b, and c.

(b) Use Heron's formula to find the area of a triangle that has sides of 5, 6, and 7.

116. Brahmagupta's Formula A cyclic quadrilateral is a quadrilateral that has each of its vertices on a circle. The area of a cyclic quadrilateral can be found with Brahmagupta's formula

$$A=\sqrt{(s-a)(s-b)(s-c)(s-d)}$$

where a, b, c, and d are the lengths of the sides of the cyclic quadrilateral and s is one-half the perimeter of the quadrilateral.

(a) Write a formula for s in terms of a, b, c, and d.

(b) Find the area of a cyclic quadrilateral with sides of 4, 6, 9, and 3.

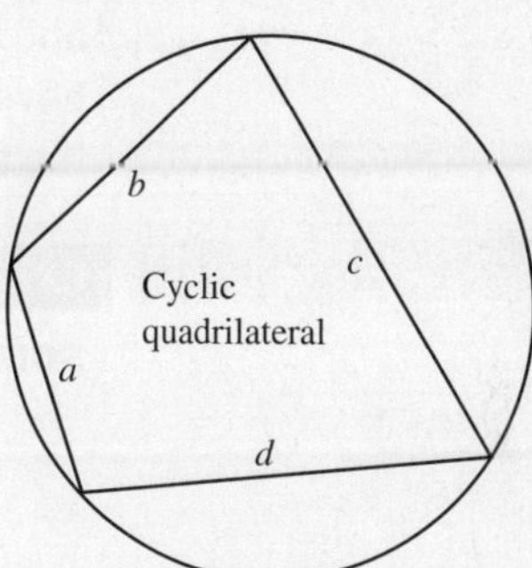

4.4 Addition and Subtraction of Radical Expressions

We have been able to add and subtract polynomials by combining similar terms. The same idea applies to addition and subtraction of radical expressions.

DEFINITION

Two radicals are said to be **similar radicals** if they have the same index and the same radicand.

The expressions $5\sqrt[3]{7}$ and $-8\sqrt[3]{7}$ are similar since the index is ^{3} in both cases and the radicands are 7. The expressions $3\sqrt[4]{5}$ and $7\sqrt[3]{5}$ are not similar since they have different indices, and the expressions $2\sqrt[5]{8}$ and $3\sqrt[5]{9}$ are not similar because the radicands are not the same.

Practice Problems

1. Combine $3\sqrt{5} - 2\sqrt{5} + 4\sqrt{5}$.

EXAMPLE 1 Combine $5\sqrt{3} - 4\sqrt{3} + 6\sqrt{3}$.

SOLUTION All three radicals are similar. We apply the distributive property to get

$$5\sqrt{3} - 4\sqrt{3} + 6\sqrt{3} = (5 - 4 + 6)\sqrt{3}$$
$$= 7\sqrt{3}$$

2. Combine $4\sqrt{50} + 3\sqrt{8}$.

EXAMPLE 2 Combine $3\sqrt{8} + 5\sqrt{18}$.

SOLUTION The two radicals do not seem to be similar. We must write each in simplified form before applying the distributive property.

$$3\sqrt{8} + 5\sqrt{18} = 3\sqrt{4 \cdot 2} + 5\sqrt{9 \cdot 2}$$
$$= 3\sqrt{4}\,\sqrt{2} + 5\sqrt{9}\,\sqrt{2}$$
$$= 3 \cdot 2\sqrt{2} + 5 \cdot 3\sqrt{2}$$
$$= 6\sqrt{2} + 15\sqrt{2}$$
$$= (6 + 15)\sqrt{2}$$
$$= 21\sqrt{2}$$

The result of Example 2 can be generalized to the following rule for sums and differences of radical expressions.

Rule

To add or subtract radical expressions, put each in simplified form and apply the distributive property if possible. We can add only similar radicals. We must write each expression in simplified form for radicals before we can tell if the radicals are similar.

3. Assume $x, y \geq 0$ and combine: $4\sqrt{18x^2y} - 3x\sqrt{50y}$

EXAMPLE 3 Combine $7\sqrt{75xy^3} - 4y\sqrt{12xy}$, where $x, y \geq 0$.

SOLUTION We write each expression in simplified form and combine similar radicals:

$$7\sqrt{75xy^3} - 4y\sqrt{12xy} = 7\sqrt{25y^2}\,\sqrt{3xy} - 4y\sqrt{4}\,\sqrt{3xy}$$
$$= 35y\sqrt{3xy} - 8y\sqrt{3xy}$$
$$= (35y - 8y)\sqrt{3xy}$$
$$= 27y\sqrt{3xy}$$

Answers

1. $5\sqrt{5}$ 2. $26\sqrt{2}$ 3. $-3x\sqrt{2y}$

EXAMPLE 4 Combine $10\sqrt[3]{8a^4b^2} + 11a\sqrt[3]{27ab^2}$.

SOLUTION Writing each radical in simplified form and combining similar terms, we have

$$\begin{aligned} 10\sqrt[3]{8a^4b^2} + 11a\sqrt[3]{27ab^2} &= 10\sqrt[3]{8a^3}\,\sqrt[3]{ab^2} + 11a\sqrt[3]{27}\,\sqrt[3]{ab^2} \\ &= 20a\sqrt[3]{ab^2} + 33a\sqrt[3]{ab^2} \\ &= 53a\sqrt[3]{ab^2} \end{aligned}$$

4. Combine $2\sqrt[3]{27a^2b^4} + 3b\sqrt[3]{125a^2b}$.

EXAMPLE 5 Combine $\frac{\sqrt{3}}{2} + \frac{1}{\sqrt{3}}$.

SOLUTION We begin by writing the second term in simplified form.

$$\begin{aligned} \frac{\sqrt{3}}{2} + \frac{1}{\sqrt{3}} &= \frac{\sqrt{3}}{2} + \frac{1}{\sqrt{3}} \cdot \frac{\sqrt{3}}{\sqrt{3}} \\ &= \frac{\sqrt{3}}{2} + \frac{\sqrt{3}}{3} \\ &= \frac{1}{2}\sqrt{3} + \frac{1}{3}\sqrt{3} \\ &= \left(\frac{1}{2} + \frac{1}{3}\right)\sqrt{3} \\ &= \frac{5}{6}\sqrt{3} = \frac{5\sqrt{3}}{6} \end{aligned}$$

5. Combine $\frac{\sqrt{5}}{3} + \frac{1}{\sqrt{5}}$.

EXAMPLE 6 Construct a golden rectangle from a square of side 4. Then show that the ratio of the length to the width is the golden ratio $\frac{1 + \sqrt{5}}{2}$.

SOLUTION Figure 1 shows the golden rectangle constructed from a square of side 4.

The length of the diagonal OB is found from the Pythagorean theorem.

$$OB = \sqrt{2^2 + 4^2} = \sqrt{4 + 16} = \sqrt{20} = 2\sqrt{5}$$

The ratio of the length to the width for the rectangle is the golden ratio.

$$\text{Golden ratio} = \frac{CE}{EF} = \frac{2 + 2\sqrt{5}}{4} = \frac{\not{2}(1 + \sqrt{5})}{\not{2} \cdot 2} = \frac{1 + \sqrt{5}}{2}$$

6. Construct a golden rectangle from a square of side 6. Then show that the ratio of the length to the width is the golden ratio.

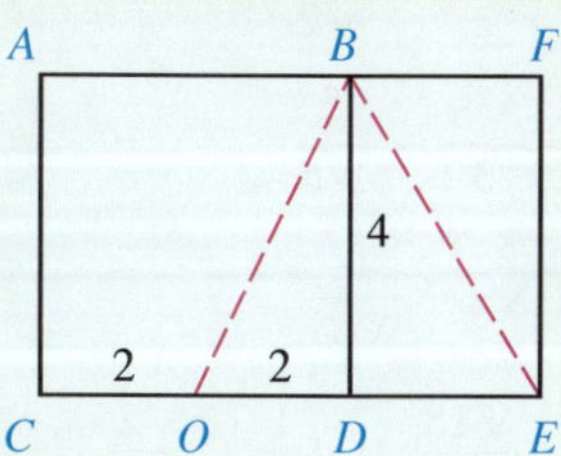

Figure 1

As you can see, showing that the ratio of length to width in this rectangle is the golden ratio depends on our ability to write $\sqrt{20}$ as $2\sqrt{5}$ and our ability to reduce to lowest terms by factoring and then dividing out the common factor 2 from the numerator and denominator.

Getting Ready for Class

After reading through the preceding section, respond in your own words and in complete sentences.

A. What are similar radicals?
B. When can we add two radical expressions?
C. What is the first step when adding or subtracting expressions containing radicals?
D. What is the golden ratio, and where does it come from?

Answers

4. $21b\sqrt[3]{a^2b}$ 5. $\frac{8\sqrt{5}}{15}$

6. See Solutions to Selected Practice Problems.

PROBLEM SET 4.4

Combine the following expressions. (Assume any variables under an even root are nonnegative.)

1. $3\sqrt{5}+4\sqrt{5}$
2. $6\sqrt{3}-5\sqrt{3}$
3. $3x\sqrt{7}-4x\sqrt{7}$
4. $6y\sqrt{a}+7y\sqrt{a}$
5. $5\sqrt[3]{10}-4\sqrt[3]{10}$
6. $6\sqrt[4]{2}+9\sqrt[4]{2}$
7. $8\sqrt[5]{6}-2\sqrt[5]{6}+3\sqrt[5]{6}$
8. $7\sqrt[6]{7}-\sqrt[6]{7}+4\sqrt[6]{7}$
9. $3x\sqrt{2}-4x\sqrt{2}+x\sqrt{2}$
10. $5x\sqrt{6}-3x\sqrt{6}-2x\sqrt{6}$
11. $\sqrt{20}-\sqrt{80}+\sqrt{45}$
12. $\sqrt{8}-\sqrt{32}-\sqrt{18}$
13. $4\sqrt{8}-2\sqrt{50}-5\sqrt{72}$
14. $\sqrt{48}-3\sqrt{27}+2\sqrt{75}$
15. $5x\sqrt{8}+3\sqrt{32x^2}-5\sqrt{50x^2}$
16. $2\sqrt{50x^2}-8x\sqrt{18}-3\sqrt{72x^2}$
17. $5\sqrt[3]{16}-4\sqrt[3]{54}$
18. $\sqrt[3]{81}+3\sqrt[3]{24}$
19. $\sqrt[3]{x^4y^2}+7x\sqrt[3]{xy^2}$
20. $2\sqrt[3]{x^8y^6}-3y^2\sqrt[3]{8x^8}$

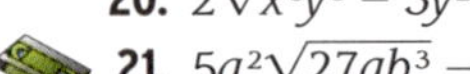
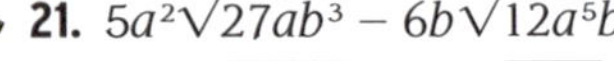

21. $5a^2\sqrt{27ab^3}-6b\sqrt{12a^5b}$
22. $9a\sqrt{20a^3b^2}+7b\sqrt{45a^5}$
23. $b\sqrt[3]{24a^5b}+3a\sqrt[3]{81a^2b^4}$
24. $7\sqrt[3]{a^4b^3c^2}-6ab\sqrt[3]{ac^2}$
25. $5x\sqrt[4]{3y^5}+y\sqrt[4]{243x^4y}+\sqrt[4]{48x^4y^5}$
26. $x\sqrt[4]{5xy^8}+y\sqrt[4]{405x^5y^4}+y^2\sqrt[4]{80x^5}$

27. $\dfrac{\sqrt{2}}{2}+\dfrac{1}{\sqrt{2}}$
28. $\dfrac{\sqrt{3}}{3}+\dfrac{1}{\sqrt{3}}$
29. $\dfrac{\sqrt{5}}{3}+\dfrac{1}{\sqrt{5}}$
30. $\dfrac{\sqrt{6}}{2}+\dfrac{1}{\sqrt{6}}$
31. $\sqrt{x}-\dfrac{1}{\sqrt{x}}$
32. $\sqrt{x}+\dfrac{1}{\sqrt{x}}$
33. $\dfrac{\sqrt{18}}{6}+\sqrt{\dfrac{1}{2}}+\dfrac{\sqrt{2}}{2}$
34. $\dfrac{\sqrt{12}}{6}+\sqrt{\dfrac{1}{3}}+\dfrac{\sqrt{3}}{3}$
35. $\sqrt{6}-\sqrt{\dfrac{2}{3}}+\sqrt{\dfrac{1}{6}}$
36. $\sqrt{15}-\sqrt{\dfrac{3}{5}}+\sqrt{\dfrac{5}{3}}$
37. $\sqrt[3]{25}+\dfrac{3}{\sqrt[3]{5}}$
38. $\sqrt[4]{8}+\dfrac{1}{\sqrt[4]{2}}$
39. Use a calculator to find a decimal approximation for $\sqrt{12}$ and for $2\sqrt{3}$.
40. Use a calculator to find decimal approximations for $\sqrt{50}$ and $5\sqrt{2}$.
41. Use a calculator to find a decimal approximation for $\sqrt{8}+\sqrt{18}$. Is it equal to the decimal approximation for $\sqrt{26}$ or $\sqrt{50}$?
42. Use a calculator to find a decimal approximation for $\sqrt{3}+\sqrt{12}$. Is it equal to the decimal approximation for $\sqrt{15}$ or $\sqrt{27}$?

Each of the following statements is false. Correct the right side of each one to make the statement true.

43. $3\sqrt{2x}+5\sqrt{2x}=8\sqrt{4x}$
44. $5\sqrt{3}-7\sqrt{3}=-2\sqrt{9}$
45. $\sqrt{9+16}=3+4$
46. $\sqrt{36+64}=6+8$

Applying the Concepts

47. **Golden Rectangle** Construct a golden rectangle from a square of side 8. Then show that the ratio of the length to the width is the golden ratio $\frac{1+\sqrt{5}}{2}$.
48. **Golden Rectangle** Construct a golden rectangle from a square of side 10. Then show that the ratio of the length to the width is the golden ratio $\frac{1+\sqrt{5}}{2}$.
49. **Golden Rectangle** Use a ruler to construct a golden rectangle from a square of side 1 inch. Then show that the ratio of the length to the width is the golden ratio.
50. **Golden Rectangle** Use a ruler to construct a golden rectangle from a square of side $\frac{2}{3}$ inch. Then show that the ratio of the length to the width is the golden ratio.
51. **Golden Rectangle** To show that all golden rectangles have the same ratio of length to width, construct a golden rectangle from a square of side $2x$. Then show that the ratio of the length to the width is the golden ratio.
52. **Golden Rectangle** To show that all golden rectangles have the same ratio of length to width, construct a golden rectangle from a square of side x. Then show that the ratio of the length to the width is the golden ratio.
53. **Isosceles Right Triangles** A triangle is isosceles if it has two equal sides, and a triangle is a right triangle if it has a right angle in it. Sketch an isosceles right triangle, and find the ratio of the hypotenuse to a leg.
54. **Equilateral Triangles** A triangle is equilateral if it has three equal sides. The triangle in the figure is equilateral with each side of length $2x$. Find the ratio of the height to a side.

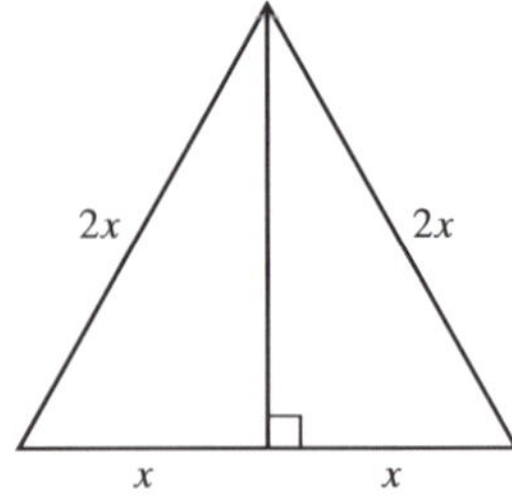

55. Pyramids The following solid is called a regular square pyramid because its base is a square and all eight edges are the same length, 5. It is also true that the vertex, V, is directly above the center of the base.

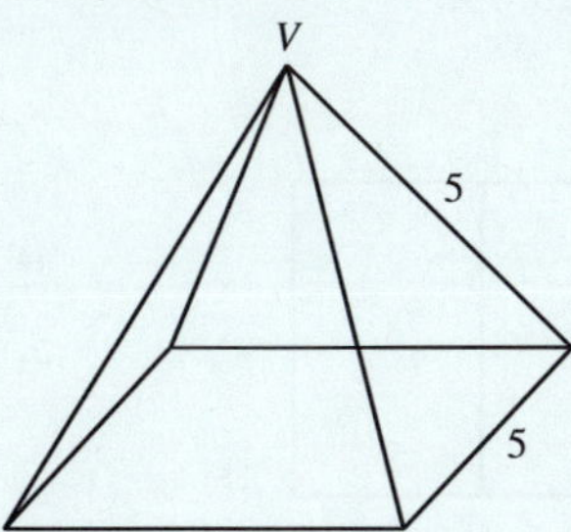

(a) Find the ratio of a diagonal of the base to the length of a side.

(b) Find the ratio of the area of the base to the diagonal of the base.

(c) Find the ratio of the area of the base to the perimeter of the base.

56. Pyramids Refer to the diagram of a square pyramid below. Find the ratio of the height h of the pyramid to the altitude a.

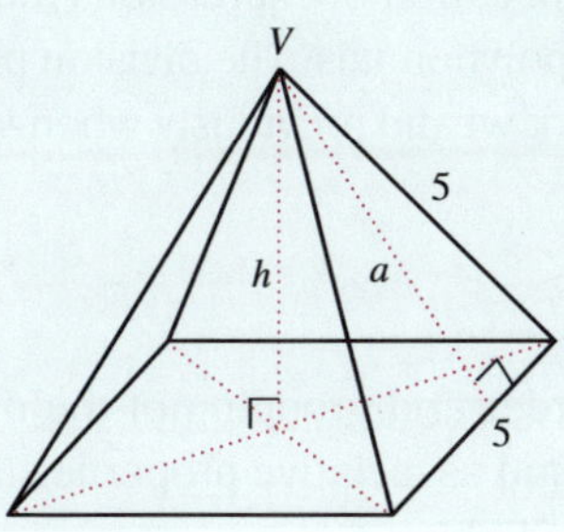

Review Problems

Add and subtract as indicated.

57. $\dfrac{2a-4}{a+2} - \dfrac{a-6}{a+2}$

58. $\dfrac{2a-3}{a-2} - \dfrac{a-1}{a-2}$

59. $3 + \dfrac{4}{3-t}$

60. $6 + \dfrac{2}{5-t}$

61. $\dfrac{3}{2x-5} - \dfrac{39}{8x^2-14x-15}$

62. $\dfrac{2}{4x-5} + \dfrac{9}{8x^2-38x+35}$

63. $\dfrac{1}{x-y} - \dfrac{3xy}{x^3-y^3}$

64. $\dfrac{1}{x+y} + \dfrac{3xy}{x^3+y^3}$

Extending the Concepts

Combine expressions, if possible. Assume all variables represent positive numbers.

65. $\dfrac{\sqrt{36a^2b^4c}}{2} - \dfrac{b\sqrt{16a^2b^2c}}{3}$

66. $\dfrac{y\sqrt{18xyz}}{5} + \dfrac{\sqrt{32xy^3z}}{4}$

67. $\dfrac{-b+\sqrt{b^2-4ac}}{2a} + \dfrac{-b-\sqrt{b^2-4ac}}{2a}$

68. $\dfrac{c-\sqrt{c^2-4ad}}{2a} + \dfrac{c+\sqrt{c^2-4ad}}{2a}$

69. $\sqrt{\dfrac{7}{4}} + \dfrac{\sqrt{7}}{2}$

70. $\sqrt{\dfrac{5}{9}} - \dfrac{-2\sqrt{5}}{3}$

71. $-5a\sqrt{75b^3} + \sqrt{18a^2b} + 7b\sqrt{108a^2b} - a\sqrt{50b}$

72. $3\sqrt{24x^3y} - y\sqrt{27x} + 11x\sqrt{54x^2y} - y\sqrt{48x}$

73. $5x\sqrt{\dfrac{3y^2}{2}} - 3y\sqrt{\dfrac{8x^2}{3}} + 2\sqrt{\dfrac{3x^2y^2}{2}}$

74. $\sqrt{\dfrac{3y}{4x^2}} + \dfrac{2y}{x}\sqrt{\dfrac{3}{4y}} - \dfrac{2}{3x}\sqrt{3y}$

4.5 Multiplication and Division of Radical Expressions

We have worked with the golden rectangle more than once in this chapter. The following is one such golden rectangle.

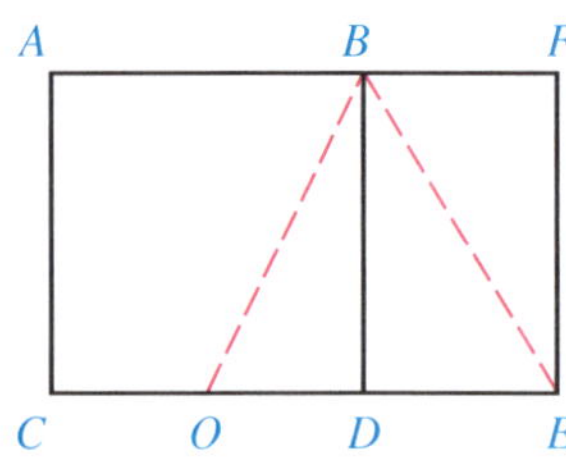

By now you know that in any golden rectangle constructed from a square (of any size) the ratio of the length to the width will be

$$\frac{1+\sqrt{5}}{2}$$

which we call the golden ratio. What is interesting is that the smaller rectangle on the right, *BFED*, is also a golden rectangle. We will use the mathematics developed in this section to confirm this fact.

In this section we will look at multiplication and division of expressions that contain radicals. As you will see, multiplication of expressions that contain radicals is very similar to multiplication of polynomials. The division problems in this section are just an extension of the work we did previously when we rationalized denominators.

Practice Problems

1. Multiply $(7\sqrt{3})(5\sqrt{11})$.

EXAMPLE 1 Multiply $(3\sqrt{5})(2\sqrt{7})$.

SOLUTION We can rearrange the order and grouping of the numbers in this product by applying the commutative and associative properties. Following this, we apply property 1 for radicals and multiply:

$$(3\sqrt{5})(2\sqrt{7}) = (3 \cdot 2)(\sqrt{5}\,\sqrt{7}) \quad \text{Commutative and associative properties}$$
$$= (3 \cdot 2)(\sqrt{5 \cdot 7}) \quad \text{Property 1 for radicals}$$
$$= 6\sqrt{35} \quad \text{Multiplication}$$

In practice, it is not necessary to show the first two steps.

2. Multiply $\sqrt{2}(3\sqrt{5} - 4\sqrt{2})$.

EXAMPLE 2 Multiply $\sqrt{3}(2\sqrt{6} - 5\sqrt{12})$.

SOLUTION Applying the distributive property, we have

$$\sqrt{3}(2\sqrt{6} - 5\sqrt{12}) = \sqrt{3} \cdot 2\sqrt{6} - \sqrt{3} \cdot 5\sqrt{12}$$
$$= 2\sqrt{18} - 5\sqrt{36}$$

Writing each radical in simplified form gives

$$2\sqrt{18} - 5\sqrt{36} = 2\sqrt{9}\,\sqrt{2} - 5\sqrt{36}$$
$$= 6\sqrt{2} - 30$$

Answers

1. $35\sqrt{33}$ 2. $3\sqrt{10} - 8$

EXAMPLE 3 Multiply $(\sqrt{3} + \sqrt{5})(4\sqrt{3} - \sqrt{5})$.

SOLUTION The same principle that applies when multiplying two binomials applies to this product. We must multiply each term in the first expression by each term in the second one. Any convenient method can be used. Let's use the FOIL method.

$$(\sqrt{3} + \sqrt{5})(4\sqrt{3} - \sqrt{5}) = \overset{F}{\sqrt{3} \cdot 4\sqrt{3}} - \overset{O}{\sqrt{3}\sqrt{5}} + \overset{I}{\sqrt{5} \cdot 4\sqrt{3}} - \overset{L}{\sqrt{5}\sqrt{5}}$$
$$= 4 \cdot 3 - \sqrt{15} + 4\sqrt{15} - 5$$
$$= 12 + 3\sqrt{15} - 5$$
$$= 7 + 3\sqrt{15}$$

3. Multiply: $(\sqrt{2} + \sqrt{7})(\sqrt{2} - 3\sqrt{7})$

EXAMPLE 4 Expand and simplify $(\sqrt{x} + 3)^2$.

SOLUTION 1 We can write this problem as a multiplication problem and proceed as we did in Example 3:

$$(\sqrt{x} + 3)^2 = (\sqrt{x} + 3)(\sqrt{x} + 3)$$
$$= \overset{F}{\sqrt{x} \cdot \sqrt{x}} + \overset{O}{3\sqrt{x}} + \overset{I}{3\sqrt{x}} + \overset{L}{3 \cdot 3}$$
$$= x + 3\sqrt{x} + 3\sqrt{x} + 9$$
$$= x + 6\sqrt{x} + 9$$

SOLUTION 2 We can obtain the same result by applying the formula for the square of a sum: $(a + b)^2 = a^2 + 2ab + b^2$.

$$(\sqrt{x} + 3)^2 = (\sqrt{x})^2 + 2(\sqrt{x})(3) + 3^2$$
$$= x + 6\sqrt{x} + 9$$

4. Expand and simplify. $(\sqrt{x} + 5)^2$

EXAMPLE 5 Expand $(3\sqrt{x} - 2\sqrt{y})^2$ and simplify the result.

SOLUTION Let's apply the formula for the square of a difference, $(a - b)^2 = a^2 - 2ab + b^2$.

$$(3\sqrt{x} - 2\sqrt{y})^2 = (3\sqrt{x})^2 - 2(3\sqrt{x})(2\sqrt{y}) + (2\sqrt{y})^2$$
$$= 9x - 12\sqrt{xy} + 4y$$

5. Expand $(5\sqrt{a} - 3\sqrt{b})^2$ and simplify the result.

EXAMPLE 6 Expand and simplify $(\sqrt{x + 2} - 1)^2$.

SOLUTION Applying the formula $(a - b)^2 = a^2 - 2ab + b^2$, we have

$$(\sqrt{x + 2} - 1)^2 = (\sqrt{x + 2})^2 - 2\sqrt{x + 2}(1) + 1^2$$
$$= x + 2 - 2\sqrt{x + 2} + 1$$
$$= x + 3 - 2\sqrt{x + 2}$$

6. Expand and simplify. $(\sqrt{x + 3} - 1)^2$

EXAMPLE 7 Multiply $(\sqrt{6} + \sqrt{2})(\sqrt{6} - \sqrt{2})$.

SOLUTION We notice the product is of the form $(a + b)(a - b)$, which always gives the difference of two squares, $a^2 - b^2$:

$$(\sqrt{6} + \sqrt{2})(\sqrt{6} - \sqrt{2}) = (\sqrt{6})^2 - (\sqrt{2})^2$$
$$= 6 - 2$$
$$= 4$$

7. Multiply. $(\sqrt{5} + \sqrt{3})(\sqrt{5} - \sqrt{3})$

Answers

3. $-19 - 2\sqrt{14}$ 4. $x + 10\sqrt{x} + 25$
5. $25a - 30\sqrt{ab} + 9b$
6. $x + 4 - 2\sqrt{x + 3}$ 7. 2

Note: We can prove that conjugates always multiply to yield a rational number as follows: If a and b are positive integers, then

$$(\sqrt{a}+\sqrt{b})(\sqrt{a}-\sqrt{b})$$
$$=\sqrt{a}\sqrt{a}-\sqrt{a}\sqrt{b}+\sqrt{a}\sqrt{b}-\sqrt{b}\sqrt{b}$$
$$=a-\sqrt{ab}+\sqrt{ab}-b$$
$$=a-b$$

which is rational if a and b are rational.

The two expressions $(\sqrt{6}+\sqrt{2})$ and $(\sqrt{6}-\sqrt{2})$ are called *conjugates.* In general, the conjugate of $\sqrt{a}+\sqrt{b}$ is $\sqrt{a}-\sqrt{b}$. If a and b are integers, multiplying conjugates of this form always produces a rational number.

Division with radical expressions is the same as rationalizing the denominator. In a previous section we were able to divide $\sqrt{3}$ by $\sqrt{2}$ by rationalizing the denominator:

$$\frac{\sqrt{3}}{\sqrt{2}}=\frac{\sqrt{3}}{\sqrt{2}}\cdot\frac{\mathbf{\sqrt{2}}}{\mathbf{\sqrt{2}}}=\frac{\sqrt{6}}{2}$$

We can accomplish the same result with expressions such as

$$\frac{6}{\sqrt{5}-\sqrt{3}}$$

by multiplying the numerator and denominator by the conjugate of the denominator.

8. Divide $\dfrac{3}{\sqrt{7}-\sqrt{3}}$.

EXAMPLE 8 Divide $\dfrac{6}{\sqrt{5}-\sqrt{3}}$. (Rationalize the denominator.)

SOLUTION Since the product of two conjugates is a rational number, we multiply the numerator and denominator by the conjugate of the denominator.

$$\frac{6}{\sqrt{5}-\sqrt{3}}=\frac{6}{\sqrt{5}-\sqrt{3}}\cdot\frac{\mathbf{(\sqrt{5}+\sqrt{3})}}{\mathbf{(\sqrt{5}+\sqrt{3})}}$$
$$=\frac{6\sqrt{5}+6\sqrt{3}}{(\sqrt{5})^2-(\sqrt{3})^2}$$
$$=\frac{6\sqrt{5}+6\sqrt{3}}{5-3}$$
$$=\frac{6\sqrt{5}+6\sqrt{3}}{2}$$

The numerator and denominator of this last expression have a factor of 2 in common. We can reduce to lowest terms by factoring 2 from the numerator and then dividing both the numerator and denominator by 2:

$$=\frac{\cancel{2}(3\sqrt{5}+3\sqrt{3})}{\cancel{2}}$$
$$=3\sqrt{5}+3\sqrt{3}$$

9. Rationalize the denominator:

$$\frac{\sqrt{10}-3}{\sqrt{10}+3}$$

EXAMPLE 9 Rationalize the denominator $\dfrac{\sqrt{5}-2}{\sqrt{5}+2}$.

SOLUTION To rationalize the denominator, we multiply the numerator and denominator by the conjugate of the denominator:

$$\frac{\sqrt{5}-2}{\sqrt{5}+2}=\frac{\sqrt{5}-2}{\sqrt{5}+2}\cdot\frac{\mathbf{(\sqrt{5}-2)}}{\mathbf{(\sqrt{5}-2)}}$$
$$=\frac{5-2\sqrt{5}-2\sqrt{5}+4}{(\sqrt{5})^2-2^2}$$
$$=\frac{9-4\sqrt{5}}{5-4}$$
$$=\frac{9-4\sqrt{5}}{1}$$
$$=9-4\sqrt{5}$$

Answers

8. $\dfrac{3\sqrt{7}+3\sqrt{3}}{4}$ 9. $19-6\sqrt{10}$

EXAMPLE 10 A golden rectangle constructed from a square of side 2 is shown in Figure 1. Show that the smaller rectangle *BDEF* is also a golden rectangle by finding the ratio of its length to its width.

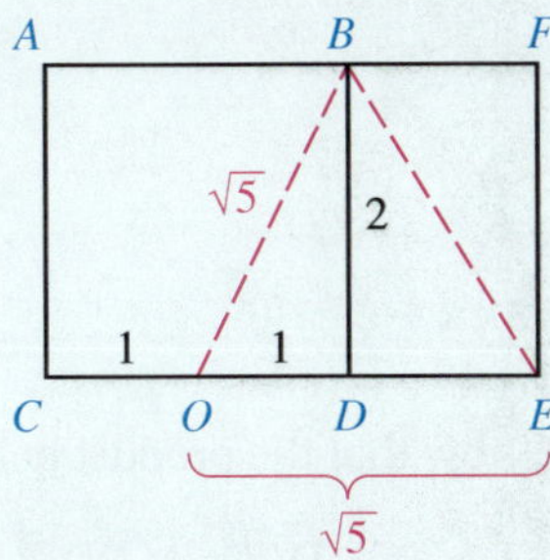

Figure 1

SOLUTION First we find expressions for the length and width of the smaller rectangle.

$$\text{Length} = EF = 2$$

$$\text{Width} = DE = \sqrt{5} - 1$$

Next, we find the ratio of length to width.

$$\text{Ratio of length to width} = \frac{EF}{DE} = \frac{2}{\sqrt{5} - 1}$$

To show that the small rectangle is a golden rectangle, we must show that the ratio of length to width is the golden ratio. We do so by rationalizing the denominator.

$$\begin{aligned}\frac{2}{\sqrt{5} - 1} &= \frac{2}{\sqrt{5} - 1} \cdot \frac{\mathbf{\sqrt{5} + 1}}{\mathbf{\sqrt{5} + 1}} \\ &= \frac{2(\sqrt{5} + 1)}{5 - 1} \\ &= \frac{2(\sqrt{5} + 1)}{4} \\ &= \frac{\sqrt{5} + 1}{2} \quad \text{Divide out common factor 2}\end{aligned}$$

Since addition is commutative, this last expression is the golden ratio. Therefore, the small rectangle in Figure 1 is a golden rectangle.

Getting Ready for Class

After reading through the preceding section, respond in your own words and in complete sentences.

A. Explain why $(\sqrt{5} + \sqrt{2})^2 \neq 5 + 2$.

B. Explain in words how you would rationalize the denominator in the expression $\dfrac{\sqrt{3}}{\sqrt{5} - \sqrt{2}}$.

C. What are conjugates?

D. What result is guaranteed when multiplying radical expressions that are conjugates?

10. If side *AC* in Figure 1 were 4 instead of 2, show that the smaller rectangle *BDEF* is also a golden rectangle by finding the ratio of its length to its width.

Answer

10. See Solutions to Selected Practice Problems section.

PROBLEM SET 4.5

Multiply. (Assume all expressions appearing under a square root symbol represent nonnegative numbers throughout this problem set.)

1. $\sqrt{6}\,\sqrt{3}$
2. $\sqrt{6}\,\sqrt{2}$
3. $(2\sqrt{3})(5\sqrt{7})$
4. $(3\sqrt{5})(2\sqrt{7})$
5. $(4\sqrt{6})(2\sqrt{15})(3\sqrt{10})$
6. $(4\sqrt{35})(2\sqrt{21})(5\sqrt{15})$
7. $(3\sqrt[3]{3})(6\sqrt[3]{9})$
8. $(2\sqrt[3]{2})(6\sqrt[3]{4})$

9. $\sqrt{3}(\sqrt{2} - 3\sqrt{3})$
10. $\sqrt{2}(5\sqrt{3} + 4\sqrt{2})$
11. $6\sqrt[3]{4}(2\sqrt[3]{2} + 1)$
12. $7\sqrt[3]{5}(3\sqrt[3]{25} - 2)$
13. $(\sqrt{3} + \sqrt{2})(3\sqrt{3} - \sqrt{2})$
14. $(\sqrt{5} - \sqrt{2})(3\sqrt{5} + 2\sqrt{2})$

15. $(\sqrt{x} + 5)(\sqrt{x} - 3)$
16. $(\sqrt{x} + 4)(\sqrt{x} + 2)$
17. $(3\sqrt{6} + 4\sqrt{2})(\sqrt{6} + 2\sqrt{2})$
18. $(\sqrt{7} - 3\sqrt{3})(2\sqrt{7} - 4\sqrt{3})$
19. $(\sqrt{3} + 4)^2$
20. $(\sqrt{5} - 2)^2$
21. $(\sqrt{x} - 3)^2$
22. $(\sqrt{x} + 4)^2$
23. $(2\sqrt{a} - 3\sqrt{b})^2$
24. $(5\sqrt{a} - 2\sqrt{b})^2$
25. $(\sqrt{x - 4} + 2)^2$
26. $(\sqrt{x - 3}) + 2)^2$
27. $(\sqrt{x - 5} - 3)^2$
28. $(\sqrt{x - 3} - 4)^2$
29. $(\sqrt{3} - \sqrt{2})(\sqrt{3} + \sqrt{2})$
30. $(\sqrt{5} - \sqrt{2})(\sqrt{5} + \sqrt{2})$
31. $(\sqrt{a} + 7)(\sqrt{a} - 7)$
32. $(\sqrt{a} + 5)(\sqrt{a} - 5)$
33. $(5 - \sqrt{x})(5 + \sqrt{x})$
34. $(3 - \sqrt{x})(3 + \sqrt{x})$
35. $(\sqrt{x - 4} + 2)(\sqrt{x - 4} - 2)$
36. $(\sqrt{x + 3} + 5)(\sqrt{x + 3} - 5)$
37. $(\sqrt{3} + 1)^3$
38. $(\sqrt{5} - 2)^3$

Rationalize the denominator in each of the following.

39. $\dfrac{\sqrt{2}}{\sqrt{6} - \sqrt{2}}$
40. $\dfrac{\sqrt{5}}{\sqrt{5} + \sqrt{3}}$
41. $\dfrac{\sqrt{5}}{\sqrt{5} + 1}$
42. $\dfrac{\sqrt{7}}{\sqrt{7} - 1}$
43. $\dfrac{\sqrt{x}}{\sqrt{x} - 3}$
44. $\dfrac{\sqrt{x}}{\sqrt{x} + 2}$
45. $\dfrac{\sqrt{5}}{2\sqrt{5} - 3}$
46. $\dfrac{\sqrt{7}}{3\sqrt{7} - 2}$
47. $\dfrac{3}{\sqrt{x} - \sqrt{y}}$
48. $\dfrac{2}{\sqrt{x} + \sqrt{y}}$
49. $\dfrac{\sqrt{6} + \sqrt{2}}{\sqrt{6} - \sqrt{2}}$
50. $\dfrac{\sqrt{5} - \sqrt{3}}{\sqrt{5} + \sqrt{3}}$
51. $\dfrac{\sqrt{7} - 2}{\sqrt{7} + 2}$
52. $\dfrac{\sqrt{11} + 3}{\sqrt{11} - 3}$
53. $\dfrac{\sqrt{a} + \sqrt{b}}{\sqrt{a} - \sqrt{b}}$
54. $\dfrac{\sqrt{a} - \sqrt{b}}{\sqrt{a} + \sqrt{b}}$
55. $\dfrac{\sqrt{x} + 2}{\sqrt{x} - 2}$
56. $\dfrac{\sqrt{x} - 3}{\sqrt{x} + 3}$
57. $\dfrac{2\sqrt{3} - \sqrt{7}}{3\sqrt{3} + \sqrt{7}}$
58. $\dfrac{5\sqrt{6} + 2\sqrt{2}}{\sqrt{6} - \sqrt{2}}$
59. $\dfrac{3\sqrt{x} + 2}{1 + \sqrt{x}}$
60. $\dfrac{5\sqrt{x} - 1}{2 + \sqrt{x}}$
61. Show that the product below

$$(\sqrt[3]{2} + \sqrt[3]{3})(\sqrt[3]{4} - \sqrt[3]{6} + \sqrt[3]{9})$$

is 5.

62. Show that the product below

$$(\sqrt[3]{x} + 2)(\sqrt[3]{x^2} - 2\sqrt[3]{x} + 4)$$

is $x + 8$.

Each of the following statements below is false. Correct the right side of each one to make it true.

63. $5(2\sqrt{3}) = 10\sqrt{15}$
64. $3(2\sqrt{x}) = 6\sqrt{3x}$
65. $(\sqrt{x} + 3)^2 = x + 9$
66. $(\sqrt{x} - 7)^2 = x - 49$
67. $(5\sqrt{3})^2 = 15$
68. $(3\sqrt{5})^2 = 15$

Applying the Concepts

69. **Gravity** If an object is dropped from the top of a 100-foot building, the amount of time t (in seconds) that it takes for the object to be h feet from the ground is given by the formula

$$t = \frac{\sqrt{100 - h}}{4}$$

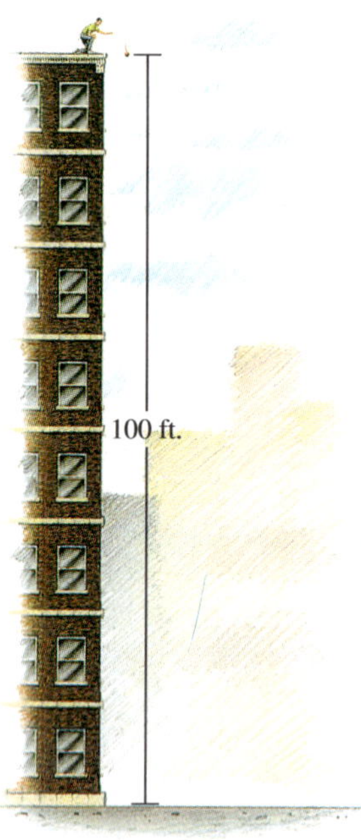

How long does it take before the object is 50 feet from the ground? How long does it take to reach the ground? (When it is on the ground, h is 0.)

70. Gravity Use the formula given in Problem 69 to determine h if t is 1.25 seconds.

71. Golden Rectangle Rectangle *ACEF* in Figure 2 is a golden rectangle. If side *AC* is 6 inches, show that the smaller rectangle *BDEF* is also a golden rectangle.

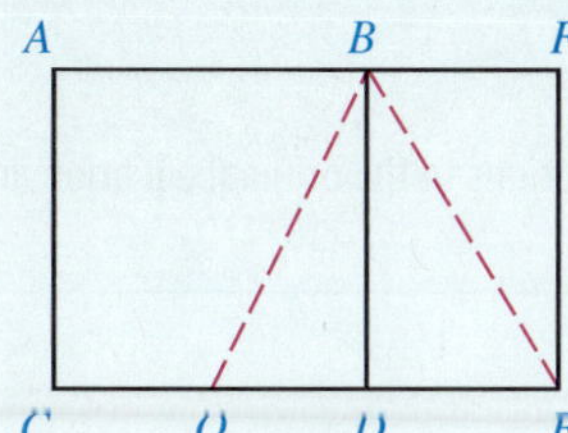

Figure 2

72. Golden Rectangle Rectangle *ACEF* in Figure 2 is a golden rectangle. If side *AC* is 1 inch, show that the smaller rectangle *BDEF* is also a golden rectangle.

73. Golden Rectangle If side *AC* in Figure 2 is $2x$, show that rectangle *BDEF* is a golden rectangle.

74. Golden Rectangle If side *AC* in Figure 2 is x, show that rectangle *BDEF* is a golden rectangle.

Review Problems

Simplify each complex fraction.

75. $\dfrac{\frac{1}{4} - \frac{1}{3}}{\frac{1}{2} + \frac{1}{6}}$

76. $\dfrac{\frac{1}{8} - \frac{1}{3}}{\frac{1}{4} - \frac{1}{3}}$

77. $\dfrac{1 - \frac{2}{y}}{1 + \frac{2}{y}}$

78. $\dfrac{1 + \frac{3}{y}}{1 - \frac{3}{y}}$

79. $\dfrac{4 + \frac{4}{x} + \frac{1}{x^2}}{4 - \frac{1}{x^2}}$

80. $\dfrac{1 - \frac{1}{x} - \frac{6}{x^2}}{1 - \frac{9}{x^2}}$

Extending the Concepts

For the rest of the problems in the problem set, assume that all variables represent positive numbers.

Multiply.

81. $\sqrt{\dfrac{2y}{x}} \cdot \sqrt{\dfrac{x^2}{8}}$

82. $\sqrt{\dfrac{3a}{b^3}} \cdot \sqrt{\dfrac{ab}{27}}$

83. $\sqrt[3]{\dfrac{16a^7}{b^4}} \cdot \sqrt[3]{\dfrac{b}{2a}}$

84. $\sqrt[3]{\dfrac{3y}{x^7}} \cdot \sqrt[3]{\dfrac{x}{81y^4}}$

85. $\sqrt[5]{\dfrac{x^2y}{z^7}} \cdot \sqrt[5]{\dfrac{y^9z^3}{x^7}}$

86. $\sqrt[5]{\dfrac{a^3b}{c^8}} \cdot \sqrt[5]{\dfrac{b^4c^4}{a^8}}$

Rationalize the denominator.

87. $\dfrac{3x}{\sqrt{x} - \sqrt{xy}}$

88. $\dfrac{2a}{\sqrt{u} - \sqrt{ub}}$

89. $\dfrac{\sqrt{a}}{\sqrt{ab} - \sqrt{a}}$

90. $\dfrac{2\sqrt{x}}{3\sqrt{x} - 2\sqrt{y}}$

91. $\dfrac{\sqrt{x-4}}{\sqrt{x-4} + 2}$

92. $\dfrac{\sqrt{x+3}}{\sqrt{x+3} + 5}$

93. $\dfrac{\sqrt{x+3} + \sqrt{x-3}}{\sqrt{x+3} - \sqrt{x-3}}$

94. $\dfrac{\sqrt{x+5} + \sqrt{x-5}}{\sqrt{x+5} - \sqrt{x-5}}$

95. $\dfrac{1}{\sqrt[3]{x} + 2}$

96. $\dfrac{1}{\sqrt[3]{x} - 2}$

97. $\dfrac{1}{\sqrt[3]{3} + \sqrt[3]{2}}$

98. $\dfrac{1}{\sqrt[3]{3} - \sqrt[3]{2}}$

4.6 Equations with Radicals

This section is concerned with solving equations that involve one or more radicals. The first step in solving an equation that contains a radical is to eliminate the radical from the equation. To do so, we need an additional property.

Squaring Property of Equality

If both sides of an equation are squared, the solutions to the original equation are solutions to the resulting equation.

We will never lose solutions to our equations by squaring both sides. We may, however, introduce *extraneous solutions.* Extraneous solutions satisfy the equation obtained by squaring both sides of the original equation, but they do not satisfy the original equation.

We know that if two real numbers a and b are equal, then so are their squares:

If $\quad a = b$

then $\quad a^2 = b^2$

However, extraneous solutions are introduced when we square opposites; that is, even though opposites are not equal, their squares are. For example,

$$5 = -5 \qquad \textbf{A false statement}$$

$$(5)^2 = (-5)^2 \qquad \textbf{Square both sides}$$

$$25 = 25 \qquad \textbf{A true statement}$$

We are free to square both sides of an equation any time it is convenient. We must be aware, however, that doing so may introduce extraneous solutions. We must, therefore, check all our solutions in the original equation if at any time we square both sides of the original equation.

Practice Problems

1. Solve for x: $\sqrt{2x + 4} = 4$

EXAMPLE 1 Solve for x: $\sqrt{3x + 4} = 5$.

SOLUTION We square both sides and proceed as usual:

$$\sqrt{3x + 4} = 5$$

$$(\sqrt{3x + 4})^2 = 5^2$$

$$3x + 4 = 25$$

$$3x = 21$$

$$x = 7$$

Checking $x = 7$ in the original equation, we have

$$\sqrt{3(7) + 4} \stackrel{?}{=} 5$$

$$\sqrt{21 + 4} \stackrel{?}{=} 5$$

$$\sqrt{25} \stackrel{?}{=} 5$$

$$5 = 5$$

The solution $x = 7$ satisfies the original equation.

Answer

1. 6

EXAMPLE 2 Solve $\sqrt{4x-7} = -3$.

SOLUTION Squaring both sides, we have

$$\sqrt{4x-7} = -3$$
$$(\sqrt{4x-7})^2 = (-3)^2$$
$$4x - 7 = 9$$
$$4x = 16$$
$$x = 4$$

Checking $x = 4$ in the original equation gives

$$\sqrt{4(4)-7} \stackrel{?}{=} -3$$
$$\sqrt{16-7} \stackrel{?}{=} -3$$
$$\sqrt{9} \stackrel{?}{=} -3$$
$$3 = -3$$

The solution $x = 4$ produces a false statement when checked in the original equation. Since $x = 4$ was the only possible solution, there is no solution to the original equation. The possible solution $x = 4$ is an extraneous solution. It satisfies the equation obtained by squaring both sides of the original equation, but it does not satisfy the original equation.

2. Solve $\sqrt{7x-3} = -5$.

Note: The fact that there is no solution to the equation in Example 2 was obvious to begin with. Notice that the left side of the equation is the *positive* square root of $4x - 7$, which must be a positive number or 0. The right side of the equation is -3. Since we cannot have a number that is either positive or zero equal to a negative number, there is no solution to the equation.

EXAMPLE 3 Solve $\sqrt{5x-1} + 3 = 7$.

SOLUTION We must isolate the radical on the left side of the equation. If we attempt to square both sides without doing so, the resulting equation will also contain a radical. Adding -3 to both sides, we have

$$\sqrt{5x-1} + 3 = 7$$
$$\sqrt{5x-1} = 4$$

We can now square both sides and proceed as usual:

$$(\sqrt{5x-1})^2 = 4^2$$
$$5x - 1 = 16$$
$$5x = 17$$
$$x = \frac{17}{5}$$

Checking $x = \frac{17}{5}$, we have

$$\sqrt{5\left(\frac{17}{5}\right) - 1} + 3 \stackrel{?}{=} 7$$
$$\sqrt{17-1} + 3 \stackrel{?}{=} 7$$
$$\sqrt{16} + 3 \stackrel{?}{=} 7$$
$$4 + 3 \stackrel{?}{=} 7$$
$$7 = 7$$

3. Solve $\sqrt{4x+5} + 2 = 7$.

Answers

2. No solution 3. 5

4. Solve $t - 6 = \sqrt{t - 4}$.

EXAMPLE 4 Solve $t + 5 = \sqrt{t + 7}$.

SOLUTION This time, squaring both sides of the equation results in a quadratic equation:

$$(t + 5)^2 = (\sqrt{t + 7})^2 \quad \textbf{Square both sides}$$
$$t^2 + 10t + 25 = t + 7$$
$$t^2 + 9t + 18 = 0 \quad \textbf{Standard form}$$
$$(t + 3)(t + 6) = 0 \quad \textbf{Factor the left side}$$
$$t + 3 = 0 \quad \text{or} \quad t + 6 = 0 \quad \textbf{Set factors equal to 0}$$
$$t = -3 \quad \text{or} \quad t = -6$$

We must check each solution in the original equation:

Check $t = -3$	Check $t = -6$
$-3 + 5 \stackrel{?}{=} \sqrt{-3 + 7}$	$-6 + 5 \stackrel{?}{=} \sqrt{-6 + 7}$
$2 \stackrel{?}{=} \sqrt{4}$	$-1 \stackrel{?}{=} \sqrt{1}$
$2 = 2$	$-1 = 1$
A true statement	A false statement

Since $t = -6$ does not check, our only solution is $t = -3$.

5. Solve $\sqrt{x - 9} = \sqrt{x} - 3$.

EXAMPLE 5 Solve $\sqrt{x - 3} = \sqrt{x} - 3$.

SOLUTION We begin by squaring both sides. Note what happens when we square the right side of the equation, and compare the square of the right side with the square of the left side. You must convince yourself that these results are correct. (The note in the margin will help if you are having trouble convincing yourself that what is written below is true.)

$$(\sqrt{x - 3})^2 = (\sqrt{x} - 3)^2$$
$$x - 3 = x - 6\sqrt{x} + 9$$

Note: It is very important that you realize that the square of $(\sqrt{x} - 3)$ is not $x + 9$. Remember, when we square a difference with two terms, we use the formula

$$(a - b)^2 = a^2 - 2ab + b^2$$

Applying this formula to $(\sqrt{x} - 3)^2$, we have

$$(\sqrt{x} - 3)^2 = (\sqrt{x})^2 - 2(\sqrt{x})(3) + 3^2$$
$$= x - 6\sqrt{x} + 9$$

Now we still have a radical in our equation, so we will have to square both sides again. Before we do, though, let's isolate the remaining radical.

$$x - 3 = x - 6\sqrt{x} + 9$$
$$-3 = -6\sqrt{x} + 9 \quad \textbf{Add } -x \textbf{ to each side}$$
$$-12 = -6\sqrt{x} \quad \textbf{Add } -9 \textbf{ to each side}$$
$$2 = \sqrt{x} \quad \textbf{Divide each side by } -6$$
$$4 = x \quad \textbf{Square each side}$$

Our only possible solution is $x = 4$, which we check in our original equation as follows:

$$\sqrt{4 - 3} \stackrel{?}{=} \sqrt{4} - 3$$
$$\sqrt{1} \stackrel{?}{=} 2 - 3$$
$$1 = -1 \quad \textbf{A false statement}$$

Substituting 4 for x in the original equation yields a false statement. Since 4 was our only possible solution, there is no solution to our equation.

Answers
4. 8 5. 9

Here is another example of an equation for which we must apply our squaring property twice before all radicals are eliminated.

EXAMPLE 6 Solve $\sqrt{x+1} = 1 - \sqrt{2x}$.

6. Solve $\sqrt{x+4} = 2 - \sqrt{3x}$.

SOLUTION This equation has two separate terms involving radical signs. Squaring both sides gives

$$x + 1 = 1 - 2\sqrt{2x} + 2x$$

$$-x = -2\sqrt{2x} \qquad \textbf{Add } -2x \textbf{ and } -1 \textbf{ to both sides}$$

$$x^2 = 4(2x) \qquad \textbf{Square both sides}$$

$$x^2 - 8x = 0 \qquad \textbf{Standard form}$$

Our equation is a quadratic equation in standard form. To solve for x, we factor the left side and set each factor equal to 0.:

$$x(x-8) = 0 \qquad \textbf{Factor left side}$$

$$x = 0 \quad \text{or} \quad x - 8 = 0 \qquad \textbf{Set factors equal to 0}$$

$$x = 8$$

Since we squared both sides of our equation, we have the possibility that one or both of the solutions are extraneous. We must check each one in the original equation:

Check $x = 8$	Check $x = 0$
$\sqrt{8+1} \stackrel{?}{=} 1 - \sqrt{2 \cdot 8}$	$\sqrt{0+1} \stackrel{?}{=} 1 - \sqrt{2 \cdot 0}$
$\sqrt{9} \stackrel{?}{=} 1 - \sqrt{16}$	$\sqrt{1} \stackrel{?}{=} 1 - \sqrt{0}$
$3 \stackrel{?}{=} 1 - 4$	$1 \stackrel{?}{=} 1 - 0$
$3 = -3$	$1 = 1$
A false statement	A true statement

Since $x = 8$ does not check, it is an extraneous solution. Our only solution is $x = 0$.

EXAMPLE 7 Solve $\sqrt{x+1} = \sqrt{x+2} - 1$.

7. Solve $\sqrt{x+2} = \sqrt{x+3} - 1$.

SOLUTION Squaring both sides we have

$$(\sqrt{x+1})^2 = (\sqrt{x+2} - 1)^2$$

$$x + 1 = x + 2 - 2\sqrt{x+2} + 1$$

Once again we are left with a radical in our equation. Before we square each side again, we must isolate the radical on the right side of the equation.

$$x + 1 = x + 3 - 2\sqrt{x+2} \qquad \textbf{Simplify the right side}$$

$$1 = 3 - 2\sqrt{x+2} \qquad \textbf{Add } -x \textbf{ to each side}$$

$$-2 = -2\sqrt{x+2} \qquad \textbf{Add } -3 \textbf{ to each side}$$

$$1 = \sqrt{x+2} \qquad \textbf{Divide each side by } -2$$

$$1 = x + 2 \qquad \textbf{Square both sides}$$

$$-1 = x \qquad \textbf{Add } -2 \textbf{ to each side}$$

Answer

6. 0

Checking our only possible solution, $x = -1$, in our original equation, we have

$$\sqrt{-1+1} \stackrel{?}{=} \sqrt{-1+2} - 1$$
$$\sqrt{0} \stackrel{?}{=} \sqrt{1} - 1$$
$$0 \stackrel{?}{=} 1 - 1$$
$$0 = 0 \quad \textbf{A true statement}$$

Our solution checks.

It is also possible to raise both sides of an equation to powers greater than 2. We only need to check for extraneous solutions when we raise both sides of an equation to an even power. Raising both sides of an equation to an odd power will not produce extraneous solutions.

8. Solve $\sqrt[3]{3x-7} = 2$.

EXAMPLE 8 Solve $\sqrt[3]{4x+5} = 3$.

SOLUTION Cubing both sides we have

$$(\sqrt[3]{4x+5})^3 = 3^3$$
$$4x + 5 = 27$$
$$4x = 22$$
$$x = \frac{22}{4}$$
$$x = \frac{11}{2}$$

We do not need to check $x = \frac{11}{2}$ since we raised both sides to an odd power.

We end this section by looking at graphs of some equations that contain radicals.

9. Graph each equation:
$y = \sqrt{x} + 3$
$y = \sqrt{x+3}$

EXAMPLE 9 Graph $y = \sqrt{x}$ and $y = \sqrt[3]{x}$.

SOLUTION The graphs are shown in Figures 1 and 2. Notice that the graph of $y = \sqrt{x}$ appears in the first quadrant only because in the equation $y = \sqrt{x}$, x and y cannot be negative.

The graph of $y = \sqrt[3]{x}$ appears in quadrants 1 and 3 since the cube root of a positive number is also a positive number and the cube root of a negative number is a negative number; that is, when x is positive, y will be positive and when x is negative, y will be negative.

The graphs of both equations will contain the origin since $y = 0$ when $x = 0$ in both equations.

Answers
7. -2 **8.** 5

x	y
−4	undefined
−1	undefined
0	0
1	1
4	2
9	3
16	4

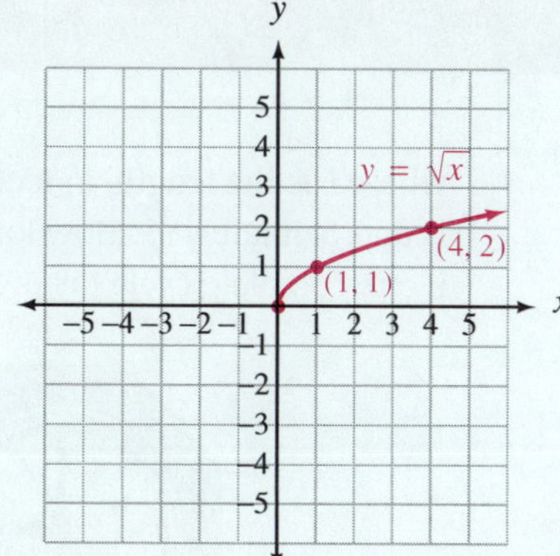

Figure 1

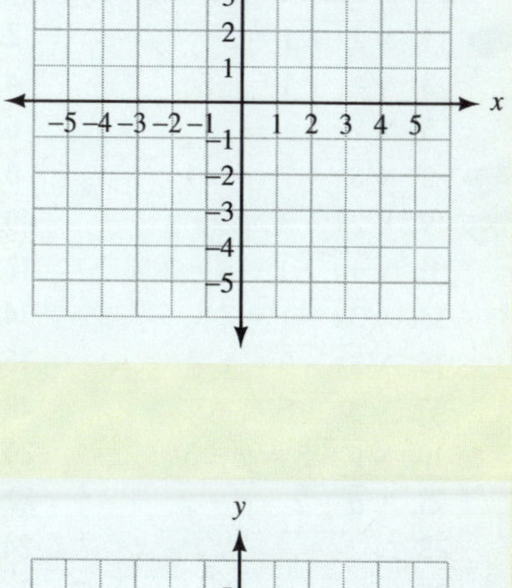

x	y
−27	−3
−8	−2
−1	−1
0	0
1	1
8	2
27	3

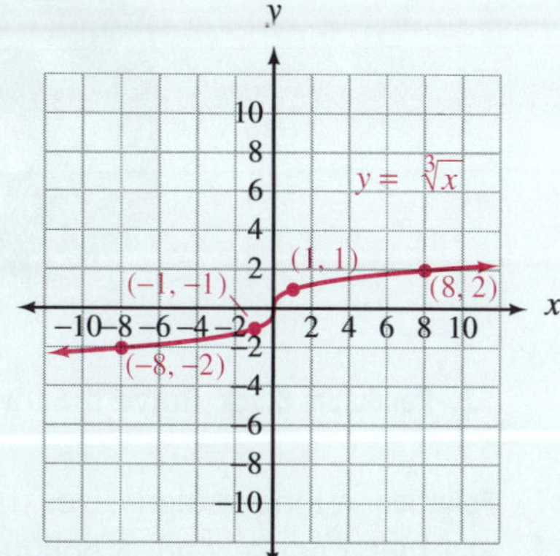

Figure 2

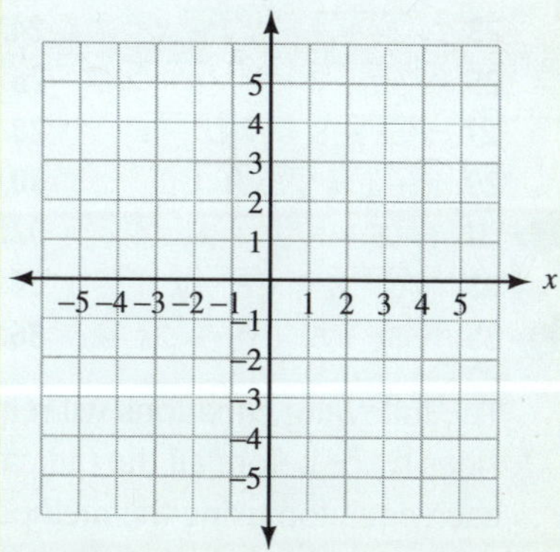

Getting Ready for Class

After reading through the preceding section, respond in your own words and in complete sentences.

A. What is the squaring property of equality?

B. Under what conditions do we obtain extraneous solutions to equations that contain radical expressions?

C. If we have raised both sides of an equation to a power, when is it not necessary to check for extraneous solutions?

D. When will you need to apply the squaring property of equality twice in the process of solving an equation containing radicals?

Answer

9. See Solutions to Selected Practice Problems.

PROBLEM SET 4.6

Solve each of the following equations.

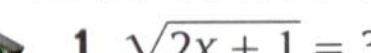

1. $\sqrt{2x+1} = 3$
2. $\sqrt{3x+1} = 4$
3. $\sqrt{4x+1} = -5$
4. $\sqrt{6x+1} = -5$
5. $\sqrt{2y-1} = 3$
6. $\sqrt{3y-1} = 2$
7. $\sqrt{5x-7} = -1$
8. $\sqrt{8x+3} = -6$
9. $\sqrt{2x-3} - 2 = 4$
10. $\sqrt{3x+1} - 4 = 1$
11. $\sqrt{4a+1} + 3 = 2$
12. $\sqrt{5a-3} + 6 = 2$
13. $\sqrt[4]{3x+1} = 2$
14. $\sqrt[4]{4x+1} = 3$
15. $\sqrt[3]{2x-5} = 1$
16. $\sqrt[3]{5x+7} = 2$
17. $\sqrt[3]{3a+5} = -3$
18. $\sqrt[3]{2a+7} = -2$
19. $\sqrt{y-3} = y-3$
20. $\sqrt{y+3} = y-3$
21. $\sqrt{a+2} = a+2$
22. $\sqrt{a+10} = a-2$
23. $\sqrt{2x+4} = \sqrt{1-x}$
24. $\sqrt{3x+4} = -\sqrt{2x+3}$
25. $\sqrt{4a+7} = -\sqrt{a+2}$
26. $\sqrt{7a-1} = \sqrt{2a+4}$
27. $\sqrt[4]{5x-8} = \sqrt[4]{4x-1}$
28. $\sqrt[4]{6x+7} = \sqrt[4]{x+2}$
29. $x+1 = \sqrt{5x+1}$
30. $x-1 = \sqrt{6x+1}$
31. $t+5 = \sqrt{2t+9}$
32. $t+7 = \sqrt{2t+13}$
33. $\sqrt{y-8} = \sqrt{8-y}$
34. $\sqrt{2y+5} = \sqrt{5y+2}$
35. $\sqrt[3]{3x+5} = \sqrt[3]{5-2x}$
36. $\sqrt[3]{4x+9} = \sqrt[3]{3-2x}$

The following equations wll require that you square both sides twice before all the radicals are eliminated. Solve each equation using the methods shown in Examples 5, 6, and 7.

37. $\sqrt{x-8} = \sqrt{x} - 2$
38. $\sqrt{x+3} = \sqrt{x} - 3$
39. $\sqrt{x+1} = \sqrt{x} + 1$
40. $\sqrt{x-1} = \sqrt{x} - 1$
41. $\sqrt{x+8} = \sqrt{x-4} + 2$
42. $\sqrt{x+5} = \sqrt{x-3} + 2$
43. $\sqrt{x-5} - 3 = \sqrt{x-8}$
44. $\sqrt{x-3} - 4 = \sqrt{x-3}$
45. $\sqrt{x+4} = 2 - \sqrt{2x}$
46. $\sqrt{5x+1} = 1 + \sqrt{5x}$
47. $\sqrt{2x+4} = \sqrt{x+3} + 1$
48. $\sqrt{2x-1} = \sqrt{x-4} + 2$

Applying the Concepts

49. **Solving a Formula** Solve the following formula for h:

$$t = \frac{\sqrt{100-h}}{4}$$

50. **Solving a Formula** Solve the following formula for h:

$$t = \sqrt{\frac{2h-40t}{g}}$$

51. **Pendulum Clock** The length of time (T) in seconds it takes the pendulum of a clock to swing through one complete cycle is given by the formula

$$T = 2\pi\sqrt{\frac{L}{32}}$$

where L is the length, in feet, of the pendulum, and π is approximately $\frac{22}{7}$. How long must the pendulum be if one complete cycle takes 2 seconds?

52. **Pendulum Clock** Solve the formula in Problem 51 for L.

Pollution A long straight river, 100 meters wide, is flowing at 1 meter per second. A pollutant is entering the river at a constant rate from one of its banks. As the pollutant disperses in the water, it forms a plume that is modeled by the equation $y = \sqrt{x}$. Use this information to answer the following questions.

53. How wide is the plume 25 meters down river from the source of the pollution?
54. How wide is the plume 100 meters down river from the source of the pollution?
55. How far down river from the source of the pollution does the plume reach halfway across the river?
56. How far down the river from the source of the pollution does the plume reach the other side of the river?
57. For the situation described in the instructions and modeled by the equation $y = \sqrt{x}$, what is the range of values that y can assume?

58. If the river was moving at 2 meters per second, would the plume be larger or smaller 100 meters downstream from the source?

Graph each equation.

59. $y = 2\sqrt{x}$

60. $y = -2\sqrt{x}$

61. $y = \sqrt{x} - 2$

62. $y = \sqrt{x} + 2$

63. $y = \sqrt{x - 2}$

64. $y = \sqrt{x + 2}$

65. $y = 3\sqrt[3]{x}$

66. $y = -3\sqrt[3]{x}$

67. $y = \sqrt[3]{x} + 3$

68. $y = \sqrt[3]{x} - 3$

69. $y = \sqrt[3]{x + 3}$

70. $y = \sqrt[3]{x - 3}$

Review Problems

Multiply.

71. $\sqrt{2}(\sqrt{3} - \sqrt{2})$

72. $(\sqrt{x} - 4)(\sqrt{x} + 5)$

73. $(\sqrt{x} + 5)^2$

74. $(\sqrt{5} + \sqrt{3})(\sqrt{5} - \sqrt{3})$

Rationalize the denominator.

75. $\dfrac{\sqrt{x}}{\sqrt{x} + 3}$

76. $\dfrac{\sqrt{5} - \sqrt{3}}{\sqrt{5} + \sqrt{3}}$

Extending the Concepts

Solve each equation.

77. $\dfrac{x}{3\sqrt{2x - 3}} - \dfrac{1}{\sqrt{2x - 3}} = \dfrac{1}{3}$

78. $\dfrac{x}{5\sqrt{2x + 10}} + \dfrac{1}{\sqrt{2x + 10}} = \dfrac{1}{5}$

79. $x + 1 = \sqrt[3]{4x + 4}$

80. $x - 1 = \sqrt[3]{4x - 4}$

Solve for y in terms of x.

81. $y + 2 = \sqrt{x^2 + (y - 2)^2}$

82. $y + \dfrac{1}{2} = \sqrt{x^2 + \left(y - \dfrac{1}{2}\right)^2}$

83. Use your Y variables list, or write a program, to graph the family of curves $Y = \sqrt{X} + B$ for B = −3, −2, −1, 0, 1, 2 and 3.

84. Use your Y variables list, or write a program, to graph the family of curves $Y = \sqrt{X + B}$ for B = −3, −2, −1,0, 1, 2, and 3.

85. Summarize the results of Problem 83 by giving a written description of the effect of b on the graph of $y = \sqrt{x} + b$.

86. Summarize the results of Problem 84 by giving a written description of the effect of b on the graph of $y = \sqrt{x + b}$.

87. Use your Y variables list, or write a program, to graph the family of curves $Y = \sqrt[3]{X} + B$ for B = −3, −2, −1, 0, 1, 2, and 3.

88. Use your Y variables list, or write a program, to graph the family of curves $Y = \sqrt[3]{X + B}$ for B = −3, −2, −1, 0, 1, 2, and 3.

89. Summarize the results of Problem 87 by giving a written description of the effect of b on the graph of $y = \sqrt[3]{x} + b$.

90. Summarize the results of Problem 88 by giving a written description of the effect of b on the graph of $y = \sqrt[3]{x + b}$.

91. Use your Y variables list, or write a program, to graph the family of curves $Y = A\sqrt{X}$ for A = −3, −2, −1, 0, 1, 2, and 3.

92. Use your Y variables list, or write a program, to graph the family of curves $Y = A\sqrt{X}$ for $A = \frac{1}{4}, \frac{1}{3}, \frac{1}{2}$, 1, 2, and 3.

93. Summarize the results of Problems 91 and 92 by giving a written description of the effect of a on the graph of $y = a\sqrt{x}$.

CHAPTER 4 SUMMARY

Examples

1. The number 49 has two square roots, 7 and −7. They are written like this:
   $\sqrt{49} = 7 \qquad -\sqrt{49} = -7$

Square Roots [4.1]

Every positive real number x has two square roots. The *positive square root* of x is written $\sqrt{x}$, and the *negative square root* of x is written $-\sqrt{x}$. Both the positive and the negative square roots of x are numbers we square to get x; that is,

$$\left.\begin{aligned}(\sqrt{x})^2 &= x\\ \text{and}\quad (-\sqrt{x})^2 &= x\end{aligned}\right\}\quad \text{for } x \geq 0$$

2. $\sqrt[3]{8} = 2$
   $\sqrt[3]{-27} = -3$

Higher Roots [4.1]

In the expression $\sqrt[n]{a}$, n is the *index*, a is the *radicand*, and $\sqrt{}$ is the *radical sign*. The expression $\sqrt[n]{a}$ is such that

$$(\sqrt[n]{a})^n = a \qquad a \geq 0 \text{ when } n \text{ is even}$$

3. $25^{1/2} = \sqrt{25} = 5$
   $8^{2/3} = (\sqrt[3]{8})^2 = 2^2 = 4$
   $9^{3/2} = (\sqrt{9})^3 = 3^3 = 27$

Rational Exponents [4.1, 4.2]

Rational exponents are used to indicate roots. The relationship between rational exponents and roots is as follows:

$$a^{1/n} = \sqrt[n]{a} \qquad \text{and} \qquad a^{m/n} = (a^{1/n})^m = (a^m)^{1/n}$$

$$a \geq 0 \text{ when } n \text{ is even}$$

4. $\sqrt{4 \cdot 5} = \sqrt{4}\,\sqrt{5} = 2\sqrt{5}$
   $\sqrt{\dfrac{7}{9}} = \dfrac{\sqrt{7}}{\sqrt{9}} = \dfrac{\sqrt{7}}{3}$

Properties of Radicals [4.3]

If a and b are nonnegative real numbers whenever n is even, then

1. $\sqrt[n]{ab} = \sqrt[n]{a}\,\sqrt[n]{b}$
2. $\sqrt[n]{\dfrac{a}{b}} = \dfrac{\sqrt[n]{a}}{\sqrt[n]{b}} \qquad (b \neq 0)$

5. $\sqrt{\dfrac{4}{5}} = \dfrac{\sqrt{4}}{\sqrt{5}}$
   $= \dfrac{2}{\sqrt{5}} \cdot \dfrac{\sqrt{5}}{\sqrt{5}}$
   $= \dfrac{2\sqrt{5}}{5}$

Simplified Form for Radicals [4.3]

A radical expression is said to be in *simplified form*

1. If there is no factor of the radicand that can be written as a power greater than or equal to the index;
2. If there are no fractions under the radical sign; and
3. If there are no radicals in the denominator.

6. $5\sqrt{3} - 7\sqrt{3} = (5 - 7)\sqrt{3}$
   $= -2\sqrt{3}$
   $\sqrt{20} + \sqrt{45} = 2\sqrt{5} + 3\sqrt{5}$
   $= (2 + 3)\sqrt{5}$
   $= 5\sqrt{5}$

Addition and Subtraction of Radical Expressions [4.4]

We add and subtract radical expressions by using the distributive property to combine similar radicals. Similar radicals are radicals with the same index and the same radicand.

7. $(\sqrt{x} + 2)(\sqrt{x} + 3)$
   $= \sqrt{x}\,\sqrt{x} + 3\sqrt{x} + 2\sqrt{x} + 2 \cdot 3$
   $= x + 5\sqrt{x} + 6$

Multiplication of Radical Expressions [4.5]

We multiply radical expressions in the same way that we multiply polynomials. We can use the distributive property and the FOIL method.

8. $\dfrac{3}{\sqrt{2}} = \dfrac{3}{\sqrt{2}} \cdot \dfrac{\sqrt{2}}{\sqrt{2}} = \dfrac{3\sqrt{2}}{2}$
   $\dfrac{3}{\sqrt{5} - \sqrt{3}} = \dfrac{3}{\sqrt{5} - \sqrt{3}} \cdot \dfrac{\sqrt{5} + \sqrt{3}}{\sqrt{5} + \sqrt{3}}$
   $= \dfrac{3\sqrt{5} + 3\sqrt{3}}{5 - 3}$
   $= \dfrac{3\sqrt{5} + 3\sqrt{3}}{2}$

Rationalizing the Denominator [4.3, 4.5]

When a fraction contains a square root in the denominator, we rationalize the denominator by multiplying numerator and denominator by

1. The square root itself if there is only one term in the denominator, or
2. The conjugate of the denominator if there are two terms in the denominator.

Rationalizing the denominator is also called division of radical expressions.

9. $\sqrt{2x + 1} = 3$
   $(\sqrt{2x + 1})^2 = 3^2$
   $2x + 1 = 9$
   $x = 4$

Squaring Property of Equality [4.6]

We may square both sides of an equation any time it is convenient to do so, as long as we check all resulting solutions in the original equation.

CHAPTER 4 REVIEW

The problems below form a comprehensive review of the material in this chapter. They can be used to study for exams. If you would like to take a practice test on this chapter, you can use the odd-numbered problems. Give yourself an hour and work as many of the odd-numbered problems as possible. When you are finished, or when an hour has passed, check your answers with the answers in the back of the book. You can use the even-numbered problems for a second practice test.

Simplify each expression as much as possible. [4.1]

1. $49^{1/2}$ **2.** $(-27)^{1/3}$ **3.** $16^{1/4}$ **4.** $9^{3/2}$ **5.** $\sqrt[5]{32x^{15}y^{10}}$ **6.** $8^{-4/3}$

Use the properties of exponents to simplify each expression. Assume all bases represent positive numbers. [4.1]

7. $x^{2/3} \cdot x^{4/3}$ **8.** $(a^{2/3}b^{4/3})^3$ **9.** $\dfrac{a^{3/5}}{a^{1/4}}$ **10.** $\dfrac{a^{2/3}b^3}{a^{1/4}b^{1/3}}$

Multiply. [4.2]

11. $(3x^{1/2} + 5y^{1/2})(4x^{1/2} - 3y^{1/2})$ **12.** $(a^{1/3} - 5)^2$

13. Divide: $\dfrac{28x^{5/6} + 14x^{7/6}}{7x^{1/3}}$. (Assume $x > 0$.) [4.2] **14.** Factor $2(x-3)^{1/4}$ from $8(x-3)^{5/4} - 2(x-3)^{1/4}$. [4.2]

15. Simplify $x^{3/4} + \dfrac{5}{x^{1/4}}$ into a single fraction. (Assume $x > 0$.) [4.2]

Write each expression in simplified form for radicals. (Assume all variables represent nonnegative numbers.) [4.3]

16. $\sqrt{12}$ **17.** $\sqrt{50}$ **18.** $\sqrt[3]{16}$ **19.** $\sqrt{18x^2}$ **20.** $\sqrt{80a^3b^4c^2}$ **21.** $\sqrt[4]{32a^4b^5c^6}$

Rationalize the denominator in each expression. [4.3]

22. $\dfrac{3}{\sqrt{2}}$ **23.** $\dfrac{6}{\sqrt[3]{2}}$

Write each expression in simplified form. (Assume all variables represent positive numbers.) [4.3]

24. $\sqrt{\dfrac{48x^3}{7y}}$ **25.** $\sqrt[3]{\dfrac{40x^2y^3}{3z}}$

Combine the following expressions. (Assume all variables represent positive numbers.) [4.4]

26. $5x\sqrt{6} + 2x\sqrt{6} - 9x\sqrt{6}$ **27.** $\sqrt{12} + \sqrt{3}$ **28.** $\dfrac{3}{\sqrt{5}} + \sqrt{5}$

29. $3\sqrt{8} - 4\sqrt{72} + 5\sqrt{50}$ **30.** $3b\sqrt{27a^5b} + 2a\sqrt{3a^3b^3}$ **31.** $2x\sqrt[3]{xy^3z^2} - 6y\sqrt[3]{x^4z^2}$

Multiply. [4.5]

32. $\sqrt{2}(\sqrt{3} - 2\sqrt{2})$ **33.** $(\sqrt{x} - 2)(\sqrt{x} - 3)$

Rationalize the denominator. [4.5]

34. $\dfrac{3}{\sqrt{5} - 2}$ **35.** $\dfrac{\sqrt{7} + \sqrt{5}}{\sqrt{7} - \sqrt{5}}$ **36.** $\dfrac{3\sqrt{7}}{3\sqrt{7} - 4}$

Solve each equation. [4.6]

37. $\sqrt{4a+1} = 1$ **38.** $\sqrt[3]{3x-8} = 1$ **39.** $\sqrt{3x+1} - 3 = 1$ **40.** $\sqrt{x+4} = \sqrt{x} - 2$

Graph each equation. [4.6]

41. $y = 3\sqrt{x}$ **42.** $y = \sqrt[3]{x} + 2$

43. Construction The roof of the house shown in Figure 1 is to extend up 13.5 feet above the ceiling, which is 36 feet across. Find the length of one side of the roof.

Figure 1

44. Surveying A surveyor is attempting to find the distance across a pond. From a point on one side of the pond he walks 25 yards to the end of the pond and then makes a 90-degree turn and walks another 60 yards before coming to a point directly across the pond from the point at which he started. What is the distance across the pond? (See Figure 2.)

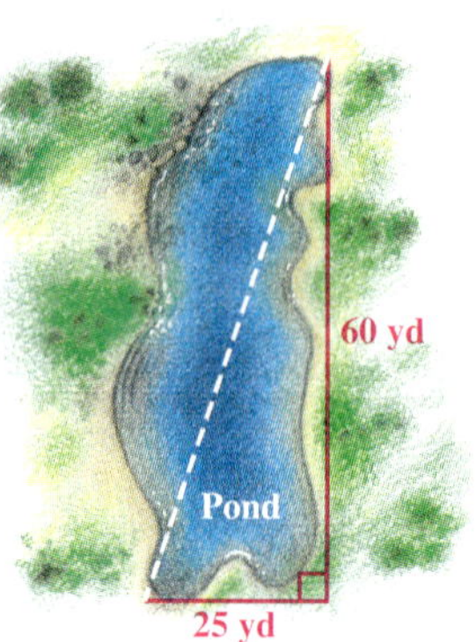

Figure 2

Quadratic Equations

Table 1 is taken from the trail map given to skiers at the Northstar at Tahoe Ski Resort in Lake Tahoe, California. The table gives the length of each chair lift at Northstar, along with the change in elevation from the beginning of the lift to the end of the lift.

Right triangles are good mathematical models for chair lifts. In this chapter, we will use our knowledge of right triangles, along with the new material developed in the chapter, to solve problems involving chair lifts and a variety of other examples.

CHAPTER OUTLINE

Table 1

FROM THE TRAIL MAP FOR THE NORTHSTAR AT TAHOE SKI RESORT

LIFT INFORMATION

LIFT	VERTICAL RISE (FEET)	LENGTH (FEET)
Big Springs Gondola	480	4,100
Bear Paw Double	120	790
Echo Triple	710	4,890
Aspen Express Quad	900	5,100
Forest Double	1,170	5,750
Lookout Double	960	4,330
Comstock Express Quad	1,250	5,900
Rendezvous Triple	650	2,900
Schaffer Camp Triple	1,860	6,150
Chipmunk Tow Lift	28	280
Bear Cub Tow Lift	120	750

5.1 Complex Numbers

The stamp shown here was issued by Germany in 1977 to commemorate the 200th anniversary of the birth of mathematician Carl Gauss. The number $-5+6i$ shown on the stamp is a complex number, as are the other numbers on the stamp. Working with complex numbers gives us a way to solve a wider variety of equations. For example, the equation $x^2 = -9$ has no real number solutions since the square of a real number is always positive. We have been unable to work with square roots of negative numbers like $\sqrt{-25}$ and $\sqrt{-16}$ for the same reason. Complex numbers allow us to expand our work with radicals to include square roots of negative numbers and solve equations like $x^2 = -9$ and $x^2 = -64$. Our work with complex numbers is based on the following definition.

DEFINITION

The number i is such that $i = \sqrt{-1}$ (which is the same as saying $i^2 = -1$).

The number i, as we have defined it here, is not a real number. Because of the way we have defined i, we can use it to simplify square roots of negative numbers.

Square Roots of Negative Numbers

If a is a positive number, then $\sqrt{-a}$ can always be written as $i\sqrt{a}$; that is,

$$\sqrt{-a} = i\sqrt{a} \quad \text{if } a \text{ is a positive number}$$

To justify our rule, we simply square the quantity $i\sqrt{a}$ to obtain $-a$. Here is what it looks like when we do so:

$$\begin{aligned}(i\sqrt{a})^2 &= i^2 \cdot (\sqrt{a})^2 \\ &= -1 \cdot a \\ &= -a\end{aligned}$$

Here are some examples that illustrate the use of our new rule.

EXAMPLES Write each square root in terms of the number i.

1. $\sqrt{-25} = i\sqrt{25} = i \cdot 5 = 5i$
2. $\sqrt{-49} = i\sqrt{49} = i \cdot 7 = 7i$
3. $\sqrt{-12} = i\sqrt{12} = i \cdot 2\sqrt{3} = 2i\sqrt{3}$
4. $\sqrt{-17} = i\sqrt{17}$

Practice Problems

1. $\sqrt{-36}$
2. $-\sqrt{-64}$
3. $\sqrt{-18}$
4. $-\sqrt{-19}$

Note: In Examples 3 and 4 we wrote i before the radical simply to avoid confusion. If we were to write the answer to 3 as $2\sqrt{3}i$, some people would think the i was under the radical sign and it is not.

Answers

1. $6i$ 2. $-8i$ 3. $3i\sqrt{2}$ 4. $-i\sqrt{19}$

If we assume all the properties of exponents hold when the base is i, we can write any power of i as i, -1, $-i$, or 1. Using the fact that $i^2 = -1$, we have

$$\begin{aligned}i^1 &= i \\ i^2 &= -1 \\ i^3 &= i^2 \cdot i = -1(i) = -i \\ i^4 &= i^2 \cdot i^2 = -1(-1) = 1\end{aligned}$$

Since $i^4 = 1$, i^5 will simplify to i, and we will begin repeating the sequence i, -1, $-i$, 1 as we simplify higher powers of i: Any power of i simplifies to i, -1, $-i$, or 1. The easiest way to simplify higher powers of i is to write them in terms of i^2. For instance, to simplify i^{21}, we would write it as

$$(i^2)^{10} \cdot i \qquad \text{because } 2 \cdot 10 + 1 = 21$$

Then, since $i^2 = -1$, we have

$$(-1)^{10} \cdot i = 1 \cdot i = i$$

EXAMPLES Simplify as much as possible.

5. $i^{30} = (i^2)^{15} = (-1)^{15} = -1$

6. $i^{11} = (i^2)^5 \cdot i = (-1)^5 \cdot i = (-1)i = -i$

7. $i^{40} = (i^2)^{20} = (-1)^{20} = 1$

Simplify.

5. i^{20}

6. i^{23}

7. i^{50}

DEFINITION

A **complex number** is any number that can be put in the form

$$a + bi$$

where a and b are real numbers and $i = \sqrt{-1}$. The form $a + bi$ is called **standard form** for complex numbers. The number a is called the **real part** of the complex number. The number b is called the **imaginary part** of the complex number.

Every real number is a complex number. For example, 8 can be written as $8 + 0i$. Likewise, $-\frac{1}{2}$, π, $\sqrt{3}$, and -9 are complex numbers because they can all be written in the form $a + bi$:

$$-\frac{1}{2} = -\frac{1}{2} + 0i \qquad \pi = \pi + 0i$$

$$\sqrt{3} = \sqrt{3} + 0i \qquad -9 = -9 + 0i$$

The real numbers occur when $b = 0$. When $b \neq 0$, we have complex numbers that contain i, such as $2 + 5i$, $6 - i$, $4i$, and $\frac{1}{2}i$. These numbers are called *imaginary numbers.* The diagram explains this further.

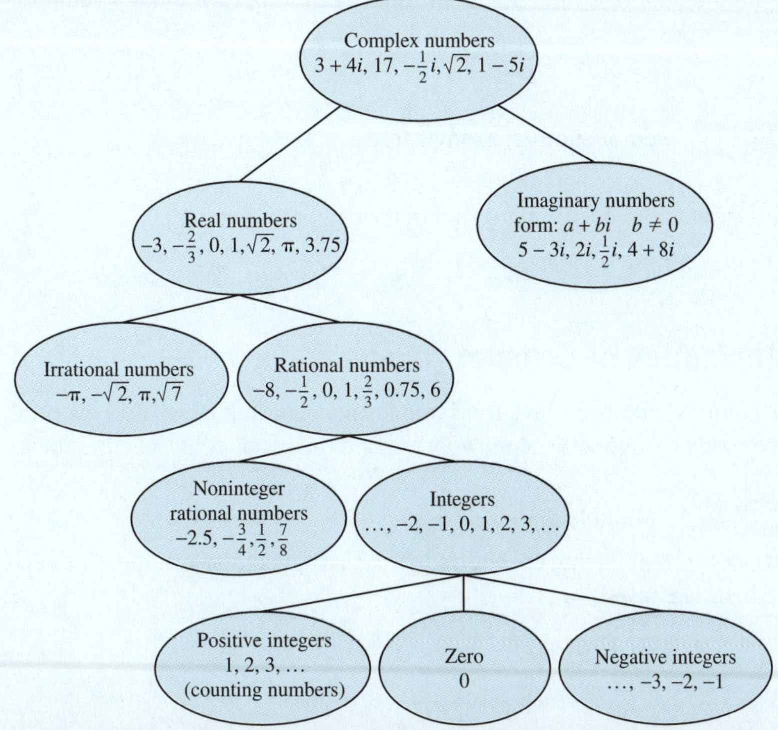

Answers

5. 1 **6.** $-i$ **7.** -1

Equality for Complex Numbers

Two complex numbers are equal if and only if their real parts are equal and their imaginary parts are equal; that is, for real numbers a, b, c, and d,

$$a + bi = c + di \quad \text{if and only if} \quad a = c \quad \text{and} \quad b = d$$

8. Find x and y if
$4x + 7i = 8 - 14yi$

EXAMPLE 8 Find x and y if $3x + 4i = 12 - 8yi$.

SOLUTION Since the two complex numbers are equal, their real parts are equal and their imaginary parts are equal:

$$3x = 12 \quad \text{and} \quad 4 = -8y$$

$$x = 4 \qquad\qquad y = -\frac{1}{2}$$

9. Find x and y if
$(2x - 1) + 9i = 5 + (4y + 1)i$

EXAMPLE 9 Find x and y if $(4x - 3) + 7i = 5 + (2y - 1)i$.

SOLUTION The real parts are $4x - 3$ and 5. The imaginary parts are 7 and $2y - 1$:

$$4x - 3 = 5 \quad \text{and} \quad 7 = 2y - 1$$

$$4x = 8 \qquad\qquad 8 = 2y$$

$$x = 2 \qquad\qquad y = 4$$

Addition and Subtraction of Complex Numbers

To add two complex numbers, add their real parts and add their imaginary parts; that is, if a, b, c, and d are real numbers, then

$$(a + bi) + (c + di) = (a + c) + (b + d)i$$

If we assume that the commutative, associative, and distributive properties hold for the number i, then the definition of addition is simply an extension of these properties.

We define subtraction in a similar manner. If a, b, c, and d are real numbers, then

$$(a + bi) - (c + di) = (a - c) + (b - d)i$$

Add or subtract as indicated.

10. $(2 + 6i) + (3 - 4i)$

11. $(6 + 5i) - (4 + 3i)$

12. $(7 - i) - (8 - 2i)$

EXAMPLES Add or subtract as indicated.

10. $(3 + 4i) + (7 - 6i) = (3 + 7) + (4 - 6)i = 10 - 2i$

11. $(7 + 3i) - (5 + 6i) = (7 - 5) + (3 - 6)i = 2 - 3i$

12. $(5 - 2i) - (9 - 4i) = (5 - 9) + (-2 + 4)i = -4 + 2i$

Multiplication of Complex Numbers

Since complex numbers have the same form as binomials, we find the product of two complex numbers the same way we find the product of two binomials.

13. Multiply $(2 + 3i)(1 - 4i)$.

EXAMPLE 13 Multiply $(3 - 4i)(2 + 5i)$.

SOLUTION Multiplying each term in the second complex number by each term in the first, we have

$$(3 - 4i)(2 + 5i) = 3 \cdot 2 + 3 \cdot 5i - 2 \cdot 4i - 5i(4i)$$

$$= 6 + 15i - 8i - 20i^2$$

Answers

8. $x = 2, y = -\frac{1}{2}$ **9.** $x = 3, y = 2$
10. $5 + 2i$ **11.** $2 + 2i$ **12.** $-1 + i$

Combining similar terms and using the fact that $i^2 = -1$, we can simplify as follows:

$$\begin{aligned} 6 + 15i - 8i - 20i^2 &= 6 + 7i - 20(-1) \\ &= 6 + 7i + 20 \\ &= 26 + 7i \end{aligned}$$

The product of the complex numbers $3 - 4i$ and $2 + 5i$ is the complex number $26 + 7i$.

EXAMPLE 14 Multiply $2i(4 - 6i)$.

SOLUTION Applying the distributive property gives us

$$\begin{aligned} 2i(4 - 6i) &= 2i \cdot 4 - 2i \cdot 6i \\ &= 8i - 12i^2 \\ &= 12 + 8i \end{aligned}$$

EXAMPLE 15 Expand $(3 + 5i)^2$.

SOLUTION We treat this like the square of a binomial. Remember, $(a + b)^2 = a^2 + 2ab + b^2$.

$$\begin{aligned} (3 + 5i)^2 &= 3^2 + 2(3)(5i) + (5i)^2 \\ &= 9 + 30i + 25i^2 \\ &= 9 + 30i - 25 \\ &= -16 + 30i \end{aligned}$$

EXAMPLE 16 Multiply $(2 - 3i)(2 + 3i)$.

SOLUTION This product has the form $(a - b)(a + b)$, which we know results in the difference of two squares, $a^2 - b^2$:

$$\begin{aligned} (2 - 3i)(2 + 3i) &= 2^2 - (3i)^2 \\ &= 4 - 9i^2 \\ &= 4 + 9 \\ &= 13 \end{aligned}$$

The product of the two complex numbers $2 - 3i$ and $2 + 3i$ is the real number 13. The two complex numbers $2 - 3i$ and $2 + 3i$ are called complex conjugates. The fact that their product is a real number is very useful.

DEFINITION

The complex numbers $a + bi$ and $a - bi$ are called **complex conjugates.** One important property they have is that their product is the real number $a^2 + b^2$. Here's why:

$$\begin{aligned} (a + bi)(a - bi) &= a^2 - (bi)^2 \\ &= a^2 - b^2i^2 \\ &= a^2 - b^2(-1) \\ &= a^2 + b^2 \end{aligned}$$

14. Multiply $-3i(2 + 3i)$.

15. Expand $(2 + 4i)^2$.

Note: We can obtain the same result by writing $(3 + 5i)^2$ as $(3 + 5i)$ times $(3 + 5i)$ and applying the FOIL method as we did with the problem in Example 13.

16. Multiply $(3 - 5i)(3 + 5i)$.

Answers

13. $14 - 5i$ **14.** $9 - 6i$
15. $-12 + 16i$ **16.** 34

Division with Complex Numbers

The fact that the product of two complex conjugates is a real number is the key to division with complex numbers.

17. Divide $\frac{3 + 2i}{2 - 5i}$.

EXAMPLE 17 Divide $\frac{2 + i}{3 - 2i}$.

SOLUTION We want a complex number in standard form that is equivalent to the quotient $\frac{2 + i}{3 - 2i}$. We need to eliminate i from the denominator. Multiplying the numerator and denominator by $3 + 2i$ will give us what we want:

$$\frac{2+i}{3-2i} = \frac{2+i}{3-2i} \cdot \frac{\mathbf{(3+2i)}}{\mathbf{(3+2i)}}$$

$$= \frac{6 + 4i + 3i + 2i^2}{9 - 4i^2}$$

$$= \frac{6 + 7i - 2}{9 + 4}$$

$$= \frac{4 + 7i}{13}$$

$$= \frac{4}{13} + \frac{7}{13}i$$

Dividing the complex number $2 + i$ by $3 - 2i$ gives the complex number $\frac{4}{13} + \frac{7}{13}i$.

18. Divide $\frac{3 + 2i}{i}$.

EXAMPLE 18 Divide $\frac{7 - 4i}{i}$.

SOLUTION The conjugate of the denominator is $-i$. Multiplying numerator and denominator by this number, we have

$$\frac{7-4i}{i} = \frac{7-4i}{i} \cdot \frac{\mathbf{-i}}{\mathbf{-i}}$$

$$= \frac{-7i + 4i^2}{-i^2}$$

$$= \frac{-7i + 4(-1)}{-(-1)}$$

$$= -4 - 7i$$

Getting Ready for Class

After reading through the preceding section, respond in your own words and in complete sentences.

A. What is the number i?
B. What is a complex number?
C. What kind of number results when we multiply complex conjugates?
D. Explain how to divide complex numbers.

Answers
17. $-\frac{4}{29} + \frac{19}{29}i$ 18. $2 - 3i$

PROBLEM SET 5.1

Write the following in terms of i, and simplify as much as possible.

1. $\sqrt{-36}$ **2.** $\sqrt{-49}$
3. $-\sqrt{-25}$ **4.** $-\sqrt{-81}$
5. $\sqrt{-72}$ **6.** $\sqrt{-48}$
7. $-\sqrt{-12}$ **8.** $-\sqrt{-75}$

Write each of the following as i, -1, $-i$, or 1.

9. i^{28} **10.** i^{31} **11.** i^{26}
12. i^{37} **13.** i^{75} **14.** i^{42}

Find x and y so each of the following equations is true.

15. $2x + 3yi - 6 - 3i$
16. $4x - 2yi = 4 + 8i$
17. $2 - 5i = -x + 10yi$
18. $4 + 7i = 6x - 14yi$
19. $2x + 10i = -16 - 2yi$
20. $4x - 5i = -2 + 3yi$
21. $(2x - 4) - 3i = 10 - 6yi$
22. $(4x - 3) - 2i = 8 + yi$
23. $(7x - 1) + 4i = 2 + (5y + 2)i$
24. $(5x + 2) - 7i = 4 + (2y + 1)i$

Combine the following complex numbers.

25. $(2 + 3i) + (3 + 6i)$
26. $(4 + i) + (3 + 2i)$
27. $(3 - 5i) + (2 + 4i)$
28. $(7 + 2i) + (3 - 4i)$
29. $(5 + 2i) - (3 + 6i)$

30. $(6 + 7i) - (4 + i)$
31. $(3 - 5i) - (2 + i)$
32. $(7 - 3i) - (4 + 10i)$
33. $[(3 + 2i) - (6 + i)] + (5 + i)$
34. $[(4 - 5i) - (2 + i)] + (2 + 5i)$
35. $[(7 - i) - (2 + 4i)] - (6 + 2i)$
36. $[(3 - i) - (4 + 7i)] - (3 - 4i)$
37. $(3 + 2i) - [(3 - 4i) - (6 + 2i)]$
38. $(7 - 4i) - [(-2 + i) - (3 + 7i)]$
39. $(4 - 9i) + [(2 - 7i) - (4 + 8i)]$
40. $(10 - 2i) - [(2 + i) - (3 - i)]$

Find the following products.

41. $3i(4 + 5i)$ **42.** $2i(3 + 4i)$
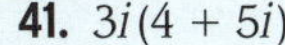
43. $6i(4 - 3i)$ **44.** $11i(2 - i)$
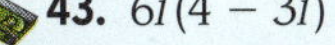
45. $(3 + 2i)(4 + i)$ **46.** $(2 - 4i)(3 + i)$
47. $(4 + 9i)(3 - i)$ **48.** $(5 - 2i)(1 + i)$
49. $(1 + i)^3$ **50.** $(1 - i)^3$
51. $(2 - i)^3$ **52.** $(2 + i)^3$
53. $(2 + 5i)^2$ **54.** $(3 + 2i)^2$
55. $(1 - i)^2$ **56.** $(1 + i)^2$
57. $(3 - 4i)^2$ **58.** $(6 - 5i)^2$
59. $(2 + i)(2 - i)$ **60.** $(3 + i)(3 - i)$
61. $(6 - 2i)(6 + 2i)$ **62.** $(5 + 4i)(5 - 4i)$
63. $(2 + 3i)(2 - 3i)$ **64.** $(2 - 7i)(2 + 7i)$
65. $(10 + 8i)(10 - 8i)$ **66.** $(11 - 7i)(11 + 7i)$

Find the following quotients. Write all answers in standard form for complex numbers.

67. $\dfrac{2 - 3i}{i}$ **68.** $\dfrac{3 + 4i}{i}$
69. $\dfrac{5 + 2i}{-i}$ **70.** $\dfrac{4 - 3i}{-i}$
71. $\dfrac{4}{2 - 3i}$ **72.** $\dfrac{3}{4 - 5i}$
73. $\dfrac{6}{-3 + 2i}$ **74.** $\dfrac{-1}{-2 - 5i}$
75. $\dfrac{2 + 3i}{2 - 3i}$ **76.** $\dfrac{4 - 7i}{4 + 7i}$
77. $\dfrac{5 + 4i}{3 + 6i}$ **78.** $\dfrac{2 + i}{5 - 6i}$

Review Problems

Solve each equation.

79. $\dfrac{t}{3} - \dfrac{1}{2} = -1$ **80.** $\dfrac{x}{x - 2} + \dfrac{2}{3} = \dfrac{2}{x - 2}$
81. $2 + \dfrac{5}{y} = \dfrac{3}{y^2}$ **82.** $1 - \dfrac{1}{y} = \dfrac{12}{y^2}$

Solve each application problem.

83. The sum of a number and its reciprocal is $\frac{41}{20}$. Find the number.

84. It takes an inlet pipe 8 hours to fill a tank. The drain can empty the tank in 6 hours. If the tank is full and both the inlet pipe and drain are open, how long will it take to drain the tank?

Extending the Concepts

85. Show that $-i$ and $\frac{1}{i}$ (the opposite and the reciprocal of i) are the same number.

86. Show that i^{2n+1} is the same as i for all positive even integers n.

87. Show that $x = 1 + i$ is a solution to the equation $x^2 - 2x + 2 = 0$.

88. Show that $x = 1 - i$ is a solution to the equation $x^2 - 2x + 2 = 0$.

89. Show that $x = 2 + i$ is a solution to the equation $x^3 - 11x + 20 = 0$.

90. Show that $x = 2 - i$ is a solution to the equation $x^3 - 11x + 20 = 0$.

5.2 Completing the Square

In this section, we will develop the first of our new methods of solving quadratic equations. The new method is called *completing the square.* Completing the square on a quadratic equation allows us to obtain solutions, regardless of whether the equation can be factored. Before we solve equations by completing the square, we need to learn how to solve equations by taking square roots of both sides.

Consider the equation

$$x^2 = 16$$

We can solve the equation by writing it in standard form, factoring, and then setting each factor equal to 0.

$$x^2 = 16$$

$$x^2 - 16 = 0 \quad \textbf{Standard form}$$

$$(x + 4)(x - 4) = 0 \quad \textbf{Factor}$$

$$x + 4 = 0 \quad \text{or} \quad x - 4 = 0 \quad \textbf{Zero-factor property}$$

$$x = -4 \quad \text{or} \quad x = 4$$

We can shorten our work considerably, however, if we simply notice that x must be either the positive square root of 16 or the negative square root of 16; that is,

$$\text{If} \quad x^2 = 16$$

$$\text{then} \quad x = \sqrt{16} \quad \text{or} \quad x = -\sqrt{16}$$

$$x = 4 \quad \text{or} \quad x = -4$$

We can generalize this result into a theorem as follows.

Theorem 1

If $a^2 = b$ where b is a real number, then $a = \sqrt{b}$ or $a = -\sqrt{b}$.

Notation The expression $a = \sqrt{b}$ or $a = -\sqrt{b}$ can be written in shorthand form as $a = \pm\sqrt{b}$. The symbol $\pm$ is read "plus or minus."

We can apply Theorem 1 to some fairly complicated quadratic equations.

EXAMPLE 1 Solve $(2x - 3)^2 = 25$.

SOLUTION

$$
\begin{aligned}
(2x - 3)^2 &= 25 \\
2x - 3 &= \pm\sqrt{25} && \textbf{Theorem 1} \\
2x - 3 &= \pm 5 && \sqrt{25} = 5 \\
2x &= 3 \pm 5 && \textbf{Add 3 to both sides} \\
x &= \frac{3 \pm 5}{2} && \textbf{Divide both sides by 2}
\end{aligned}
$$

The last equation can be written as two separate statements:

$$
\begin{aligned}
x &= \frac{3+5}{2} \quad \text{or} \quad x = \frac{3-5}{2} \\
&= \frac{8}{2} \qquad\qquad\;\; = \frac{-2}{2} \\
&= 4 \quad\quad \text{or} \qquad = -1
\end{aligned}
$$

The solution set is $\{4, -1\}$.

Notice that we could have solved the equation in Example 1 by expanding the left side, writing the resulting equation in standard form, and then factoring. The problem would look like this:

$$
\begin{aligned}
(2x - 3)^2 &= 25 && \textbf{Original equation} \\
4x^2 - 12x + 9 &= 25 && \textbf{Expand the left side} \\
4x^2 - 12x - 16 &= 0 && \textbf{Add } -25 \textbf{ to each side} \\
4(x^2 - 3x - 4) &= 0 && \textbf{Begin factoring} \\
4(x - 4)(x + 1) &= 0 && \textbf{Factor completely} \\
x - 4 = 0 \quad \text{or} \quad x + 1 &= 0 && \textbf{Set variable factors equal to 0} \\
x = 4 \quad \text{or} \quad x &= -1
\end{aligned}
$$

As you can see, solving the equation by factoring leads to the same two solutions.

EXAMPLE 2 Solve for x: $(3x - 1)^2 = -12$.

SOLUTION

$$
\begin{aligned}
(3x - 1)^2 &= -12 \\
3x - 1 &= \pm\sqrt{-12} && \textbf{Theorem 1} \\
3x - 1 &= \pm 2i\sqrt{3} && \sqrt{-12} = 2i\sqrt{3} \\
3x &= 1 \pm 2i\sqrt{3} && \textbf{Add 1 to both sides} \\
x &= \frac{1 \pm 2i\sqrt{3}}{3} && \textbf{Divide both sides by 3}
\end{aligned}
$$

The solution set is $\left\{\frac{1 + 2i\sqrt{3}}{3}, \frac{1 - 2i\sqrt{3}}{3}\right\}$

Practice Problems

1. Solve $(3x + 2)^2 = 16$.

2. Solve $(4x - 3)^2 = -50$.

Note: We cannot solve the equation in Example 2 by factoring. If we expand the left side and write the resulting equation in standard form, we are left with a quadratic equation that does not factor:

$(3x - 1)^2 = -12$ Equation from Example 2

$9x^2 - 6x + 1 = -12$ Expand the left side

$9x^2 - 6x + 13 = 0$ Standard form, but not factorable

Answer

1. $\frac{2}{3}, -2$

3. Solve $x^2 + 10x + 25 = 20$.

Both solutions are complex. Here is a check of the first solution:

When $$x = \frac{1 + 2i\sqrt{3}}{3}$$

the equation $$(3x - 1)^2 = -12$$

becomes $$\left(3 \cdot \frac{1 + 2i\sqrt{3}}{3} - 1\right)^2 \stackrel{?}{=} -12$$

or
$$(1 + 2i\sqrt{3} - 1)^2 \stackrel{?}{=} -12$$
$$(2i\sqrt{3})^2 \stackrel{?}{=} -12$$
$$4 \cdot i^2 \cdot 3 \stackrel{?}{=} -12$$
$$12(-1) \stackrel{?}{=} -12$$
$$-12 = -12$$

EXAMPLE 3 Solve $x^2 + 6x + 9 = 12$.

SOLUTION We can solve this equation as we have the equations in Examples 1 and 2 if we first write the left side as $(x + 3)^2$.

$x^2 + 6x + 9 = 12$	**Original equation**
$(x + 3)^2 = 12$	**Write $x^2 + 6x + 9$ as $(x + 3)^2$**
$x + 3 = \pm 2\sqrt{3}$	**Theorem 1**
$x = -3 \pm 2\sqrt{3}$	**Add −3 to each side**

We have two irrational solutions: $-3 + 2\sqrt{3}$ and $-3 - 2\sqrt{3}$. What is important about this problem, however, is the fact that the equation was easy to solve because the left side was a perfect square trinomial.

Completing the Square

The method of completing the square is simply a way of transforming any quadratic equation into an equation of the form found in the preceding three examples.

The key to understanding the method of completing the square lies in recognizing the relationship between the last two terms of any perfect square trinomial whose leading coefficient is 1.

Consider the following list of perfect square trinomials and their corresponding binomial squares:

$$x^2 - 6x + 9 = (x - 3)^3$$
$$x^2 + 8x + 16 = (x + 4)^2$$
$$x^2 - 10x + 25 = (x - 5)^2$$
$$x^2 + 12x + 36 = (x + 6)^2$$

In each case the leading coefficient is 1. A more important observation comes from noticing the relationship between the linear and constant terms (middle and last terms) in each trinomial. Observe that the constant term in each case is the square of half the coefficient of x in the middle term. For example, in the last expression, the constant term 36 is the square of half of 12, where 12 is the coefficient of x in the middle term. (Notice also that the second terms in all the binomials on the right side are half the coefficients of the middle terms of the trinomials

Answers

2. $\frac{3 \pm 5i\sqrt{2}}{4}$ 3. $-5 \pm 2\sqrt{5}$

on the left side.) We can use these observations to build our own perfect square trinomials and, in doing so, solve some quadratic equations.

Consider the following equation:

$$x^2 + 6x = 3$$

We can think of the left side as having the first two terms of a perfect square trinomial. We need only add the correct constant term. If we take half the coefficient of x, we get 3. If we then square this quantity, we have 9. Adding the 9 to both sides, the equation becomes

$$x^2 + 6x + \mathbf{9} = 3 + \mathbf{9}$$

Note: This is the step in which we actually complete the square.

The left side is the perfect square $(x + 3)^2$; the right side is 12:

$$(x + 3)^2 = 12$$

The equation is now in the correct form. We can apply Theorem 1 and finish the solution:

$$(x + 3)^2 = 12$$

$$x + 3 = \pm\sqrt{12} \qquad \textbf{Theorem 1}$$

$$x + 3 = \pm 2\sqrt{3}$$

$$x = -3 \pm 2\sqrt{3}$$

The solution set is $\{-3 + 2\sqrt{3}, -3 - 2\sqrt{3}\}$. The method just used is called *completing the square* since we complete the square on the left side of the original equation by adding the appropriate constant term.

EXAMPLE 4 Solve by completing the square: $x^2 + 5x - 2 = 0$.

4. Solve $x^2 + 3x - 4 = 0$ by completing the square.

SOLUTION We must begin by adding 2 to both sides. (The left side of the equation, as it is, is not a perfect square because it does not have the correct constant term. We will simply "move" that term to the other side and use our own constant term.)

$$x^2 + 5x = 2 \qquad \textbf{Add 2 to each side}$$

We complete the square by adding the square of half the coefficient of the linear term to both sides:

$$x^2 + 5x + \mathbf{\frac{25}{4}} = 2 + \mathbf{\frac{25}{4}} \qquad \textbf{Half of 5 is } \tfrac{5}{2}\textbf{, the square of which is } \tfrac{25}{4}$$

$$\left(x + \frac{5}{2}\right)^2 = \frac{33}{4} \qquad 2 + \tfrac{25}{4} = \tfrac{8}{4} + \tfrac{25}{4} = \tfrac{33}{4}$$

$$x + \frac{5}{2} = \pm\sqrt{\frac{33}{4}} \qquad \textbf{Theorem 1}$$

$$= \pm\frac{\sqrt{33}}{2} \qquad \textbf{Simplify the radical}$$

$$x = -\frac{5}{2} \pm \frac{\sqrt{33}}{2} \qquad \textbf{Add } -\tfrac{5}{2} \textbf{ to both sides}$$

$$= \frac{-5 \pm \sqrt{33}}{2}$$

The solution set is $\left\{\frac{-5 + \sqrt{33}}{2}, \frac{-5 - \sqrt{33}}{2}\right\}$.

Note: We can use a calculator to get decimal approximations to these solutions. If $\sqrt{33} \approx 5.74$, then

$$\frac{-5 + 5.74}{2} = 0.37$$

$$\frac{-5 - 5.74}{2} = -5.37$$

Answer

4. 1, −4

5. Solve for x: $5x^2 - 3x + 2 = 0$.

EXAMPLE 5 Solve for x: $3x^2 - 8x + 7 = 0$.

SOLUTION

$$3x^2 - 8x + 7 = 0$$

$$3x^2 - 8x = -7 \quad \textbf{Add } -7 \textbf{ to both sides}$$

We cannot complete the square on the left side because the leading coefficient is not 1. We take an extra step and divide both sides by 3:

$$\frac{3x^2}{3} - \frac{8x}{3} = -\frac{7}{3}$$

$$x^2 - \frac{8}{3}x = -\frac{7}{3}$$

Half of $\frac{8}{3}$ is $\frac{4}{3}$, the square of which is $\frac{16}{9}$.

$$x^2 - \frac{8}{3}x + \mathbf{\frac{16}{9}} = -\frac{7}{3} + \mathbf{\frac{16}{9}} \quad \textbf{Add } \tfrac{16}{9} \textbf{ to both sides}$$

$$\left(x - \frac{4}{3}\right)^2 = -\frac{5}{9} \quad \textbf{Simplify right side}$$

$$x - \frac{4}{3} = \pm\sqrt{-\frac{5}{9}} \quad \textbf{Theorem 1}$$

$$x - \frac{4}{3} = \pm\frac{i\sqrt{5}}{3} \quad \sqrt{-\frac{5}{9}} = \frac{\sqrt{-5}}{3} = \frac{i\sqrt{5}}{3}$$

$$x = \frac{4}{3} \pm \frac{i\sqrt{5}}{3} \quad \textbf{Add } \tfrac{4}{3} \textbf{ to both sides}$$

$$x = \frac{4 \pm i\sqrt{5}}{3}$$

The solution set is $\left\{\frac{4 + i\sqrt{5}}{3}, \frac{4 - i\sqrt{5}}{3}\right\}$.

To Solve a Quadratic Equation by Completing the Square

To summarize the method used in the preceding two examples, we list the following steps:

Step 1: Write the equation in the form $ax^2 + bx = c$.

Step 2: If the leading coefficient is not 1, divide both sides by the coefficient so that the resulting equation has a leading coefficient of 1; that is, if $a \neq 1$, then divide both sides by a.

Step 3: Add the square of half the coefficient of the linear term to both sides of the equation.

Step 4: Write the left side of the equation as the square of a binomial, and simplify the right side if possible.

Step 5: Apply Theorem 1, and solve as usual.

Answer

5. $\frac{3 \pm i\sqrt{31}}{10}$

Facts from Geometry: More Special Triangles

The triangles shown in Figures 1 and 2 occur frequently in mathematics.

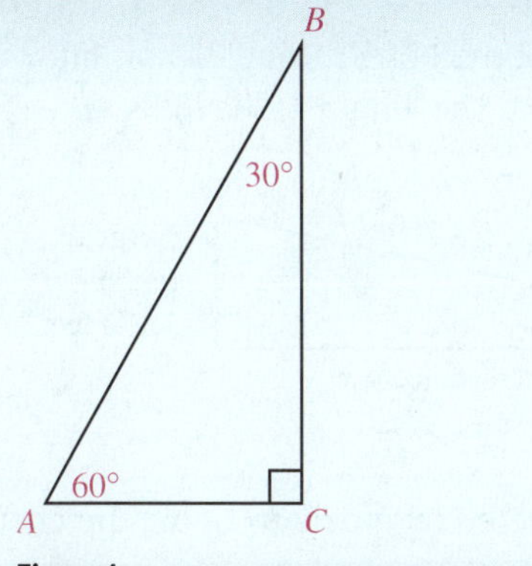

Figure 1

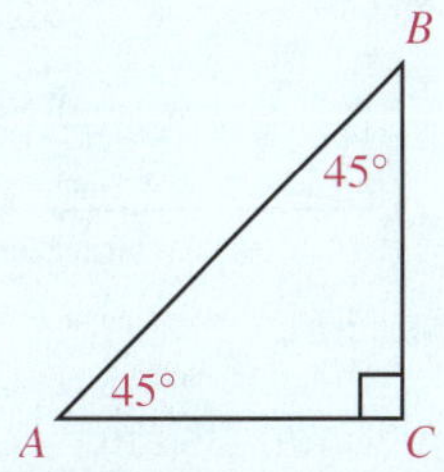

Figure 2

Note that both of the triangles are right triangles. We refer to the triangle in Figure 1 as a 30°–60°–90° triangle and the triangle in Figure 2 as a 45°–45°–90° triangle.

EXAMPLE 6 If the shortest side in a 30°–60°–90° triangle is 1 inch, find the lengths of the other two sides.

SOLUTION In Figure 3 triangle *ABC* is a 30°–60°–90° triangle in which the shortest side *AC* is 1 inch long. Triangle *DBC* is also a 30°–60°–90° triangle in which the shortest side *DC* is 1 inch long.

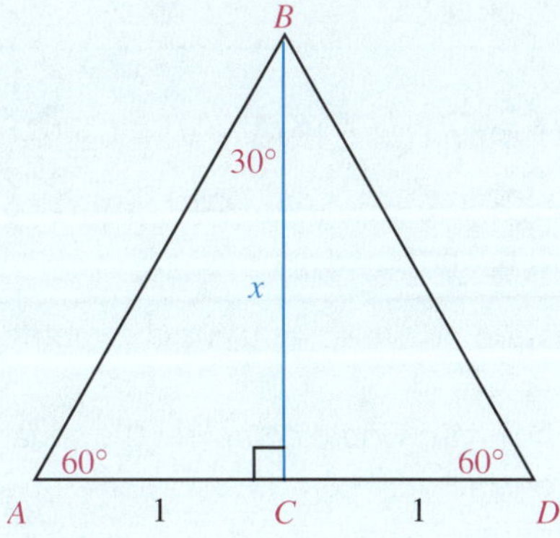

Figure 3

Notice that the large triangle *ABD* is an equilateral triangle because each of its interior angles is 60°. Each side of triangle *ABD* is 2 inches long. Side *AB* in triangle *ABC* is therefore 2 inches. To find the length of side *BC*, we use the Pythagorean theorem.

$$BC^2 + AC^2 = AB^2$$

$$x^2 + 1^2 = 2^2$$

$$x^2 + 1 = 4$$

$$x^2 = 3$$

$$x = \sqrt{3} \text{ inches}$$

Note that we write only the positive square root because x is the length of a side in a triangle and is therefore a positive number.

6. If the shortest side in a 30°–60°–90° triangle is 2 inches long, find the lengths of the other two sides.

Answer

6. 4 inches and $2\sqrt{3}$ inches

7. Table 1 in the introduction to this chapter gives the vertical rise of the Lookout Double chair lift as 960 feet and the length of the chair lift as 4,330 feet. To the nearest foot, find the horizontal distance covered by a person riding this lift.

EXAMPLE 7 Table 1 in the introduction to this chapter gives the vertical rise of the Forest Double chair lift as 1,170 feet and the length of the chair lift as 5,750 feet. To the nearest foot, find the horizontal distance covered by a person riding this lift.

SOLUTION Figure 4 is a model of the Forest Double chair lift. A rider gets on the lift at point A and exits at point B. The length of the lift is AB.

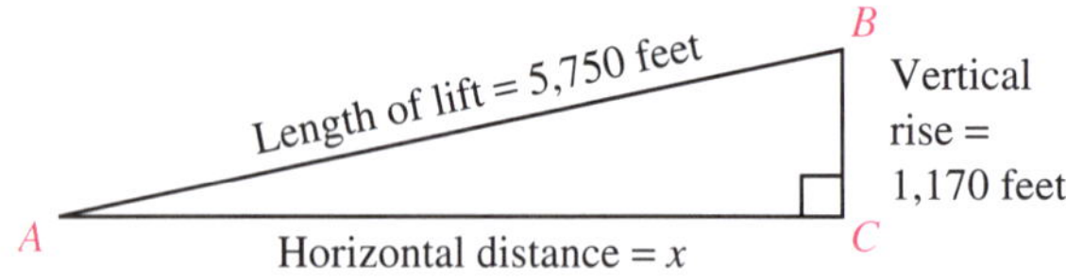

Figure 4

To find the horizontal distance covered by a person riding the chair lift, we use the Pythagorean theorem.

$5{,}750^2 = x^2 + 1{,}170^2$	**Pythagorean theorem**
$33{,}062{,}500 = x^2 + 1{,}368{,}900$	**Simplify squares**
$x^2 = 33{,}062{,}500 - 1{,}368{,}900$	**Solve for x^2**
$x^2 = 31{,}693{,}600$	**Simplify the right side**
$x = \sqrt{31{,}693{,}600}$	**Theorem 1**
$= 5{,}630$ feet (to the nearest foot)	

A rider getting on the lift at point A and riding to point B will cover a horizontal distance of approximately 5,630 feet.

Getting Ready for Class

After reading through the preceding section, respond in your own words and in complete sentences.

A. What kind of equation do we solve using the method of completing the square?

B. Explain in words how you would complete the square on $x^2 - 16x = 4$.

C. What is the relationship between the shortest side and the longest side in a 30°–60°–90° triangle?

D. What two expressions together are equivalent to $x = \pm 4$?

Answer

7. 4,222 feet

PROBLEM SET 5.2

Solve the following equations.

1. $x^2 = 25$
2. $x^2 = 16$
3. $a^2 = -9$
4. $a^2 = -49$
5. $y^2 = \frac{3}{4}$
6. $y^2 = \frac{5}{9}$
7. $x^2 + 12 = 0$
8. $x^2 + 8 = 0$
9. $4a^2 - 45 = 0$
10. $9a^2 - 20 = 0$
11. $(2y - 1)^2 = 25$
12. $(3y + 7)^2 = 1$
13. $(2a + 3)^2 = -9$
14. $(3a - 5)^2 = -49$
15. $(5x + 2)^2 = -8$
16. $(6x - 7)^2 = -75$
17. $x^2 + 8x + 16 = -27$
18. $x^2 - 12x + 36 = -8$
19. $4a^2 - 12a + 9 = -4$
20. $9a^2 - 12a + 4 = -9$

Copy each of the following, and fill in the blanks so the left side of each is a perfect square trinomial. That is, complete the square.

21. $x^2 + 12x + \_\_ = (x + \_\_)^2$
22. $x^2 + 6x + \_\_ = (x + \_\_)^2$
23. $x^2 - 4x + \_\_ = (x - \_\_)^2$
24. $x^2 - 2x + \_\_ = (x - \_\_)^2$
25. $a^2 - 10a + \_\_ = (a - \_\_)^2$
26. $a^2 - 8a + \_\_ = (a - \_\_)^2$
27. $x^2 + 5x + \_\_ = (x + \_\_)^2$
28. $x^2 + 3x + \_\_ = (x + \_\_)^2$
29. $y^2 - 7y + \_\_ = (y - \_\_)^2$
30. $y^2 - y + \_\_ = (y - \_\_)^2$
31. $x^2 + \frac{1}{2}x + \_\_ = (x + \_\_)^2$
32. $x^2 - \frac{3}{4}x + \_\_ = (x - \_\_)^2$
33. $x^2 + \frac{2}{3}x + \_\_ = (x + \_\_)^2$
34. $x^2 - \frac{4}{5}x + \_\_ = (x - \_\_)^2$

Solve each of the following quadratic equations by completing the square.

35. $x^2 + 4x = 12$
36. $x^2 - 2x = 8$
37. $x^2 + 12x = -27$
38. $x^2 - 6x = 16$
39. $a^2 - 2a + 5 = 0$
40. $a^2 + 10a + 22 = 0$
41. $y^2 - 8y + 1 = 0$
42. $y^2 + 6y - 1 = 0$
43. $x^2 - 5x - 3 = 0$
44. $x^2 - 5x - 2 = 0$

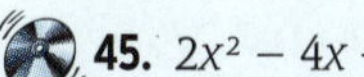

45. $2x^2 - 4x - 8 = 0$
46. $3x^2 - 9x - 12 = 0$
47. $3t^2 - 8t + 1 = 0$
48. $5t^2 + 12t - 1 = 0$
49. $4x^2 - 3x + 5 = 0$
50. $7x^2 - 5x + 2 = 0$
51. $3x^2 + 4x - 1 = 0$
52. $2x^2 + 6x - 1 = 0$
53. $2x^2 - 10x = 11$
54. $25x^2 - 20x = 1$
55. $4x^2 - 10x + 11 = 0$
56. $4x^2 - 6x + 1 = 0$

Applying the Concepts

57. Geometry If the shortest side in a 30°–60°–90° triangle is $\frac{1}{2}$ inch long, find the length of the other two sides.

58. Geometry If the shortest side in a 30°–60°–90° triangle is 3 feet long, find the length of the other two sides.

59. Geometry If the length of the shortest side of a 30°–60°–90° triangle is x, find the length of the other two sides in terms of x.

60. Geometry If the length of the longest side of a 30°–60°–90° triangle is x, find the lengths of the other two sides in terms of x.

61. Geometry If the length of the shorter sides of a 45°–45°–90° triangle is 1 inch, find the length of the hypotenuse.

62. Geometry If the length of the shorter sides of a 45°–45°–90° triangle is 3 feet, find the length of the hypotenuse.

63. Geometry If the length of the hypotenuse of a 45°–45°–90° triangle is 1 inch, find the length of the shorter sides.

64. Geometry If the length of the hypotenuse of a 45°–45°–90° triangle is 2 feet, find the length of the shorter sides.

65. **Geometry** If the length of the shorter sides of a 45°–45°–90° triangle is x, find the length of the hypotenuse, in terms of x.
66. **Geometry** If the length of the hypotenuse of a 45°–45°–90° triangle is x, find the length of the shorter sides, in terms of x.
67. **Chair Lift** Use Table 1 from the introduction to this chapter to find the horizontal distance covered by a person riding the Bear Paw Double chair lift. Round your answer to the nearest foot.
68. **Chair Lift** Use Table 1 from the introduction to this chapter to find the horizontal distance covered by a person riding the Big Springs Gondola lift. Round your answer to the nearest foot.
69. **Chair Lift** Using a right triangle to model the Forest Double chair lift, find the slope of the lift to the nearest hundredth.
70. **Chair Lift** Using a right triangle to model the Echo Triple chair lift, find the slope of the lift to the nearest hundredth.
71. **Length of an Escalator** An escalator in a department store is to carry people a vertical distance of 20 feet between floors. How long is the escalator if it makes an angle of 45° with the ground? (See Figure 5.)

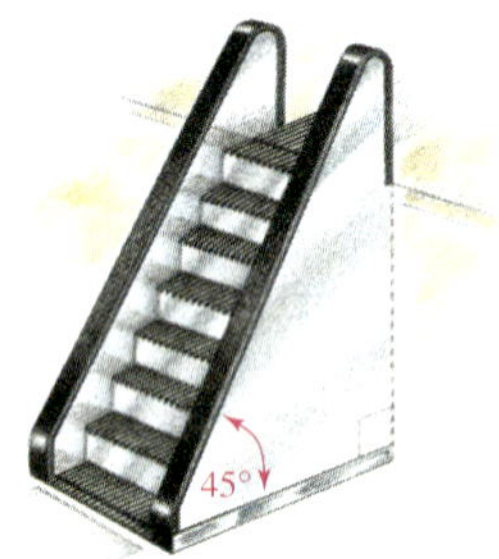

Figure 5

72. **Dimensions of a Tent** A two-person tent is to be made so the height at the center is 4 feet. If the sides of the tent are to meet the ground at an angle of 60° and the tent is to be 6 feet in length, how many square feet of material will be needed to make the tent? (Figure 6; assume that the tent has a floor and is closed at both ends.) Give your answer to the nearest tenth of a square foot.

Figure 6

73. **Interest Rate** Suppose a deposit of \$3,000 in a savings account that paid an annual interest rate r (compounded yearly) is worth \$3,456 after 2 years. Using the formula $A = P(1 + r)^t$, we have

$$3{,}456 = 3{,}000(1 + r)^2$$

Solve for r to find the annual interest rate.

74. **Special Triangles** In Figure 7, triangle ABC has angles 45° and 30°, and height x. Find the lengths of sides AB, BC, and AC, in terms of x.

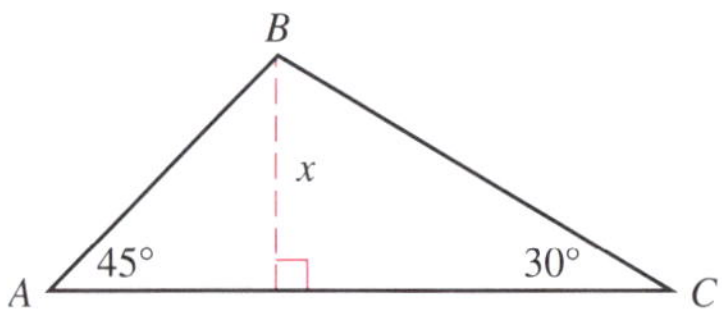

Figure 7

75. **Special Triangles** In Figure 8, AB has a length of 12 inches. Find the length of AD.

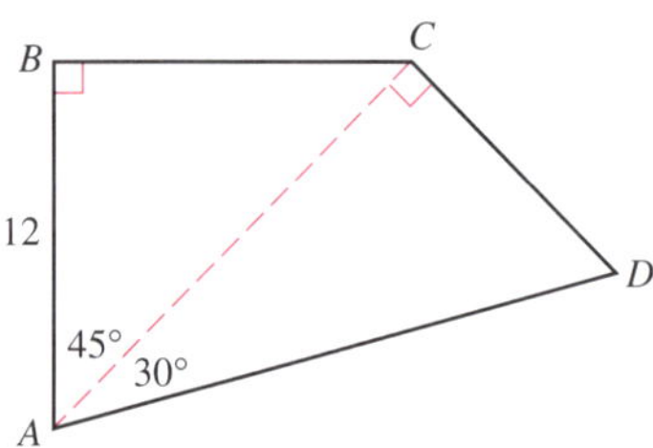

Figure 8

76. **Wheelchair Access** Most local laws state that ramps for wheelchairs may not have more than a 1/10 slope;

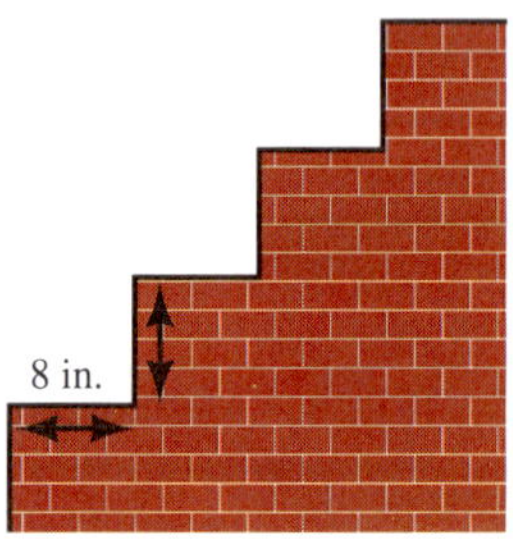

Figure 9

that is, for every foot in the vertical direction the ramp must travel at least 10 feet in the horizontal direction. A wheelchair ramp will take the place of four steps that are 8 inches deep and 8 inches high (Figure 9). What is the minimum length of such a ramp?

Review Problems

Write each of the following in simplified form for radicals.

77. $\sqrt{45}$
78. $\sqrt{24}$
79. $\sqrt{27y^5}$
80. $\sqrt{8y^3}$
81. $\sqrt[3]{54x^6y^5}$
82. $\sqrt[3]{16x^9y^7}$

83. Simplify $\sqrt{b^2 - 4ac}$ when $a = 6$, $b = 7$, and $c = -5$.

84. Simplify $\sqrt{b^2 - 4ac}$ when $a = 2$, $b = -6$, and $c = 3$.

Rationalize the denominator.

85. $\frac{3}{\sqrt{2}}$

86. $\frac{5}{\sqrt{3}}$

87. $\frac{2}{\sqrt[3]{4}}$

88. $\frac{3}{\sqrt[3]{2}}$

Extending the Concepts

Solve for x.

89. $(x + a)^2 + (x - a)^2 = 10a^2$

90. $(ax + 1)^2 + (ax - 1)^2 = 10$

Assume a is a positive number and solve for x by completing the square on x.

91. $x^2 + 2ax = -a^2$

92. $x^2 + 2ax = -4a^2$

93. $x^2 + 2ax = 0$

94. $x^2 + ax = 0$

Assume p and q are positive numbers and solve for x by completing the square on x.

95. $x^2 + px + q = 0$

96. $x^2 - px + q = 0$

97. $3x^2 + px + q = 0$

98. $3x^2 + 2px + q = 0$

5.3 The Quadratic Formula

In this section we will use the method of completing the square from the preceding section to derive the quadratic formula. The *quadratic formula* is a very useful tool in mathematics. It allows us to solve all types of quadratic equations.

The Quadratic Theorem

For any quadratic equation in the form $ax^2 + bx + c = 0$, where $a \neq 0$, the two solutions are

$$x = \frac{-b + \sqrt{b^2 - 4ac}}{2a} \quad \text{and} \quad x = \frac{-b - \sqrt{b^2 - 4ac}}{2a}$$

Proof We will prove the quadratic theorem by completing the square on $ax^2 + bx + c = 0$;

$$ax^2 + bx + c = 0$$

$$ax^2 + bx = -c \qquad \textbf{Add } -c \textbf{ to both sides}$$

$$x^2 + \frac{b}{a}x = -\frac{c}{a} \qquad \textbf{Divide both sides by } a$$

To complete the square on the left side, we add the square of $\frac{1}{2}$ of $\frac{b}{a}$ to both sides. $\left(\frac{1}{2} \text{ of } \frac{b}{a} \text{ is } \frac{b}{2a}.\right)$

$$x^2 + \frac{b}{a}x + \left(\frac{b}{2a}\right)^2 = -\frac{c}{a} + \left(\frac{b}{2a}\right)^2$$

We now simplify the right side as a separate step. We square the second term and combine the two terms by writing each with the least common denominator $4a^2$.

$$-\frac{c}{a} + \left(\frac{b}{2a}\right)^2 = -\frac{c}{a} + \frac{b^2}{4a^2} = \frac{\mathbf{4a}}{\mathbf{4a}}\left(\frac{-c}{a}\right) + \frac{b^2}{4a^2} = \frac{-4ac + b^2}{4a^2}$$

It is convenient to write this last expression as

$$\frac{b^2 - 4ac}{4a^2}$$

Continuing with the proof, we have

$$x^2 + \frac{b}{a}x + \left(\frac{b}{2a}\right)^2 = \frac{b^2 - 4ac}{4a^2}$$

$$\left(x + \frac{b}{2a}\right)^2 = \frac{b^2 - 4ac}{4a^2} \qquad \textbf{Write left side as a binomial square}$$

$$x + \frac{b}{2a} = \pm\frac{\sqrt{b^2 - 4ac}}{2a} \qquad \textbf{Theorem 1}$$

$$x = -\frac{b}{2a} \pm \frac{\sqrt{b^2 - 4ac}}{2a} \qquad \textbf{Add } -\frac{b}{2a} \textbf{ to both sides}$$

$$= \frac{-b \pm \sqrt{b^2 - 4ac}}{2a}$$

Our proof is now complete. What we have is this: If our equation is in the form $ax^2 + bx + c = 0$ (standard form), where $a \neq 0$, the two solutions are always given by the formula

$$x = \frac{-b \pm \sqrt{b^2 - 4ac}}{2a}$$

This formula is known as the *quadratic formula.* If we substitute the coefficients a, b, and c of any quadratic equation in standard form into the formula, we need only perform some basic arithmetic to arrive at the solution set.

EXAMPLE 1 Use the quadratic formula to solve $6x^2 + 7x - 5 = 0$.

SOLUTION Using $a = 6$, $b = 7$, and $c = -5$ in the formula

$$x = \frac{-b \pm \sqrt{b^2 - 4ac}}{2a}$$

we have $$x = \frac{-7 \pm \sqrt{49 - 4(6)(-5)}}{2(6)}$$

or $$x = \frac{-7 \pm \sqrt{49 + 120}}{12}$$

$$= \frac{-7 \pm \sqrt{169}}{12}$$

$$= \frac{-7 \pm 13}{12}$$

We separate the last equation into the two statements

$$x = \frac{-7 + 13}{12} \quad \text{or} \quad x = \frac{-7 - 13}{12}$$

$$x = \frac{1}{2} \quad \text{or} \quad x = -\frac{5}{3}$$

The solution set is $\{\frac{1}{2}, -\frac{5}{3}\}$.

Whenever the solutions to a quadratic equation are rational numbers, as they are in Example 1, it means that the original equation was solvable by factoring. To illustrate, let's solve the equation from Example 1 again but this time by factoring:

$$6x^2 + 7x - 5 = 0 \quad \textbf{Equation in standard form}$$

$$(3x + 5)(2x - 1) = 0 \quad \textbf{Factor the left side}$$

$$3x + 5 = 0 \quad \text{or} \quad 2x - 1 = 0 \quad \textbf{Set factors equal to 0}$$

$$x = -\frac{5}{3} \quad \text{or} \quad x = \frac{1}{2}$$

When an equation can be solved by factoring, then factoring is usually the faster method of solution. It is best to try to factor first, and then if you have trouble factoring, go to the quadratic formula. It always works.

EXAMPLE 2 Solve $\frac{x^2}{3} - x = -\frac{1}{2}$.

SOLUTION Multiplying through by 6 and writing the result in standard form, we have

$$2x^2 - 6x + 3 = 0$$

Practice Problems

1. Solve $6x^2 + 7x + 2 = 0$ using the quadratic formula.

2. Solve $\frac{x^2}{2} + x = \frac{1}{3}$.

Answer

1. $-\frac{1}{2}, -\frac{2}{3}$

the left side of which is not factorable. Therefore, we use the quadratic formula with $a = 2$, $b = -6$, and $c = 3$. The two solutions are given by

$$x = \frac{-(-6) \pm \sqrt{36 - 4(2)(3)}}{2(2)}$$

$$= \frac{6 \pm \sqrt{12}}{4}$$

$$= \frac{6 \pm 2\sqrt{3}}{4} \qquad \sqrt{12} = \sqrt{4 \cdot 3} = \sqrt{4}\sqrt{3} = 2\sqrt{3}$$

We can reduce this last expression to lowest terms by factoring 2 from the numerator and denominator and then dividing the numerator and denominator by 2:

$$x = \frac{\not{2}(3 \pm \sqrt{3})}{\not{2} \cdot 2} = \frac{3 \pm \sqrt{3}}{2}$$

3. Solve $\frac{1}{x+4} - \frac{1}{x} = \frac{1}{2}$.

EXAMPLE 3 Solve $\frac{1}{x+2} - \frac{1}{x} = \frac{1}{3}$.

SOLUTION To solve this equation, we must first put it in standard form. To do so, we must clear the equation of fractions by multiplying each side by the LCD for all the denominators, which is $3x(x + 2)$. Multiplying both sides by the LCD, we have

$$\mathbf{3x(x+2)}\left(\frac{1}{x+2} - \frac{1}{x}\right) = \frac{1}{3} \cdot \mathbf{3x(x+2)} \qquad \textbf{Multiply each by the LCD}$$

$$3x(\cancel{x+2}) \cdot \frac{1}{\cancel{x+2}} - 3\cancel{x}(x+2) \cdot \frac{1}{\cancel{x}} = \frac{1}{\cancel{3}} \cdot \cancel{3}x(x+2)$$

$$3x - 3(x+2) = x(x+2)$$

$$3x - 3x - 6 = x^2 + 2x \qquad \textbf{Multiplication}$$

$$-6 = x^2 + 2x \qquad \textbf{Simplify left side}$$

$$0 = x^2 + 2x + 6 \qquad \textbf{Add 6 to each side}$$

Since the right side of our last equation is not factorable, we use the quadratic formula. From our last equation, we have $a = 1$, $b = 2$, and $c = 6$. Using these numbers for a, b, and c in the quadratic formula gives us

$$x = \frac{-2 \pm \sqrt{4 - 4(1)(6)}}{2(1)}$$

$$= \frac{-2 \pm \sqrt{4 - 24}}{2} \qquad \textbf{Simplify inside the radical}$$

$$= \frac{-2 \pm \sqrt{-20}}{2} \qquad \mathbf{4 - 24 = -20}$$

$$= \frac{-2 \pm 2i\sqrt{5}}{2} \qquad \mathbf{-20 = i\sqrt{20} = i\sqrt{4}\sqrt{5} = 2i\sqrt{5}}$$

$$= \frac{\not{2}(-1 \pm i\sqrt{5})}{\not{2})} \qquad \textbf{Factor 2 from the numerator}$$

$$= -1 \pm i\sqrt{5} \qquad \textbf{Divide numerator and denominator by 2}$$

Since neither of the two solutions, $-1 + i\sqrt{5}$ nor $-1 - i\sqrt{5}$, will make any of the denominators in our original equation 0, they are both solutions.

Although the equation in our next example is not a quadratic equation, we solve it by using both factoring and the quadratic formula.

Answers

2. $\frac{-3 \pm \sqrt{15}}{3}$ 3. $-2 \pm 2i$

EXAMPLE 4 Solve $27t^3 - 8 = 0$.

SOLUTION It would be a mistake to add 8 to each side of this equation and then take the cube root of each side because we would lose two of our solutions. Instead, we factor the left side, and then set the factors equal to 0:

$$27t^3 - 8 = 0 \quad \textbf{Equation in standard form}$$

$$(3t - 2)(9t^2 + 6t + 4) = 0 \quad \textbf{Factor as the difference of two cubes}$$

$$3t - 2 = 0 \quad \text{or} \quad 9t^2 + 6t + 4 = 0 \quad \textbf{Set each factor equal to 0}$$

The first equation leads to a solution of $t = \frac{2}{3}$. The second equation does not factor, so we use the quadratic formula with $a = 9$, $b = 6$, and $c = 4$:

$$t = \frac{-6 \pm \sqrt{36 - 4(9)(4)}}{2(9)}$$

$$= \frac{-6 \pm \sqrt{36 - 144}}{18}$$

$$= \frac{-6 \pm \sqrt{-108}}{18}$$

$$= \frac{-6 \pm 6i\sqrt{3}}{18} \quad \sqrt{-108} = i\sqrt{36 \cdot 3} = 6i\sqrt{3}$$

$$= \frac{\cancel{6}(-1 \pm i\sqrt{3})}{\cancel{6} \cdot 3} \quad \textbf{Factor 6 from the numerator and denominator}$$

$$= \frac{-1 \pm i\sqrt{3}}{3} \quad \textbf{Divide out common factor 6}$$

The three solutions to our original equation are

$$\frac{2}{3}, \quad \frac{-1 + i\sqrt{3}}{3}, \quad \text{and} \quad \frac{-1 - i\sqrt{3}}{3}$$

4. Solve $8t^3 - 27 = 0$.

EXAMPLE 5 If an object is thrown downward with an initial velocity of 20 feet per second, the distance $s(t)$, in feet, it travels in t seconds is given by the function $s(t) = 20t + 16t^2$. How long does it take the object to fall 40 feet?

SOLUTION We let $s(t) = 40$, and solve for t:

When $s(t) = 40$

the function $s(t) = 20t + 16t^2$

becomes $40 = 20t + 16t^2$

or $16t^2 + 20t - 40 = 0$

$$4t^2 + 5t - 10 = 0 \quad \textbf{Divide by 4}$$

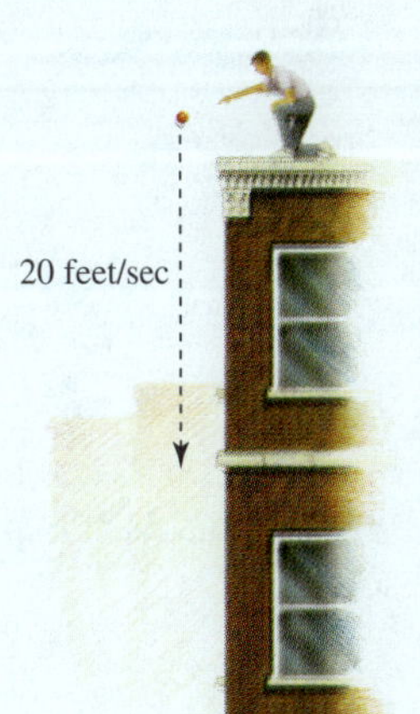

Using the quadratic formula, we have

$$t = \frac{-5 \pm \sqrt{25 - 4(4)(-10)}}{2(4)}$$

$$= \frac{-5 \pm \sqrt{185}}{8}$$

$$= \frac{-5 + \sqrt{185}}{8} \quad \text{or} \quad \frac{-5 - \sqrt{185}}{8}$$

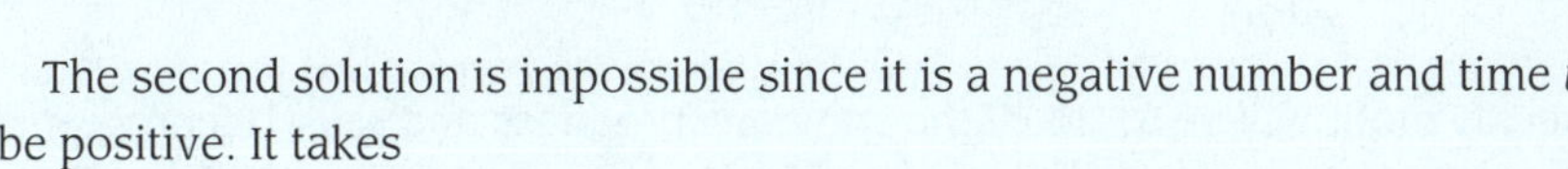

The second solution is impossible since it is a negative number and time t must be positive. It takes

$$t = \frac{-5 + \sqrt{185}}{8} \quad \text{or approximately} \quad \frac{-5 + 13.60}{8} \approx 1.08 \text{ seconds}$$

for the object to fall 40 feet.

5. An object thrown upward with an initial velocity of 32 feet per second rises and falls according to the equation

$$s = 32t - 16t^2$$

where s is the height of the object above the ground at any time t. At what times t will the object be 12 feet above the ground?

Answers

4. $\frac{-3 \pm 3i\sqrt{3}}{4}, \frac{3}{2}$

5. $\frac{3}{2}$ second, $\frac{1}{2}$ second

Recall that the relationship between profit, revenue, and cost is given by the formula

$$P(x) = R(x) - C(x)$$

where $P(x)$ is the profit, $R(x)$ is the total revenue, and $C(x)$ is the total cost of producing and selling x items.

6. Use the information in Example 6 to find the number of programs the company must sell each week for its weekly profit to be $1,320.

EXAMPLE 6 A company produces and sells copies of an accounting program for home computers. The total weekly cost (in dollars) to produce x copies of the program is $C(x) = 8x + 500$, and the weekly revenue for selling all x copies of the program is $R(x) = 35x - 0.1x^2$. How many programs must be sold each week for the weekly profit to be $1,200?

SOLUTION Substituting the given expressions for $R(x)$ and $C(x)$ in the equation $P(x) = R(x) - C(x)$, we have a polynomial in x that represents the weekly profit $P(x)$:

$$\begin{aligned} P(x) &= R(x) - C(x) \\ &= 35x - 0.1x^2 - (8x + 500) \\ &= 35x - 0.1x^2 - 8x - 500 \\ &= -500 + 27x - 0.1x^2 \end{aligned}$$

Setting this expression equal to 1,200, we have a quadratic equation to solve that gives us the number of programs x that need to be sold each week to bring in a profit of $1,200:

$$1{,}200 = -500 + 27x - 0.1x^2$$

We can write this equation in standard form by adding the opposite of each term on the right side of the equation to both sides of the equation. Doing so produces the following equation:

$$0.1x^2 - 27x + 1{,}700 = 0$$

Applying the quadratic formula to this equation with $a = 0.1$, $b = -27$, and $c = 1{,}700$, we have

$$\begin{aligned} x &= \frac{27 \pm \sqrt{(-27)^2 - 4(0.1)(1{,}700)}}{2(0.1)} \\ &= \frac{27 \pm \sqrt{729 - 680}}{0.2} \\ &= \frac{27 \pm \sqrt{49}}{0.2} \\ &= \frac{27 \pm 7}{0.2} \end{aligned}$$

Writing this last expression as two separate expressions, we have our two solutions:

$$x = \frac{27 + 7}{0.2} = \frac{34}{0.2} = 170 \quad \text{or} \quad x = \frac{27 - 7}{0.2} = \frac{20}{0.2} = 100$$

The weekly profit will be $1,200 if the company produces and sells 100 programs or 170 programs.

Answer
6. 130 or 140 programs

What is interesting about the equation we solved in Example 6 is that it has rational solutions, meaning it could have been solved by factoring. But looking back at the equation, factoring does not seem like a reasonable method of solution because the coefficients are either very large or very small. So, there are times when using the quadratic formula is a faster method of solution, even though the equation you are solving is factorable.

USING TECHNOLOGY

Graphing Calculators

More About Example 5

We can solve the problem discussed in Example 5 by graphing the function $Y_1 = 20X + 16X^2$ in a window with X from 0 to 2 (because X is taking the place of *t* and we know *t* is a positive quantity) and Y from 0 to 50 (because we are looking for X when Y_1 is 40). Graphing Y_1 gives a graph similar to the graph in Figure 1. Using the Zoom and Trace features at $Y_1 = 40$ gives us X = 1.08 to the nearest hundredth, matching the results we obtained by solving the original equation algebraically.

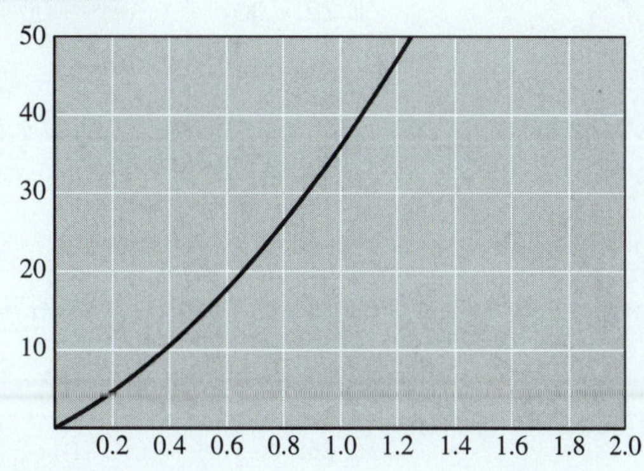

Figure 1

More About Example 6

To visualize the functions in Example 6, we set up our calculator this way:

$Y_1 = 35X - .1X^2$ **Revenue function**

$Y_2 = 8X + 500$ **Cost function**

$Y_3 = Y_1 - Y_2$ **Profit function**

Window: X from 0 to 350, Y from 0 to 3500

Graphing these functions produces graphs similar to the ones shown in Figure 2. The lower graph is the graph of the profit function. Using the Zoom and Trace features on the lower graph at $Y_3 = 1{,}200$ produces two corresponding values of X, 170 and 100, which match the results in Example 6.

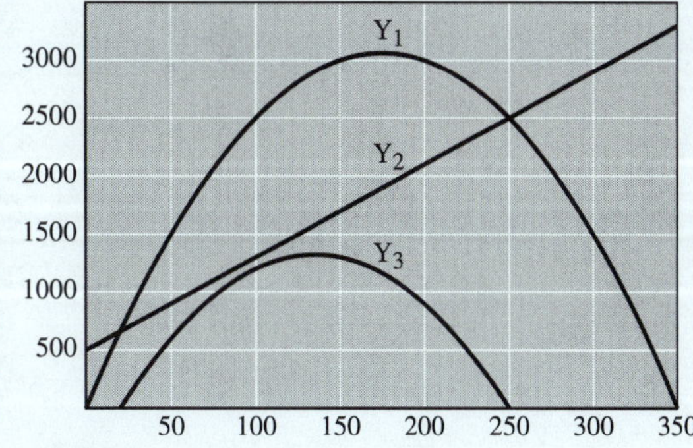

Figure 2

We will continue this discussion of the relationship between graphs of functions and solutions to equations in the Using Technology material in the next section.

Getting Ready for Class

After reading through the preceding section, respond in your own words and in complete sentences.

A. What is the quadratic formula?

B. Under what circumstances should the quadratic formula be applied?

C. When would the quadratic formula result in complex solutions?

D. When will the quadratic formula result in only one solution?

PROBLEM SET 5.3

Solve each equation in each problem using the quadratic formula.

1. (a) $3x^2 + 4x - 2 = 0$
   (b) $3x^2 - 4x - 2 = 0$
   (c) $3x^2 + 4x + 2 = 0$
   (d) $2x^2 + 4x - 3 = 0$
   (e) $2x^2 - 4x + 3 = 0$
2. (a) $3x^2 + 6x - 2 = 0$
   (b) $3x^2 - 6x - 2 = 0$
   (c) $3x^2 + 6x + 2 = 0$
   (d) $2x^2 + 6x + 3 = 0$
   (e) $2x^2 + 6x - 3 = 0$

Solve each equation. Use factoring or the quadratic formula, whichever is appropriate. (Try factoring first. If you have any difficulty factoring, then go right to the quadratic formula.)

3. $x^2 + 5x + 6 = 0$
4. $x^2 + 5x - 6 = 0$
5. $a^2 - 4a + 1 = 0$
6. $a^2 + 4a + 1 = 0$
7. $\frac{1}{6}x^2 - \frac{1}{2}x + \frac{1}{3} = 0$
8. $\frac{1}{4}x^2 + \frac{1}{4}x - \frac{1}{2} = 0$
9. $y^2 - 5y = 0$
10. $2y^2 + 10y = 0$
11. $30x^2 + 40x = 0$
12. $50x^2 - 20x = 0$
13. $\frac{2t^2}{3} - t = -\frac{1}{6}$
14. $\frac{t^2}{3} - \frac{t}{2} = -\frac{3}{2}$
15. $0.01x^2 + 0.06x - 0.08 = 0$
16. $0.02x^2 - 0.03x + 0.05 = 0$
17. $2x + 3 = -2x^2$
18. $2x - 3 = 3x^2$
19. $100x^2 - 200x + 100 = 0$
20. $100x^2 - 600x + 900 = 0$
21. $\frac{1}{2}r^2 = \frac{1}{6}r - \frac{2}{3}$
22. $\frac{1}{4}r^2 = \frac{2}{5}r + \frac{1}{10}$
23. $(x - 3)(x - 5) = 1$
24. $(x - 3)(x + 1) = -6$
25. $(x + 3)^2 + (x - 8)(x - 1) = 16$
26. $(x - 4)^2 + (x + 2)(x + 1) = 9$
27. $\frac{x^2}{3} - \frac{5x}{6} = \frac{1}{2}$
28. $\frac{x^2}{6} + \frac{5}{6} = -\frac{x}{3}$

Multiply both sides of each equation by its LCD. Then solve the resulting equation.

29. $\frac{1}{x+1} - \frac{1}{x} = \frac{1}{2}$
30. $\frac{1}{x+1} + \frac{1}{x} = \frac{1}{3}$
31. $\frac{1}{y-1} + \frac{1}{y+1} = 1$
32. $\frac{2}{y+2} + \frac{3}{y-2} = 1$
33. $\frac{1}{x+2} + \frac{1}{x+3} = 1$
34. $\frac{1}{x+3} + \frac{1}{x+4} = 1$
35. $\frac{6}{r^2 - 1} - \frac{1}{2} = \frac{1}{r+1}$
36. $2 + \frac{5}{r-1} = \frac{12}{(r-1)^2}$

Solve each equation. In each case you will have three solutions.

37. $x^3 - 8 = 0$
38. $x^3 - 27 = 0$
39. $8a^3 + 27 = 0$
40. $27a^3 + 8 = 0$
41. $125t^3 - 1 = 0$
42. $64t^3 + 1 = 0$

Each of the following equations has three solutions. Look for the greatest common factor; then use the quadratic formula to find all solutions.

43. $2x^3 + 2x^2 + 3x = 0$

44. $6x^3 - 4x^2 + 6x = 0$

45. $3y^4 = 6y^3 - 6y^2$

46. $4y^4 = 16y^3 - 20y^2$

47. $6t^5 + 4t^4 = -2t^3$

48. $8t^5 + 2t^4 = -10t^3$

49. One solution to a quadratic equation is $\dfrac{-3 + 2i}{5}$. What is the other solution?

50. One solution to a quadratic equation is $\dfrac{-2 + 3i\sqrt{2}}{5}$. What is the other solution?

Applying the Concepts

51. Falling Object An object is thrown downward with an initial velocity of 5 feet per second. The relationship between the distance s it travels and time t is given by $s = 5t + 16t^2$. How long does it take the object to fall 74 feet?

52. Falling Object The distance an object falls from rest is given by the equation $s = 16t^2$, where s = distance in feet and t = time in seconds. How long does it take an object dropped from a 100-foot cliff to hit the ground?

53. Ball Toss A ball is thrown upward with an initial velocity of 20 feet per second. The equation that gives the height h of the ball at any time t is $h = 20t - 16t^2$. At what times will the ball be 4 feet off the ground?

54. Coin Toss A coin is tossed upward with an initial velocity of 32 feet per second from a height of 16 feet above the ground. The equation giving the object's height h at any time t is $h = 16 + 32t - 16t^2$. When does the object reach a height of 32 feet?

55. Profit The total cost (in dollars) for a company to manufacture and sell x items per week is $C = 60x + 300$, whereas the revenue brought in by selling all x items is $R = 100x - 0.5x^2$. How many items must be sold to obtain a weekly profit of \$300?

56. Profit The total cost (in dollars) for a company to produce and sell x items per week is $C = 200x + 1{,}600$, whereas the revenue brought in by selling all x items is $R = 300x - 0.5x^2$. How many items must be sold in order for the weekly profit to be \$2,150?

57. Profit Suppose it costs a company selling patterns $C = 800 + 6.5x$ dollars to produce and sell x patterns a month. If the revenue obtained by selling x patterns is $R = 10x - 0.002x^2$, how many patterns must it sell each month if it wants a monthly profit of \$700?

58. Profit Suppose a company manufactures and sells x picture frames each month with a total cost of $C = 1{,}200 + 3.5x$ dollars. If the revenue obtained by selling x frames is $R = 9x - 0.002x^2$, find the number of frames it must sell each month if its monthly profit is to be \$2,300.

59. Photograph Cropping The following figure shows a photographic image on a 10.5-centimeter by 8.2-centimeter background. The overall area of the background is to be reduced to 80% of its original area by cutting off (cropping) equal strips on all four sides. What is the width of the strip that is cut from each side?

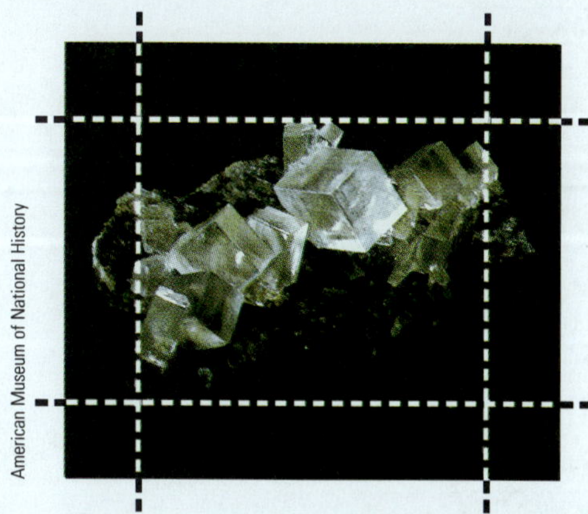

60. Area of a Garden A garden measures 20.3 meters by 16.4 meters. To double the area of the garden, strips of equal width are added to all four sides.

(a) Draw a diagram that illustrates these conditions.

(b) What are the new overall dimensions of the garden?

61. Area and Perimeter A rectangle has a perimeter of 20 yards and an area of 15 square yards.
(a) Write two equations that state these facts in terms of the rectangle's length, l, and its width, w.
(b) Solve the two equations from part (a) to determine the actual length and width of the rectangle.
(c) Explain why two answers are possible to part (b).

62. Population Size Writing in 1829, former President James Madison made some predictions about the growth of the population of the United States. The populations he predicted fit the equation

$$y = 0.029x^2 - 1.39x + 42$$

where y is the population in millions of people x years from 1829.
(a) Use the equation to determine the approximate year President Madison would have predicted that the U.S. population would reach 100,000,000
(b) If the U.S. population in 1990 was approximately 200,000,000, were President Madison's predictions accurate in the long term? Explain why or why not.

Review Problems

Divide, using long division.

63. $\dfrac{8y^2 - 26y - 9}{2y - 7}$

64. $\dfrac{6y^2 + 7y - 18}{3y - 4}$

65. $\dfrac{x^3 + 9x^2 + 26x + 24}{x + 2}$

66. $\dfrac{x^3 + 6x^2 + 11x + 6}{x + 3}$

Simplify each expression. (Assume $x, y > 0$.)

67. $25^{1/2}$

68. $8^{1/3}$

69. $\left(\dfrac{9}{25}\right)^{3/2}$

70. $\left(\dfrac{16}{81}\right)^{3/4}$

71. $8^{-2/3}$

72. $4^{-3/2}$

73. $\dfrac{(49x^8y^{-4})^{1/2}}{(27x^{-3}y^9)^{-1/3}}$

74. $\dfrac{(x^{-2}y^{1/3})^6}{x^{-10}y^{3/2}}$

Extending the Concepts

So far, all the equations we have solved have had coefficients that were rational numbers. Here are some equations that have irrational coefficients and some that have complex coefficients. Solve each equation. (Remember, $i^2 = -1$.)

75. $x^2 + \sqrt{3}x - 6 = 0$

76. $x^2 - \sqrt{5}x - 5 = 0$

77. $\sqrt{2}x^2 + 2x - \sqrt{2} = 0$

78. $\sqrt{7}x^2 + 2\sqrt{2}x - \sqrt{7} = 0$

79. $x^2 + ix + 2 = 0$

80. $x^2 + 3ix - 2 = 0$

81. $ix^2 + 3x + 4i = 0$

82. $4ix^2 + 5x + 9i = 0$

CHAPTER 5 SUMMARY

Complex Numbers [5.1]

A *complex number* is any number that can be put in the form

$$a + bi$$

where a and b are real numbers and $i = \sqrt{-1}$. The *real part* of the complex number is a, and b is the *imaginary part.*

If a, b, c, and d are real numbers, then we have the following definitions associated with complex numbers:

1. Equality

$$a + bi = c + di \quad \text{if and only if} \quad a = c \text{ and } b = d$$

2 Addition and subtraction

$$(a + bi) + (c + di) = (a + c) + (b + d)i$$

$$(a + bi) - (c + di) = (a - c) + (b - d)i$$

3. Multiplication

$$(a + bi)(c + di) = (ac - bd) + (ad + bc)i$$

4. Division is similar to rationalizing the denominator.

Examples

1. $3 + 4i$ is a complex number.

Addition

$$(3 + 4i) + (2 - 5i) = 5 - i$$

Multiplication

$$(3 + 4i)(2 - 5i) = 6 - 15i + 8i - 20i^2 = 6 - 7i + 20 = 26 - 7i$$

Division

$$\frac{2}{3 + 4i} = \frac{2}{3 + 4i} \cdot \frac{\mathbf{3 - 4i}}{\mathbf{3 - 4i}} = \frac{6 - 8i}{9 + 16} = \frac{6}{25} - \frac{8}{25}i$$

Theorem 1 [5.2]

If $a^2 = b$, where b is a real number, then

$$a = \sqrt{b} \qquad \text{or} \qquad a = -\sqrt{b}$$

which can be written as $a = \pm\sqrt{b}$.

2. If $(x - 3)^2 = 25$

then $x - 3 = \pm 5$

$x = 3 \pm 5$

$x = 8$ or $x = -2$

To Solve a Qaudratic Equation by Completing the Square [5.2]

Step 1: Write the equation in the form $ax^2 + bx = c$.

Step 2: If $a \neq 1$, divide through by the constant a so the coefficient of x^2 is 1.

Step 3: Complete the square on the left side by adding the square of $\frac{1}{2}$ the coefficient of x to both sides.

Step 4: Write the left side of the equation as the square of a binomial. Simplify the right side if possible.

Step 5: Apply Theorem 1, and solve as usual.

3. Solve $x^2 - 6x - 6 = 0$

$$x^2 - 6x = 6$$

$$x^2 - 6x + \mathbf{9} = 6 + \mathbf{9}$$

$$(x - 3)^2 = 15$$

$$x - 3 = \pm\sqrt{15}$$

$$x = 3 \pm \sqrt{15}$$

The Quadratic Theorem [5.3]

For any quadratic equation in the form $ax^2 + bx + c = 0$, $a \neq 0$, the two solutions are

$$x = \frac{-b \pm \sqrt{b^2 - 4ac}}{2a}$$

This last expression is known as the *quadratic formula.*

4. If $2x^2 + 3x - 4 = 0$, then

$$x = \frac{-3 \pm \sqrt{9 - 4(2)(-4)}}{2(2)} = \frac{-3 \pm \sqrt{41}}{4}$$

CHAPTER 5 REVIEW

The problems below form a comprehensive review of the material in this chapter. They can be used to study for exams. If you would like to take a practice test on this chapter, you can use the odd-numbered problems. Give yourself an hour and work as many of the odd-numbered problems as possible. When you are finished, or when an hour has passed, check your answers with the answers in the back of the book. You can use the even-numbered problems for a second practice test.

Write each of the following as i, -1, $-i$, or 1. [5.1]

1. i^{24}

2. i^{27}

Find x and y so that each of the following equations is true. [5.1]

3. $3 - 4i = -2x + 8yi$

4. $(3x + 2) - 8i = -4 + 2yi$

Combine the following complex numbers. [5.1]

5. $(3 + 5i) + (6 - 2i)$

6. $(2 + 5i) - [(3 + 2i) + (6 - i)]$

Multiply. [5.1]

7. $3i(4 + 2i)$

8. $(2 + 3i)(4 + i)$

9. $(4 + 2i)^2$

10. $(4 + 3i)(4 - 3i)$

Divide. Write all answers in standard form for complex numbers. [5.1]

11. $\frac{3 + i}{i}$

12. $\frac{-3}{2 + i}$

Solve each equation. [5.2]

13. $(2t - 5)^2 = 25$

14. $(3t - 2)^2 = 4$

15. $(3y - 4)^2 = -49$

16. $(2x + 6)^2 = 12$

Solve by completing the square. [5.2]

17. $2x^2 + 6x - 20 = 0$

18. $3x^2 + 15x = -18$

19. $a^2 + 9 = 6a$

20. $a^2 + 4 = 4a$

21. $2y^2 + 6y = -3$

22. $3y^2 + 3 = 9y$

Solve each equation. [5.3]

23. $\frac{1}{6}x^2 + \frac{1}{2}x - \frac{5}{3} = 0$

24. $8x^2 - 18x = 0$

25. $4t^2 - 8t + 19 = 0$

26. $100x^2 - 200x = 100$

27. $0.06a^2 + 0.05a = 0.04$

28. $9 - 6x = -x^2$

29. $(2x + 1)(x - 5) - (x + 3)(x - 2) = -17$

30. $2y^3 + 2y = 10y^2$

31. $5x^2 = -2x + 3$

32. $x^3 - 27 = 0$

33. $3 - \frac{2}{x} + \frac{1}{x^2} = 0$

34. $\frac{1}{x - 3} + \frac{1}{x + 2} = 1$

35. Profit The total cost (in dollars) for a company to produce x items per week is $C = 7x + 400$. The revenue for selling all x items is $R = 34x - 0.1x^2$. How many items must it produce and sell each week for its weekly profit to be $1,300? [5.3]

36. Profit The total cost (in dollars) for a company to produce x items per week is $C = 70x + 300$. The revenue for selling all x items is $R = 110x - 0.5x^2$. How many items must it produce and sell each week for its weekly profit to be $300? [5.3]

Solutions to Selected Practice Problems

Solutions to all practice problems that require more than one step are shown here. Before you look at these solutions to see where you have made a mistake, you should try the problem you are working on twice. If you do not get the correct answer the second time you work the problem, then the solution shown here should show you where you have gone wrong.

Chapter 1

Section 1.1

1. $3x^8 - 27x^6 = 3x^6(x^2 - 9)$
$= 3x^6(x + 3)(x - 3)$

2. $4x^4 + 40x^3 + 100x^2 = 4x^2(x^2 + 10x + 25)$
$= 4x^2(x + 5)^2$

3. $y^4 + 36y^2 = y^2(y^2 + 36)$

4. $6x^2 - x - 15 = (3x - 5)(2x + 3)$

5. $3x^5 - 81x^2 = 3x^2(x^3 - 27)$
$= 3x^2(x - 3)(x^2 + 3x + 9)$

6. $3a^2b^3 + 6a^2b^2 - 3a^2b = 3a^2b(b^2 + 2b - 1)$

7. $x^2 - 10x + 25 - b^2 = (x - 5)^2 - b^2$
$= (x - 5 + b)(x - 5 - b)$

8. $15x^4 + x^2 - 2 = (5x^2 + 2)(3x^2 - 1)$

9. $3x^2(x - 2) - 7x(x - 2) + 2(x - 2)$
$= (x - 2)(3x^2 - 7x + 2)$
$= (x - 2)(3x - 1)(x - 2)$
$= (x - 2)^2(3x - 1)$

Section 1.2

1. $3(5) + 1 \stackrel{?}{=} 16$
$15 + 1 = 16$
$16 = 16$
$4(5) - 6 \stackrel{?}{=} 14$
$20 - 6 = 14$
$14 = 14$

2. $3a - 3 = -5a + 9$
$3a + \mathbf{5a} - 3 = -5a + \mathbf{5a} + 9$
$8a - 3 = 9$
$8a - 3 + \mathbf{3} = 9 + \mathbf{3}$
$8a = 12$
$\frac{\mathbf{1}}{\mathbf{8}} \cdot 8a = \frac{\mathbf{1}}{\mathbf{8}} \cdot 12$
$a = \frac{12}{8} = \frac{3}{2}$

The solution set is $\left\{\frac{3}{2}\right\}$.

3. **a.** $-x = \frac{2}{3}$
$-1(-x) = -1\left(\frac{2}{3}\right)$
$x = -\frac{2}{3}$

b. $-y = -4$
$-1(-y) = -1(-4)$
$y = 4$

4. $\frac{3}{5}x + \frac{1}{3} = -\frac{5}{6}$
$\frac{3}{5}x + \frac{1}{3} + \left(-\frac{\mathbf{1}}{\mathbf{3}}\right) = -\frac{5}{6} + \left(-\frac{\mathbf{1}}{\mathbf{3}}\right)$
$\frac{3}{5}x = -\frac{7}{6}$
$\frac{\mathbf{5}}{\mathbf{3}} \cdot \frac{3}{5}x = \frac{\mathbf{5}}{\mathbf{3}}\left(-\frac{7}{6}\right)$
$x = -\frac{35}{18}$

Or eliminating the fractions in the beginning we can solve it this way:

$30 \cdot \frac{3}{5}x + 30 \cdot \frac{1}{3} = 30\left(-\frac{5}{6}\right)$
$18x + 10 = -25$
$18x = -35$
$x = -\frac{35}{18}$

5. a. Working with the decimals.

$$\begin{aligned}0.08x + 0.10(8{,}000 - x) &= 680\\ 0.08x + 800 - 0.10x &= 680\\ -0.02x + 800 &= 680\\ -0.02x + 800 + (\mathbf{-800}) &= 680 + (\mathbf{-800})\\ -0.02x &= -120\\ x &= \frac{-120}{-0.02} = 6{,}000\end{aligned}$$

b. Eliminating the decimals by multiplying each side by 100.

$$\begin{aligned}100(0.08x) + 100(0.10)(8{,}000 - x) &= 100(680)\\ 8x + 10(8{,}000 - x) &= 68{,}000\\ 8x + 80{,}000 - 10x &= 68{,}000\\ -2x + 80{,}000 &= 68{,}000\\ -2x &= -12{,}000\\ x &= \frac{-12{,}000}{-2}\\ &= 6{,}000\end{aligned}$$

6.
$$\begin{aligned}6 - 2(5x - 1) + 4x &= 20\\ 6 - 10x + 2 + 4x &= 20\\ -6x + 8 &= 20\\ -6x + 8 + (\mathbf{-8}) &= 20 + (\mathbf{-8})\\ -6x &= 12\\ -\frac{\mathbf{1}}{\mathbf{6}}(-6x) &= -\frac{\mathbf{1}}{\mathbf{6}}(12)\\ x &= -2\end{aligned}$$

7.
$$\begin{aligned}3(5x + 1) &= 10 + 15x\\ 15x + 3 &= 10 + 15x\\ 15x + (\mathbf{-15x}) + 3 &= 10 + 15x + (\mathbf{-15x})\\ 3 &= 10\end{aligned}$$

A false statement indicates there is no solution to our original equation.

8.
$$\begin{aligned}-4 + 8x &= 2(4x - 2)\\ -4 + 8x &= 8x - 4\\ -4 + 8x + (\mathbf{-8x}) &= 8x + (\mathbf{-8x}) - 4\\ -4 &= -4\end{aligned}$$

A true statement indicates all real numbers are solutions to our equation.

9.
$$\begin{aligned}x^2 - x - 6 &= 0\\ (x - 3)(x + 2) &= 0\end{aligned}$$
$x - 3 = 0$ or $x + 2 = 0$
$x = 3$ or $x = -2$

10.
$$\begin{aligned}6 \cdot \frac{1}{2}x^3 &= 6 \cdot \frac{5}{6}x^2 + 6 \cdot \frac{1}{3}x\\ 3x^3 &= 5x^2 + 2x\\ 3x^3 - 5x^2 - 2x &= 0\\ x(3x^2 - 5x - 2) &= 0\\ x(3x + 1)(x - 2) &= 0\end{aligned}$$
$x = 0$ or $3x + 1 = 0$ or $x - 2 = 0$
$x = 0$ or $x = -\frac{1}{3}$ or $x = 2$

11.
$$\begin{aligned}100x^2 &= 500x\\ 100x^2 - 500x &= 0\\ 100x(x - 5) &= 0\end{aligned}$$
$100x = 0$ or $x - 5 = 0$
$x = 0$ or $x = 5$

12.
$$\begin{aligned}(x + 1)(x + 2) &= 12\\ x^2 + 3x + 2 &= 12\\ x^2 + 3x - 10 &= 0\\ (x + 5)(x - 2) &= 0\end{aligned}$$
$x + 5 = 0$ or $x - 2 = 0$
$x = -5$ or $x = 2$

13.
$$\begin{aligned}x^3 + 5x^2 - 4x - 20 &= 0\\ x^2(x + 5) - 4(x + 5) &= 0\\ (x + 5)(x^2 - 4) &= 0\\ (x + 5)(x + 2)(x - 2) &= 0\end{aligned}$$
$x + 5 = 0$ or $x + 2 = 0$ or $x - 2 = 0$
$x = -5$ or $x = -2$ or $x = 2$

Section 1.3

1. If $|x| = 3$
then $x = 3$ or $x = -3$

2. If $|3x - 6| = 9$
then $3x - 6 = 9$ or $3x - 6 = -9$
$3x = 15$ or $3x = -3$
$x = 5$ or $x = -1$

3. If $|4x - 3| + 2 = 3$
then $|4x - 3| = 1$
$4x - 3 = 1$ or $4x - 3 = -1$
$4x = 4$ or $4x = 2$
$x = 1$ or $x = \frac{1}{2}$

4. $|7a - 1| = -2$ has no solution since the left side is positive or zero and the right side is negative.

5. $|x + 3| = |x + 8|$

$x + 3 = x + 8$ or $x + 3 = -(x + 8)$

$3 = 8$ — No solution here

$x + 3 = -x - 8$

$2x + 3 = -8$

$2x = -11$

$x = -\frac{11}{2}$

Section 1.4

1. When $x = -3$, we have

$$2(-3) - 3y = 6$$
$$-6 - 3y = 6$$
$$-3y = 12$$
$$y = -4$$

2. When $x = 375$, we have

$$375 = 900 - 300p$$
$$-525 = -300p$$
$$p = \frac{-525}{-300} = \$1.75$$

3.

$$486.7 = (3.14)(6^2) + 2(3.14)(6)h$$
$$486.7 = 113.04 + 37.68h$$
$$373.66 = 37.68h$$

$h = 9.9$ centimeters to the nearest tenth

4. If $v = 64$ and $h = 64$

then $h = vt - 16t^2$

becomes $64 = 64t - 16t^2$

$$16t^2 - 64t + 64 = 0$$
$$t^2 - 4t + 4 = 0$$
$$(t - 2)^2 = 0$$

$t = 2$ seconds

5.

$$R = (1{,}300 - 100p)p$$

If $R = 3{,}600$, then

$$3{,}600 = (1{,}300 - 100p)p$$
$$3{,}600 = 1{,}300p - 100p^2$$
$$100p^2 - 1{,}300p + 3{,}600 = 0$$
$$p^2 - 13p + 36 = 0$$
$$(p - 4)(p - 9) = 0$$

$p - 4 = 0$ or $p - 9 = 0$

$p = 4$ or $p = 9$

The price can be set at either \$4 or \$9.

6.

$$P = 2w + 2l$$
$$P - 2w = 2l$$
$$\frac{P - 2w}{2} = l$$

or $\frac{P}{2} - w = l$

7.

$$ax + 5 = cx + 3$$
$$ax - cx = 3 - 5$$
$$(a - c)x = -2$$
$$x = \frac{-2}{a - c}$$

8. $x = 0.25(74)$

$x = 18.5$

9. $x \cdot 84 = 21$

$$x = \frac{21}{84}$$
$$x = 0.25$$
$$x = 25\%$$

10.

$$35 = 0.40x$$
$$\frac{35}{0.40} = x$$
$$87.5 = x$$

11.

$$a_1 = 3(1) - 2 = 1$$
$$a_2 = 3(2) - 2 = 4$$
$$a_3 = 3(3) - 2 = 7$$
$$a_4 = 3(4) - 2 = 10$$

12.

$$a_1 = \frac{1}{1 + 1} = \frac{1}{2}$$
$$a_2 = \frac{2}{2 + 1} = \frac{2}{3}$$
$$a_3 = \frac{3}{3 + 1} = \frac{3}{4}$$
$$a_4 = \frac{4}{4 + 1} = \frac{4}{5}$$

Section 1.5

1.

$$2x + 2(4x - 10) = 12.5$$
$$2x + 8x - 20 = 12.5$$
$$10x - 20 = 12.5$$
$$10x = 32.5$$
$$x = 3.25$$

The length is $x = 3.25$ feet; the width is $4x - 10 = 4(3.25) - 10 = 3$ feet.

2.

$$x + 0.0725x = 23{,}466.30$$
$$1.0725x = 23{,}466.30$$
$$x = \frac{23{,}466.30}{1.0725}$$
$$= 21{,}880$$

The price of the car is \$21,880.00.

3.

$$x + 5x = 180$$
$$6x = 180$$
$$x = 30$$

The angles are 30° and 150°.

4. 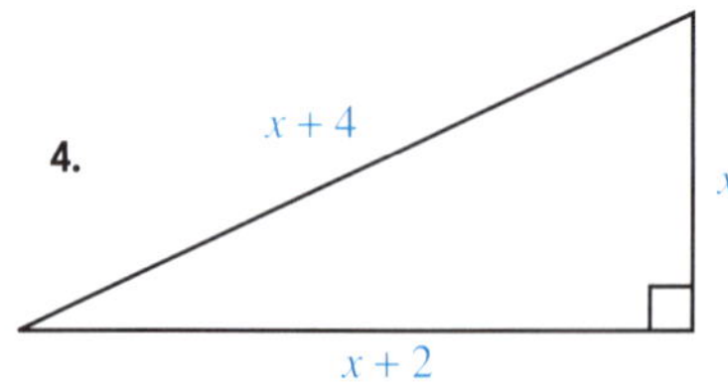

$$(x+4)^2 = (x+2)^2 + x^2$$
$$x^2 + 8x + 16 = x^2 + 4x + 4 + x^2$$
$$x^2 + 8x + 16 = 2x^2 + 4x + 4$$
$$0 = x^2 - 4x - 12$$
$$0 = (x-6)(x+2)$$
$x - 6 = 0$ or $x + 2 = 0$

$x = 6$ or $x = -2$

Since the length of a side cannot be negative, the shortest side is 6. The other two sides are $6 + 2 = 8$ and $6 + 4 = 10$.

5.

	DOLLARS AT 8%	DOLLARS AT 10%	TOTAL
Number	x	$8{,}000 - x$	8,000
Interest	$0.08x$	$0.10(8{,}000 - x)$	680

$$0.08x + 0.10(8{,}000 - x) = 680$$
$$8x + 10(8{,}000 - x) = 68{,}000$$
$$8x + 80{,}000 - 10x = 68{,}000$$
$$-2x + 80{,}000 = 68{,}000$$
$$-2x = -12{,}000$$
$$x = 6{,}000$$
\$6,000 at 8% and \$2,000 at 10%

6.

WIDTH	LENGTH	AREA (IN²)
1	$\ell = 7 - 1 = 6$	6
2	$\ell = 7 - 2 = 5$	10
3	$\ell = 7 - 3 = 4$	12
4	$\ell = 7 - 4 = 3$	12
5	$\ell = 7 - 5 = 2$	10
6	$\ell = 7 - 6 = 1$	6

Section 1.6

1.
$$4x - 2 > 3x + 4$$
$$4x + (\mathbf{-3x}) - 2 > 3x + (\mathbf{-3x}) + 4$$
$$x - 2 > 4$$
$$x - 2 + \mathbf{2} > 4 + \mathbf{2}$$
$$x > 6$$

Interval notation $(6, \infty)$

2.
$$-3y - 2 < 7$$
$$-3y < 9$$
$$-\frac{1}{3}(-3y) > -\frac{1}{3}(9)$$
$$y > -3$$

Interval notation $(-3, \infty)$

3. 

4. 

5.
$$-7 \le 2x + 1 \le 7$$
$$-8 \le 2x \le 6$$
$$-4 \le x \le 3$$

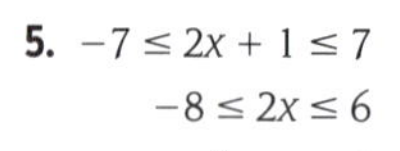
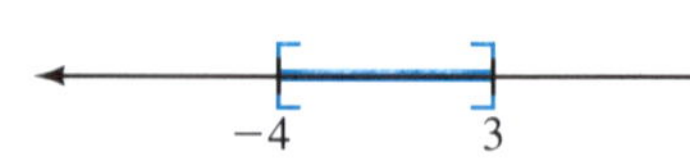

Interval notation $[-4, 3]$

6. $3t - 6 \le -3$ or $3t - 6 \ge 3$

$3t \le 3$ or $3t \ge 9$

$t \le 1$ or $t \ge 3$

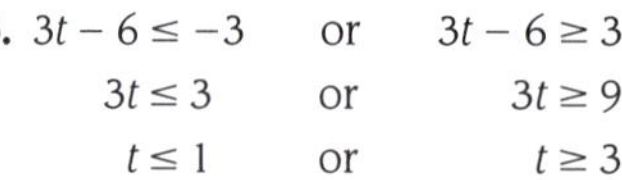
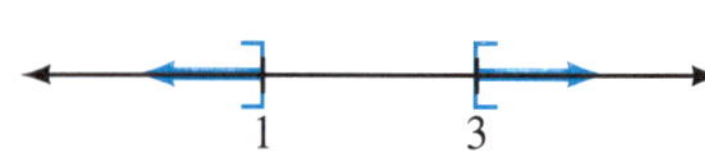

Interval notation $(-\infty, 1] \cup [3, \infty)$

Chapter 2

Section 2.1

1.

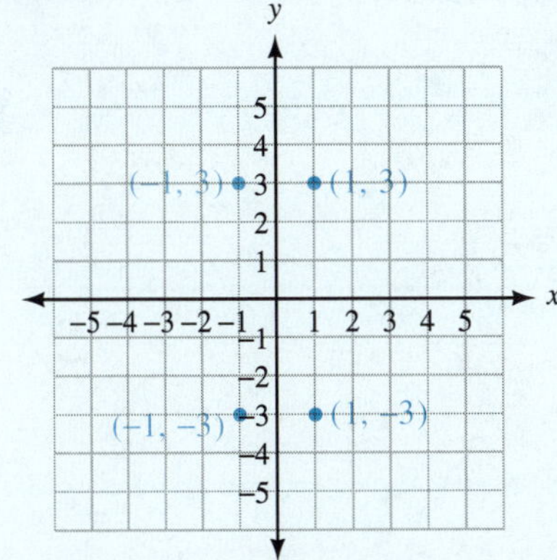

2.

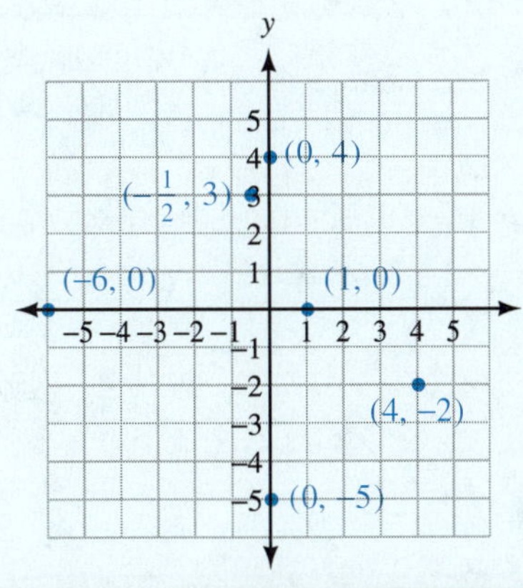

3.

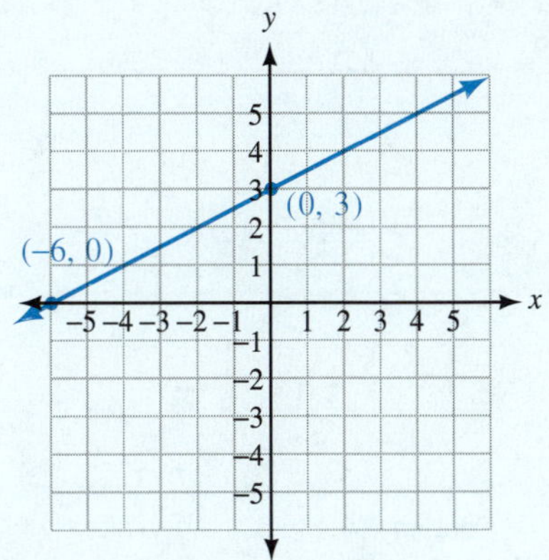

4.

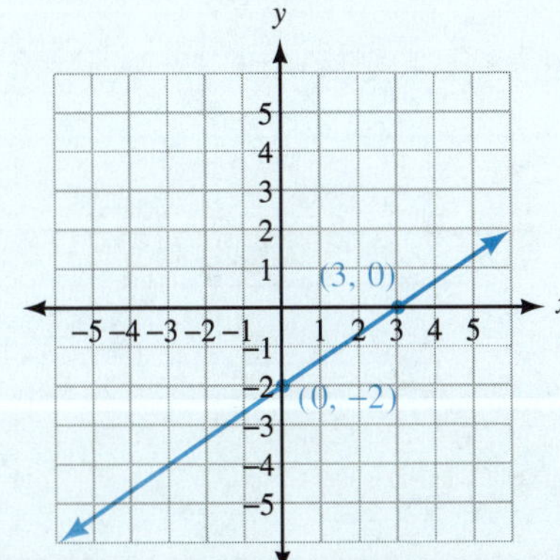

5.

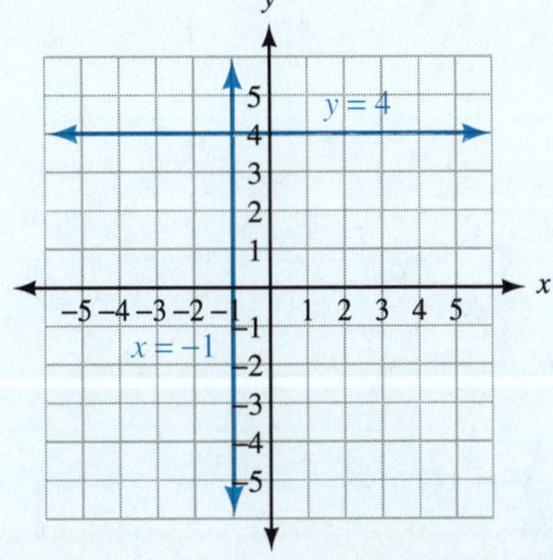

Section 2.2

1. When $x = 0, y = -5$, so the point $(0, -5)$ is on the graph. When $x = 2, y = 1$, so the point $(2, 1)$ is a second point on the graph. The ratio of rise to run going from $(0, -5)$ to $(2, 1)$ is $\frac{6}{2} = 3$. The slope of $y = 3x - 5$ is 3.

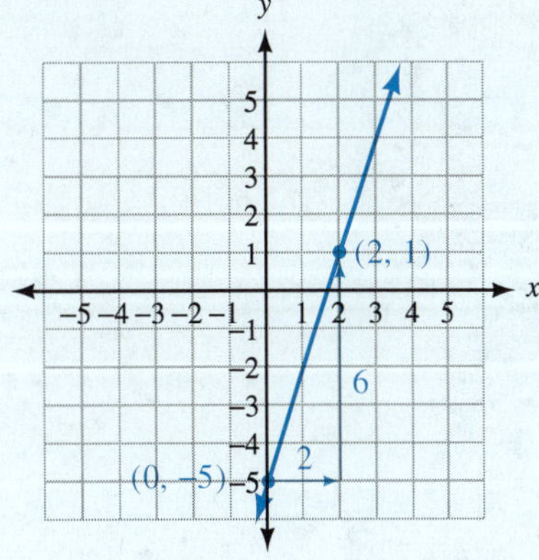

2. $m = \frac{4-(-2)}{3-1} = \frac{6}{2} = 3$ **3.** $m = \frac{-3-(-3)}{2-(-1)} = \frac{0}{3} = 0$

Section 2.3

1.

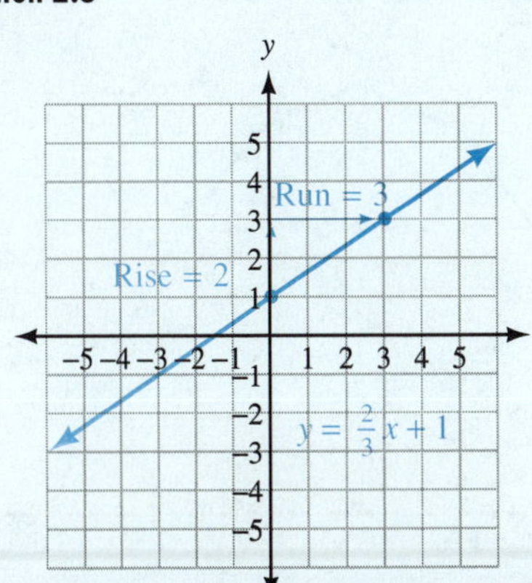

2.

$$4x - 5y = 7$$
$$-5y = -4x + 7$$
$$y = \frac{4}{5}x - \frac{7}{5}$$

Slope $= \frac{4}{5}$, y-intercept $= -\frac{7}{5}$

3.

$$-3x + 2y = -6$$
$$2y = 3x - 6$$
$$y = \frac{3}{2}x - 3$$

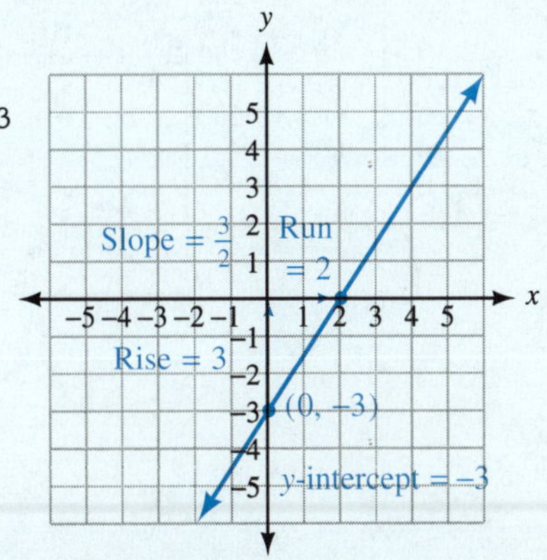

4. $m = 3$, $(x_1, y_1) = (-1, 2)$

$y - y_1 = m(x - x_1)$

$y - 2 = 3(x + 1)$

$y - 2 = 3x + 3$

$y = 3x + 5$

5. $m = \dfrac{5 - (-3)}{2 - 6} = \dfrac{8}{-4} = -2$

let $(x_1, y_1) = (2, 5)$

then $y - 5 = -2(x - 2)$

$y - 5 = -2x + 4$

$y = -2x + 9$

6. $3x - y = 2$

$-y = -3x + 2$

$y = 3x - 2$

The slope of a line perpendicular to this line is $-\frac{1}{3}$.

Using $m = -\frac{1}{3}$ and $(x_1, y_1) = (3, 2)$ we have

$$y - 2 = -\frac{1}{3}(x - 3)$$

$$y - 2 = -\frac{1}{3}x + 1$$

$$y = -\frac{1}{3}x + 3$$

$$3y = -x + 9$$

$$x + 3y = 9$$

Section 2.4

1.

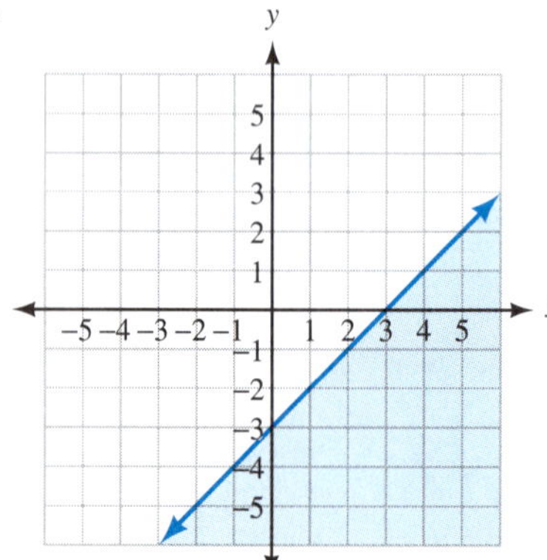

2.

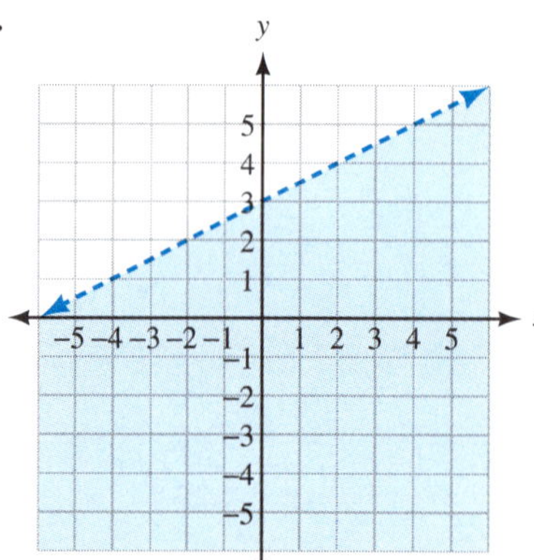

3.

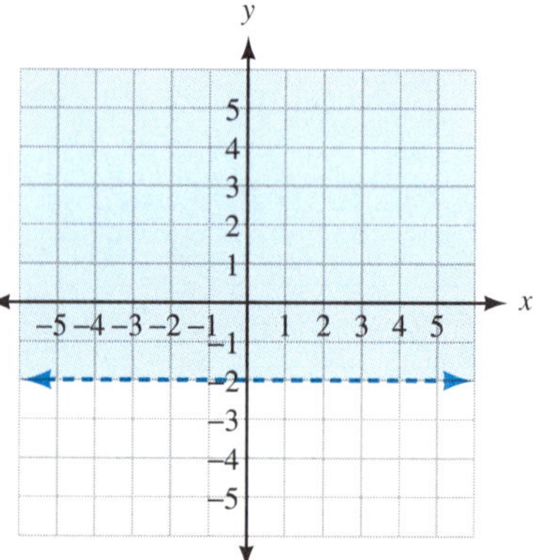

4.

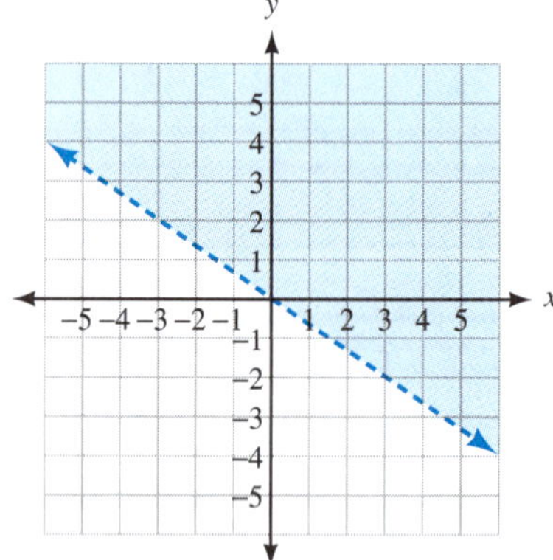

Section 2.5

1.

x	y
0	0
10	80
20	160
30	240
40	320

4.

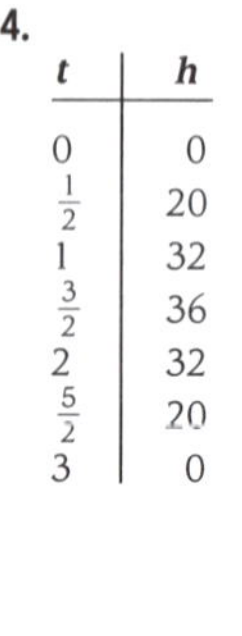

t	h
0	0
$\frac{1}{2}$	20
1	32
$\frac{3}{2}$	36
2	32
$\frac{5}{2}$	20
3	0

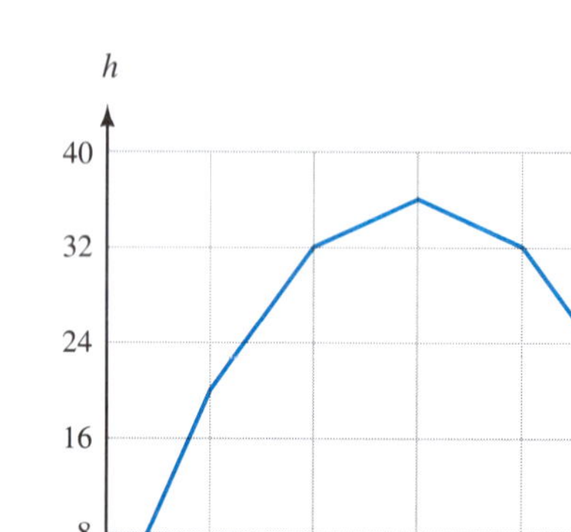

5.

x	y
5	−3
0	−2
−3	−1
−4	0
−3	1
0	2
5	3

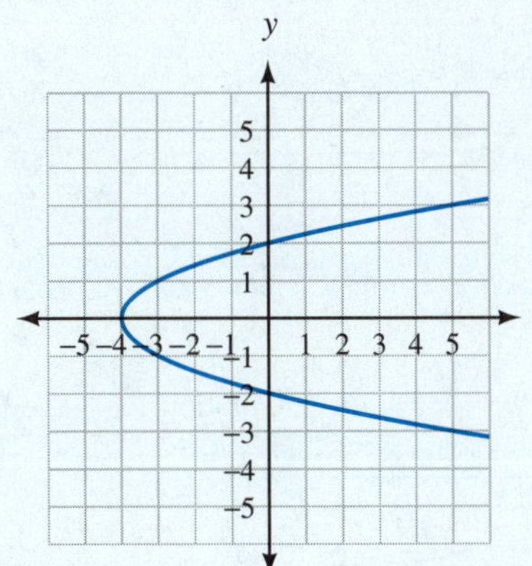

6.

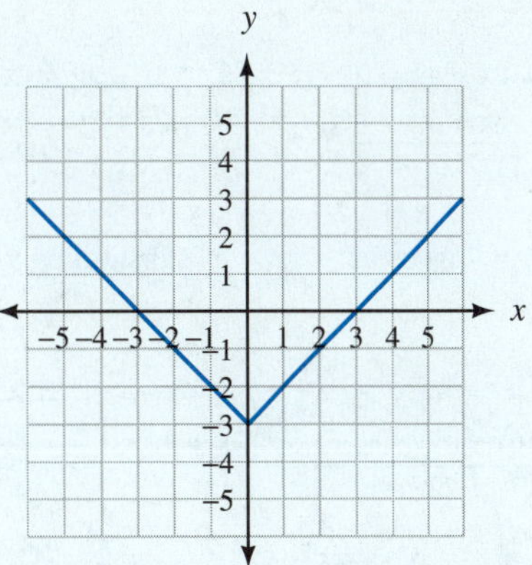

Section 2.6

1. (a) $f(0) = 8(0)$
$= 0$

(b) $f(5) = 8(5)$
$= 40$

(c) $f(10.5) = 8(10.5)$
$= 84$

2. (a) $s(8) = \dfrac{88}{8}$
$= 11$

(b) $s(11) = \dfrac{88}{11}$
$= 8$

3. (a) $C(5) = 80\left(\dfrac{1}{2}\right)^{5/5}$
$= 80\left(\dfrac{1}{2}\right)^{1}$
$= 40$

(b) $C(10) = 80\left(\dfrac{1}{2}\right)^{10/5}$
$= 80\left(\dfrac{1}{2}\right)^{2}$
$= 80\left(\dfrac{1}{4}\right)$
$= 20$

4. (a) $C(5) = 2\pi(5)$
$= 10\pi$
≈ 31.4 inches

(b) $A(5) = \pi(5)^2$
$= 25\pi$
≈ 78.5 inches^2

5. (a) $f(0) = 4(0)^2 - 3$
$= -3$

(b) $f(3) = 4(3)^2 - 3$
$= 4(9) - 3$
$= 36 - 3$
$= 33$

(c) $f(-2) = 4(-2)^2 - 3$
$= 4(4) - 3$
$= 16 - 3$
$= 13$

6. (a) $f(5) = 2(5) + 1$
$= 10 + 1$
$= 11$

(b) $g(5) = (5)^2 - 3$
$= 25 - 3$
$= 22$

(c) $f(-2) = 2(-2) + 1$
$= -4 + 1$
$= -3$

(d) $g(-2) = (-2)^2 - 3$
$= 4 - 3$
$= 1$

(e) $f(a) = 2(a) + 1$
$= 2a + 1$

(f) $g(a) = (a)^2 - 3$
$= a^2 - 3$

8. (a) $f[g(2)] = f[4(2) + 1]$
$= f[8 + 1]$
$= f(9)$
$= 3(9)^2$
$= 3(81)$
$= 243$

(b) $g[f(2)] = g[3(2)^2]$
$= g[3(4)]$
$= g(12)$
$= 4(12) + 1$
$= 48 + 1$
$= 49$

Section 2.7

1. $y = Kx$
$24 = K \cdot 8$
$\Rightarrow K = 3$
$y = 3x$
$y = 3 \cdot 2$
$= 6$

2. $d(2.5) = 16(2.5)^2$
$= 16(6.25)$
$= 100$ feet

3. $V = \dfrac{2{,}400}{P}$
$150 = \dfrac{2{,}400}{P}$
$150P = 2{,}400$
$P = \dfrac{2{,}400}{150}$
$= 16$ pounds per square inch

4. $y = Kxz^2$
$81 = K(2)\,9^2$
$81 = 162K$
$\Rightarrow K = \dfrac{1}{2}$
$y = \dfrac{1}{2}xz^2$
$= \dfrac{1}{2}(4)(4^2)$
$= 32$

5. $R = \dfrac{0.0005l}{d^2}$
$= \dfrac{0.0005(300)}{(0.25)^2}$
$= 2.4$ ohms

Section 2.8

1.
$$\begin{aligned} 2x - y &= 7 \xrightarrow{\text{4 times each side}} & 8x - 4y &= 28 \\ 3x + 4y &= -6 \xrightarrow[\text{no change}]{} & 3x + 4y &= -6 \\ & & \overline{11x \qquad} &\overline{= 22} \\ & & x &= 2 \end{aligned}$$

Substituting $x = 2$ into $2x - y = 7$ gives us $y = -3$. The solution is $(2, -3)$.

2.
$$\begin{aligned} 3x - 2y &= -8 \xrightarrow{\text{2 times each side}} & 6x - 4y &= -16 \\ -2x + 3y &= 7 \xrightarrow[\text{3 times each side}]{} & -6x + 9y &= 21 \\ & & 5y &= 5 \\ & & y &= 1 \end{aligned}$$

Substituting $y = 1$ into $3x - 2y = -8$ gives us $x = -2$. The solution is $(-2, 1)$.

3.
$$\begin{aligned} 3x - 5y &= 2 \xrightarrow{\text{2 times each side}} & 6x - 10y &= 4 \\ 2x + 4y &= 1 \xrightarrow[\text{-3 times each side}]{} & -6x - 12y &= -3 \\ & & -22y &= 1 \\ & & y &= -\frac{1}{22} \end{aligned}$$

$$\begin{aligned} 3x - 5y &= 2 \xrightarrow{\text{4 times each side}} & 12x - 20y &= 8 \\ 2x + 4y &= 1 \xrightarrow[\text{5 times each side}]{} & 10x + 20y &= 5 \\ & & 22x &= 13 \\ & & x &= \frac{13}{22} \end{aligned}$$

Solution: $\left(\frac{13}{22}, -\frac{1}{22}\right)$

4.
$$\begin{aligned} 2x + 7y &= 3 \xrightarrow{\text{-2 times each side}} & -4x - 14y &= -6 \\ 4x + 14y &= 1 \xrightarrow[\text{no change}]{} & 4x + 14y &= 1 \\ & & 0 &= -5 \end{aligned}$$

We have eliminated both variables and are left with a false statement, indicating that the lines are parallel. There is no solution to the system.

5.
$$\begin{aligned} 2x + 7y &= 3 \xrightarrow{\text{-2 times each side}} & -4x - 14y &= -6 \\ 4x + 14y &= 6 \xrightarrow[\text{no change}]{} & 4x + 14y &= 6 \\ & & 0 &= 0 \end{aligned}$$

We have eliminated both variables and are left with a true statement. The lines coincide. Any ordered pair that satisfies one of the equations will satisfy the other.

6.
$$\begin{aligned} \frac{1}{3}x + \frac{1}{2}y &= 4 \xrightarrow{\text{times 6}} & 2x + 3y &= 24 \\ \frac{2}{3}x - \frac{1}{4}y &= 3 \xrightarrow[\text{times 12}]{} & 8x - 3y &= 36 \\ & & 10x &= 60 \\ & & x &= 6 \end{aligned}$$

Substituting $x = 6$ into any equation with both variables gives $y = 4$. The solution is $(6, 4)$.

7. $4x - 2y = -2$

$y = x + 3$

Substituting $x + 3$ for y in the first equation gives us

$$\begin{aligned} 4x - 2(x + 3) &= -2 \\ 4x - 2x - 6 &= -2 \\ 2x - 6 &= -2 \\ 2x &= 4 \\ x &= 2 \end{aligned}$$

When $x = 2, y = 2 + 3 = 5$. The solution is $(2, 5)$.

8. $5x - 3y = -4$

$x + 2y = 7$

Solving the second equation for x gives $x = -2y + 7$. Substituting this for x in the first equation we have

$$\begin{aligned} 5(-2y + 7) - 3y &= -4 \\ -10y + 35 - 3y &= -4 \\ -13y + 35 &= -4 \\ -13y &= -39 \\ y &= 3 \end{aligned}$$

Putting $y = 3$ into either of the first two equations gives $x = 1$. The solution is $(1, 3)$.

Section 2.9

1.

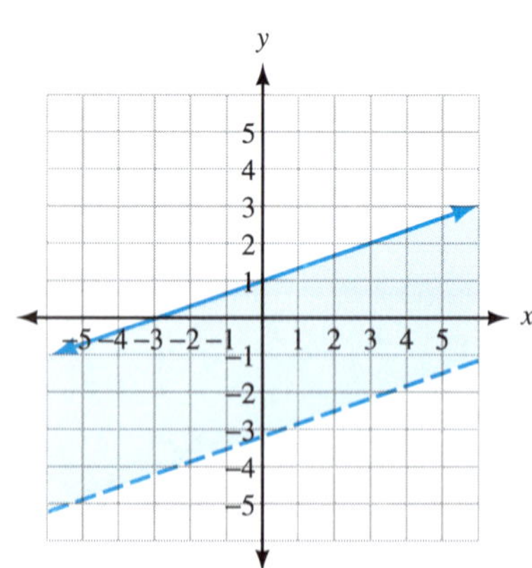

2.

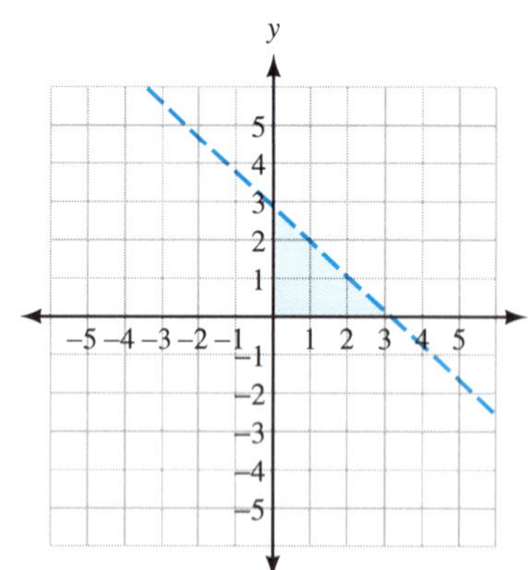

3.

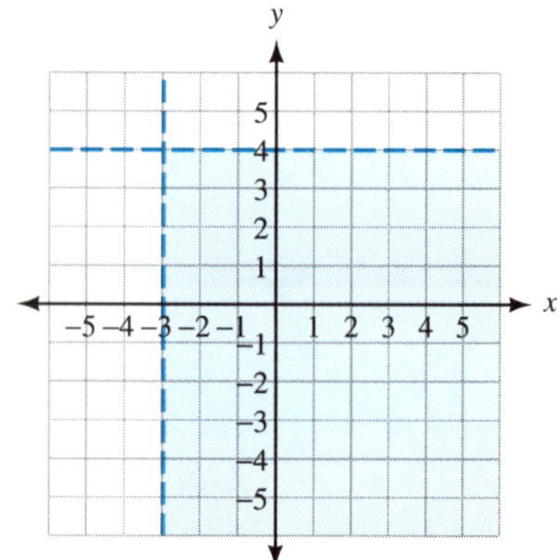

4.

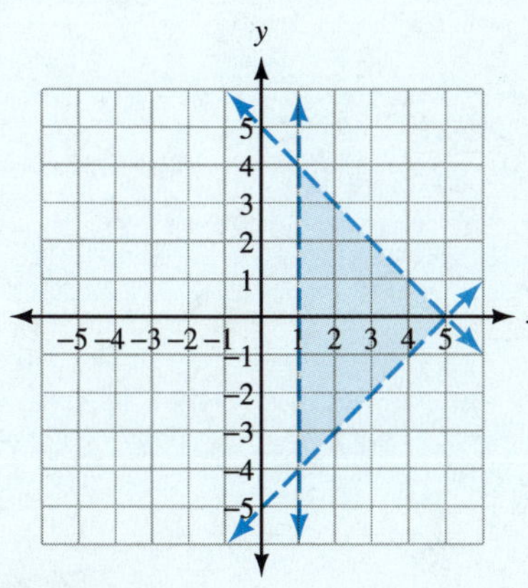

5. The horizontal line would move down to $y = 300$ from $y = 500$.

Chapter 3

Section 3.1

1. $\dfrac{x^2-9}{x+3} = \dfrac{\cancel{(x+3)}(x-3)}{\cancel{x+3}}$

$= x - 3$

2. $\dfrac{y^2-y-6}{y^2-4} = \dfrac{(y-3)\cancel{(y+2)}}{(y-2)\cancel{(y+2)}}$

$= \dfrac{y-3}{y-2}$

3. $\dfrac{3a^3+3}{6a^2-6a+6} = \dfrac{3(a^3+1)}{6(a^2-a+1)}$

$= \dfrac{3(a+1)\cancel{(a^2-a+1)}}{6\cancel{(a^2-a+1)}}$

$= \dfrac{a+1}{2}$

4. $\dfrac{x^2+4x+ax+4a}{x^2+ax+4x+4a} = \dfrac{x(x+4)+a(x+4)}{x(x+a)+4(x+a)}$

$= \dfrac{\cancel{(x+4)}\cancel{(x+a)}}{\cancel{(x+a)}\cancel{(x+4)}}$

$= 1$

5. When $a = 7$ and $b = 4$

the expression $\dfrac{a-b}{b-a}$ becomes

$\dfrac{7-4}{4-7} = \dfrac{3}{-3} = -1$

6. $\dfrac{7-x}{x^2-49} = \dfrac{-1\cancel{(x-7)}}{(x+7)\cancel{(x-7)}} = \dfrac{-1}{x+7}$

7. $C = 165(3.14) = 518$ feet to the nearest foot

$r = \dfrac{d}{t} = \dfrac{518}{40} = 13.0$ feet per second

9. (a) $f(0) = \dfrac{0+6}{0-3}$

$= \dfrac{6}{-3}$

$= -2$

(b) $f(-6) = \dfrac{-6+6}{-6-3}$

$= \dfrac{0}{-9}$

$= 0$

(c) $f(6) = \dfrac{6+6}{6-3}$

$= \dfrac{12}{3}$

$= 4$

(d) $f(3) = \dfrac{3+6}{3-3}$

$= \dfrac{9}{0}$

Undefined

10.

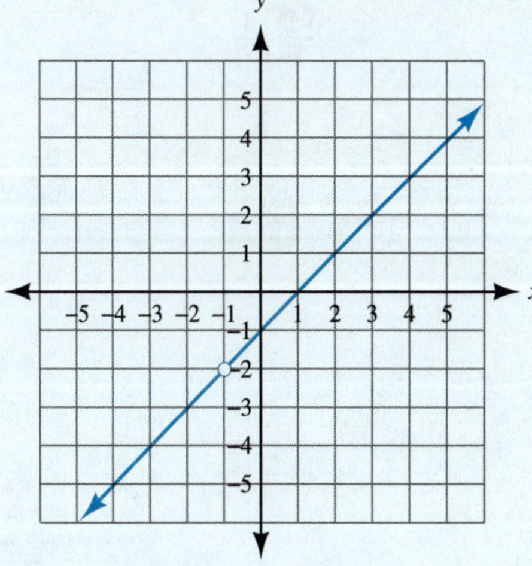

This graph does not contain the point $(-1, -2)$, whereas the graph of $y = x - 1$ does.

Section 3.2

1. $\dfrac{12x^4-18x^3+24x^2}{6x} = \dfrac{12x^4}{6x} - \dfrac{18x^3}{6x} + \dfrac{24x^2}{6x}$

$= 2x^3 - 3x^2 + 4x$

2. $\dfrac{27x^4y^7-81x^5y^3}{-9x^3y^2} = \dfrac{27x^4y^7}{-9x^3y^2} - \dfrac{81x^5y^3}{-9x^3y^2}$

$= -3xy^5 + 9x^2y$

3. $\dfrac{12a^5+8a^4+16a^3+4a^2}{8a^4} = \dfrac{12a^5}{8a^4} + \dfrac{8a^4}{8a^4} + \dfrac{16a^3}{8a^4} + \dfrac{4a^2}{8a^4}$

$= \dfrac{3a}{2} + 1 + \dfrac{2}{a} + \dfrac{1}{2a^2}$

4. $\dfrac{2x^2-5xy+3y^2}{x-y} = \dfrac{(2x-3y)\cancel{(x-y)}}{\cancel{x-y}}$

$= 2x - 3y$

5.
$$\frac{f(x)-f(a)}{x-a}=\frac{(2x+3)-(2a+3)}{x-a}=\frac{2x-2a}{x-a}=\frac{2(x-a)}{x-a}=2$$

6.
$$\frac{f(x)-f(a)}{x-a}=\frac{(x^2-1)-(a^2-1)}{x-a}=\frac{x^2-a^2}{x-a}=\frac{(x+a)(x-a)}{x-a}=x+a$$

7.
$$\begin{array}{r} 215 \\ 35\overline{)7{,}546} \\ \underline{7\,0} \\ 54 \\ \underline{35} \\ 196 \\ \underline{175} \\ 21 \end{array}$$

Answer: $215+\frac{3}{5}$

8.
$$\begin{array}{r} 3x+1 \\ x-3\overline{)\;3x^2-8x-1} \\ -\quad + \\ \underline{\cancel{+}3x^2\cancel{-}9x} \\ +\;x-1 \\ -\quad+ \\ \underline{\cancel{+}\;x\cancel{-}3} \\ 2 \end{array}$$

Answer: $3x+1+\dfrac{2}{x-3}$

9.
$$\begin{array}{r} 3x^2+6x+15 \\ x-2\overline{)\;3x^3+0x^2+3x+1} \\ -\quad+ \\ \underline{\cancel{+}3x^3\cancel{-}6x^2} \\ +6x^2+\;3x \\ -\quad+ \\ \underline{\cancel{+}6x^2\cancel{-}12x} \\ +15x+1 \\ -\quad+ \\ \underline{\cancel{+}15x\cancel{-}30} \\ 31 \end{array}$$

Answer: $3x^2+6x+15+\dfrac{31}{x-2}$

10.
$$\begin{array}{r} 2x-3y \\ x-y\overline{)\;2x^2-5xy+3y^2} \\ -\quad+ \\ \underline{\cancel{+}2x^2\cancel{-}2xy} \\ -3xy+3y^2 \\ +\quad- \\ \underline{\cancel{-}3xy\cancel{+}3y^2} \\ 0 \end{array}$$

Answer: $2x-3y$

Section 3.3

1.
$$\frac{3}{4}\cdot\frac{12}{27}=\frac{3\cdot 12}{4\cdot 27}=\frac{\cancel{3}\cdot\cancel{2}\cdot\cancel{2}\cdot\cancel{3}}{\cancel{2}\cdot\cancel{2}\cdot\cancel{3}\cdot\cancel{3}\cdot 3}=\frac{1}{3}$$

2.
$$\frac{6x^4}{4y^9}\cdot\frac{12y^5}{3x^2}=\frac{\overset{2}{\cancel{6}}\cdot\overset{3}{\cancel{12}}x^4y^5}{\cancel{3}\cdot\cancel{4}x^2y^9}=\frac{6x^2}{y^4}$$

3.
$$\frac{x+5}{x^2-25}\cdot\frac{x-5}{x^2-10x+25}=\frac{\cancel{(x+5)}\cancel{(x-5)}}{\cancel{(x+5)}\cancel{(x-5)}(x-5)^2}=\frac{1}{(x-5)^2}$$

4.
$$\frac{3y^2-3y}{3y-12}\cdot\frac{y^2-2y-8}{y^2+3y+2}=\frac{\cancel{3}y(y-1)\cancel{(y-4)}\cancel{(y+2)}}{\cancel{3}\cancel{(y-4)}(y+1)\cancel{(y+2)}}=\frac{y(y-1)}{y+1}$$

5.
$$\frac{5}{9}\div\frac{10}{27}=\frac{5}{9}\cdot\frac{27}{10}=\frac{\cancel{5}\cdot\cancel{3}\cdot\cancel{3}\cdot 3}{\cancel{3}\cdot\cancel{3}\cdot 2\cdot\cancel{5}}=\frac{3}{2}$$

6.
$$\frac{9x^4}{4y^3}\div\frac{3x^2}{8y^5}=\frac{9x^4}{4y^3}\cdot\frac{8y^5}{3x^2}=\frac{\overset{3}{\cancel{9}}\cdot\overset{2}{\cancel{8}}x^4y^5}{\cancel{3}\cdot\cancel{4}x^2y^3}=6x^2y^2$$

7.
$$\frac{xy^2-y^3}{x^2-y^2}\div\frac{x^3+y^3}{x^2+2xy+y^2}=\frac{xy^2-y^3}{x^2-y^2}\cdot\frac{x^2+2xy+y^2}{x^3+y^3}$$
$$=\frac{y^2\cancel{(x-y)}\cancel{(x+y)}\cancel{(x+y)}}{\cancel{(x+y)}\cancel{(x-y)}\cancel{(x+y)}(x^2-xy+y^2)}=\frac{y^2}{x^2-xy+y^2}$$

8.
$$\frac{a^2+3a-4}{a-4}\cdot\frac{a+3}{a^2-4a+3}\div\frac{a+1}{a^2-2a-3}=\frac{(a^2+3a-4)(a+3)(a^2-2a-3)}{(a-4)(a^2-4a+3)(a+1)}$$
$$=\frac{(a+4)\cancel{(a-1)}(a+3)\cancel{(a-3)}\cancel{(a+1)}}{(a-4)\cancel{(a-3)}\cancel{(a-1)}\cancel{(a+1)}}=\frac{(a+4)(a+3)}{a-4}$$

9. $$\frac{xa + xb - ya - yb}{xa + 2x + ya + 2y} \cdot \frac{xa + 2x + ya + 2y}{xa + xb + ya + yb}$$

$$= \frac{x(a + b) - y(a + b)}{x(a + 2) + y(a + 2)} \cdot \frac{x(a + 2) + y(a + 2)}{x(a + b) + y(a + b)}$$

$$= \frac{\cancel{(a + b)}\,(x - y)\cancel{(a + 2)}\,\cancel{(x + y)}}{\cancel{(a + 2)}\,(x + y)\cancel{(a + b)}\,\cancel{(x + y)}}$$

$$= \frac{x - y}{x + y}$$

10. $$(5x^2 - 45) \cdot \frac{3}{5x - 15} = \frac{5x^2 - 45}{1} \cdot \frac{3}{5x - 15}$$

$$= \frac{\cancel{5}(x + 3)\cancel{(x - 3)}\,3}{\cancel{5}\cancel{(x - 3)}}$$

$$= 3(x + 3)$$

Section 3.4

1. $$\frac{3}{8} + \frac{1}{8} = \frac{3 + 1}{8}$$

$$= \frac{4}{8}$$

$$= \frac{1}{2}$$

2. $$\frac{x}{x^2 - 9} + \frac{3}{x^2 - 9} = \frac{x + 3}{x^2 - 9}$$

$$= \frac{\cancel{x + 3}}{\cancel{(x + 3)}\,(x - 3)}$$

$$= \frac{1}{x - 3}$$

3. $$\frac{2x - 7}{x - 2} - \frac{x - 5}{x - 2} = \frac{2x - 7 - (x - 5)}{x - 2}$$

$$= \frac{2x - 7 - x + 5}{x - 2}$$

$$= \frac{x - 2}{x - 2}$$

$$= 1$$

4. $$\frac{3}{10} + \frac{11}{42} = \frac{3}{2 \cdot 5} + \frac{11}{2 \cdot 3 \cdot 7}$$

$$= \frac{3}{2 \cdot 5} \cdot \frac{\mathbf{3 \cdot 7}}{\mathbf{3 \cdot 7}} + \frac{11}{2 \cdot 3 \cdot 7} \cdot \frac{\mathbf{5}}{\mathbf{5}}$$

$$= \frac{63}{2 \cdot 3 \cdot 5 \cdot 7} + \frac{55}{2 \cdot 3 \cdot 5 \cdot 7}$$

$$= \frac{118}{2 \cdot 3 \cdot 5 \cdot 7}$$

$$= \frac{\cancel{2} \cdot 59}{\cancel{2} \cdot 3 \cdot 5 \cdot 7}$$

$$= \frac{59}{105}$$

5. $$\frac{-3}{x^2 - 2x - 8} + \frac{4}{x^2 - 16} = \frac{-3}{(x - 4)(x + 2)} + \frac{4}{(x - 4)(x + 4)}$$

$$= \frac{-3}{(x - 4)(x + 2)} \cdot \frac{\mathbf{x + 4}}{\mathbf{x + 4}} + \frac{4}{(x - 4)(x + 4)} \cdot \frac{\mathbf{x + 2}}{\mathbf{x + 2}}$$

$$= \frac{-3x - 12 + 4x + 8}{(x - 4)(x + 4)(x + 2)}$$

$$= \frac{\cancel{x - 4}}{\cancel{(x - 4)}\,(x + 4)(x + 2)}$$

$$= \frac{1}{(x + 4)(x + 2)}$$

6. $$\frac{x - 4}{2x - 6} + \frac{3}{x^2 - 9} = \frac{x - 4}{2(x - 3)} + \frac{3}{(x + 3)(x - 3)}$$

$$= \frac{x - 4}{2(x - 3)} \cdot \frac{\mathbf{x + 3}}{\mathbf{x + 3}} + \frac{3}{(x + 3)(x - 3)} \cdot \frac{\mathbf{2}}{\mathbf{2}}$$

$$= \frac{(x - 4)(x + 3) + 3 \cdot 2}{2(x + 3)(x - 3)}$$

$$= \frac{x^2 - x - 12 + 6}{2(x + 3)(x - 3)}$$

$$= \frac{x^2 - x - 6}{2(x + 3)(x - 3)}$$

$$= \frac{\cancel{(x - 3)}\,(x + 2)}{2(x + 3)\cancel{(x - 3)}}$$

$$= \frac{x + 2}{2(x + 3)}$$

7. $$\frac{2x-4}{x^2+5x+4} - \frac{x-4}{x^2+6x+8} = \frac{2x-4}{(x+4)(x+1)} - \frac{x-4}{(x+4)(x+2)}$$
$$= \frac{2x-4}{(x+4)(x+1)} \cdot \frac{\mathbf{x+2}}{\mathbf{x+2}} - \frac{x-4}{(x+4)(x+2)} \cdot \frac{\mathbf{x+1}}{\mathbf{x+1}}$$
$$= \frac{(2x-4)(x+2)-(x-4)(x+1)}{(x+4)(x+1)(x+2)}$$
$$= \frac{(2x^2-8)-(x^2-3x-4)}{(x+4)(x+1)(x+2)}$$
$$= \frac{x^2+3x-4}{(x+4)(x+1)(x+2)}$$
$$= \frac{\cancel{(x+4)}(x-1)}{\cancel{(x+4)}(x+1)(x+2)}$$
$$= \frac{x-1}{(x+1)(x+2)}$$

8. $$\frac{x^2}{x-4} + \frac{x+12}{4-x} = \frac{x^2}{x-4} + \frac{x+12}{4-x} \cdot \frac{-1}{-1}$$
$$= \frac{x^2}{x-4} + \frac{-x-12}{x-4}$$
$$= \frac{x^2-x-12}{x-4}$$
$$= \frac{\cancel{(x-4)}(x+3)}{\cancel{x-4}}$$
$$= x+3$$

9. $$2 + \frac{25}{5x-1} = \frac{2}{1} + \frac{25}{5x-1}$$
$$= \frac{2}{1} \cdot \frac{\mathbf{5x-1}}{\mathbf{5x-1}} + \frac{25}{5x-1}$$
$$= \frac{10x-2+25}{5x-1}$$
$$= \frac{10x+23}{5x-1}$$

10. $$\frac{1}{x} + \frac{1}{3x} = \frac{3}{3x} + \frac{1}{3x} = \frac{4}{3x}$$

Section 3.5

1. $$\frac{\frac{2}{3}}{\frac{5}{6}} = \frac{2}{3} \cdot \frac{6}{5} = \frac{12}{15} = \frac{4}{5}$$

2. $$\frac{\frac{1}{x}-\frac{1}{3}}{\frac{1}{x}+\frac{1}{3}} = \frac{\left(\frac{1}{x}-\frac{1}{3}\right)\mathbf{3x}}{\left(\frac{1}{x}+\frac{1}{3}\right)\mathbf{3x}}$$
$$= \frac{\frac{1}{x}(3x)-\frac{1}{3}(3x)}{\frac{1}{x}(3x)+\frac{1}{3}(3x)}$$
$$= \frac{3-x}{3+x}$$

3. $$\frac{\frac{x+5}{x^2-16}}{\frac{x^2-25}{x-4}} = \frac{x+5}{x^2-16} \cdot \frac{x-4}{x^2-25}$$
$$= \frac{(x+5)(x-4)}{(x+4)(x-4)(x+5)(x-5)}$$
$$= \frac{1}{(x+4)(x-5)}$$

4. $$\frac{1-\frac{9}{x^2}}{1-\frac{1}{x}-\frac{6}{x^2}} = \frac{\mathbf{x^2}\cdot\left(1-\frac{9}{x^2}\right)}{\mathbf{x^2}\cdot\left(1-\frac{1}{x}-\frac{6}{x^2}\right)}$$
$$= \frac{x^2\cdot 1 - x^2\cdot\frac{9}{x^2}}{x^2\cdot 1 - x^2\cdot\frac{1}{x} - x^2\cdot\frac{6}{x^2}}$$
$$= \frac{x^2-9}{x^2-x-6}$$
$$= \frac{(x+3)\cancel{(x-3)}}{\cancel{(x-3)}(x+2)}$$
$$= \frac{x+3}{x+2}$$

5. $$2 + \frac{5}{x-\frac{1}{5}} = 2 + \frac{5}{x-\frac{1}{5}} \cdot \frac{\mathbf{5}}{\mathbf{5}}$$
$$= \frac{2}{1} + \frac{25}{5x-1}$$
$$= \frac{2}{1} \cdot \frac{\mathbf{5x-1}}{\mathbf{5x-1}} + \frac{25}{5x-1}$$
$$= \frac{10x-2+25}{5x-1}$$
$$= \frac{10x+23}{5x-1}$$

Section 3.6

1. $\frac{x}{3} + 1 = \frac{1}{2}$ LCD = 6

$$\mathbf{6}\left(\frac{x}{3} + 1\right) = \mathbf{6} \cdot \frac{1}{2}$$
$$6 \cdot \frac{x}{3} + 6 \cdot 1 = 6 \cdot \frac{1}{2}$$
$$2x + 6 = 3$$
$$2x = -3$$
$$x = -\frac{3}{2}$$

2. $\frac{2}{a+5} = \frac{1}{3}$ LCD = $3(a+5)$

$$\mathbf{3(a+5)} \cdot \frac{2}{a+5} = \mathbf{3(a+5)} \cdot \frac{1}{3}$$
$$6 = a + 5$$
$$1 = a$$

3. $\frac{x}{x+1} - \frac{1}{2} = \frac{-1}{x+1}$ LCD = $2(x+1)$

$$\mathbf{2(x+1)}\left[\frac{x}{x+1} - \frac{1}{2}\right] = \mathbf{2(x+1)} \cdot \frac{-1}{x+1}$$
$$2(x+1) \cdot \frac{x}{x+1} - 2(x+1) \cdot \frac{1}{2} = 2(x+1) \cdot \frac{-1}{x+1}$$
$$2x - (x+1) = 2(-1)$$
$$2x - x - 1 = -2$$
$$x - 1 = -2$$
$$x = -1$$

The only possible solution is $x = -1$, but when $x = -1$, the original equation has two undefined terms. There is no solution to the equation.

4.
$$\frac{x}{x^2 - 9} - \frac{1}{x+3} = \frac{1}{4x - 12}$$
$$\frac{x}{(x+3)(x-3)} - \frac{1}{x+3} = \frac{1}{4(x-3)} \quad \text{LCD} = 4(x+3)(x-3)$$
$$\mathbf{4(x+3)(x-3)} \cdot \frac{x}{(x+3)(x-3)} - \mathbf{4(x+3)(x-3)} \cdot \frac{1}{x+3} = \mathbf{4(x+3)(x-3)} \cdot \frac{1}{4(x-3)}$$
$$4x - 4(x-3) = x + 3$$
$$4x - 4x + 12 = x + 3$$
$$12 = x + 3$$
$$9 = x$$

5. $1 - \frac{2}{x} = \frac{8}{x^2}$ LCD = x^2

$$\mathbf{x^2}\left(1 - \frac{2}{x}\right) = \mathbf{x^2} \cdot \frac{8}{x^2}$$
$$x^2 \cdot 1 - x^2 \cdot \frac{2}{x} = x^2 \cdot \frac{8}{x^2}$$
$$x^2 - 2x = 8$$
$$x^2 - 2x - 8 = 0$$
$$(x-4)(x+2) = 0$$
$$x - 4 = 0 \quad \text{or} \quad x + 2 = 0$$
$$x = 4 \quad \text{or} \quad x = -2$$

6. $\frac{y+1}{3(y+4)} = \frac{8}{(y+4)(y-4)}$

The LCD is $3(y+4)(y-4)$. Multiplying each side by the LCD gives us

$$(y+1)(y-4) = 8 \cdot 3$$
$$y^2 - 3y - 4 = 24$$
$$y^2 - 3y - 28 = 0$$
$$(y-7)(y+4) = 0$$
$$y = 7 \quad \text{or} \quad y = -4$$

The only solution is 7 because the original equation is undefined when y is -4.

7. $x = \frac{y+2}{y-1}$

$$x(y-1) = y + 2$$
$$xy - x = y + 2$$
$$xy - y = x + 2$$
$$y(x-1) = x + 2$$
$$y = \frac{x+2}{x-1}$$

8. $\frac{1}{a} = \frac{1}{x} + \frac{1}{b}$

$$axb \cdot \frac{1}{a} = axb \cdot \frac{1}{x} + axb \cdot \frac{1}{b}$$
$$xb = ab + ax$$
$$xb - ax = ab$$
$$x(b - a) = ab$$
$$x = \frac{ab}{b-a}$$

Section 3.7

1. Let x = one of the numbers and $3x$ = the other number.

$$\frac{1}{x} + \frac{1}{3x} = \frac{4}{3} \qquad \text{LCD} = 3x$$

$$\mathbf{3x} \cdot \frac{1}{x} + \mathbf{3x} \cdot \frac{1}{3x} = \mathbf{3x} \cdot \frac{4}{3}$$

$$3 + 1 = 4x$$
$$4 = 4x$$
$$1 = x$$

The two numbers are 1 and 3.

2.

	d	*r*	*t*
Upstream	1	$15 - x$	$\frac{1}{15 - x}$
Downstream	2	$15 + x$	$\frac{2}{15 + x}$

$$\frac{1}{15 - x} = \frac{2}{15 + x} \qquad \text{LCD} = (15 - x)(15 + x)$$

$$1(15 + x) = 2(15 - x)$$
$$15 + x = 30 - 2x$$
$$3x = 15$$
$$x = 5 \text{ miles per hour}$$

The speed of the current is 5 miles per hour.

3.

	d	*r*	*t*
Upstream	8	$x - 2$	$\frac{8}{x - 2}$
Downstream	8	$x + 2$	$\frac{8}{x + 2}$

$$\text{LCD} = (x + 2)(x - 2) \qquad \frac{8}{x - 2} + \frac{8}{x + 2} = 3$$

$$\mathbf{(x + 2)(x - 2)} \cdot \frac{8}{x - 2} + \mathbf{(x + 2)(x - 2)} \cdot \frac{8}{x + 2} = \mathbf{(x + 2)(x - 2)} \cdot 3$$

$$(x + 2) \cdot 8 + (x - 2) \cdot 8 = (x + 2)(x - 2) \cdot 3$$
$$8x + 16 + 8x - 16 = 3x^2 - 12$$
$$16x = 3x^2 - 12$$
$$0 = 3x^2 - 16x - 12$$
$$0 = (3x + 2)(x - 6)$$
$$3x + 2 = 0 \quad \text{or} \quad x - 6 = 0$$
$$x = -\frac{2}{3} \quad \text{or} \quad x = 6$$

The speed of the boat in still water is 6 miles per hour. (The $-\frac{2}{3}$ cannot be a solution because it is negative.)

4. Let x = the length of time it takes to fill the sink with both the drain and faucet open.

$$\frac{1}{3} - \frac{1}{4} = \frac{1}{x} \qquad \text{LCD} = 12x$$
$$4x - 3x = 12$$
$$x = 12 \text{ minutes}$$

5. $5 \text{ gal} = 5 \cancel{\text{gal}} \times \frac{4 \cancel{\text{qt}}}{1 \cancel{\text{gal}}} \times \frac{2 \text{ pt}}{1 \cancel{\text{qt}}}$

$= 5 \times 4 \times 2 \text{ qt}$

$= 40 \text{ pt}$

6. $2{,}000 \text{ qt} = 2{,}000 \cancel{\text{qt}} \times \frac{1 \text{ gal}}{4 \cancel{\text{qt}}}$

$= \frac{2{,}000}{4} \text{ gal}$

$= 500 \text{ gal}$

The number of 10-gal containers in 500 gal is $\frac{500}{10} = 50$ containers.

7. $13.0 \text{ feet per second} = \frac{13.0 \cancel{\text{feet}}}{1 \cancel{\text{second}}} \cdot \frac{1 \text{ mile}}{5{,}280 \cancel{\text{feet}}} \cdot \frac{60 \cancel{\text{second}}}{1 \cancel{\text{minute}}} \cdot \frac{60 \cancel{\text{minutes}}}{1 \text{ hour}}$

$= \frac{13.0 \cdot 60 \cdot 60 \text{ miles}}{5{,}280 \text{ hours}}$

$= 8.9 \text{ miles per hour}$

8. $1{,}100 \text{ feet per minute} = \frac{1{,}100 \cancel{\text{feet}}}{1 \cancel{\text{minute}}} \cdot \frac{1 \text{ mile}}{5{,}280 \cancel{\text{feet}}} \cdot \frac{60 \cancel{\text{minutes}}}{1 \text{ hour}}$

$= \frac{1{,}100 \cdot 60 \text{ miles}}{5{,}280 \text{ hours}}$

$= 12.5 \text{ miles per hour}$

Chapter 4

Section 4.1

1. The positive square root of 36 is 6 because 6 is the positive number with the property $6^2 = 36$. The negative square root of 36 is -6 since -6 is the negative number whose square is 36. The square roots of 36 are 6 and -6.

2. $\sqrt[3]{-64} = -4$ because $(-4)^3 = (-4)(-4)(-4) = -64$.

3. $\sqrt{-25}$ is not a real number since there is no real number whose square is -25.

4. $-\sqrt{4} = -2$; this is the negative square root of 4.

5. $\sqrt[5]{-1} = -1$ because $(-1)^5 = (-1)(-1)(-1)(-1)(-1) = -1$.

6. $\sqrt[4]{-16}$ is not a real number since there is no real number we can raise to the fourth power and obtain -16.

10. $9^{1/2} = \sqrt{9} = 3$ **11.** $27^{1/3} = \sqrt[3]{27} = 3$ **12.** $-49^{1/2} = -\sqrt{49} = -7$ **13.** $(-49)^{1/2} = \sqrt{-49}$ which is not a real number.

14. $\left(\frac{16}{25}\right)^{1/2} = \sqrt{\frac{16}{25}} = \frac{4}{5}$

15. $\sqrt[3]{8x^3y^9} = (8x^3y^9)^{1/3}$
$= 8^{1/3}(x^3)^{1/3}(y^9)^{1/3}$
$= 2xy^3$

16. $\sqrt[4]{81a^4b^8} = (81a^4b^8)^{1/4}$
$= 81^{1/4}(a^4)^{1/4}(b^8)^{1/4}$
$= 3ab^2$

17. $9^{3/2} = (9^{1/2})^3 = 3^3 = 27$

18. $16^{3/4} = (16^{1/4})^3 = 2^3 = 8$

19. $8^{-2/3} = (8^{1/3})^{-2} = 2^{-2} = \frac{1}{4}$

20. $\left(\frac{16}{81}\right)^{-3/4} = \left(\frac{81}{16}\right)^{3/4} = \left[\left(\frac{81}{16}\right)^{1/4}\right]^3 = \left(\frac{3}{2}\right)^3 = \frac{27}{8}$

21. $x^{1/2} \cdot x^{1/4} = x^{1/2+1/4} = x^{3/4}$

22. $(y^{3/5})^{5/6} = y^{(3/5)(5/6)} = y^{1/2}$

23. $\frac{z^{3/4}}{z^{2/3}} = z^{3/4-2/3} = z^{1/12}$

24. $\frac{(x^{1/3}y^{-3})^6}{x^4y^{10}} = \frac{x^2y^{-18}}{x^4y^{10}}$
$= x^{-2}y^{-28}$
$= \frac{1}{x^2y^{28}}$

Section 4.2

1. $x^{1/3}(x^{2/3} - x^{1/3}) = x^{1/3} \cdot x^{2/3} - x^{1/3} \cdot x^{1/3}$
$= x^1 - x^{2/3}$
$= x - x^{2/3}$

2. $(x^{3/5} + 2)(x^{3/5} - 7) = x^{3/5}x^{3/5} - 7x^{3/5} + 2x^{3/5} - 14$
$= x^{6/5} - 5x^{3/5} - 14$

3. $(5a^{1/2} - 4b^{1/2})(3a^{1/2} - b^{1/2}) = 15a - 5a^{1/2}b^{1/2} - 12a^{1/2}b^{1/2} + 4b$
$= 15a - 17a^{1/2}b^{1/2} + 4b$

4. $(t^{1/3} + 3)^2 = (t^{1/3})^2 + 6t^{1/3} + 9$
$= t^{2/3} + 6t^{1/3} + 9$

5. $(x^{5/2} - 3^{1/2})(x^{5/2} + 3^{1/2}) = (x^{5/2})^2 - (3^{1/2})^2$
$= x^5 - 3$

6.
$$\begin{array}{rl} & a^{2/3} - a^{1/3}b^{1/3} + b^{2/3} \\ & a^{1/3} + b^{1/3} \\ \hline a & - a^{2/3}b^{1/3} + a^{1/3}b^{2/3} \\ & + a^{2/3}b^{1/3} - a^{1/3}b^{2/3} + b \\ \hline a & + b \end{array}$$

7. $\frac{36x^{3/4}y^{1/4} - 18x^{5/4}y^{1/4}}{6x^{1/4}y^{1/4}} = \frac{36x^{3/4}y^{1/4}}{6x^{1/4}y^{1/4}} - \frac{18x^{5/4}y^{1/4}}{6x^{1/4}y^{1/4}}$
$= 6x^{1/2} - 3x$

8. $8(x-3)^{3/2} - 6(x-3)^{1/2} = 2(x-3)^{1/2}[4(x-3) - 3]$
$= 2(x-3)^{1/2}(4x - 15)$

9. $x^{2/3} - 4x^{1/3} - 21$
$= (x^{1/3})^2 - 4x^{1/3} - 21$
$= (x^{1/3} - 7)(x^{1/3} + 3)$

10. $6x^{2/3} + 19x^{1/3} + 10$
$= 6(x^{1/3})^2 + 19x^{1/3} + 10$
$= (3x^{1/3} + 2)(2x^{1/3} + 5)$

11. $(x^2-3)^{1/2} - \frac{x^2}{(x^2-3)^{1/2}} = \frac{(x^2-3)^{1/2}}{1} \cdot \frac{\mathbf{(x^2-3)^{1/2}}}{\mathbf{(x^2-3)^{1/2}}} - \frac{x^2}{(x^2-3)^{1/2}}$
$= \frac{x^2 - 3 - x^2}{(x^2-3)^{1/2}} = \frac{-3}{(x^2-3)^{1/2}}$

12. $r = \left(\frac{800}{600}\right)^{1/3} - 1$
$= 1.101 - 1$
$= 0.101$ or 10.1%

Section 4.3

1. $\sqrt{18} = \sqrt{9 \cdot 2}$
$= \sqrt{9}\sqrt{2}$
$= 3\sqrt{2}$

2. $\sqrt{50x^2y^3} = \sqrt{25x^2y^2 \cdot 2y}$
$= \sqrt{25x^2y^2}\sqrt{2y}$
$= 5xy\sqrt{2y}$

3. $\sqrt[3]{54a^4b^3} = \sqrt[3]{27a^3b^3 \cdot 2a}$
$= \sqrt[3]{27a^3b^3}\sqrt[3]{2a}$
$= 3ab\sqrt[3]{2a}$

4. $\sqrt{75x^5y^8} = \sqrt{25x^4y^8 \cdot 3x}$
$= \sqrt{25x^4y^8}\sqrt{3x}$
$= 5x^2y^4\sqrt{3x}$

5. $\sqrt[4]{48a^8b^5c^4} = \sqrt[4]{16a^8b^4c^4 \cdot 3b}$
$= \sqrt[4]{16a^8b^4c^4}\sqrt[4]{3b}$
$= 2a^2bc\sqrt[4]{3b}$

6. $\sqrt{\frac{5}{9}} = \frac{\sqrt{5}}{\sqrt{9}}$
$= \frac{\sqrt{5}}{3}$

7. $\sqrt{\frac{2}{3}} = \frac{\sqrt{2}}{\sqrt{3}}$
$= \frac{\sqrt{2}}{\sqrt{3}} \cdot \frac{\mathbf{\sqrt{3}}}{\mathbf{\sqrt{3}}}$
$= \frac{\sqrt{6}}{3}$

8. $\frac{5}{\sqrt{2}} = \frac{5}{\sqrt{2}} \cdot \frac{\mathbf{\sqrt{2}}}{\mathbf{\sqrt{2}}}$
$= \frac{5\sqrt{2}}{2}$

9. $\frac{3\sqrt{5x}}{\sqrt{2y}} = \frac{3\sqrt{5x}}{\sqrt{2y}} \cdot \frac{\mathbf{\sqrt{2y}}}{\mathbf{\sqrt{2y}}}$
$= \frac{3\sqrt{10xy}}{2y}$

10. $\frac{5}{\sqrt[3]{9}} = \frac{5}{\sqrt[3]{3^2}}$
$= \frac{5}{\sqrt[3]{3^2}} \cdot \frac{\mathbf{\sqrt[3]{3}}}{\mathbf{\sqrt[3]{3}}}$
$= \frac{5\sqrt[3]{3}}{\sqrt[3]{3^3}}$
$= \frac{5\sqrt[3]{3}}{3}$

11. $\sqrt{\frac{48x^3y^4}{7z}} = \frac{\sqrt{48x^3y^4}}{\sqrt{7z}}$
$= \frac{\sqrt{16x^2y^4}\sqrt{3x}}{\sqrt{7z}}$
$= \frac{4xy^2\sqrt{3x}}{\sqrt{7z}}$
$= \frac{4xy^2\sqrt{3x}}{\sqrt{7z}} \cdot \frac{\mathbf{\sqrt{7z}}}{\mathbf{\sqrt{7z}}}$
$= \frac{4xy^2\sqrt{21xz}}{7z}$

12. $\sqrt{16x^2} = 4|x|$

13. $\sqrt{25x^3} = 5|x|\sqrt{x}$

14. $\sqrt{x^2 + 10x + 25}$
$= \sqrt{(x+5)^2}$
$= |x+5|$

15. $\sqrt{2x^3 + 7x^2}$
$= \sqrt{x^2(2x+7)}$
$= |x|\sqrt{2x+7}$

16. $\sqrt[3]{(-3)^3}$
$= \sqrt[3]{-27}$
$= -3$

17. $\sqrt[3]{(-1)^3}$
$= \sqrt[3]{-1}$
$= -1$

Section 4.4

1. $3\sqrt{5} - 2\sqrt{5} + 4\sqrt{5} = (3 - 2 + 4)\sqrt{5}$
$= 5\sqrt{5}$

2. $4\sqrt{50} + 3\sqrt{8} = 4\sqrt{25 \cdot 2} + 3\sqrt{4 \cdot 2}$
$= 4\sqrt{25}\,\sqrt{2} + 3\sqrt{4}\,\sqrt{2}$
$= 4 \cdot 5\sqrt{2} + 3 \cdot 2\sqrt{2}$
$= 20\sqrt{2} + 6\sqrt{2}$
$= 26\sqrt{2}$

3. $4\sqrt{18x^2y} - 3x\sqrt{50y} = 4\sqrt{9x^2 \cdot 2y} - 3x\sqrt{25 \cdot 2y}$
$= 4\sqrt{9x^2}\,\sqrt{2y} - 3x\sqrt{25}\,\sqrt{2y}$
$= 4 \cdot 3x\sqrt{2y} - 3x \cdot 5\sqrt{2y}$
$= 12x\sqrt{2y} - 15x\sqrt{2y}$
$= -3x\sqrt{2y}$

4. $2\sqrt[3]{27a^2b^4} + 3b\sqrt[3]{125a^2b} = 2\sqrt[3]{27b^3 \cdot a^2b} + 3b\sqrt[3]{125 \cdot a^2b}$
$= 2\sqrt[3]{27b^3}\,\sqrt[3]{a^2b} + 3b\sqrt[3]{125}\,\sqrt[3]{a^2b}$
$= 2 \cdot 3b\sqrt[3]{a^2b} + 3b \cdot 5\sqrt[3]{a^2b}$
$= 6b\sqrt[3]{a^2b} + 15b\sqrt[3]{a^2b}$
$= 21b\sqrt[3]{a^2b}$

5. $\dfrac{\sqrt{5}}{3} + \dfrac{1}{\sqrt{5}} = \dfrac{\sqrt{5}}{3} + \dfrac{1}{\sqrt{5}} \cdot \dfrac{\boldsymbol{\sqrt{5}}}{\boldsymbol{\sqrt{5}}}$
$= \dfrac{\sqrt{5}}{3} + \dfrac{\sqrt{5}}{5}$
$= \left(\dfrac{1}{3} + \dfrac{1}{5}\right)\sqrt{5}$
$= \left(\dfrac{5}{15} + \dfrac{3}{15}\right)\sqrt{5}$
$= \dfrac{8}{15}\sqrt{5}$
$= \dfrac{8\sqrt{5}}{15}$

6. First we construct a golden rectangle from a square of side 6.

The length of the diagonal OB is found from the Pythagorean theorem.

$$OB = \sqrt{3^2 + 6^2} = \sqrt{9 + 36} = \sqrt{45} = 3\sqrt{5}$$

The ratio of the length to the width for the rectangle is the golden ratio.

$$\text{Golden ratio} = \frac{CE}{EF} = \frac{3 + 3\sqrt{5}}{6}$$
$$= \frac{\cancel{3}(1 + \sqrt{5})}{\cancel{3} \cdot 2}$$
$$= \frac{1 + \sqrt{5}}{2}$$

Section 4.5

1. We can rearrange the order and grouping of the numbers by applying the commutative and associative properties of multiplication.
$(7\sqrt{3})(5\sqrt{11}) = (7 \cdot 5)\,(\sqrt{3} \cdot \sqrt{11})$
$= 35\sqrt{33}$

2. $\sqrt{2}(3\sqrt{5} - 4\sqrt{2}) = \sqrt{2} \cdot 3\sqrt{5} - \sqrt{2} \cdot 4\sqrt{2}$
$= 3\sqrt{10} - 4\sqrt{4}$
$= 3\sqrt{10} - 4 \cdot 2$
$= 3\sqrt{10} - 8$

3. $(\sqrt{2} + \sqrt{7})(\sqrt{2} - 3\sqrt{7}) = \sqrt{2}\sqrt{2} - \sqrt{2} \cdot 3\sqrt{7} + \sqrt{7}\sqrt{2} - \sqrt{7} \cdot 3\sqrt{7}$
$= 2 - 3\sqrt{14} + \sqrt{14} - 21$
$= -19 - 2\sqrt{14}$

4. $(\sqrt{x} + 5)^2 = (\sqrt{x} + 5)(\sqrt{x} + 5)$
$= \sqrt{x}\sqrt{x} + 5\sqrt{x} + 5\sqrt{x} + 25$
$= x + 10\sqrt{x} + 25$

5. $(5\sqrt{a} - 3\sqrt{b})^2 = (5\sqrt{a})^2 - 2 \cdot 5\sqrt{a} \cdot 3\sqrt{b} + (3\sqrt{b})^2$
$= 25a - 30\sqrt{ab} + 9b$

6. $(\sqrt{x + 3} - 1)^2 = (\sqrt{x + 3} - 1)(\sqrt{x + 3} - 1)$
$= x + 3 - 2\sqrt{x + 3} + 1$
$= x + 4 - 2\sqrt{x + 3}$

7. $(\sqrt{5} + \sqrt{3})(\sqrt{5} - \sqrt{3}) = (\sqrt{5})^2 - (\sqrt{3})^2$
$= 5 - 3$
$= 2$

8. $\dfrac{3}{\sqrt{7} - \sqrt{3}} = \dfrac{3}{\sqrt{7} - \sqrt{3}} \cdot \dfrac{\boldsymbol{\sqrt{7} + \sqrt{3}}}{\boldsymbol{\sqrt{7} + \sqrt{3}}}$
$= \dfrac{3\sqrt{7} + 3\sqrt{3}}{7 - 3}$
$= \dfrac{3\sqrt{7} + 3\sqrt{3}}{4}$

9. $\dfrac{\sqrt{10} - 3}{\sqrt{10} + 3} = \dfrac{\sqrt{10} - 3}{\sqrt{10} + 3} \cdot \dfrac{\boldsymbol{\sqrt{10} - 3}}{\boldsymbol{\sqrt{10} - 3}}$
$= \dfrac{10 - 6\sqrt{10} + 9}{10 - 9}$
$= \dfrac{19 - 6\sqrt{10}}{1}$
$= 19 - 6\sqrt{10}$

10. First we construct a golden rectangle from a square of side 4.

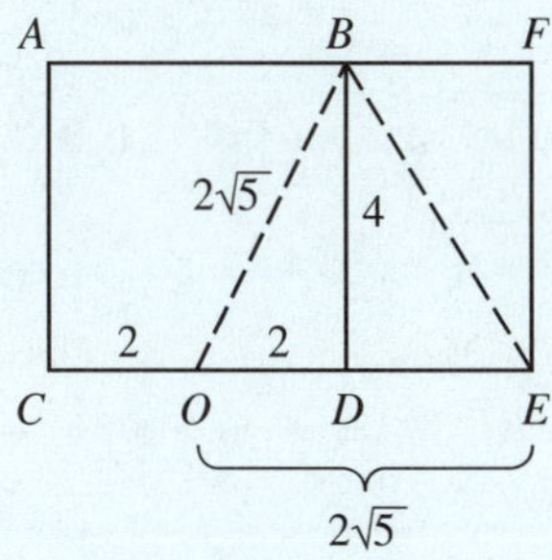

Next, we find expressions for the length and width of the smaller rectangle.

$$\text{Length} = EF = 4$$
$$\text{Width} = DE = 2\sqrt{5} - 2$$

Next, we find the ratio of length to width.

$$\text{Ratio of length to width} = \frac{EF}{DE} = \frac{4}{2\sqrt{5} - 2} = \frac{2 \cdot 2}{2(\sqrt{5} - 1)} = \frac{2}{\sqrt{5} - 1}$$

From this point on the rest of the problem looks like the solution to Example 10 in the text.

Section 4.6

1.
$$\begin{aligned} \sqrt{2x + 4} &= 4 \\ (\sqrt{2x + 4})^2 &= 4^2 \\ 2x + 4 &= 16 \\ 2x &= 12 \\ x &= 6 \end{aligned}$$

2. $\sqrt{7x - 3} = -5$ has no solution since the left side is a positive number or 0 and the right side is a negative number.

3.
$$\begin{aligned} \sqrt{4x + 5} + 2 &= 7 \\ \sqrt{4x + 5} &= 5 \\ (\sqrt{4x + 5})^2 &= 5^2 \\ 4x + 5 &= 25 \\ 4x &= 20 \\ x &= 5 \end{aligned}$$

4.
$$\begin{aligned} t - 6 &= \sqrt{t - 4} \\ (t - 6)^2 &= (\sqrt{t - 4})^2 \\ t^2 - 12t + 36 &= t - 4 \\ t^2 - 13t + 40 &= 0 \\ (t - 5)(t - 8) &= 0 \end{aligned}$$
$$t - 5 = 0 \quad \text{or} \quad t - 8 = 0$$
$$t = 5 \quad \text{or} \quad t = 8$$

Only 8 checks in the original equation.

5.
$$\begin{aligned} \sqrt{x - 9} &= \sqrt{x} - 3 \\ (\sqrt{x - 9})^2 &= (\sqrt{x} - 3)^2 \\ x - 9 &= x - 6\sqrt{x} + 9 \\ -9 &= -6\sqrt{x} + 9 \\ -18 &= -6\sqrt{x} \\ 3 &= \sqrt{x} \\ 3^2 &= (\sqrt{x})^2 \\ 9 &= x \end{aligned}$$

6.
$$\begin{aligned} \sqrt{x + 4} &= 2 - \sqrt{3x} \\ (\sqrt{x + 4})^2 &= (2 - \sqrt{3x})^2 \\ x + 4 &= 4 - 4\sqrt{3x} + 3x \\ -2x &= -4\sqrt{3x} \\ x &= 2\sqrt{3x} \\ x^2 &= (2\sqrt{3x})^2 \\ x^2 &= 4 \cdot 3x \\ x^2 - 12x &= 0 \\ x(x - 12) &= 0 \end{aligned}$$
$$x = 0 \quad \text{or} \quad x - 12 = 0$$
$$x = 12$$

Checking each solution in the original equation shows that $x = 12$ is extraneous. The only solution is $x = 0$.

7.
$$\begin{aligned} \sqrt{x + 2} &= \sqrt{x + 3} - 1 \\ (\sqrt{x + 2})^2 &= (\sqrt{x + 3} - 1)^2 \\ x + 2 &= x + 3 - 2\sqrt{x + 3} + 1 \\ 2 &= 3 - 2\sqrt{x + 3} + 1 \\ 2 &= 4 - 2\sqrt{x + 3} \\ -2 &= -2\sqrt{x + 3} \\ 1 &= \sqrt{x + 3} \\ 1^2 &= (\sqrt{x + 3})^2 \\ 1 &= x + 3 \\ -2 &= x \end{aligned}$$

8.
$$\begin{aligned} \sqrt[3]{3x - 7} &= 2 \\ (\sqrt[3]{3x - 7})^3 &= 2^3 \\ 3x - 7 &= 8 \\ 3x &= 15 \\ x &= 5 \end{aligned}$$

9. $y = \sqrt{x} + 3$

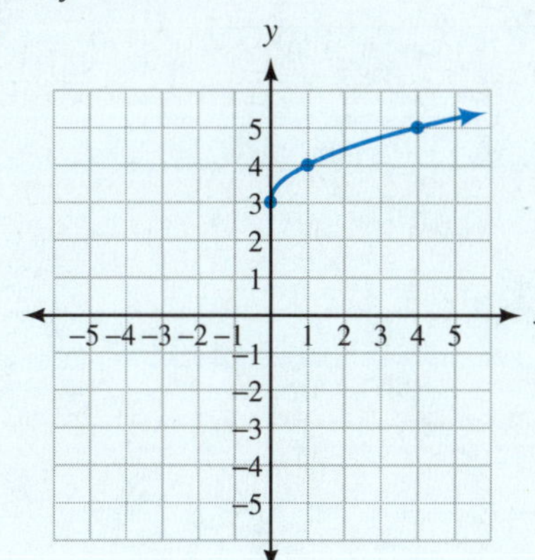

$y = \sqrt{x + 3}$

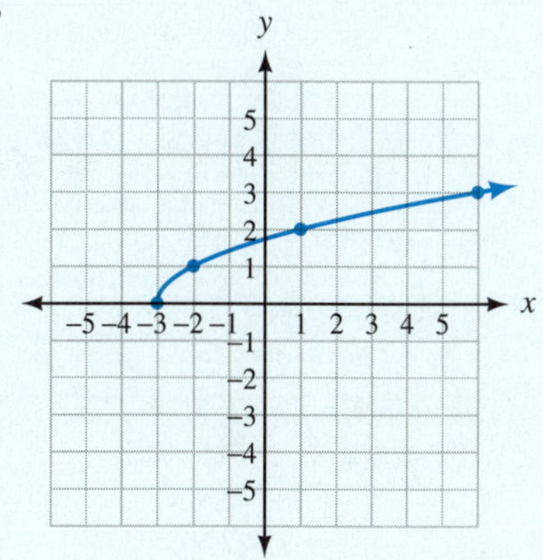

Chapter 5

Section 5.1

1. $\sqrt{-36} = i\sqrt{36} = i \cdot 6 = 6i$ **2.** $-\sqrt{-64} = -i\sqrt{64} = -i \cdot 8 = -8i$ **3.** $\sqrt{-18} = i\sqrt{18} = i \cdot 3\sqrt{2} = 3i\sqrt{2}$ **4.** $-\sqrt{-19} = -i\sqrt{19}$

5. $i^{20} = (i^2)^{10} = (-1)^{10} = 1$ **6.** $i^{23} = (i^2)^{11} \cdot i = (-1)^{11} \cdot i = -i$ **7.** $i^{50} = (i^2)^{25} = (-1)^{25} = -1$

8. $4x + 7i = 8 - 14yi$

$\Rightarrow 4x = 8$ and $7 = -14y$

$x = 2$ $\quad y = -\dfrac{1}{2}$

9. $(2x - 1) + 9i = 5 + (4y + 1)i$

$\Rightarrow 2x - 1 = 5$ and $9 = 4y + 1$

$2x = 6$ $\quad 8 = 4y$

$x = 3$ $\quad 2 = y$

10. $(2 + 6i) + (3 - 4i) = (2 + 3) + (6i - 4i)$

$= 5 + 2i$

11. $(6 + 5i) - (4 + 3i) = 6 + 5i - 4 - 3i$

$= 2 + 2i$

12. $(7 - i) - (8 - 2i) = 7 - i - 8 + 2i$

$= -1 + i$

13.
$$\begin{aligned}(2 + 3i)(1 - 4i) &= 2 \cdot 1 - 2 \cdot 4i + 3i \cdot 1 - 3i \cdot 4i\\ &= 2 - 8i + 3i - 12i^2\\ &= 2 - 5i - 12(-1)\\ &= 2 - 5i + 12\\ &= 14 - 5i\end{aligned}$$

14.
$$\begin{aligned}-3i(2 + 3i) &= -3i \cdot 2 - 3i \cdot 3i\\ &= -6i - 9i^2\\ &= -6i - 9(-1)\\ &= -6i + 9\\ &= 9 - 6i\end{aligned}$$

15.
$$\begin{aligned}(2 + 4i)^2 &= 2^2 + 2 \cdot 2 \cdot 4i + (4i)^2\\ &= 4 + 16i - 16\\ &= -12 + 16i\end{aligned}$$

16.
$$\begin{aligned}(3 - 5i)(3 + 5i) &= 3^2 - (5i)^2\\ &= 9 - 25i^2\\ &= 9 + 25\\ &= 34\end{aligned}$$

17.
$$\begin{aligned}\frac{3 + 2i}{2 - 5i} &= \frac{3 + 2i}{2 - 5i} \cdot \frac{\mathbf{2 + 5i}}{\mathbf{2 + 5i}}\\ &= \frac{6 + 15i + 4i + 10i^2}{4 - 25i^2}\\ &= \frac{6 + 19i - 10}{4 + 25}\\ &= \frac{-4 + 19i}{29}\\ &= -\frac{4}{29} + \frac{19}{29}i\end{aligned}$$

18.
$$\begin{aligned}\frac{3 + 2i}{i} &= \frac{3 + 2i}{i} \cdot \frac{\mathbf{-i}}{\mathbf{-i}}\\ &= \frac{-3i - 2i^2}{-i^2}\\ &= \frac{-3i + 2}{1}\\ &= 2 - 3i\end{aligned}$$

Section 5.2

1.
$$\begin{aligned}(3x + 2)^2 &= 16\\ 3x + 2 &= \pm 4\\ 3x &= -2 \pm 4\\ x &= \frac{-2 \pm 4}{3}\end{aligned}$$

$x = \dfrac{-2 + 4}{3}$ or $x = \dfrac{-2 - 4}{3}$

$x = \dfrac{2}{3}$ or $x = -2$

2.
$$\begin{aligned}(4x - 3)^2 &= -50\\ 4x - 3 &= \pm\sqrt{-50}\\ 4x - 3 &= \pm 5i\sqrt{2}\\ 4x &= 3 \pm 5i\sqrt{2}\\ x &= \frac{3 \pm 5i\sqrt{2}}{4}\end{aligned}$$

3.
$$\begin{aligned}x^2 + 10x + 25 &= 20\\ (x + 5)^2 &= 20\\ x + 5 &= \pm\sqrt{20} = \pm 2\sqrt{5}\\ x &= -5 \pm 2\sqrt{5}\end{aligned}$$

4.
$$\begin{aligned}x^2 + 3x - 4 &= 0\\ x^2 + 3x &= 4\\ x^2 + 3x + \mathbf{\frac{9}{4}} &= 4 + \mathbf{\frac{9}{4}}\\ \left(x + \frac{3}{2}\right)^2 &= \frac{25}{4}\\ x + \frac{3}{2} &= \pm\frac{5}{2}\\ x &= -\frac{3}{2} \pm \frac{5}{2}\end{aligned}$$

$x = -\dfrac{3}{2} + \dfrac{5}{2}$ or $x = -\dfrac{3}{2} - \dfrac{5}{2}$

$x = 1$ or $x = -4$

5.
$$\begin{aligned}5x^2 - 3x + 2 &= 0\\ 5x^2 - 3x &= -2\\ x^2 - \frac{3}{5}x &= -\frac{2}{5}\\ x^2 - \frac{3}{5}x + \mathbf{\frac{9}{100}} &= -\frac{2}{5} + \mathbf{\frac{9}{100}}\\ \left(x - \frac{3}{10}\right)^2 &= -\frac{31}{100}\\ x - \frac{3}{10} &= \pm\sqrt{-\frac{31}{100}}\\ x - \frac{3}{10} &= \pm\frac{i\sqrt{31}}{10}\\ x &= \frac{3}{10} \pm \frac{i\sqrt{31}}{10}\\ x &= \frac{3 \pm i\sqrt{31}}{10}\end{aligned}$$

6. Using Figure 3 in the text as a guide, we have $AC = 2$, so $AB = 4$. If $BC = x$, then by the Pythagorean theorem we have

$$\begin{aligned}x^2 + 2^2 &= 4^2\\ x^2 + 4 &= 16\\ x^2 &= 12\\ x &= 2\sqrt{3}\end{aligned}$$

The sides are 2, $2\sqrt{3}$, and 4 inches.

7.
$$\begin{aligned}4{,}330^2 &= x^2 + 960^2\\ 18{,}748{,}900 &= x^2 + 921{,}600\\ x^2 &= 18{,}748{,}900 - 921{,}600\\ x^2 &= 17{,}827{,}300\\ x &= \sqrt{17{,}827{,}300}\end{aligned}$$

$x = 4{,}222$ feet to the nearest foot

Section 5.3

1. $6x^2 + 7x + 2 = 0$

$a = 6, b = 7, c = 2$

$$x = \frac{-7 \pm \sqrt{7^2 - 4(6)(2)}}{2(6)}$$
$$= \frac{-7 \pm \sqrt{49 - 48}}{12}$$
$$= \frac{-7 \pm \sqrt{1}}{12}$$
$$= \frac{-7 \pm 1}{12}$$
$$x = \frac{-7 + 1}{12} \quad \text{or} \quad x = \frac{-7 - 1}{12}$$
$$x = -\frac{1}{2} \quad \text{or} \quad x = -\frac{2}{3}$$

2. $\frac{x^2}{2} + x = \frac{1}{3}$ LCD = 6

$$\mathbf{6} \cdot \frac{x^2}{2} + \mathbf{6} \cdot x = \mathbf{6} \cdot \frac{1}{3}$$
$$3x^2 + 6x = 2$$
$$3x^2 + 6x - 2 = 0$$
$$a = 3, b = 6, c = -2$$
$$x = \frac{-6 \pm \sqrt{6^2 - 4(3)(-2)}}{2(3)}$$
$$= \frac{-6 \pm \sqrt{36 + 24}}{6}$$
$$= \frac{-6 \pm \sqrt{60}}{6}$$
$$= \frac{-6 \pm 2\sqrt{15}}{6}$$
$$= \frac{\cancel{2}(-3 \pm \sqrt{15})}{\cancel{2} \cdot 3}$$
$$= \frac{-3 \pm \sqrt{15}}{3}$$

3. $\frac{1}{x+4} - \frac{1}{x} = \frac{1}{2}$ LCD = $2x(x + 4)$

$$\mathbf{2x(x+4)}\left(\frac{1}{x+4} - \frac{1}{x}\right) = \mathbf{2x(x+4)} \cdot \frac{1}{2}$$
$$\mathbf{2x(x+4)} \cdot \frac{1}{x+4} - \mathbf{2x(x+4)} \cdot \frac{1}{x} = \mathbf{2x(x+4)} \cdot \frac{1}{2}$$
$$2x - 2(x + 4) = x(x + 4)$$
$$2x - 2x - 8 = x^2 + 4x$$
$$-8 = x^2 + 4x$$
$$0 = x^2 + 4x + 8$$

$a = 1, b = 4, c = 8$

$$x = \frac{-4 \pm \sqrt{4^2 - 4(1)(8)}}{2(1)}$$
$$= \frac{-4 \pm \sqrt{16 - 32}}{2}$$
$$= \frac{-4 \pm \sqrt{-16}}{2}$$
$$= \frac{-4 \pm 4i}{2}$$
$$= \frac{2(-2 \pm 2i)}{2}$$
$$= -2 \pm 2i$$

4. $8t^3 - 27 = 0$

$$(2t - 3)(4t^2 + 6t + 9) = 0$$

$2t - 3 = 0$ or $4t^2 + 6t + 9 = 0$

$$2t = 3$$
$$t = \frac{3}{2}$$

$$t = \frac{-6 \pm \sqrt{36 - 4(4)(9)}}{2(4)}$$
$$= \frac{-6 \pm \sqrt{-108}}{8}$$
$$= \frac{-6 \pm 6i\sqrt{3}}{8}$$
$$= \frac{-3 \pm 3i\sqrt{3}}{4}$$

5. $12 = 32t - 16t^2$

$16t^2 - 32t + 12 = 0$

$4t^2 - 8t + 3 = 0$ Divide by 4.

$$t = \frac{8 \pm \sqrt{64 - 4(4)(3)}}{2(4)}$$
$$= \frac{8 \pm \sqrt{16}}{8}$$
$$= \frac{8 \pm 4}{8}$$
$$t = \frac{8 + 4}{8} = \frac{12}{8} = \frac{3}{2} \quad \text{or} \quad t = \frac{8 - 4}{8} = \frac{4}{8} = \frac{1}{2}$$

Note: Since the solutions are rational numbers, we could have solved the equation by factoring.

6. $P = \$1,320$

$P = -500 + 27x - 0.1x^2$

$1,320 = -500 + 27x - 0.1x^2$

$0.1x^2 - 27x + 1,820 = 0$

$$x = \frac{27 \pm \sqrt{(-27)^2 - 4(0.1)(1,820)}}{(2)(0.1)}$$
$$= \frac{27 \pm \sqrt{729 - 728}}{0.2}$$
$$= \frac{27 \pm \sqrt{1}}{0.2}$$
$$= \frac{27 + 1}{0.2} \quad \text{or} \quad = \frac{27 - 1}{0.2}$$
$$= \frac{28}{0.2} \quad \text{or} \quad = \frac{26}{0.2}$$
$$x = 140 \quad \text{or} \quad x = 130$$

Answers to Odd-Numbered Problems

Chapter 1

Problem Set 1.1

1. $(x-6)(x+4)$ **3.** $(x-6)(x+1)$ **5.** $(x-3)(x-2)$ **7.** $(x-5)^2$ **9.** $(2x+1)(x-3)$ **11.** $(7x-3)(3x-2)$ **13.** $(x+4)(x-4)$ **15.** $(a+1)(a-1)$ **17.** $(a+4b)(a-4b)$ **19.** $(3x+7)(3x-7)$ **21.** $(4x^2+7)(4x^2-7)$ **23.** $(t+3)(t-3)(t^2+9)$ **25.** $(a+b)(a^2-ab+b^2)$ **27.** $(x-2)(x^2+2x+4)$ **29.** $(x+1)(x^2-x+1)$ **31.** $(2x+1)(4x^2-2x+1)$ **33.** $10(3x-2)(2x-3)$ **35.** $x(x+3)(x+2)$ **37.** $x(2x+1)(x-3)$ **39.** $x(x-6)(x+4)$ **41.** $6(x+4)$ **43.** $100x(x-3)$ **45.** $5(2a+3)(2a-3)$ **47.** $a(3a+4)(3a-4)$ **49.** $2y(3+x)(2-x)$ **51.** $(a+2)(x+3)$ **53.** $(x-3a)(x-2)$ **55.** $(x+2)(x+3)(x-3)$ **57.** $(x+3)(x+2)(x-2)$ **59.** $(x+3)(2x+3)(2x-3)$ **61.** $(2x+1)(x+3)(x-3)$ **63.** $(4x+1)(x-8)$ **65.** prime **67.** $5x(6x-7)(5x+8)$ **69.** $(12x-5)(2x+1)$ **71.** $(x-1)(x^2+x+1)(x+1)(x^2-x+1)$ **73.** $(a+5)(x+3)^2$ **75.** $3(x-7)(2a+5)(2a-5)$ **77.** $\left(r+\frac{1}{3}\right)\left(r-\frac{1}{3}\right)$ **79.** $(t+2+y)(t+2-y)$ **81.** $\left(5t+\frac{1}{3}\right)\left(25t^2-\frac{5}{3}t+\frac{1}{9}\right)$ **83.** $(15t+16)(t-1)$ **85.** $100(x-3)(x+2)$ **87.** $4x(x^2+4y^2)$ **89.** $(5-t+x)(5-t-x)$ **91.** $(5x+7)(6x+11)$

Problem Set 1.2

1. 5 **3.** $-\frac{9}{2}$ **5.** $-\frac{4}{3}$ **7.** -10 **9.** -2 **11.** $\frac{3}{4}$ **13.** 3 **15.** -3 **17.** $-8, 8$ **19.** 2 **21.** 6 **23.** $-1, 6$ **25.** 0, 2, 3 **27.** $\frac{2}{3}, \frac{3}{2}$ **29.** $-4, \frac{5}{2}$ **31.** $-10, 0$ **33.** $-5, 1$ **35.** $-2, 3$ **37.** $-3, -2, 2$ **39.** $\frac{4}{5}$ **41.** $-5, 5$ **43.** 9 **45.** $-\frac{4}{3}$ **47.** 1 **49.** $-2, \frac{1}{4}$ **51.** $-\frac{4}{3}, 0, \frac{4}{3}$ **53.** $-3, -\frac{3}{2}, \frac{3}{2}$ **55.** $0, \frac{1}{6}$ **57.** $-\frac{3}{5}, \frac{1}{2}$ **59.** $0, \frac{3}{2}$ **61.** 7000 **63.** $-\frac{3}{2}, \frac{3}{2}$ **65.** $-3, -\frac{1}{2}, 3$ **67.** no solution **69.** all real numbers **71.** no solution **73.** no solution **75.** all real numbers **77.** **(a)** $\$6.60 = \$0.4n + \$1.80$ **(b)** 12 miles **79.** **(a)** $3{,}522{,}037 = 1{,}025A$ **(b)** 3,436 square miles **81.** -6 **83.** 66 **85.** -13 **87.** -4 **89.** 2 **91.** $-\frac{3}{2}$ **93.** $-\frac{21}{2}$ **95.** no solution **97.** $-3, -2, 2, 3$

Problem Set 1.3

1. $-4, 4$ **3.** $-2, 2$ **5.** no solution, $\varnothing$ **7.** $-1, 1$ **9.** no solution, $\varnothing$ **11.** $\frac{7}{3}, \frac{17}{3}$ **13.** $-\frac{5}{2}, \frac{5}{6}$ **15.** $-1, 5$ **17.** no solution, $\varnothing$ **19.** $-4, 20$ **21.** $-4, 8$ **23.** 1, 4 **25.** $-\frac{1}{7}, \frac{9}{7}$ **27.** $-3, 12$ **29.** $-\frac{10}{3}, \frac{2}{3}$ **31.** no solution, $\varnothing$ **33.** $-1, \frac{3}{2}$ **35.** 5, 25 **37.** $-30, 26$ **39.** $-12, 28$ **41.** $-2, 0$ **43.** $-\frac{1}{2}, \frac{7}{6}$ **45.** 0, 15 **47.** $-\frac{23}{7}, -\frac{11}{7}$ **49.** $-5, \frac{3}{5}$ **51.** $\frac{1}{9}, 1$ **53.** $-\frac{1}{2}$ **55.** 0 **57.** $-\frac{7}{4}, -\frac{1}{6}$ **59.** all real numbers **61.** all real numbers **63.** $-\frac{3}{10}, \frac{3}{2}$ **65.** $-0.6, -0.1$ **67.** 1987 and 1995 **69.** 1986 and 1996 **71.** 1993 and 1996

73.

n	a_n
1	2
2	1
3	0
4	1
5	2

75.

n	a_n
1	4
2	2
3	0
4	2
5	4

77. $10x^5 + 8x^3 - 6x^2$ **79.** $12a^2 + 11a - 5$ **81.** $x^4 - 81$ **83.** $16y^2 - 40y + 25$ **85.** $12xy + 28y - 6x - 14$ **87.** $9 - 6t^2 + t^4$ **89.** $x = a - b$ or $x = a + b$ **91.** $x = \frac{-b-c}{a}$ or $x = \frac{-b+c}{a}$ **93.** $x = -\frac{a}{b}y - a$ or $x = -\frac{a}{b}y + a$

Problem Set 1.4

1. -3 **3.** 0 **5.** $\frac{3}{2}$ **7.** 4 **9.** $\frac{25}{3}$ **11.** $\frac{14}{3}$ **13.** 7 ft **15.** 8 in. **17.** 8 m **19.** $l = \frac{A}{w}$ **21.** $t = \frac{I}{pr}$ **23.** $T = \frac{PV}{nR}$ **25.** $x = \frac{y-b}{m}$ **27.** $F = \frac{9}{5}C + 32$ **29.** $v = \frac{h-16t^2}{t}$ **31.** $d = \frac{A-a}{n-1}$ **33.** $y = -\frac{2}{3}x + 2$ **35.** $y = \frac{3}{5}x + 3$ **37.** $y = \frac{1}{3}x + 2$ **39.** $x = \frac{5}{a-b}$ **41.** $x = \frac{8}{b-c}$ **43.** $x = \frac{10}{c-a}$ **45.** $x = \frac{d-b}{a-c}$ **47.** $y = -\frac{1}{4}x + 2$ **49.** $y = \frac{3}{5}x - 3$ **51.** 20.52 **53.** 25% **55.** 925 **57.** 4, 7, 10, 13, 16 **59.** 4, 7, 12, 19, 28 **61.** $\frac{1}{4}, \frac{2}{5}, \frac{1}{2}, \frac{4}{7}, \frac{5}{8}$ **63.** $1, \frac{1}{4}, \frac{1}{9}, \frac{1}{16}, \frac{1}{25}$ **65.** 2, 4, 8, 16, 32 **67.** $2, \frac{3}{2}, \frac{4}{3}, \frac{5}{4}, \frac{6}{5}$ **69.** 1991

71.

PITCHER, TEAM	W	L	SAVES	BLOWN SAVES	ROLAIDS POINTS
John Franco, New York	0	8	38	8	82
Billy Wagner, Houston	4	3	30	5	82
Gregg Olson, Arizona	3	4	30	4	80
Bob Wickman, Milwaukee	6	9	25	7	55

73. 1.8 tons **75.** 1 sec and 2 sec

77. 0 sec and $\frac{3}{2}$ sec **79.** 2 sec and 3 sec **81.** \$4 and \$8 **83.** \$7 and \$10 **85.** 13,330 kilobytes **87.** \$673.68 **89.** \$674.92 **91.** \$760.05 **93.** \$18,878,664 **95.** $2(x + 3)$ **97.** $2(x + 3) = 16$ **99.** $5(x - 3)$ **101.** $3x + 2 = x - 4$ **103.** $x = -\frac{a}{b}y + a$ **105.** $a = \frac{bc}{b - c}$ **107.** Shar: 128.4 beats per minute Sara: 140.4 beats per minute

Problem Set 1.5

1. 10 feet by 20 feet **3.** 7 feet **5.** 5 inches **7.** 4 meters **9.** \$92.00 **11.** \$9,339.00 **13.** 860 items **15.** 41,667 workers **17.** \$3,260.66 **19.** \$99.6 million **21.** **(a)** 24.4% **(b)** \$14,300 **(c)** \$7,400 **23.** 8.3 grams **25.** 20°, 160° **27.** **(a)** 20.4°, 69.6° **(b)** 38.4°, 141.6° **29.** 27°, 72°, 81° **31.** 34°, 44°, 102° **33.** 43°, 43°, 94° **35.** 24 feet **37.** 6, 8, 10 **39.** 2 feet by 8 feet **41.** 5.5 feet by 12 feet **43.** Base 8 feet; height 25 feet **45.** base = 18 inches, height = 4 inches **47.** \$6,000 at 8%, \$3,000 at 9% **49.** \$5,000 at 12%, \$10,000 at 10% **51.** \$4,000 at 8%, \$2,000 at 9% **53.** 30 fathers, 45 sons **55.** \$54 **57.** 44 minutes

59.

w	l	A
2	22	44
4	20	80
6	18	108
8	16	128
10	14	140
12	12	144

61.

TIME (seconds)	HEIGHT (feet)
1	112
2	192
3	240
4	256
5	240
6	192

63.

AGE (years)	MAXIMUM HEART RATE (beats per minute)
18	202
19	201
20	200
21	199
22	198
23	197

65.

RESTING HEART RATE (beats per minute)	TRAINING HEART RATE (beats per minute)
60	144
62	145
64	146
68	147
70	148
72	149

67. 9 square inches

WIDTH	LENGTH	AREA
0.0	6.0	0
0.5	5.5	2.75
1.0	5.0	5
1.5	4.5	6.75
2.0	4.0	8
2.5	3.5	8.75
3.0	3.0	9
3.5	2.5	8.75
4.0	2.0	8
4.5	1.5	6.75
5.0	1.0	5
5.5	0.5	2.75
6.0	0.0	0

69.

TIME IN MONTHS	AMOUNT IN ACCOUNT
0	\$100.00
1	\$100.50
2	\$101.00
3	\$101.51
4	\$102.02
5	\$102.53
6	\$103.04
7	\$103.55
8	\$104.07
9	\$104.59
10	\$105.11
11	\$105.64
12	\$106.17
13	\$106.70
14	\$107.23
15	\$107.77
16	\$108.31
17	\$108.85
18	\$109.39
19	\$109.94
20	\$110.49
21	\$111.04
22	\$111.60
23	\$112.16
24	\$112.72

71. −4, 0, 2,3 **73.** $-4, -\frac{2}{5}, 0, 2, 3$ **75.** {3, 4, 5, 6, 7, 9} **77.** {7, 9}

Problem Set 1.6

1. $x \le \frac{3}{2}$

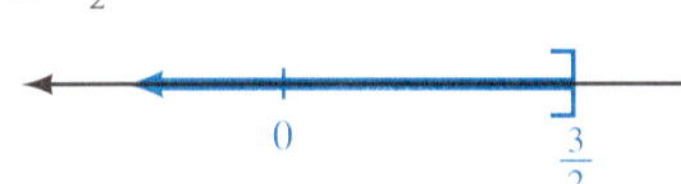

3. $x > 4$

5. $x \ge -5$

7. $x < 4$

9. $x \ge -6$

11. $x \ge 4$

13. $x < -3$

15. $m \ge -1$

17. $x \ge -3$

19. $y \le \frac{7}{2}$

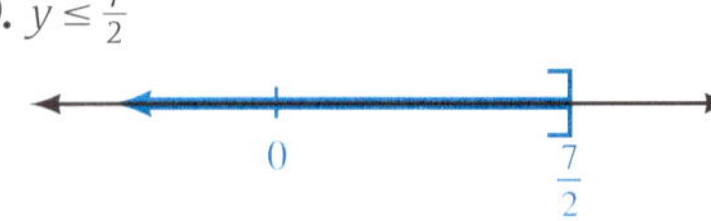

21. $x < 6$

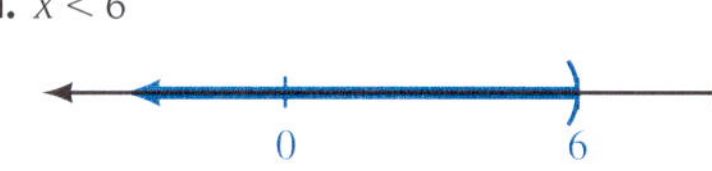

23. $y \ge -52$

25. $x > \frac{54}{7}$

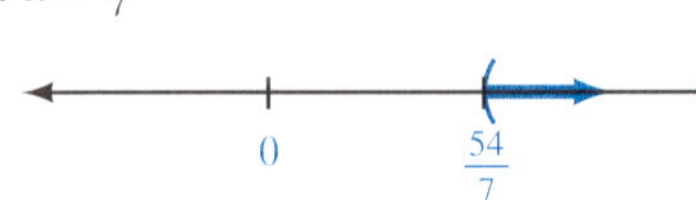

27. $x \ge -32$

29. $x < 40$

31. $(-\infty, -2]$ **33.** $[1, \infty)$ **35.** $(-\infty, 3)$ **37.** $(-\infty, -1]$ **39.** $[-17, \infty)$ **41.** $(-\infty, 10]$  **43.** $(1, \infty)$ **45.** $(-\infty, -5)$

47. $[3, 7]$

49. $(-4, 2)$

51. $[4, 6]$

53. $(-4, 2)$

55. $(-3, 3)$

57. $(-\infty, -7] \cup [-3, \infty)$

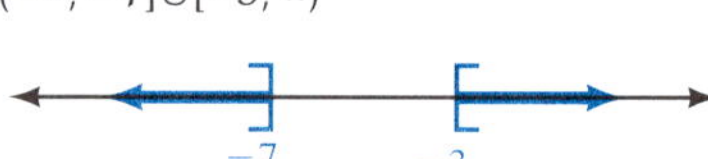

59. $(-\infty, -1] \cup [\frac{3}{5}, \infty)$

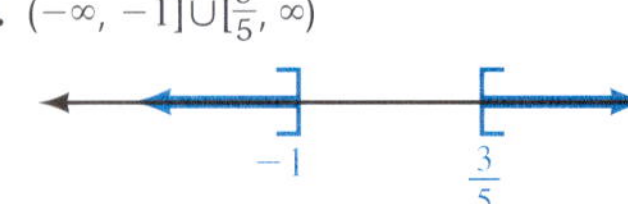

61. $(-\infty, -10) \cup (6, \infty)$

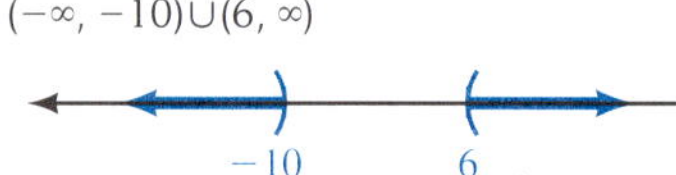

63. $(-\infty, -3) \cup [2, \infty)$

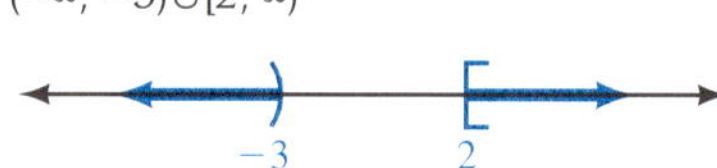

65. $(\infty, -4] \cup (-\frac{5}{2}, \infty)$

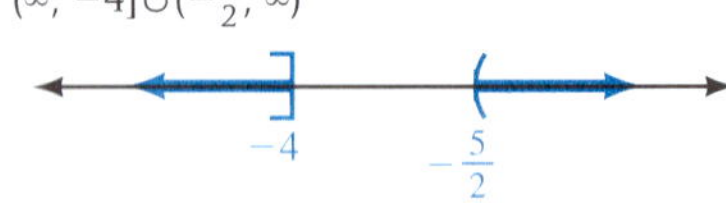

67. $-2 < x \le 4$ **69.** $x < -4$ or $x \ge 1$ **71.** \$2.00 per pad or less **73.** more than \$1.25 per pad **75.** $35° \le C \le 45°$ **77.** $-25° \le C \le -10°$ **79.** **(a)** from 1980 to 1983 **(b)** in 1991 and after **81.** **(a)** year 1999 **(b)** 2009 and later **83.** **(a)** $46\% \le R \le 52\%$, $31\% \le P \le 37\%$ **(b)** 9% **(c)** No, the pollsters are not 100% certain of the outcome. **85.** **(a)** $15 \le D < 20$ and $20{,}000 < D \le 50{,}000$ **(b)** $85 \le H < 452$ and $1{,}080 < H \le 1{,}100$ **(c)** A dog can hear low-frequency and high-frequency sounds that a human cannot. **87.** less than or equal to \$160 **89.** $3xy(2x^3 - 3y^3 + 6x^2y^2)$ **91.** $(x - 2)(x - 3)$ **93.** $(2x - 5)^2$ **95.** $(x^2 + 4)(x + 2)(x - 2)$ **97.** $(x - 6 - y)(x - 6 + y)$ **99.** does not factor (prime polynomial) **101.** $x < \frac{c - b}{a}$ **103.** $\frac{-c - b}{a} < x < \frac{c - b}{a}$ **105.** $x > \frac{d - b}{a - c}$ **107.** $\frac{-b - dc}{a} < x < \frac{-b + dc}{a}$ **109.** $x < \frac{ac(b - d)}{bd(c - a)}$

Chapter 1 Review

1. $(x^2 + 4)(x + 2)(x - 2)$ **2.** $3(a^2 + 3)^2$ **3.** $(a - 2)(a^2 + 2a + 4)$ **4.** $5x(x + 3y)^2$ **5.** $3ab(a - 3b)(a + 3b)$ **6.** $(x - 5 + y)(x - 5 - y)$ **7.** $(6 - 5a)(6 + 5a)$ **8.** $(x + 3)(x - 3)(x + 4)$ **9.** -3 **10.** $\frac{10}{13}$ **11.** $-\frac{5}{11}$ **12.** $-\frac{4}{9}$ **13.** $-\frac{3}{2}, 4$ **14.** $-\frac{1}{9}, \frac{1}{9}$ **15.** 1, 4 **16.** $-2, -\frac{2}{3}, \frac{2}{3}$ **17.** 2, 4 **18.** $-1, 5$ **19.** $-1, 4$ **20.** $-\frac{5}{3}, 3$ **21.** $-\frac{3}{2}, 3$ **22.** $-\frac{5}{3}, \frac{7}{3}$ **23.** 5, 9 **24.** 0 **25.** $h = 17$ **26.** $t = 20$ **27.** $p = \frac{I}{rt}$ **28.** $x = \frac{y - b}{m}$ **29.** $y = \frac{4}{3}x - 4$ **30.** $v = \frac{d - 16t^2}{t}$ **31.** 3, 6, 9, 12; increasing **32.** 4, 5, 6, 7; increasing **33.** $-2, 4, -8, 16$ **34.** $1, \frac{1}{2}, \frac{1}{3}, \frac{1}{4}$; decreasing **35.** 4 feet by 12 feet **36.** 3 meters, 4 meters, 5 meters

37. **(a)** 3,780 bricks **(b)** 625 feet **38.** \$24,875.24 **39.** $\left(-\infty, \frac{1}{2}\right)$ **40.** $(-\infty, 8]$ **41.** $(-\infty, 12]$ **42.** $(-1, \infty)$ **43.** $[2, 6]$
44. $[0, 1]$ **45.** $(-\infty, -\frac{3}{2}]\cup[3, \infty)$ **46.** $(-\infty, 1)\cup[4, \infty)$

Chapter 2

Problem Set 2.1

1.

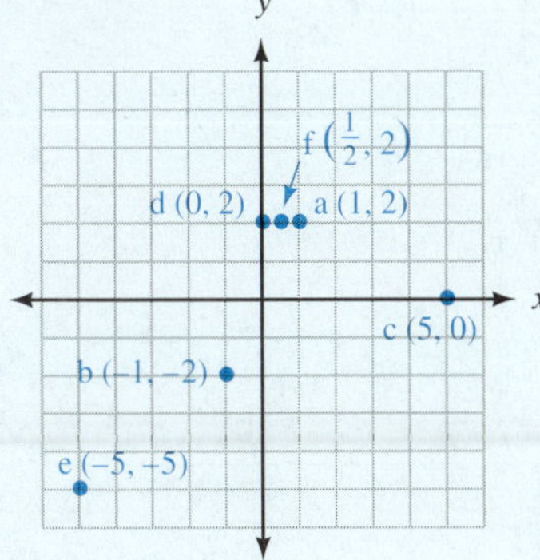

3. A (4, 1) B (−4, 3) C (−2, −5) D (2, −2) E (0, 5) F (−4, 0) G (1, 0)

5. x-intercept = 3, y-intercept = −2

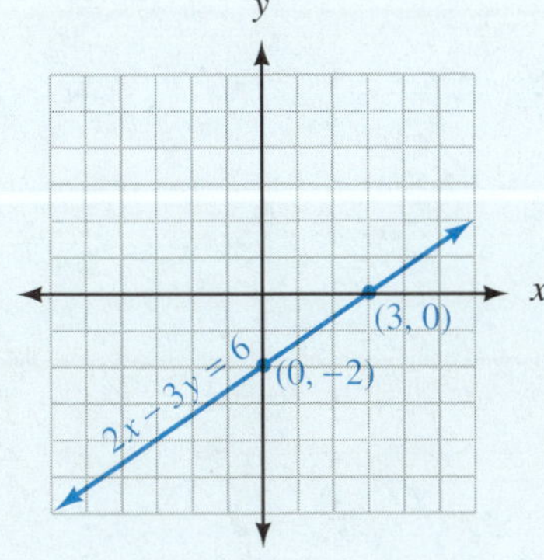

7. x-intercept = 5, y-intercept = −4

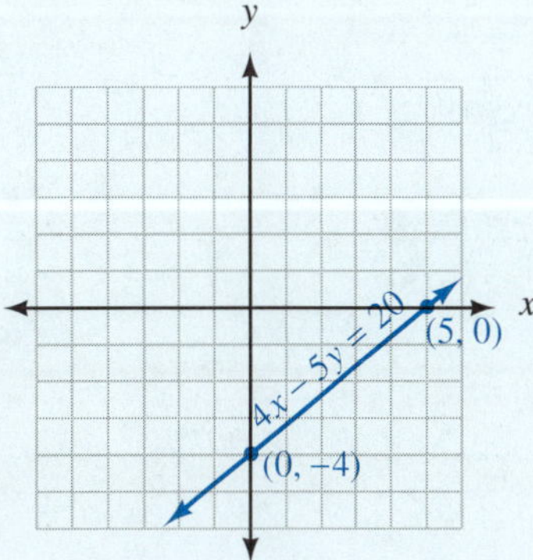

9. x-intercept = $-\frac{3}{2}$, y-intercept = 3

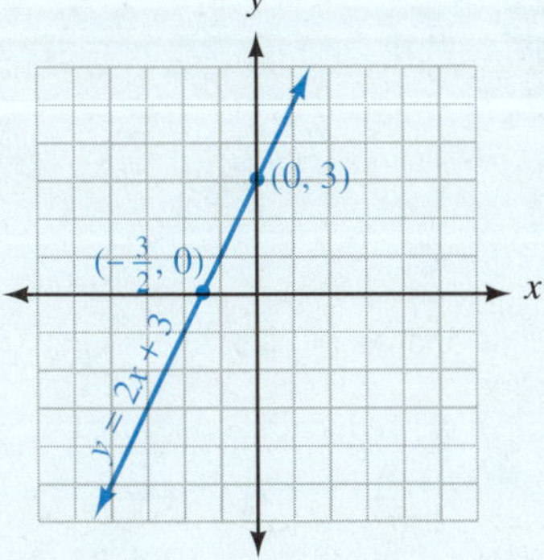

11. x-intercept = −4, y-intercept = 6

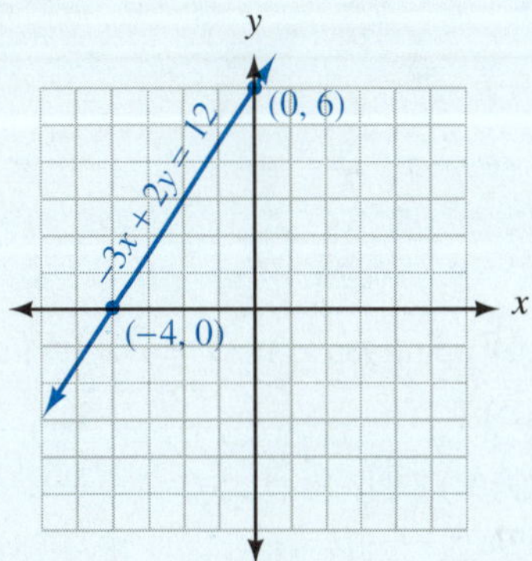

13. x-intercept $= \frac{10}{3}$, y-intercept $= -4$

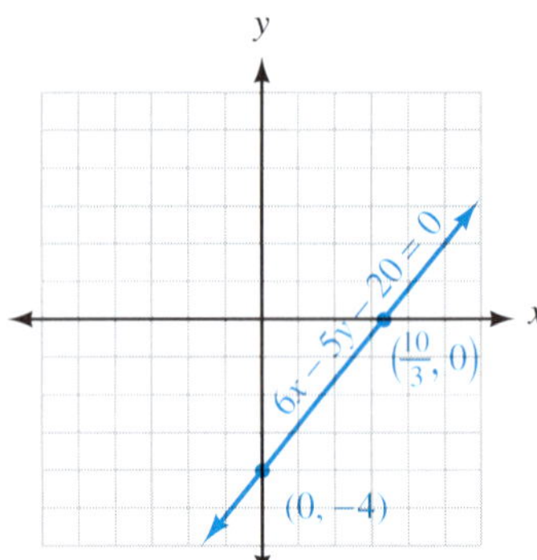

15. x-intercept $= \frac{5}{3}$, y-intercept $= -5$

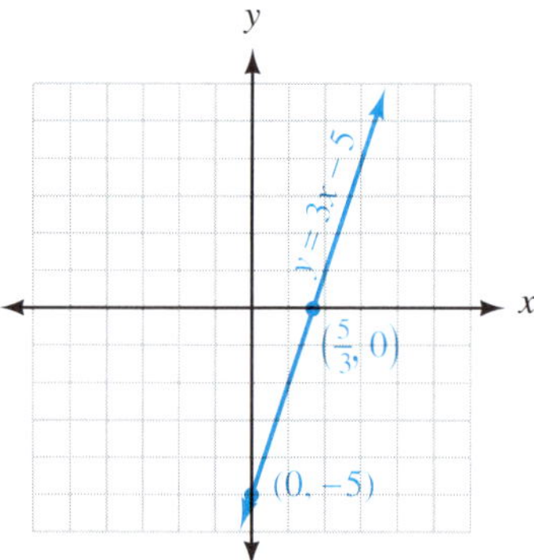

17. x-intercept $= 2$, y-intercept $= 3$

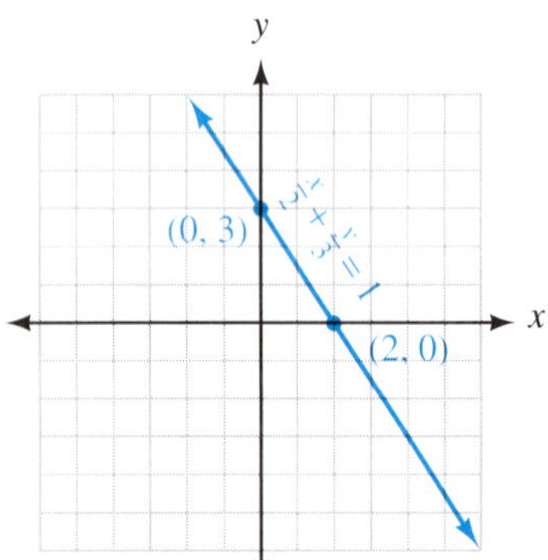

19. b

21.

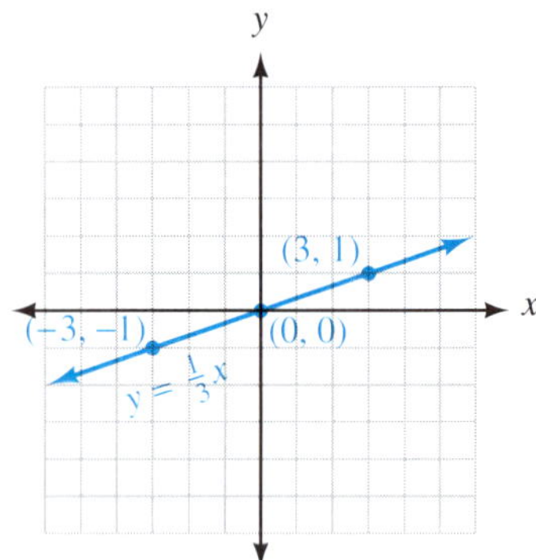

23.

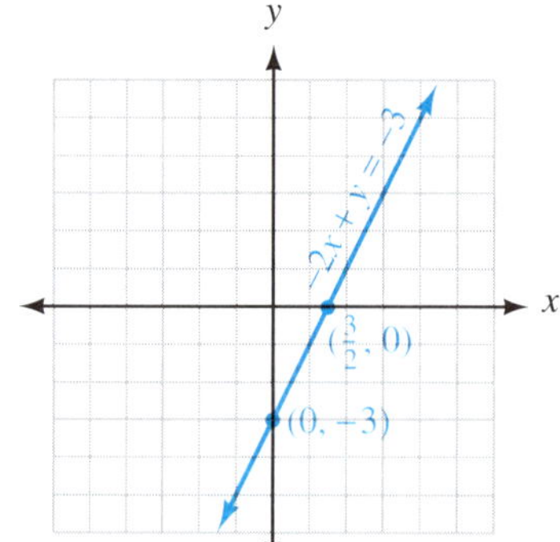

25.

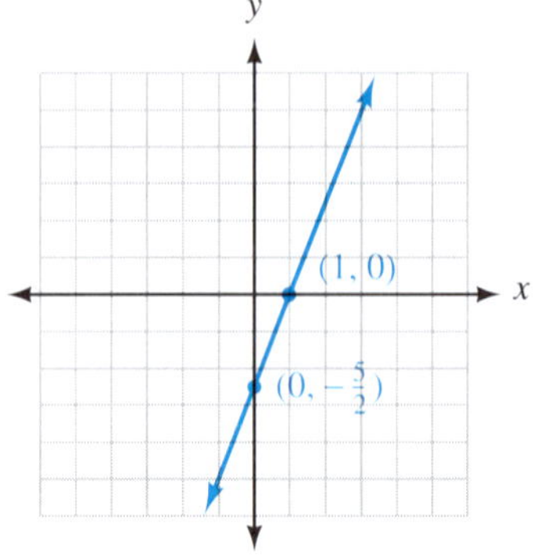

27.

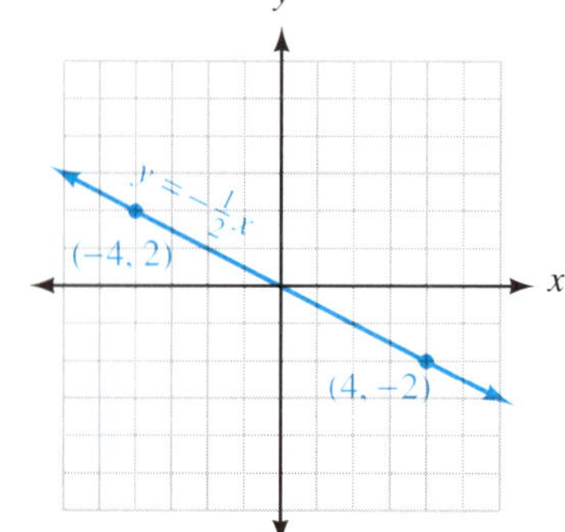

29.

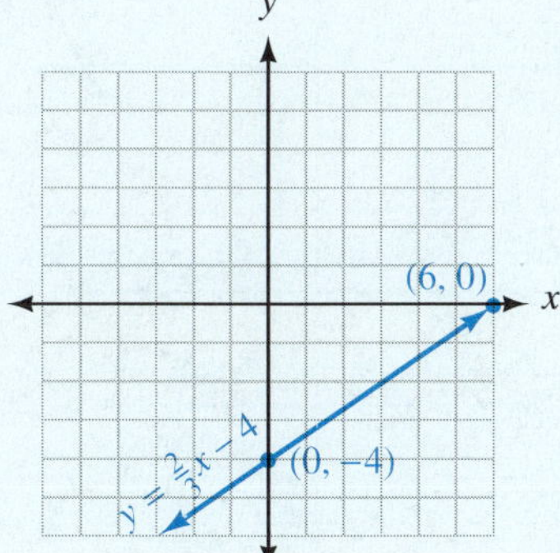

31.

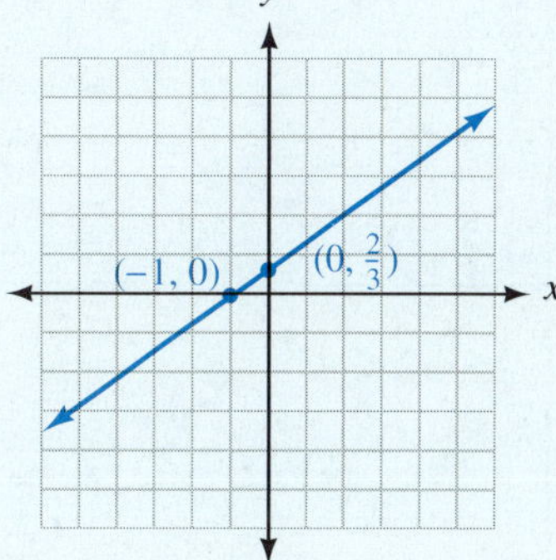

33.

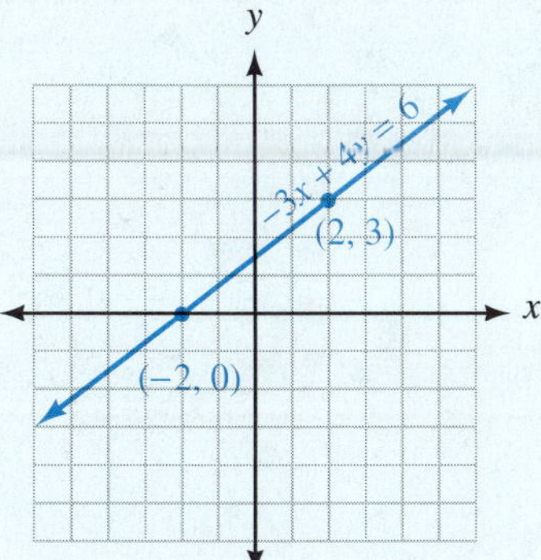

35.

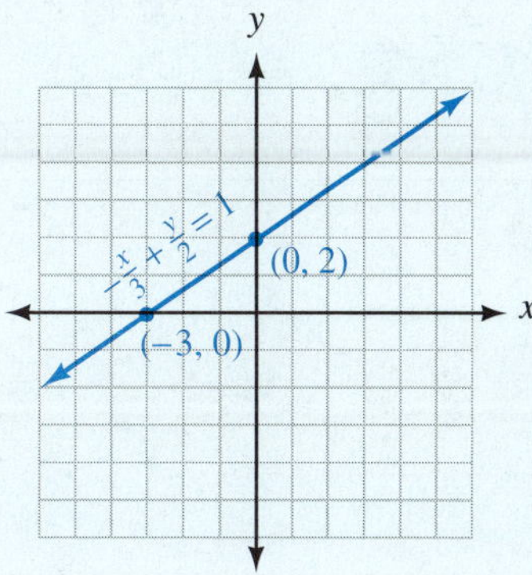

37. b

39. (a)

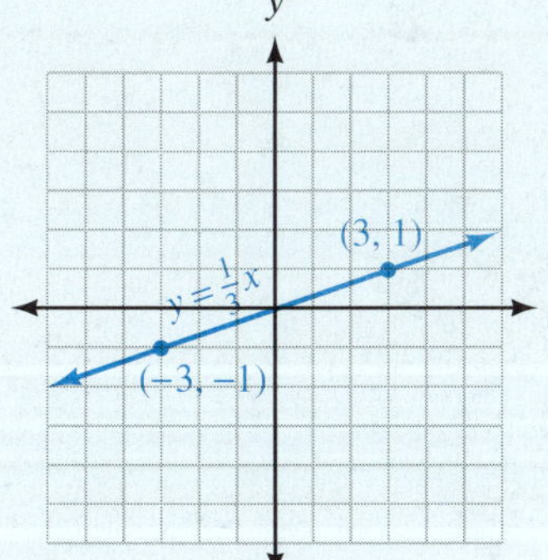

(b)

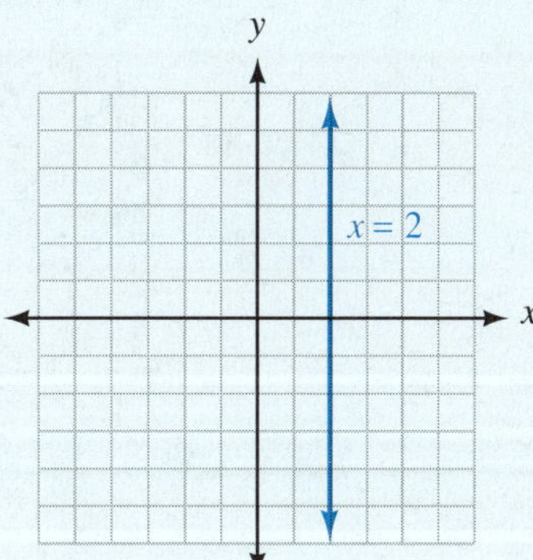

(c)

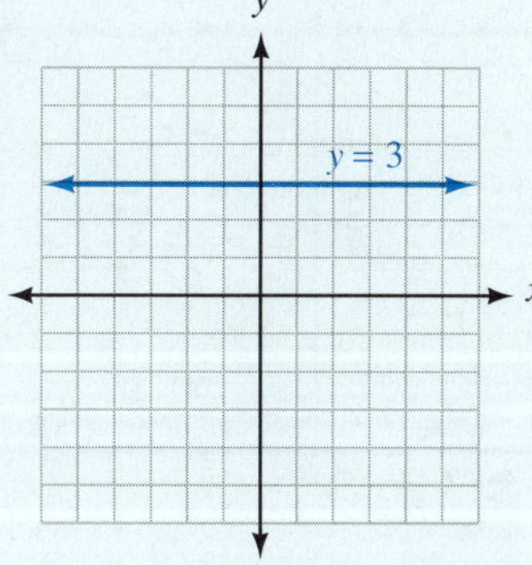

41. (a)

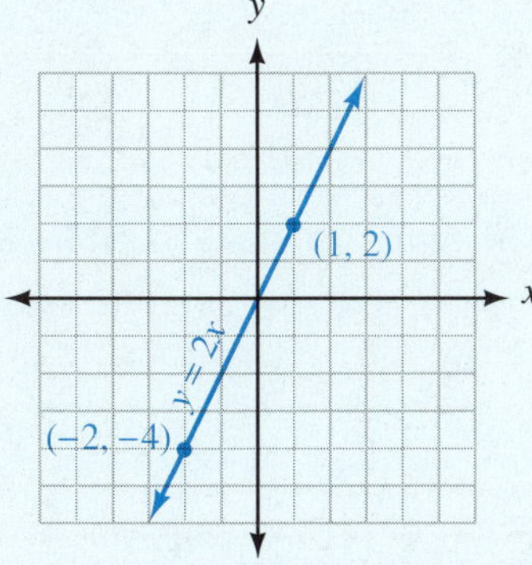

(b)

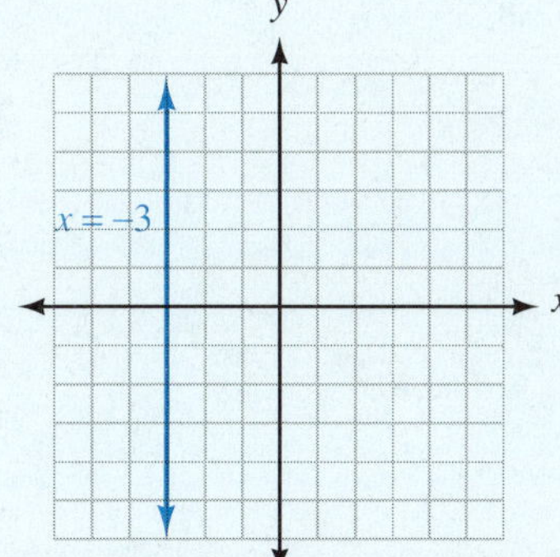

(c)

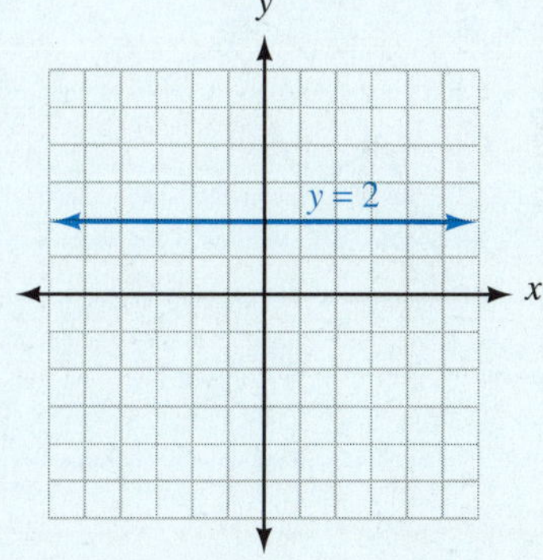

43. (a)

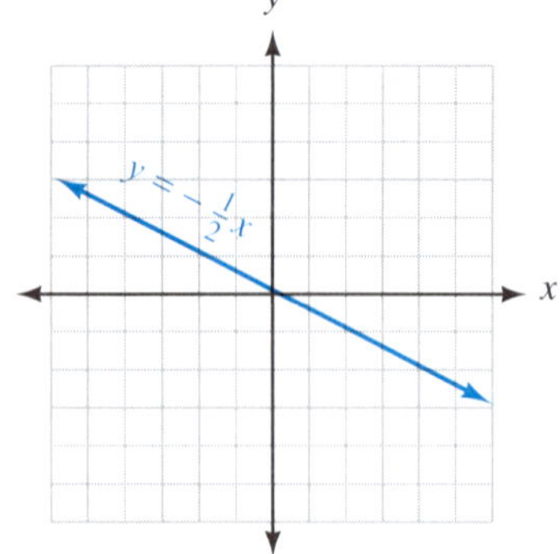

(b)

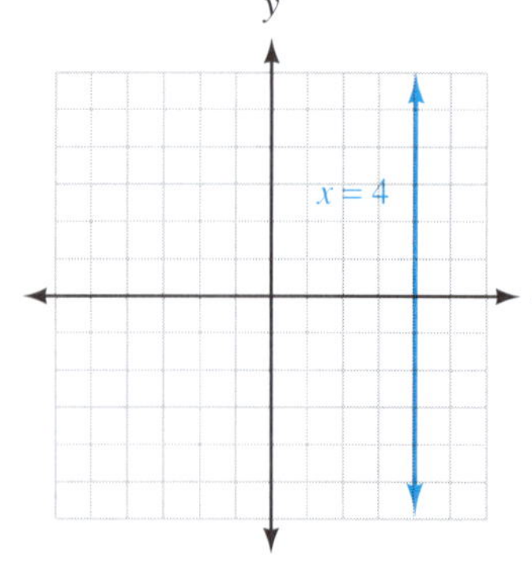

(c)

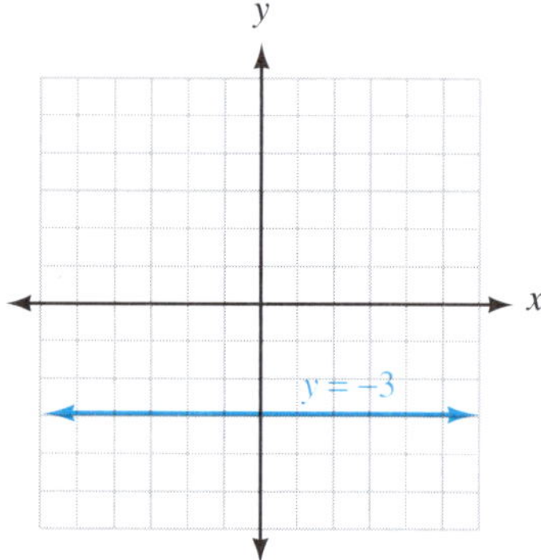

45.

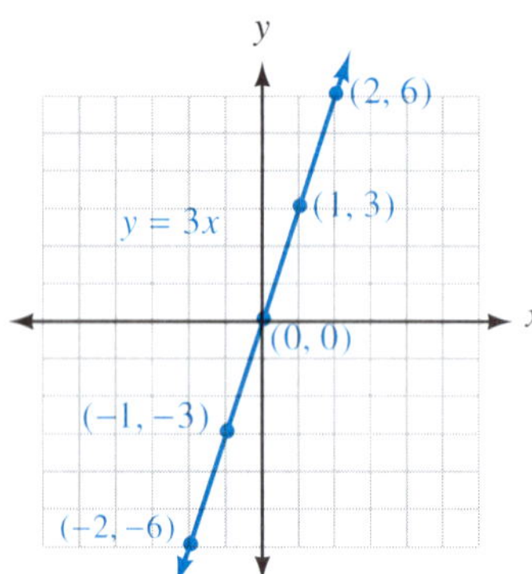

47. (a) $x > 0, y > 0$ **(b)** $x < 0, y > 0$ **(c)** $x < 0, y < 0$ **(d)** $x > 0, y < 0$

49.

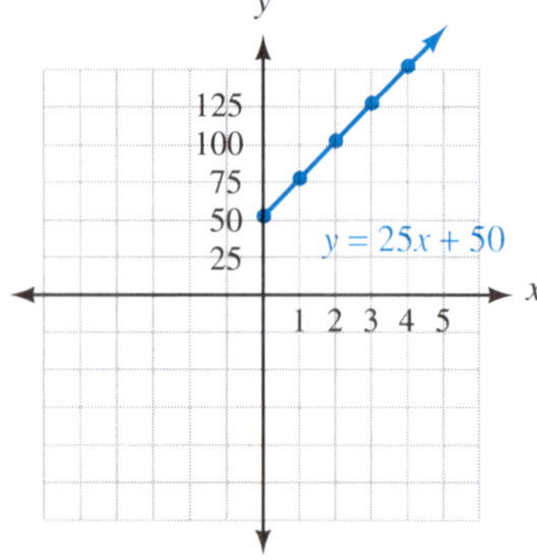

51.

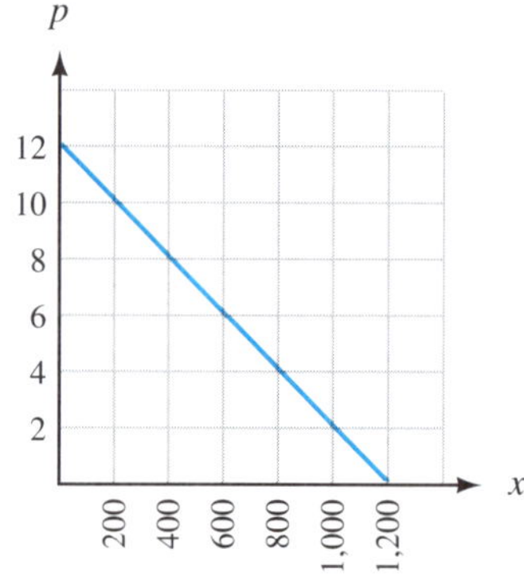

53.

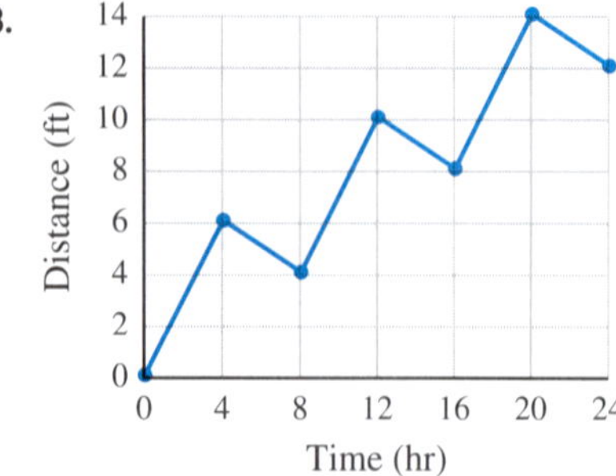

55. (a) 159 **(b)** 19,000 **(c)** 11,845 **(d)** No, because for the year 2050, $D = -80$, and there cannot be a negative number of accident-related deaths.

57. (a)

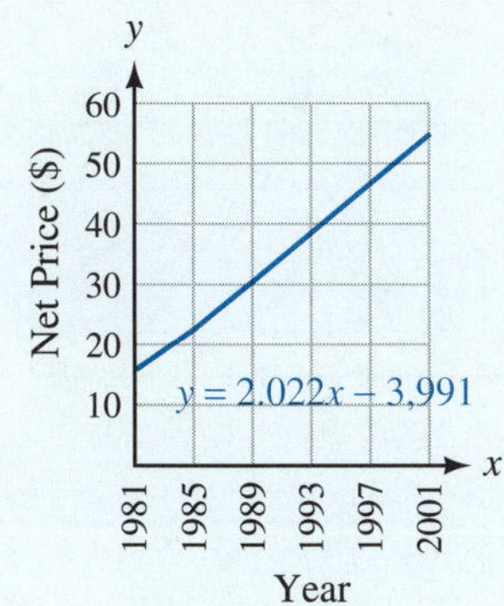

(b)

x	y
1982	16.60
1985	22.67
1989	30.76
1993	38.85
1997	46.93
2001	55.02

59. 2 **61.** $-\frac{3}{2}, 4$ **63.** -2 **65.** $-3, 0$

67.

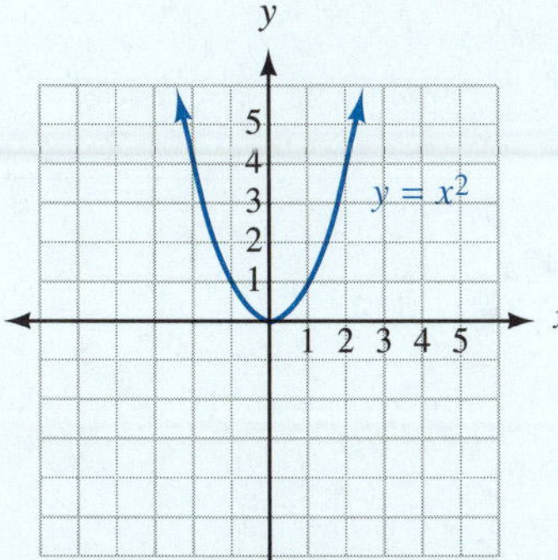

69.

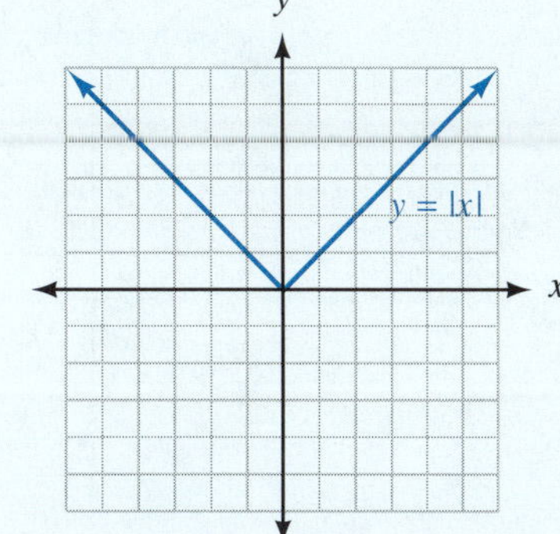

71.

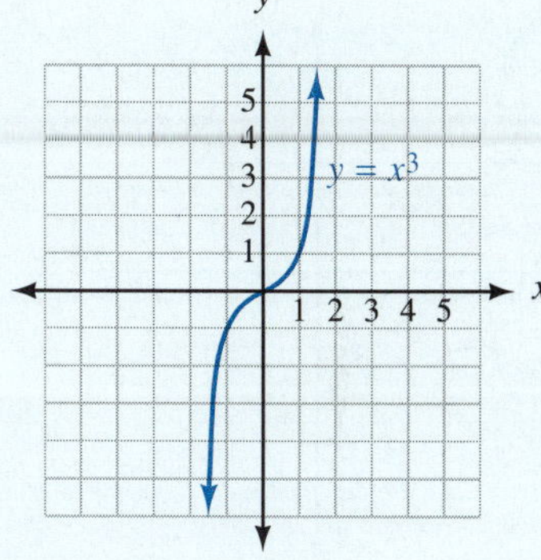

73. x-intercept $= \frac{c}{a}$, y-intercept $= \frac{c}{b}$ **75.** x-intercept $= a$, y-intercept $= b$

77.

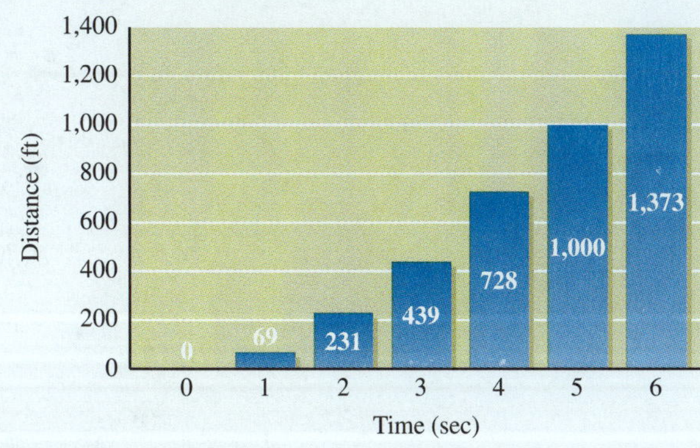

79.

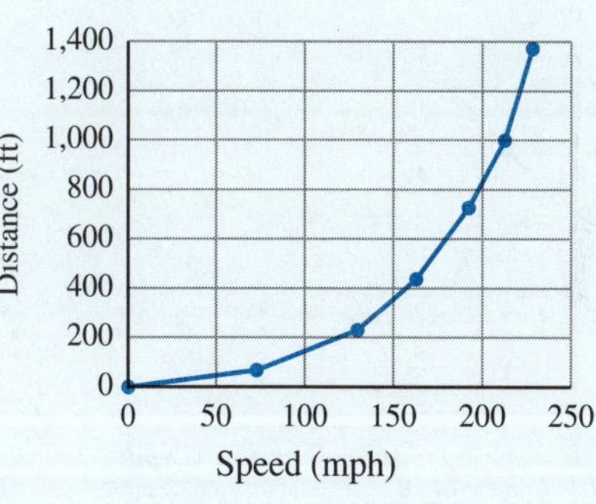

Problem Set 2.2

1. $\frac{3}{2}$ **3.** no slope (undefined) **5.** $\frac{2}{3}$ **7.** $\frac{3}{2}$ **9.** $-\frac{1}{2}$ **11.** $\frac{5}{3}$ **13.** $-\frac{5}{3}$ **15.** $-\frac{3}{2}$ **17.** 0 **19.** no slope (undefined) **21.** 5

23. 2 **25.** 3 **27.** -4 **29.** $m = -\frac{2}{3}$

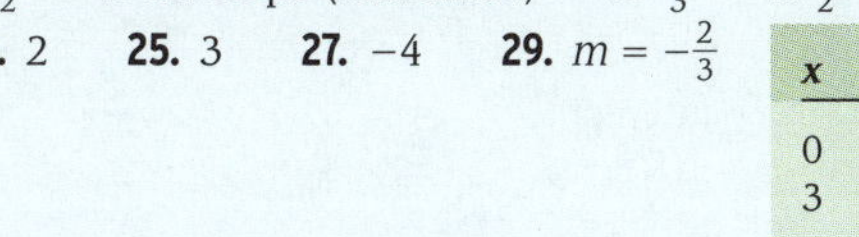

x	y
0	2
3	0

31. $m = \frac{2}{3}$

x	y
0	-5
3	-3

33. $m = \frac{2}{3}$

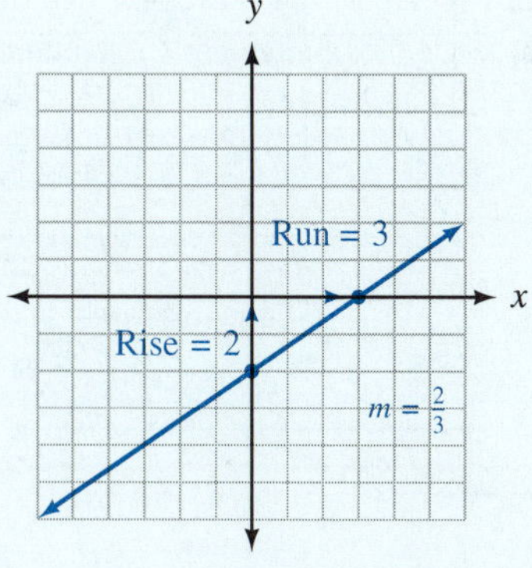

35. $\frac{1}{5}$ **37.** 0 **39.** -1 **41.** $-\frac{3}{2}$ **43. (a)** yes **(b)** no **45.** 24 feet **47.** 10 minutes **49.** 20; °C per minute **51.** first minute **53.** $-1{,}250$ dollars per year **55.** 2 to 3 years **57. (a)** 15,800 feet **(b)** $-\frac{7}{100} = -0.07$ **59.** \$1,875 per year; \$175,000 per year **61. (a)** 12,630 employees per year **(b)** 5,771 employees per year **(c)** Faster growth was occurring between 1980 and 1990. **63.** 0 **65.** $y = -\frac{3}{2}x + 6$ **67.** $t = \frac{A - P}{Pr}$ **69.** $x + 2$ **71.** $(\frac{5}{4}, \frac{7}{4})$ **73.** (0, 2) and (2, 6)

75. 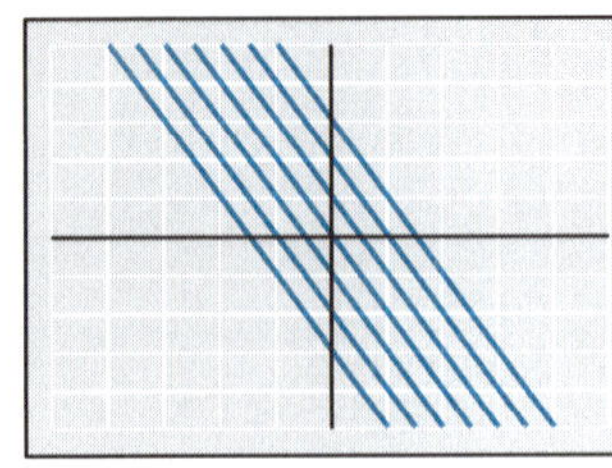

77. 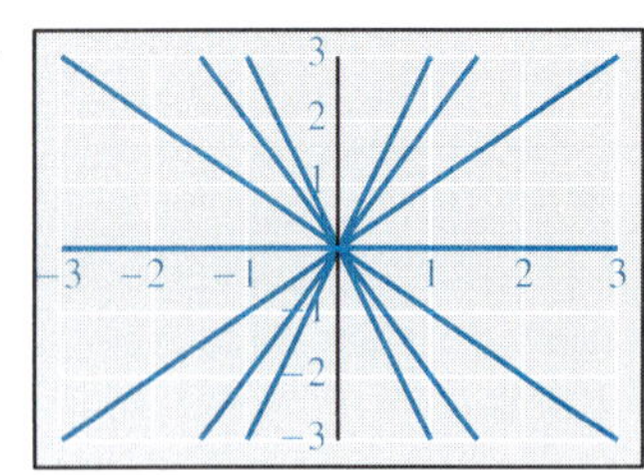

79. 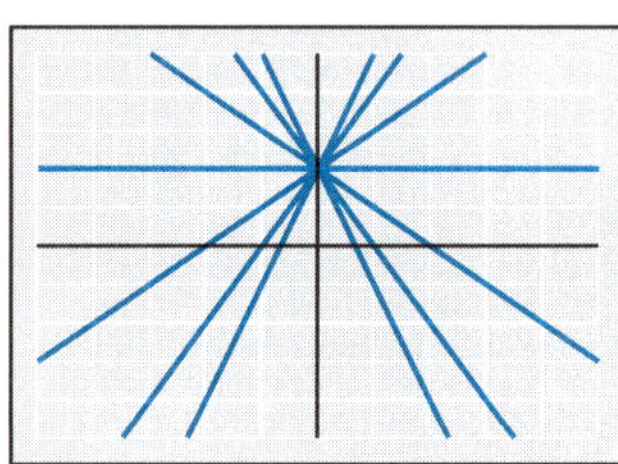

Problem Set 2.3

1. $y = -4x - 3$ **3.** $y = -\frac{2}{3}x$ **5.** $y = -\frac{2}{3}x + \frac{1}{4}$ **7. (a)** 3 **(b)** $-\frac{1}{3}$ **9. (a)** -3 **(b)** $\frac{1}{3}$ **11. (a)** $-\frac{2}{5}$ **(b)** $\frac{5}{2}$

13. slope = 3, y-intercept = -2, perpendicular slope = $-\frac{1}{3}$ **15.** slope = $\frac{2}{3}$, y-intercept = -4, perpendicular slope = $-\frac{3}{2}$

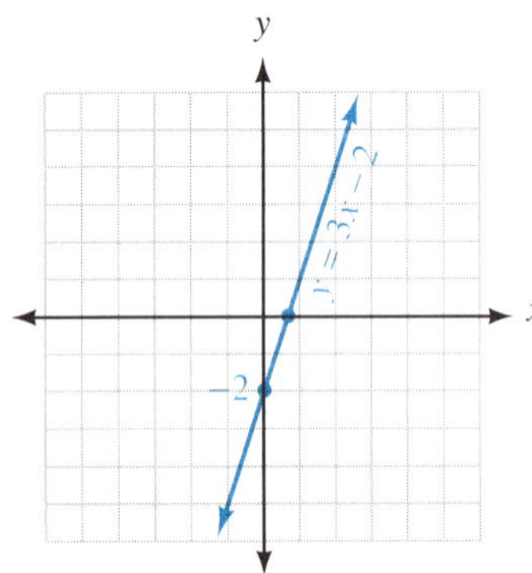

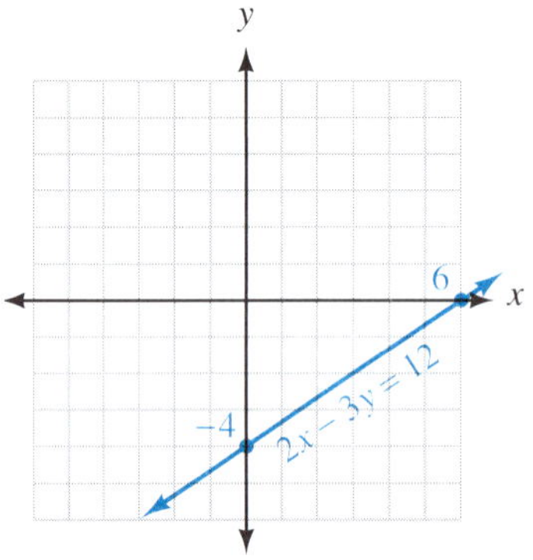

17. slope = $-\frac{4}{5}$, y-intercept = 4, perpendicular slope = $\frac{5}{4}$

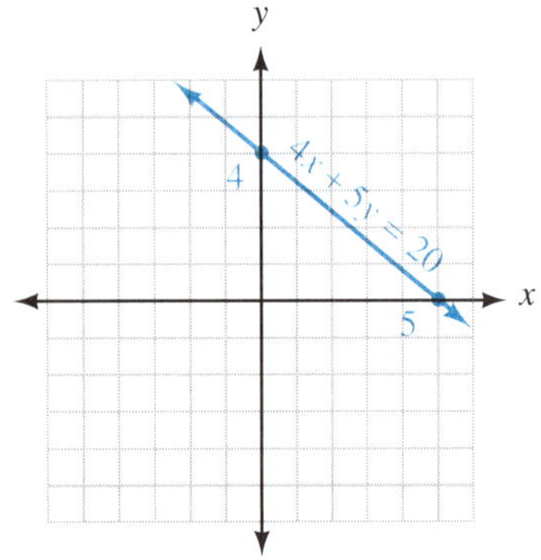

19. slope = $\frac{1}{2}$, y-intercept = -4, $y = \frac{1}{2}x - 4$ **21.** slope = $-\frac{2}{3}$, y-intercept = 3, $y = -\frac{2}{3}x + 3$ **23.** $y = 2x - 1$ **25.** $y = -\frac{1}{2}x - 1$

27. $y = -3x + 1$ **29.** $y = \frac{2}{3}x + \frac{14}{3}$ **31.** $y = -\frac{1}{4}x - \frac{13}{4}$ **33.** $3x + 5y = -1$ **35.** $x - 12y = -8$ **37.** $6x - 5y = 3$

39. $(0, -4)$ and $(2, 0)$; $y = 2x - 4$ **41.** $(0, 4)$ and $(-2, 0)$; $y = 2x + 4$

43. (a) slope = $\frac{1}{2}$, x-intercept = 0, y-intercept = 0 **(b)** no slope (undefined), x-intercept = 3, no y-intercept

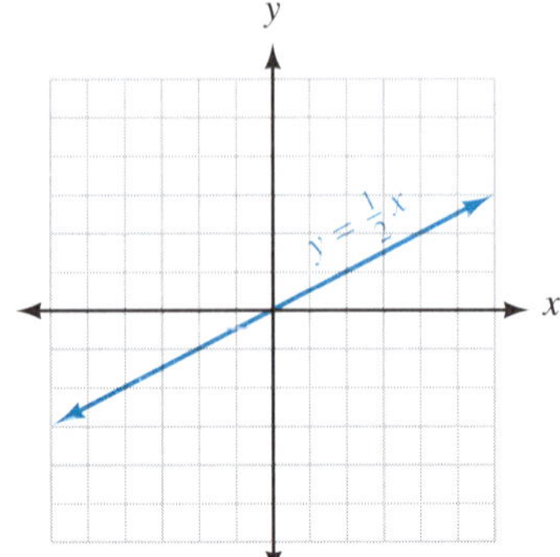

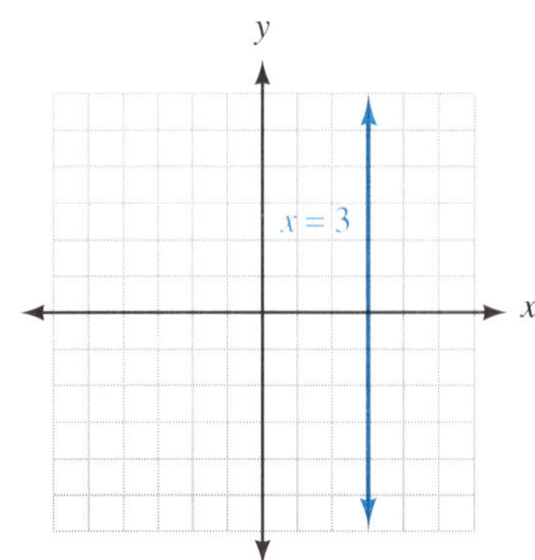

(c) slope = 0, no x-intercept, y-intercept = −2

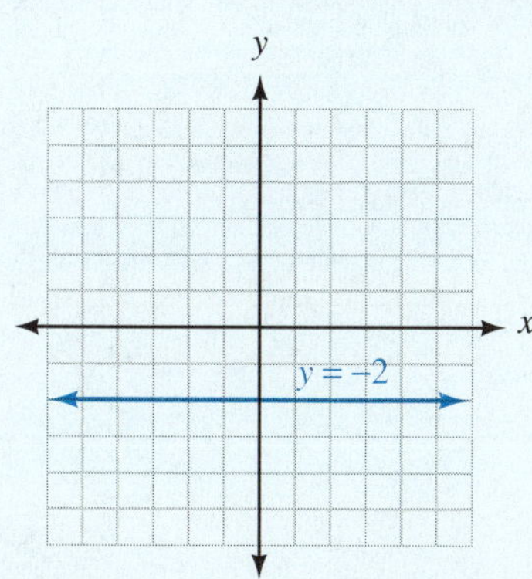

45. $y = 3x + 7$ **47.** $y = -\frac{5}{2}x - 13$ **49.** $y = \frac{1}{4}x + \frac{1}{4}$ **51.** $y = -\frac{2}{3}x + 2$ **53. (b)** 86° **55. (a)** $y = \frac{1}{114}x + \frac{8}{57}$ **(b)** $40 \le x \le 220$

57. (a) $y = 4250x - 8{,}429{,}000$ **(b)** 11,500 cases **59. (a)**

(b) $N = -0.74x + 58.5$ **(c)** 1975: 54.8 thousand; 2000: 36.3 thousand **(d)** Yes, because the points in the scatter plot tend to follow the line closely.

61. width = 5 inches, length = 23 inches **63.** $46.50 **65.** $y = -\frac{3}{2}x + 12$ **67.** $y = -\frac{4}{3}x - \frac{3}{2}$

69. $y = -\frac{3}{2}x + 3$; slope = $-\frac{3}{2}$; x-intercept = 2; y-intercept = 3 **71.** $y = \frac{3}{2}x + 3$; slope = $\frac{3}{2}$; x-intercept = −2; y-intercept = 3

73. x-intercept = a; y-intercept = b; slope = $-\frac{b}{a}$

Problem Set 2.4

1.

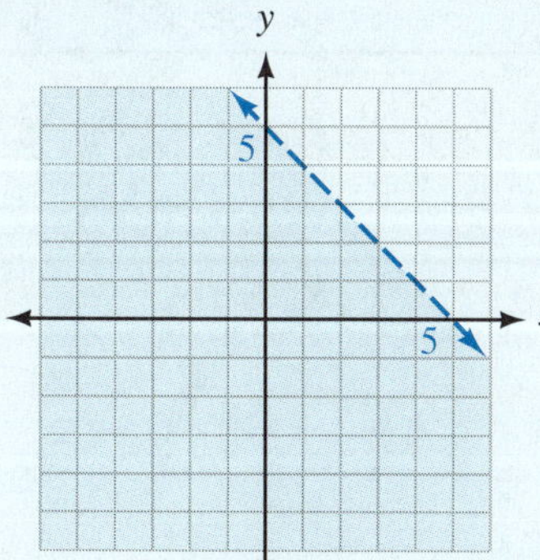

3.

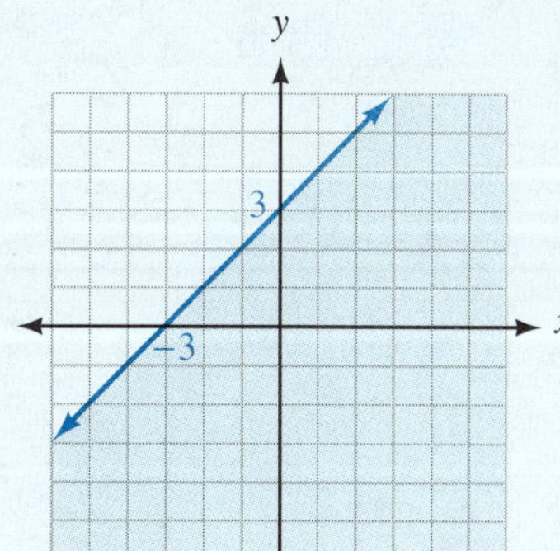

5.

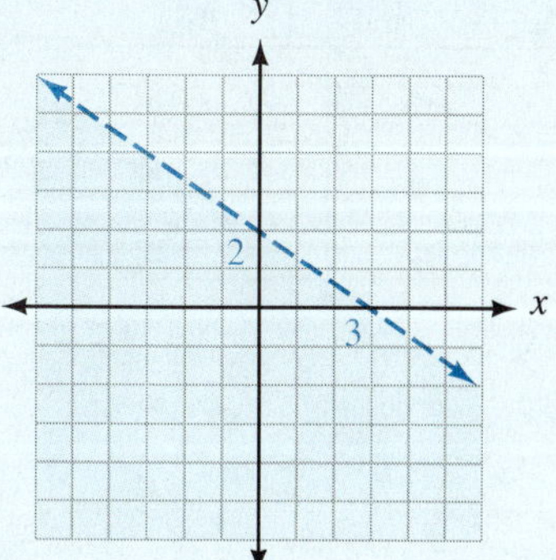

7.

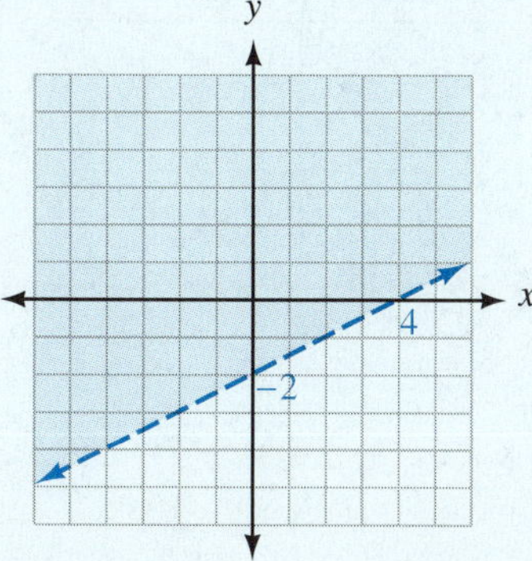

9.

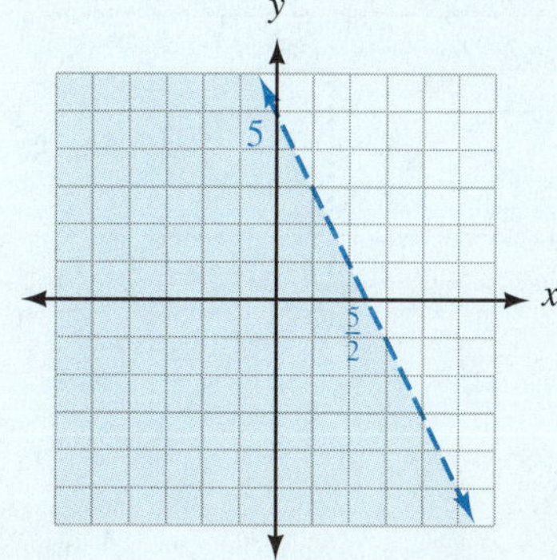

11.

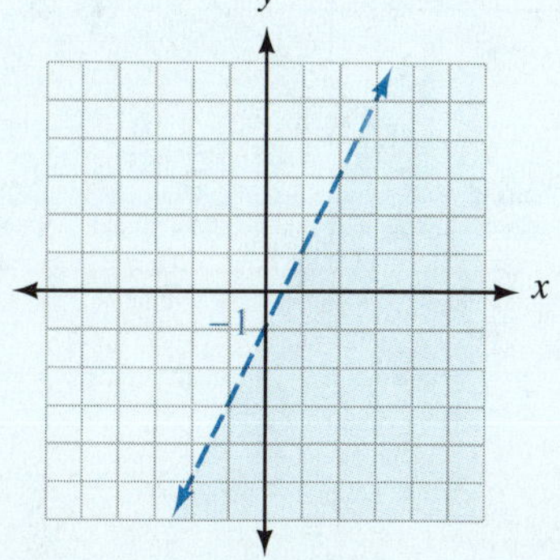

13.

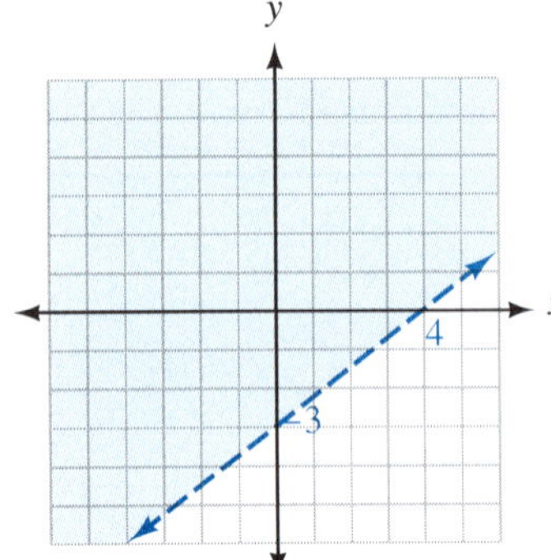

15.

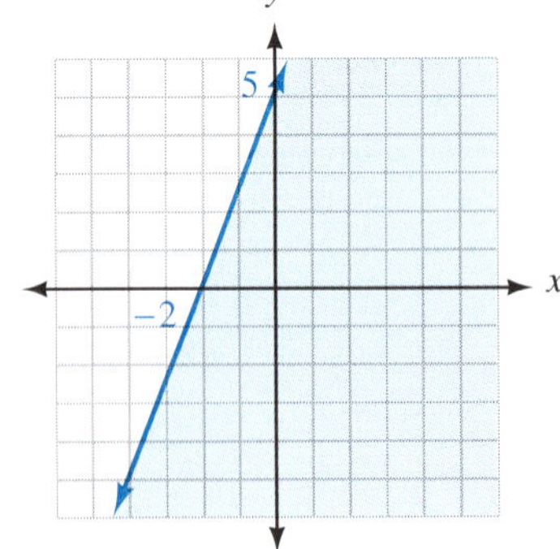

17. $x + y > 4$ 19. $-x + 2y \leq 4$

21.

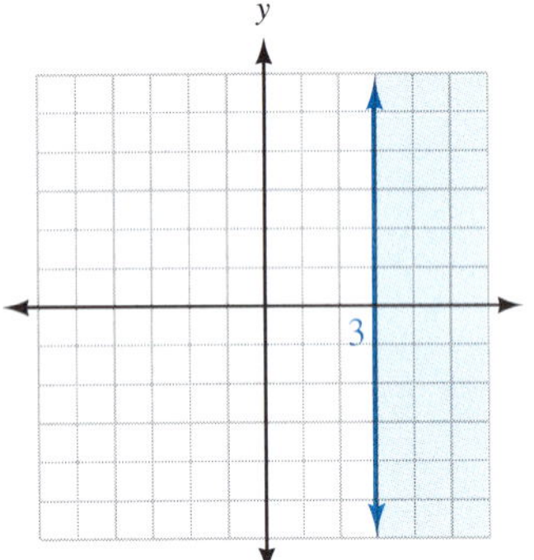

23.

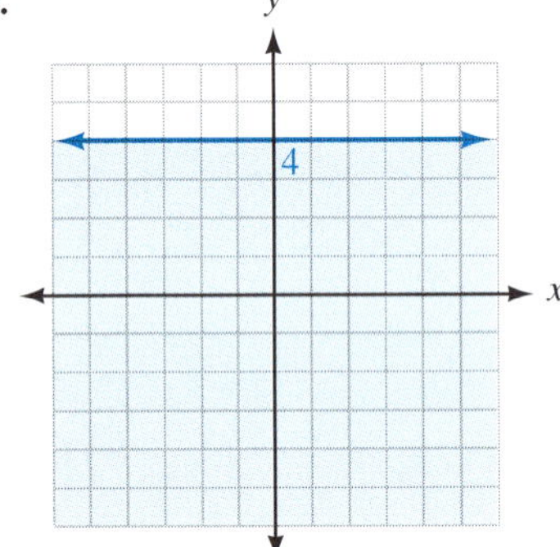

25.

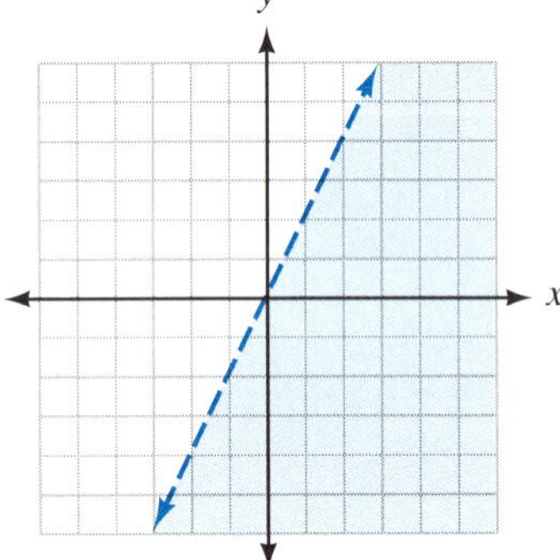

27.

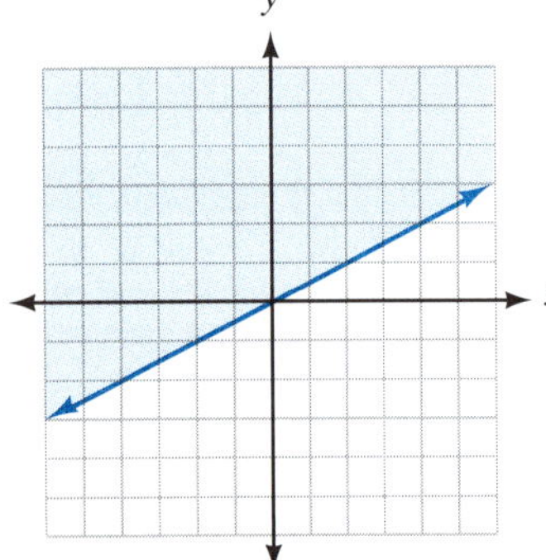

29.

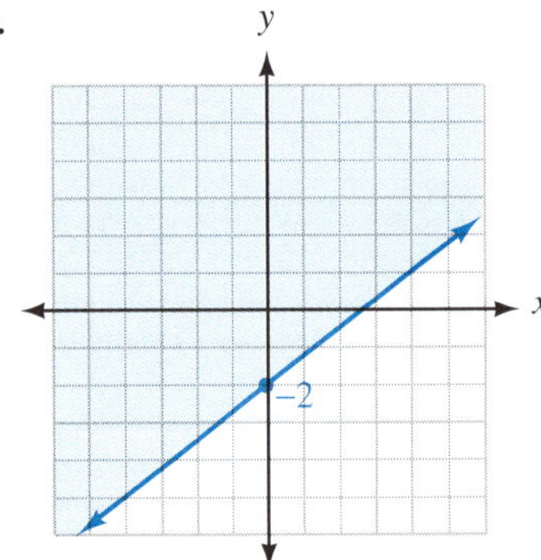

31.

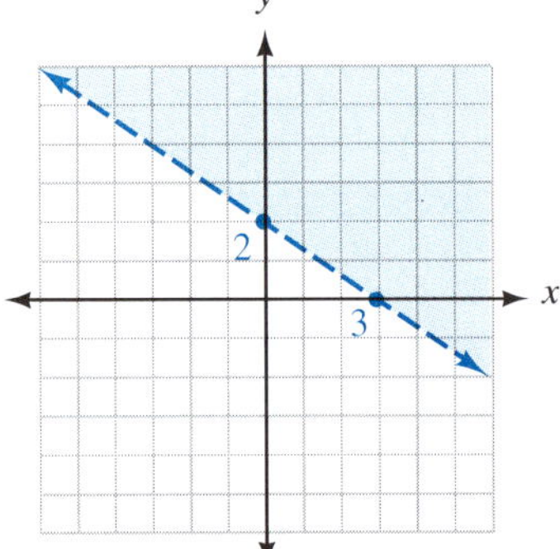

33.

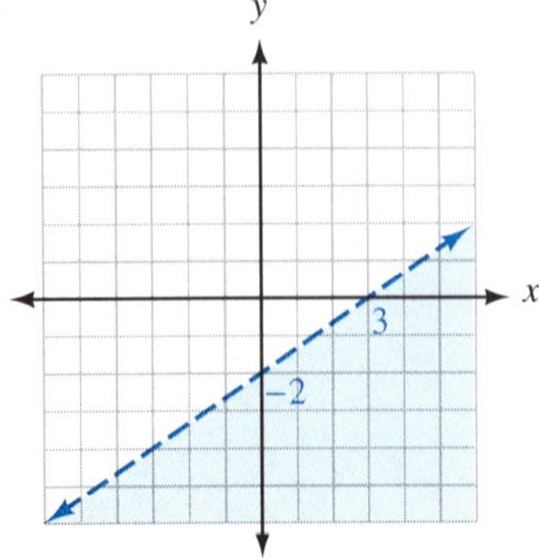

35.

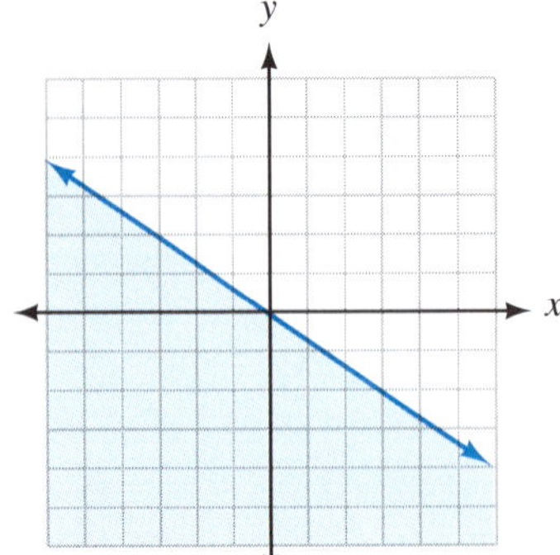

37.

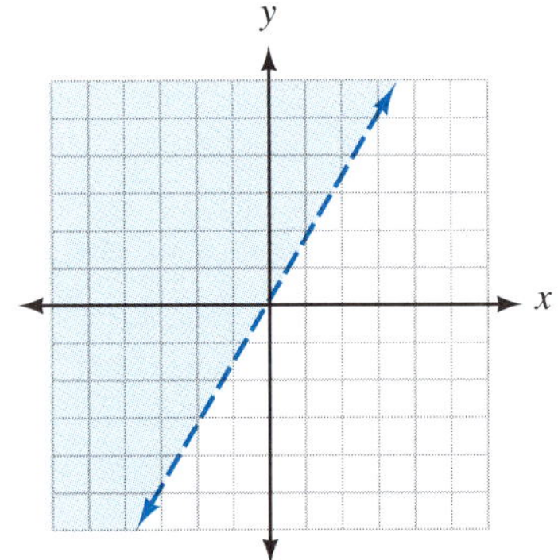

39.

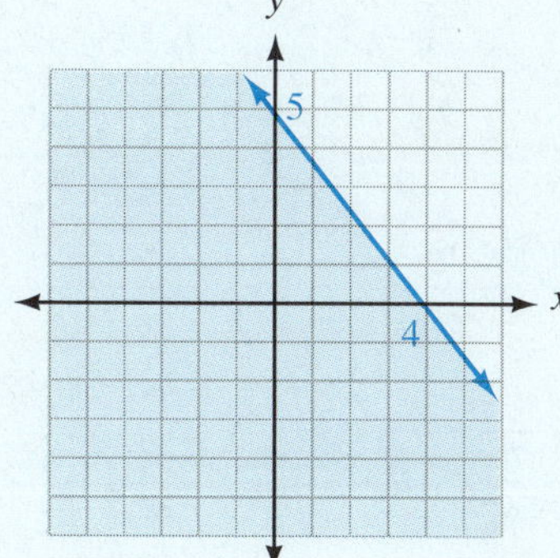

41.

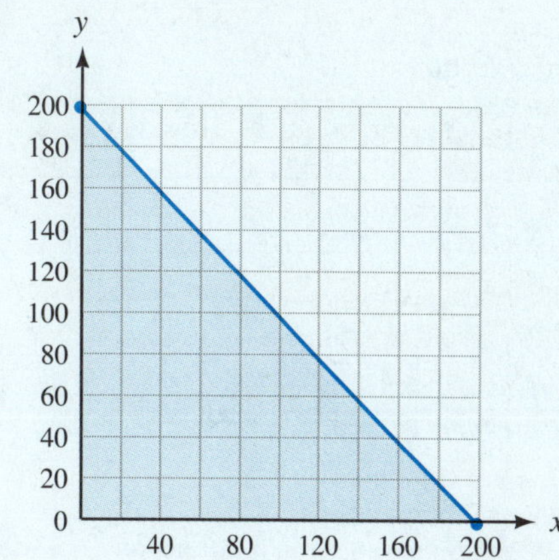

43.

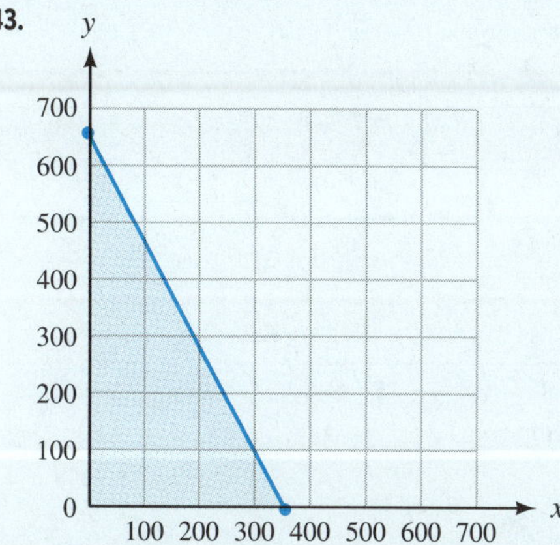

45.

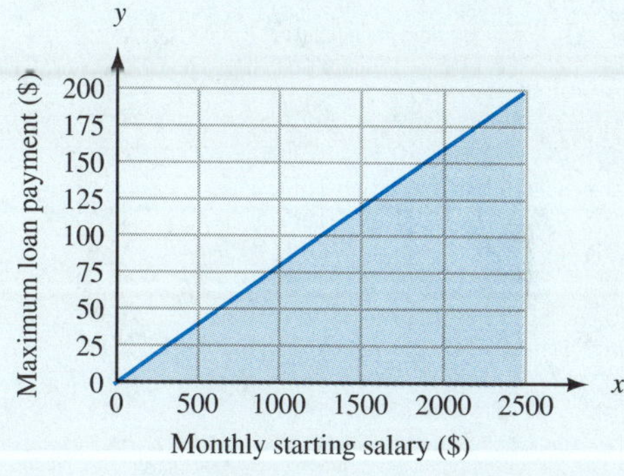

47. $y \le 7$ **49.** $t > -2$ **51.** $-1 < t < 2$

53.

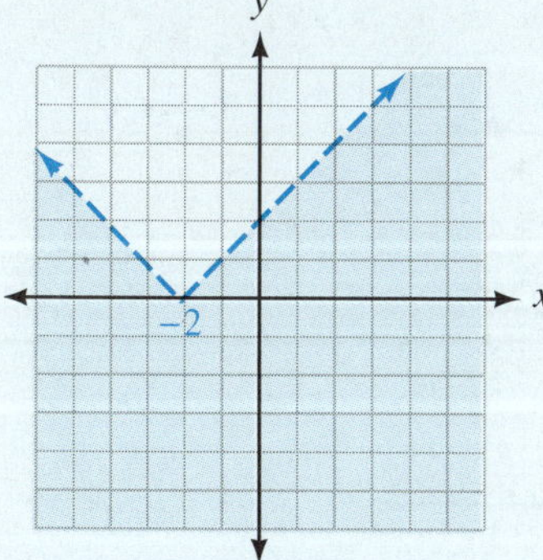

55.

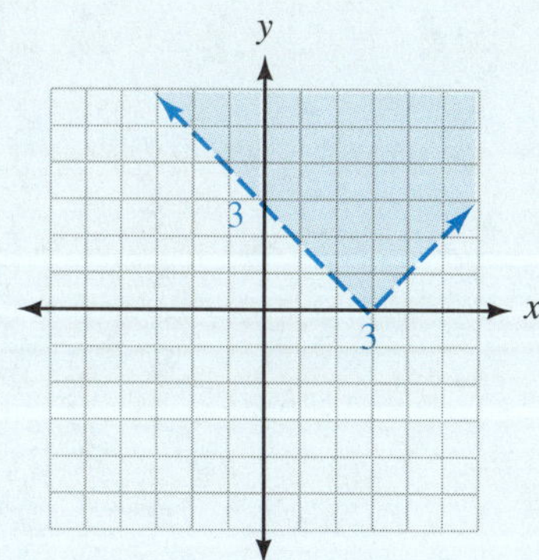

57.

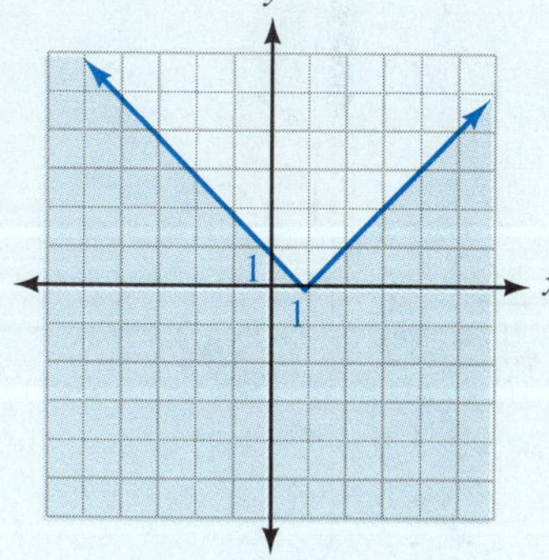

59.

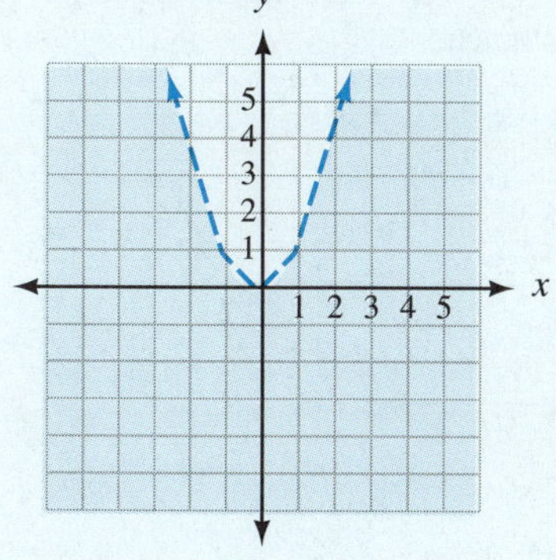

61.

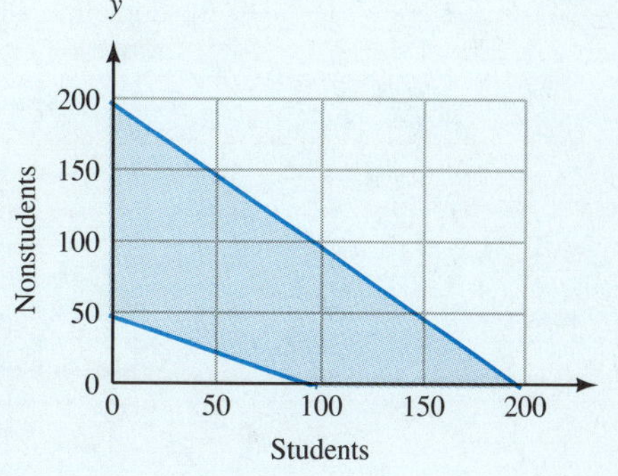

Problem Set 2.5

1. (a) $y = 8.5x$ for $10 \le x \le 40$

(b)

HOURS WORKED x	FUNCTION RULE $y = 8.5x$	GROSS PAY ($) y
10	$y = 8.5(10) = 85$	85
20	$y = 8.5(20) = 170$	170
30	$y = 8.5(30) = 255$	255
40	$y = 8.5(40) = 340$	340

(c)

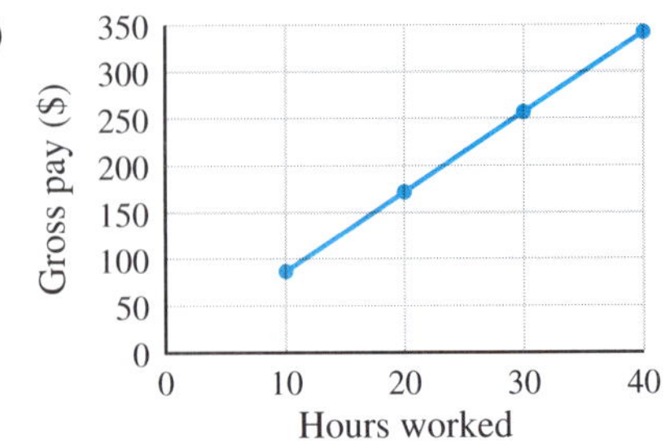

(d) domain = $\{x \mid 10 \le x \le 40\}$; range = $\{y \mid 85 \le y \le 340\}$

(e) minimum = \$85; maximum = \$340

3. domain = $\{0, 1, 2, 3\}$; range = $\{4, 5, 6\}$; is a function **5.** domain = $\{a, b, c, d\}$; range = $\{3, 4, 5\}$; is a function

7. domain = $\{a\}$; range = $\{1, 2, 3, 4\}$; not a function **9.** yes **11.** no **13.** no **15.** yes **17.** yes

19. (a)

TIME (sec) t	FUNCTION RULE $h = 16t - 16t^2$	DISTANCE (ft) h
0	$h = 16(0) - 16(0)^2$	0
0.1	$h = 16(0.1) - 16(0.1)^2$	1.44
0.2	$h = 16(0.2) - 16(0.2)^2$	2.56
0.3	$h = 16(0.3) - 16(0.3)^2$	3.36
0.4	$h = 16(0.4) - 16(0.4)^2$	3.84
0.5	$h = 16(0.5) - 16(0.5)^2$	4
0.6	$h = 16(0.6) - 16(0.6)^2$	3.84
0.7	$h = 16(0.7) - 16(0.7)^2$	3.36
0.8	$h = 16(0.8) - 16(0.8)^2$	2.56
0.9	$h = 16(0.9) - 16(0.9)^2$	1.44
1	$h = 16(1) - 16(1)^2$	0

(b) domain = $\{t \mid 0 \le t \le 1\}$; range = $\{h \mid 0 \le h \le 4\}$

(c)

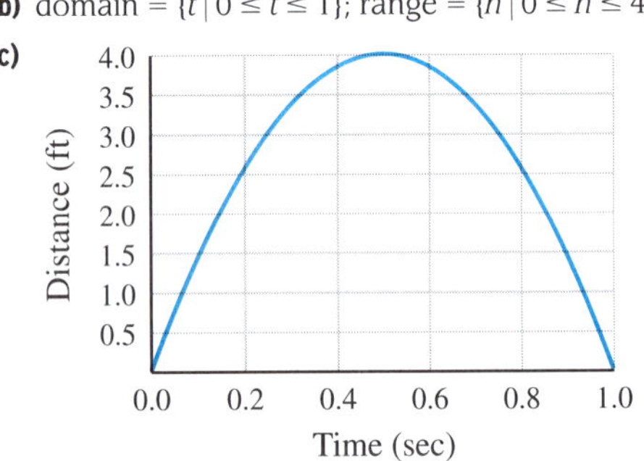

21. domain = $\{x \mid -5 \le x \le 5\}$, range = $\{y \mid 0 \le y \le 5\}$ **23.** domain = $\{x \mid -5 \le x \le 3\}$, range = $\{y \mid y = 3\}$

25. domain = all real numbers,
range = $\{y \mid y \ge -1\}$, is a function

27. domain = all real numbers,
range = $\{y \mid y \ge 4\}$, is a function

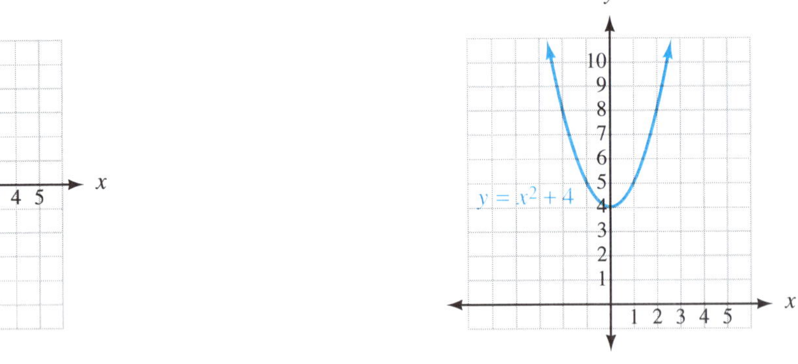

29. domain = $\{x \mid x \ge -1\}$,
range = all real numbers, not a function

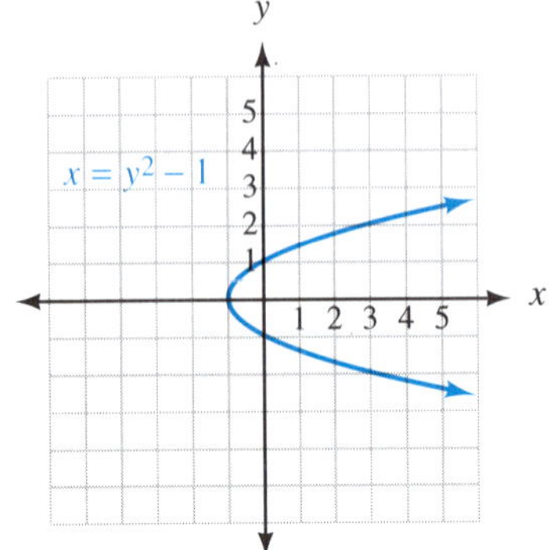

31. domain = all real numbers,
range = $\{y \mid y \ge 0\}$, is a function

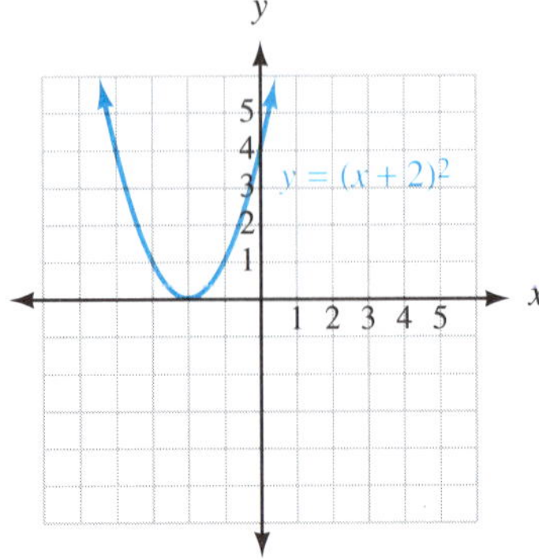

33. domain = $\{x \mid x \geq 0\}$,
range = all real numbers, not a function

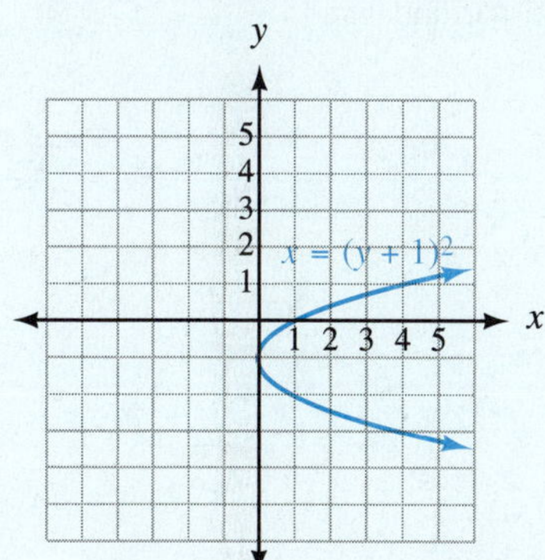

35.

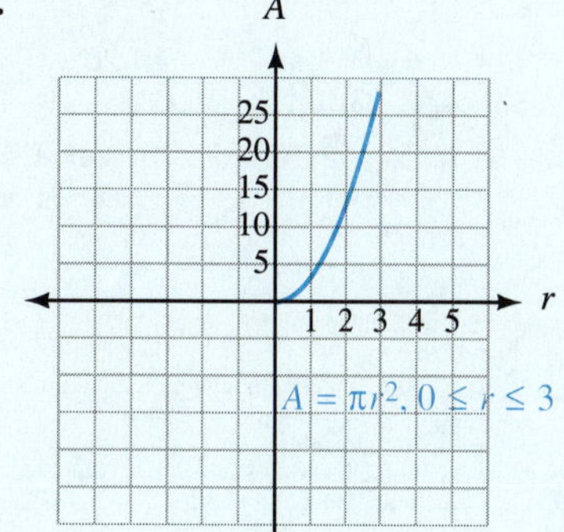

37. $P = 4x + 4, x > 0$ **39.** $A = x(x + 2), x > 0$

41. **(a)** yes **(b)** domain = $\{t \mid 0 \leq t \leq 6\}$, range = $\{h \mid 0 \leq h \leq 60\}$ **(c)** $t = 3$ **(d)** $h = 60$ **(e)** $t = 6$

43. **(a)**

YEAR	x	DEATH RATE PER 100,000 POPULATION (Y)
1930	30	430
1940	40	482
1950	50	512
1960	60	520
1970	70	506
1980	80	470
1990	90	412
1995	95	375

(c) (60, 520); The maximum death rate for 100,000 population occurred in 1960, and it was 520 per 100,000.

(d) domain = $\{x \mid 30 \leq x \leq 95\}$,
range = $\{Y \mid 375 \leq Y \leq 520\}$

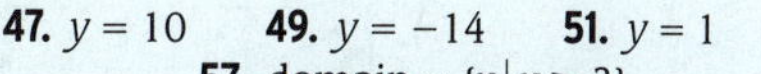

45. **(a)** Figure 9 **(b)** Figure 10 **(c)** Figure 8 **(d)** Figure 7 **47.** $y = 10$ **49.** $y = -14$ **51.** $y = 1$ **53.** $y = -3$

55. domain = all real numbers,
range = $\{y \mid y \leq 5\}$, is a function

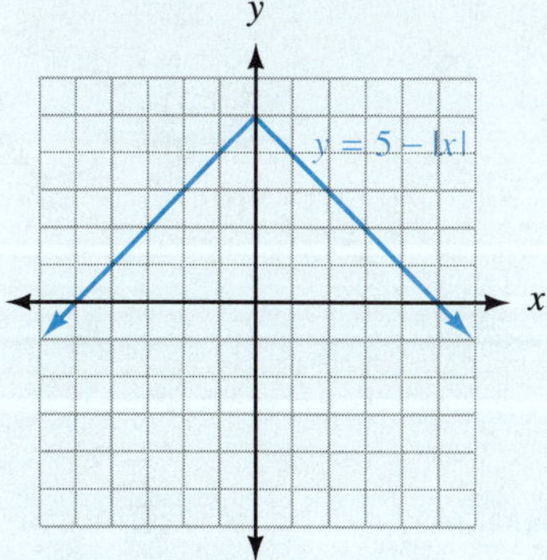

57. domain = $\{x \mid x \geq 3\}$,
range = all real numbers, not a function

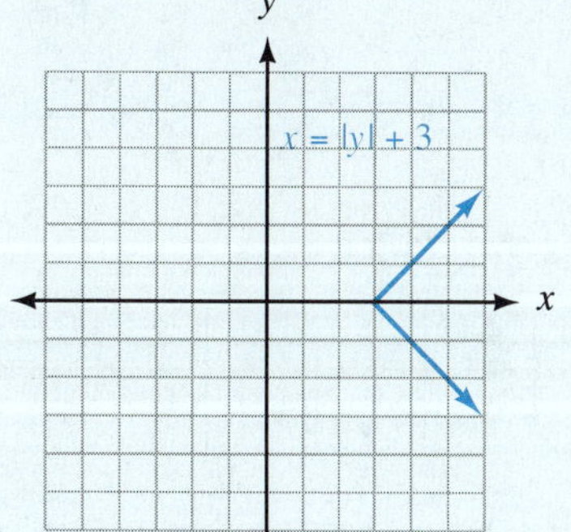

59. domain = $\{x \mid -4 \leq x \leq 4\}$,
range = $\{y \mid -4 \leq y \leq 4\}$, not a function

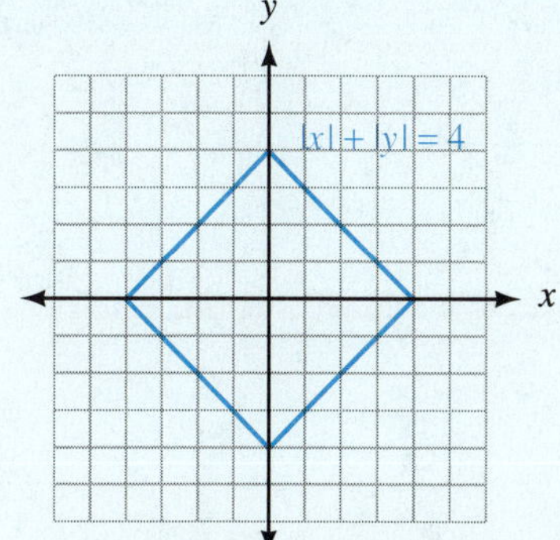

Problem Set 2.6

1. -1 **3.** -11 **5.** 2 **7.** 4 **9.** $a^2 + 3a + 4$ **11.** $2a + 7$ **13.** 1 **15.** -9 **17.** 8 **19.** 0 **21.** $3a^2 - 4a + 1$ **23.** $3a^2 + 8a + 5$ **25.** 4 **27.** 0 **29.** 2 **31.** 24 **33.** -1 **35.** $2x^2 - 19x + 12$ **37.** 99 **39.** $\frac{3}{10}$ **41.** $\frac{2}{5}$ **43.** undefined
45. **(a)** $a^2 - 7$ **(b)** $a^2 - 6a + 5$ **(c)** $x^2 - 2$ **(d)** $x^2 + 4x$ **(e)** $a^2 + 2ab + b^2 - 4$ **(f)** $x^2 + 2xh + h^2 - 4$

47.

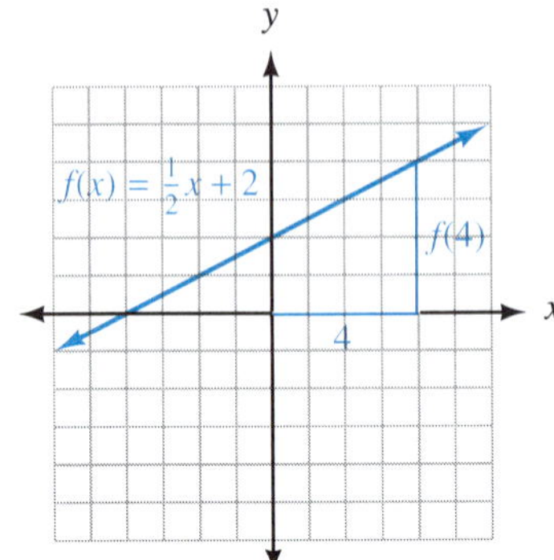

49. $x = 4$ **51.**

53. $V(3) = 300$; The painting is worth \$300 in 3 years. $V(6) = 600$; The painting is worth \$600 in 6 years.
55. **(a)** $P(x) = 6x + 6$, where $x > 0$ **(b)** $A(x) = 2x^2 + 3x$, where $x > 0$ **57.** **(a)** For 1950, find $L(50)$, for 1960, evaluate $L(60)$, for 1975, evaluate $L(75)$, and for 1995, evaluate $L(95)$. **(b)** 1950: 63.9 years; 1960: 66 years; 1975: 69.2 years; 1995: 73.4 years **(c)** 1989
59. **(a)** $F(10) = 12{,}199$; In 1910 the cotton production was 12,199 thousand bales.
$F(25) = 12{,}957$; In 1925 the cotton production was 12,957 thousand bales.
$F(40) = 11{,}978$; In 1940 the cotton production was 11,978 thousand bales.
$F(65) = 10{,}610$; In 1965 the cotton production was 10,610 thousand bales.
$F(80) = 12{,}538$; In 1980 the cotton production was 12,538 thousand bales.
$F(95) = 18{,}557$; In 1995 the cotton production was 18,557 thousand bales.
(b) 22,496 thousand bales
61. **(a)** \$5,625 **(b)** \$1,500 **(c)** $\{t \mid 0 \le t \le 5\}$ **(d)**

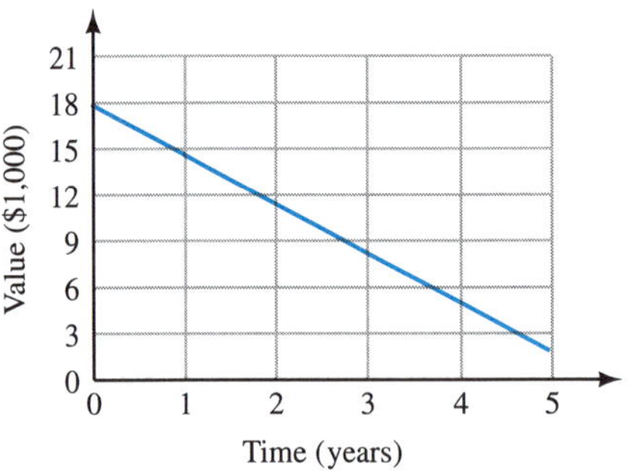

(e) $\{V(t) \mid 1{,}500 \le V(t) \le 18{,}000\}$ **(f)** 2.42 years
63. **(a)**

Weight (ounces)	0.6	1.0	1.1	2.5	3.0	4.8	5.0	5.3
Cost (cents)	32	32	55	78	78	124	124	147

(b) In words: Over 2 ounces, but not over 3 ounces. Inequality: $2 < x \le 3$ **(c)** $\{x \mid 0 < x \le 6\}$ **(d)** $\{C(x) \mid C(x) = 32, 55, 78, 101, 124, 147\}$
65. $-\frac{2}{3}, 4$ **67.** $-2, 1$ **69.** $\varnothing$ **71.** **(a)** 2 **(b)** 0 **(c)** 1 **(d)** 4 **73.** 0, 1 **75.** -6 **77.** 2 **79.** $-\frac{1}{3}$ **81.** $h + 4$ **83.** $-\frac{1}{2h + 4}$

Problem Set 2.7

1. 30 **3.** -6 **5.** 40 **7.** $\frac{81}{5}$ **9.** 64 **11.** 108 **13.** 300 **15.** ± 2 **17.** 1600 **19.** ± 8 **21.** $\frac{50}{3}$ pounds
23. **(a)** $T = 4P$ **(b)**

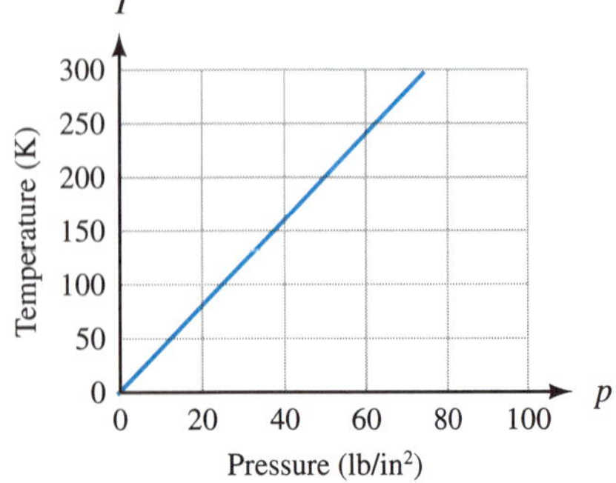

(c) 70 pounds per square inch
25. 12 pounds per square inch

27. (a) $f = \frac{80}{d}$ **(b)**

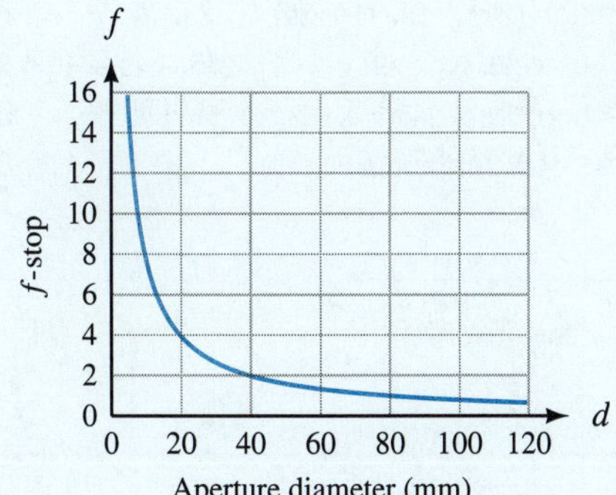

(c) 8 **29.** $\frac{1504}{15}$ square inches **31.** 1.5 ohms

33. (a) $P = 0.21\sqrt{l}$ **(b)**

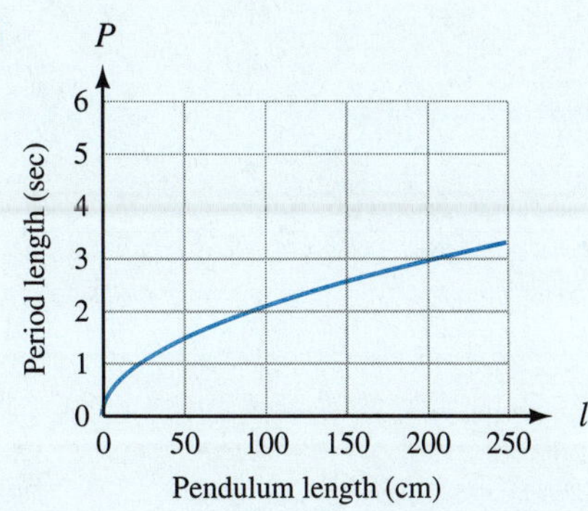

(c) 3.15 seconds

35. 1.28 meters **37.** $F = \frac{Gm_1m_2}{d^2}$ **39. (a)** $K = 1198.5$, $P = \frac{1198.5}{W}$ **(b)** \$2.42 per bushel **(c)** 368.8 million bushels
41. (a) constant = 0.65, $S = 0.65K$ **(b)** 2,630,550 seniors **(c)** 4,923,000 kindergarten students **43.** $(7a + 12)(7a - 12)$ **45.** $(x - \frac{1}{2})^2$
47. $(a - 5)(2x + 3)(2x - 3)$ **49.** y is $\frac{1}{9}$ of its original value **51.** z is 18 times it original value

Problem Set 2.8

1. (4, 3)

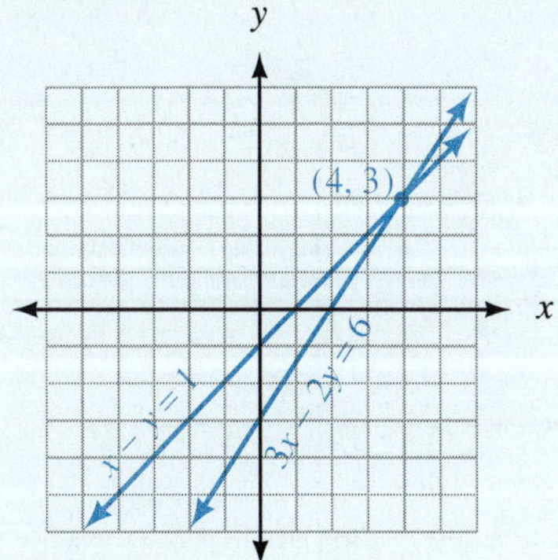

3. (−5, −6)

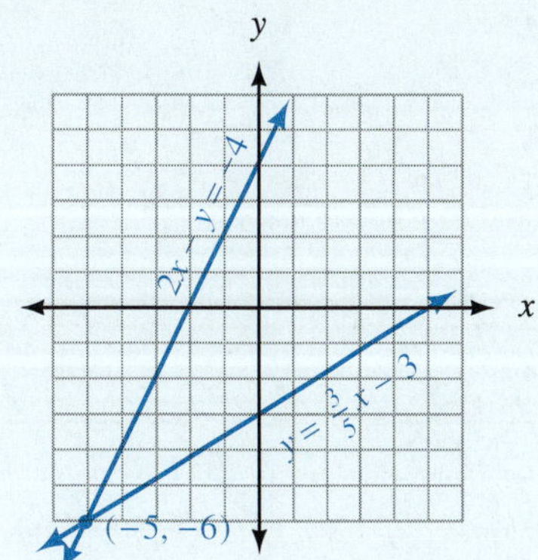

5. (4, 2)

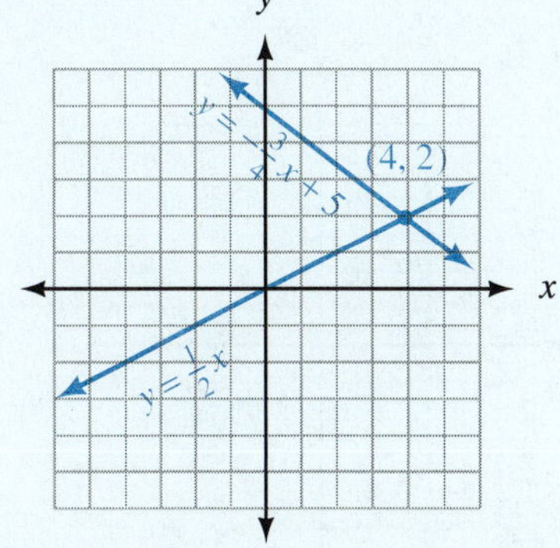

7. parallel lines; no solution

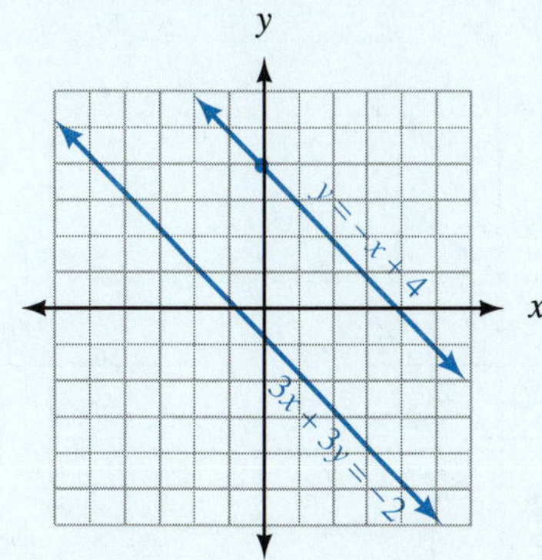

9. $(2, 3)$ **11.** $(1, -\frac{1}{2})$ **13.** $(\frac{1}{2}, -3)$ **15.** $(-\frac{8}{3}, 5)$ **17.** $(2, 2)$ **19.** no solution $(\varnothing)$ **21.** $(10, 24)$ **23.** $(4, \frac{10}{3})$ **25.** $(3, -3)$ **27.** $(\frac{4}{3}, -2)$ **29.** $(2, 4)$ **31.** lines coincide; $\{(x, y) \mid 2x - y = 5\}$ **33.** lines coincide: $\{(x, y) \mid x = \frac{3}{2}y\}$ **35.** $(-\frac{15}{43}, -\frac{27}{43})$ **37.** $(\frac{60}{43}, \frac{46}{43})$ **39.** $(\frac{9}{41}, -\frac{11}{41})$ **41.** $(-\frac{11}{7}, -\frac{20}{7})$ **43.** $(2, \frac{4}{3})$ **45.** $(-12, -12)$ **47.** no solution $(\varnothing)$ **49.** $(3, -2)$ **51.** $(-8, 1)$ **53.** $(\frac{3}{2}, \frac{3}{8})$ **55.** $(-4, -\frac{8}{3})$ **57.** $(6000, 4000)$ **59.** 2 **61. (a)** \$3.29 **(b)** $C(x) = 30x + 45$ **(c)** 2 minutes

63. (a) $0 \le x \le 12$

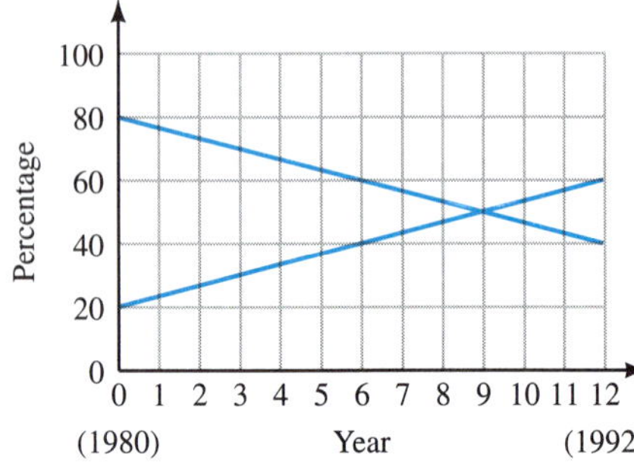

(c) $(9, 50)$; In 1989, both inpatient and outpatient surgeries were 50%.

65. (a)

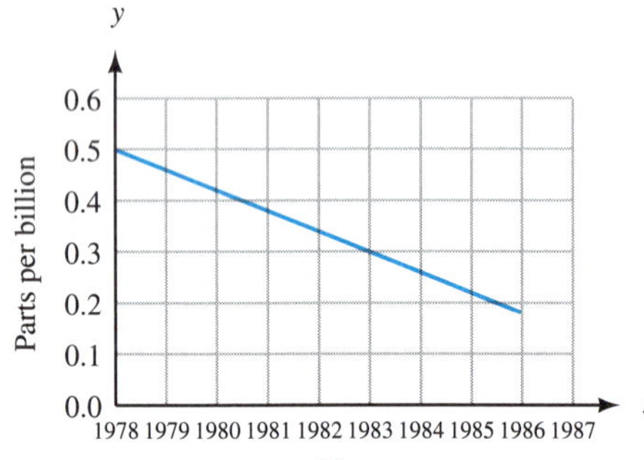

(b) $y = -0.04x + 79.62$ **(c)** 0.42 ppb **(d)** 0.02 ppb **67. (a)** women: 12.8%, men: 18.76% **(b)** No **69.** 3 **71.** $m = \frac{2}{3}, b = -2$ **73.** $y = \frac{2}{3}x + 6$ **75.** $y = \frac{2}{3}x - 2$ **77.** $a = -\frac{8}{3}, b = 16$ **79.** 24 square units **81.** $(\frac{1}{2}, \frac{1}{2})$

Problem Set 2.9

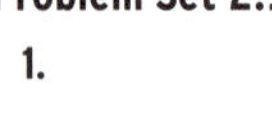

1.

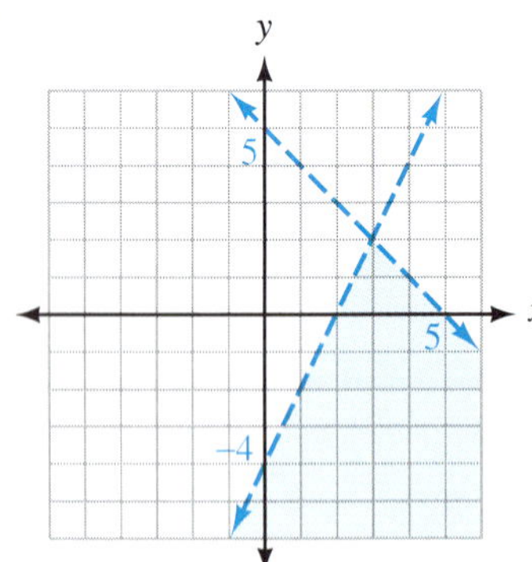

3.

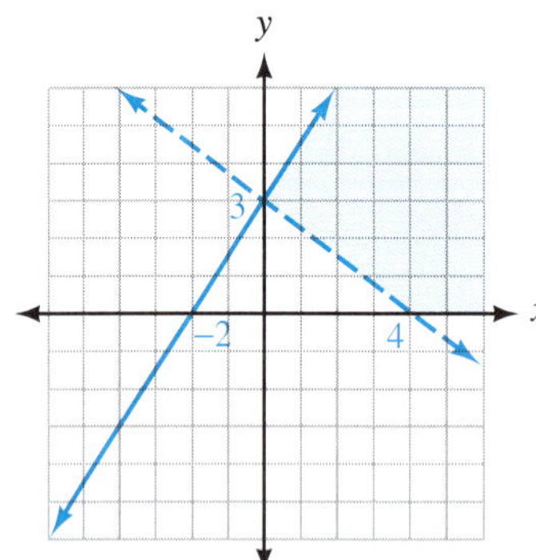

5.

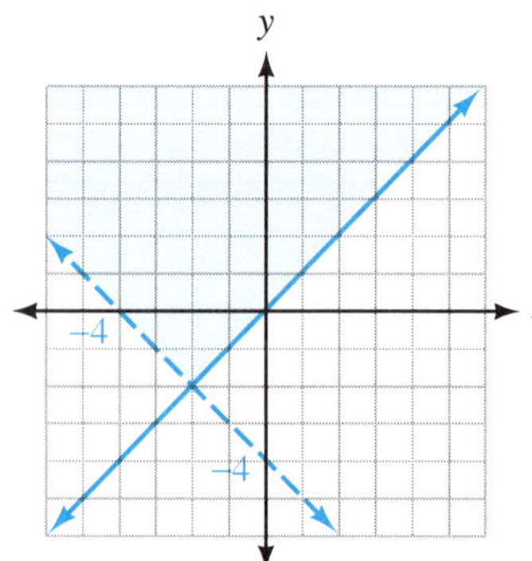

7.

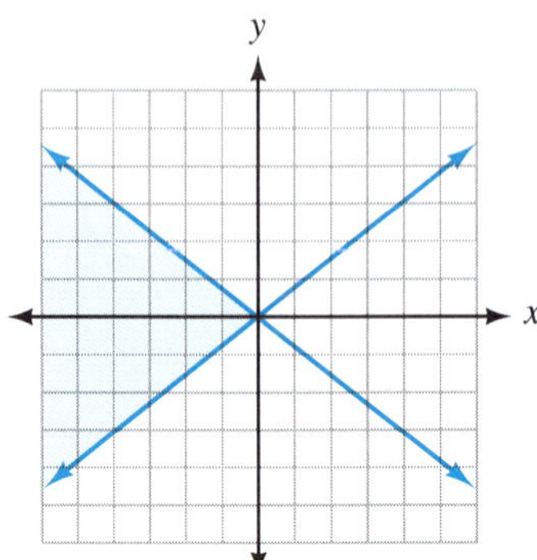

9.

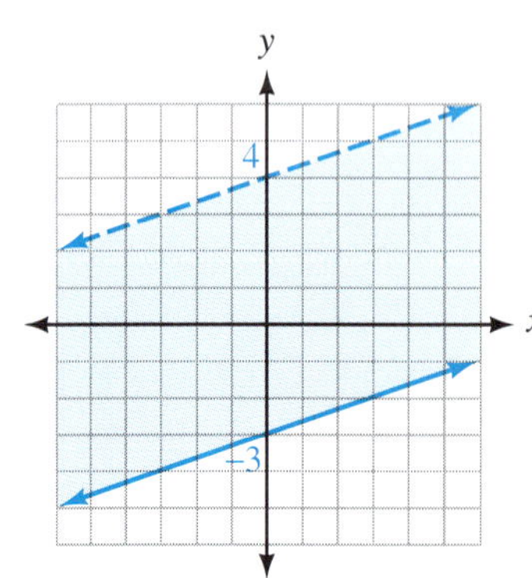

11.

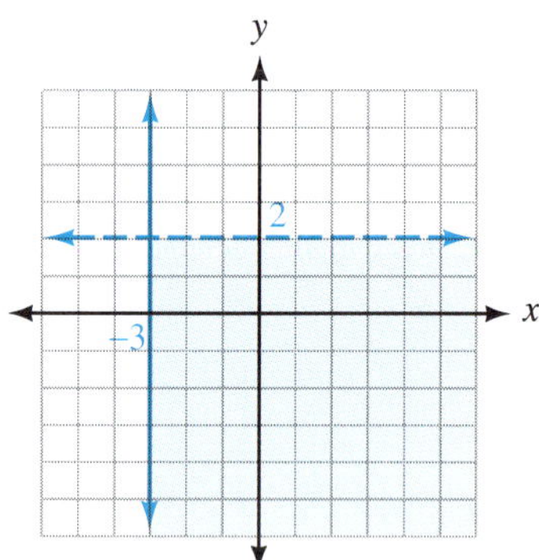

13.

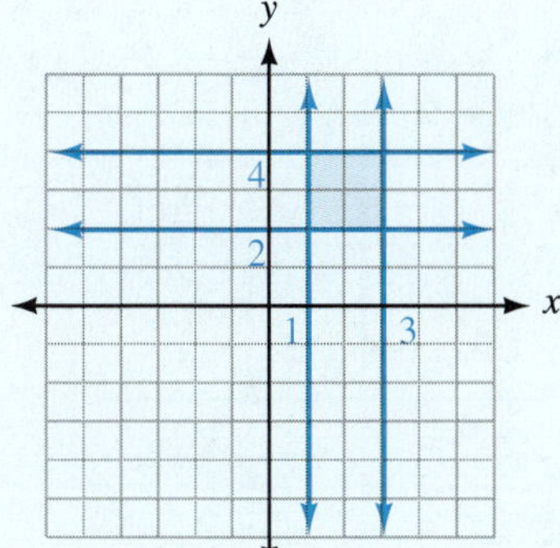

15.

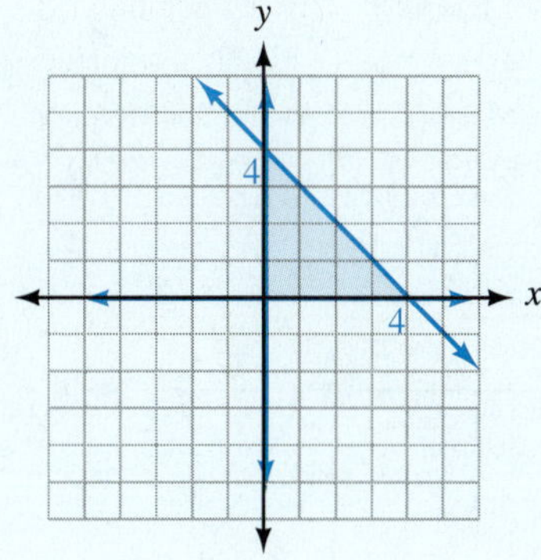

17.

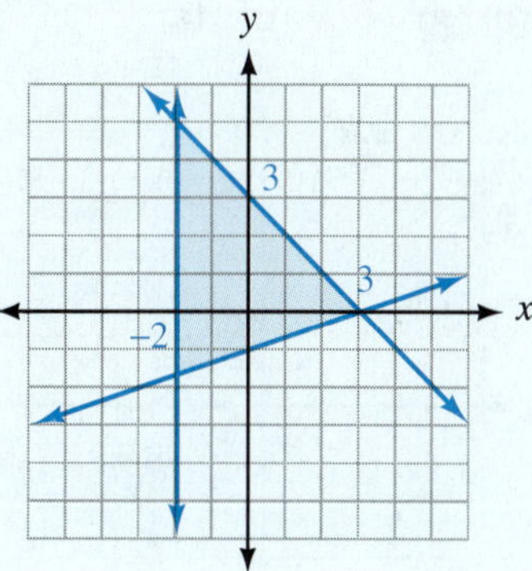

19.

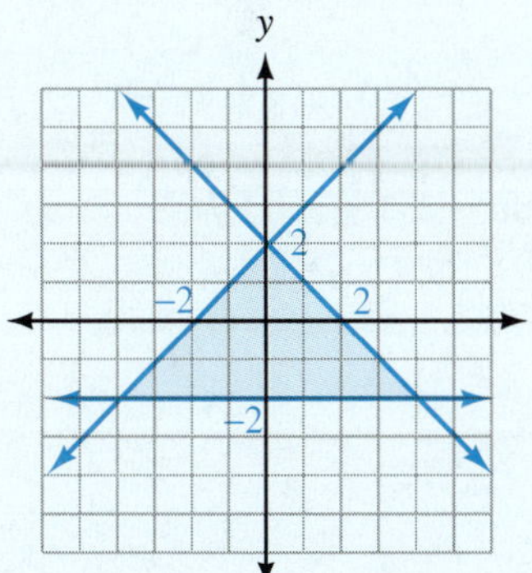

21.

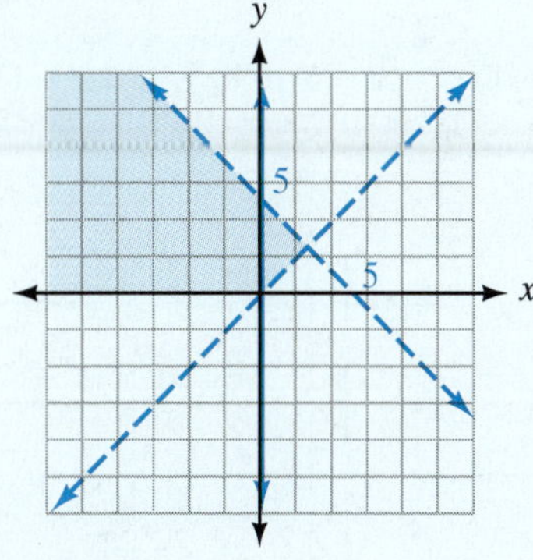

23.

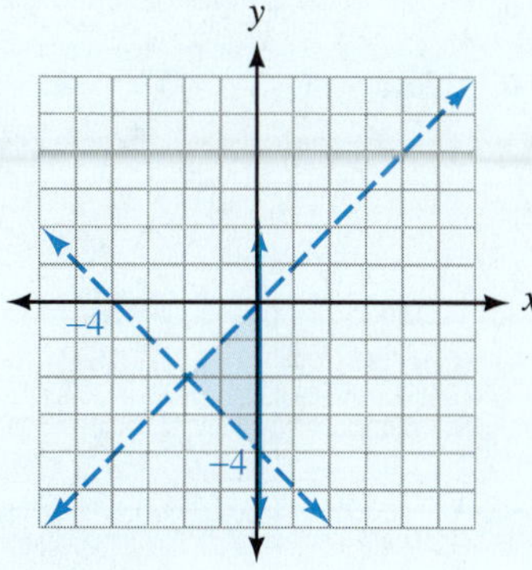

25.

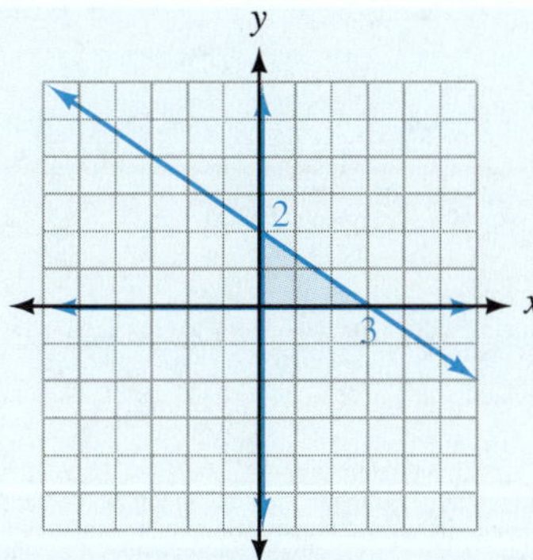

27. $x + y \leq 4$
$-x + y < 4$

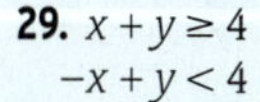
29. $x + y \geq 4$
$-x + y < 4$

31. (a) $0.55x + 0.65y \leq 40$
$x \geq 2y$
$x > 15$
$y \geq 0$

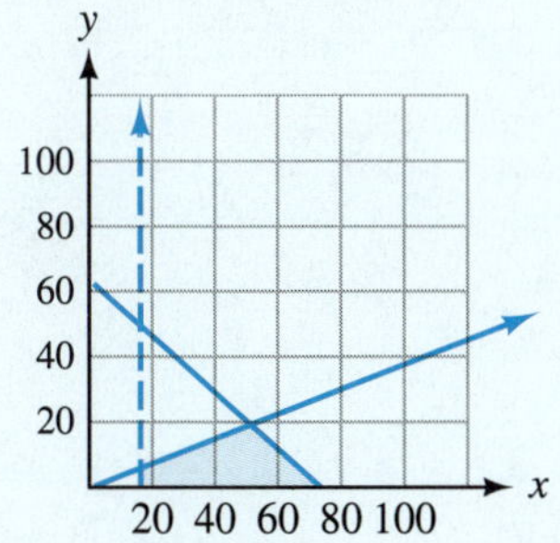

(b) 10 65-cent stamps

33. $x + y \leq 30{,}000$
$y \geq 3x$
$x \geq 4{,}000$
$y \geq 4{,}000$

35. (a) $10x + 4y \leq 80$
$7x + 5y \leq 74$
$x \geq 0$
$y \geq 0$

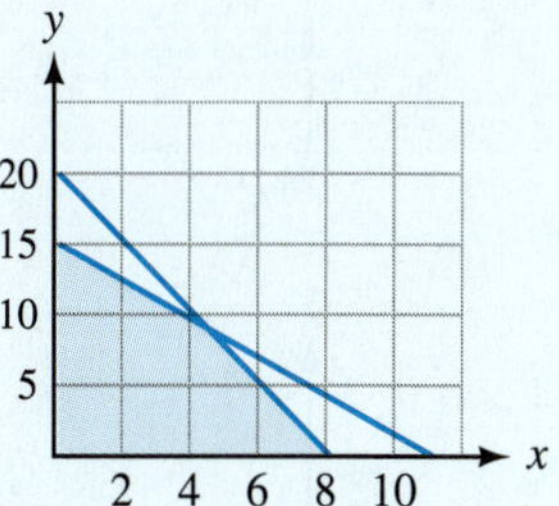

(b) yes **(c)** no

37. $(x + 4)(x - 4)$ **39.** Prime **41.** Prime **43.** $(x - 8)^2$

Chapter 2 Review

1.

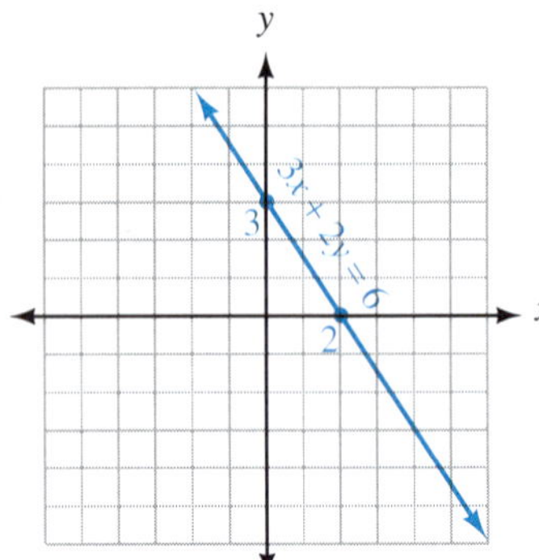

2.

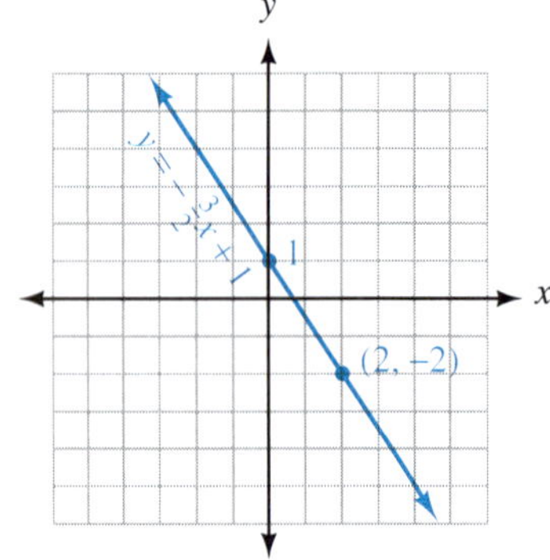

3.

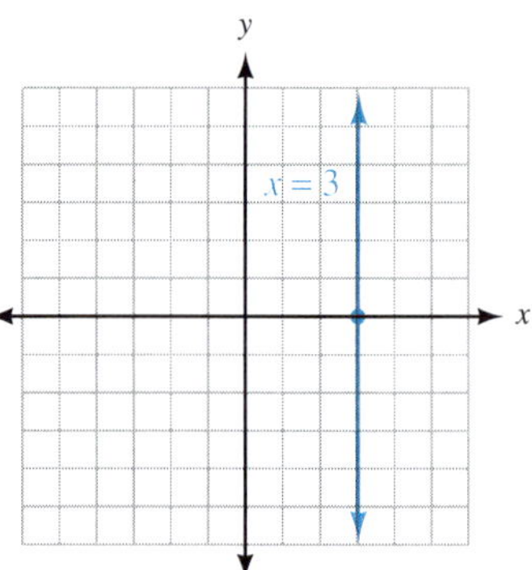

4. −2 **5.** 0 **6.** 3 **7.** 5 **8.** −5 **9.** 6 **10.** $y = 3x + 5$ **11.** $y = -2x$ **12.** $m = 3, b = -6$ **13.** $m = \frac{2}{3}, b = -3$
14. $y = 2x$ **15.** $y = -\frac{1}{3}x$ **16.** $y = 2x + 1$ **17.** $y = 7$ **18.** $y = -\frac{3}{2}x - \frac{17}{2}$ **19.** $y = 2x - 7$ **20.** $y = \frac{1}{3}x - \frac{2}{3}$

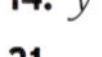

21.

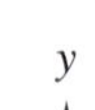

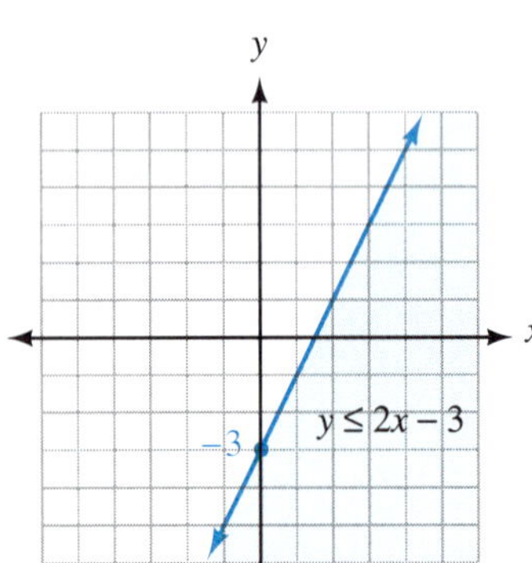

22.

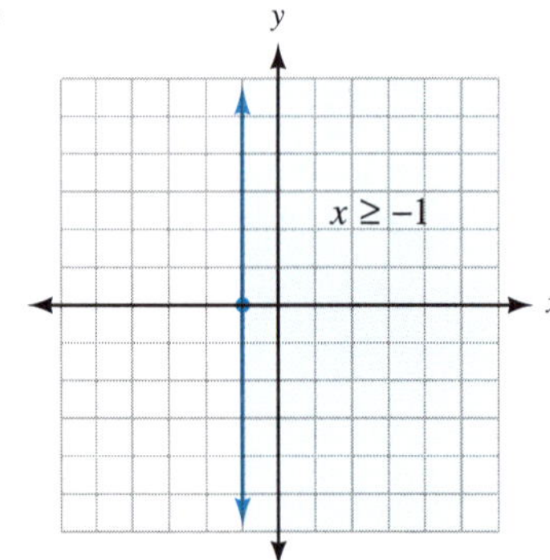

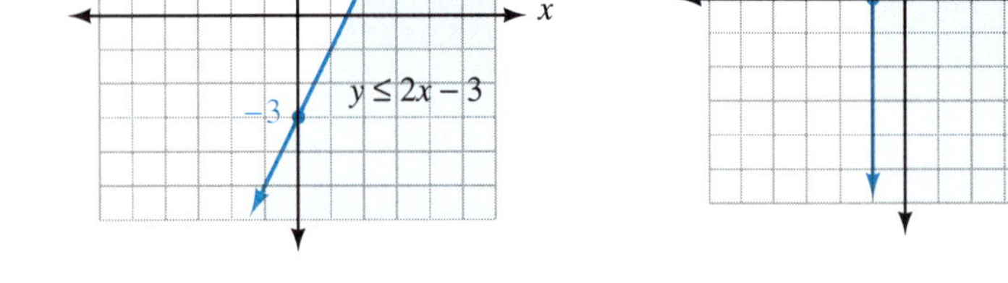

23. Domain = {2, 3, 4}; Range = {4, 3, 2}; a function

24. Domain = {6, −4, −2}; Range = {3, 0}; a function **25.** 0 **26.** 1 **27.** 1 **28.** $3a + 2$ **29.** 1 **30.** 31 **31.** 24 **32.** 6
33. 4 **34.** 25 **35.** 84 pounds **36.** 16 footcandles
37. (6, −2) **38.** no solution (parallel lines) **39.** (0, 1) **40.** no unique solution (lines coincide) **41.** (3, 12) **42.** (3, −2)
43. (4, 1) **44.** (6, −2) **45.** (−5, 3) **46.** no solution (parallel lines)

47.

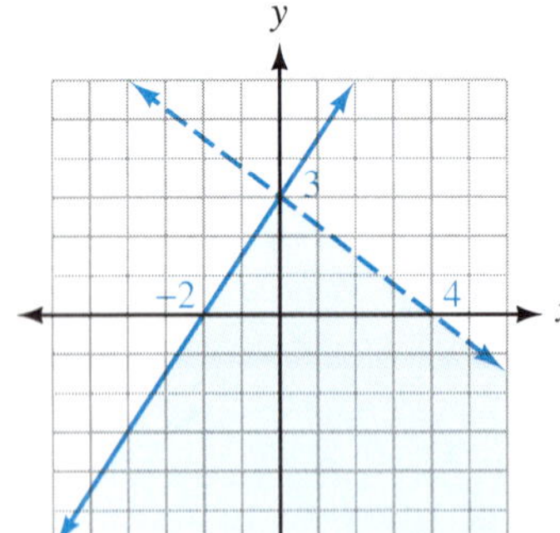

48.

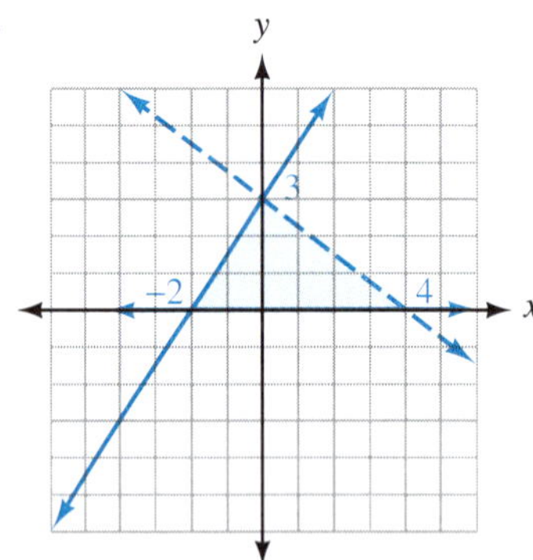

49.

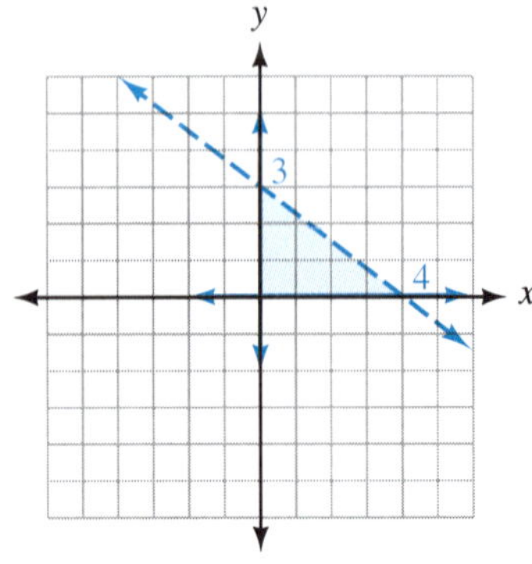

50.

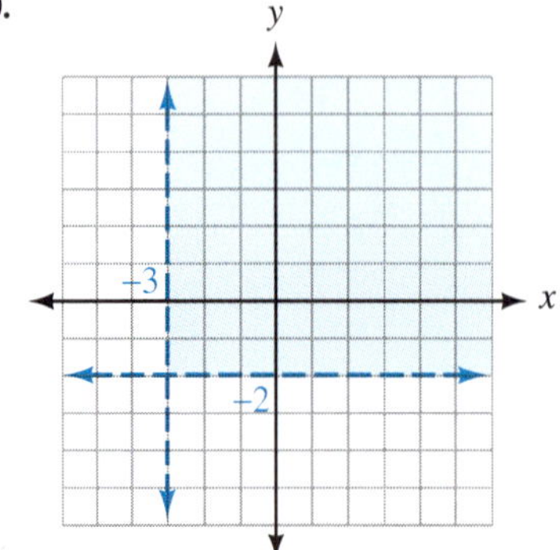

Chapter 3

Problem Set 3.1

1. $g(0) = -3, g(-3) = 0, g(3) = 3, g(-1) = -1, g(1)$ is undefined **3.** $h(0) = -3, h(-3) = 3, h(3) = 0, h(-1)$ is undefined, $h(1) = -1$
5. $\{x \mid x \neq 1\}$ **7.** $\{x \mid x \neq 2\}$ **9.** $\{t \mid t \neq -4, t \neq 4\}$ **11.** $\frac{x-4}{6}$ **13.** $(a^2+9)(a+3)$ **15.** $\frac{2y+3}{y+1}$ **17.** $\frac{x-2}{x-1}$ **19.** $\frac{x-3}{x+2}$ **21.** $\frac{x^2-x+1}{x-1}$
23. $-\frac{4a}{3}$ **25.** $\frac{b-1}{b+1}$ **27.** $\frac{7x-3}{7x+5}$ **29.** $\frac{4x+3}{4x-3}$ **31.** $\frac{x+5}{2x-7}$ **33.** $\frac{a^2-ab+b^2}{a-b}$ **35.** $\frac{2(x-1)}{x}$ **37.** $\frac{x+3}{y-4}$ **39.** $x+2$ **41.** $\frac{1}{x-3}$
43. $\frac{2x^2-5}{3x-2}$ **45.** -1 **47.** $-(y+6)$ **49.** $-\frac{3a+1}{3a-1}$ **51.** $\frac{1-y}{3+y}$ **53.** -1 **55.** $\frac{5-z}{4+z}$ **57.** 0 **59.** $4x$ **61.** $f(0) = 2, g(0) = 2$
63. $f(2)$ is undefined, $g(2) = 4$ **65.** $f(0) = 1, g(0) = 1$ **67.** $f(2) = 3, g(2) = 3$
69. The graph of $y = x + 2$ contains the point (2, 4) while the other graph does not.

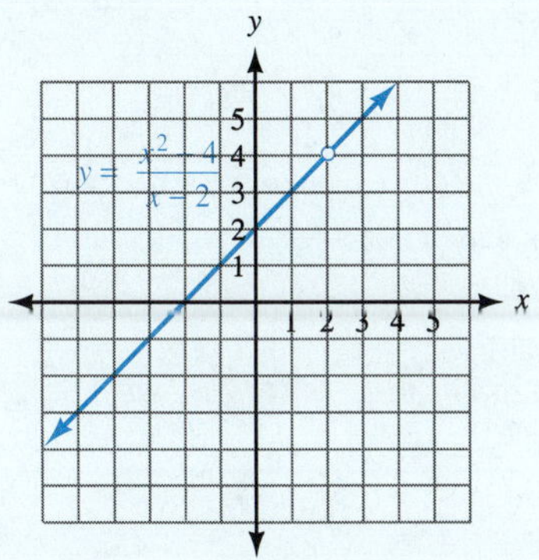

71.

WEEKS x	WEIGHT (lb) $W(x)$
0	200
1	194
4	184
12	173
24	168

73. 0.1 mile per minute **75.** 10.8 miles per gallon **77.** 6.8 feet per second
79. 2 inches: 3,768 inches/minute; 1.5 inches: 2,826 inches/minute
81. **(a)** $\{t \mid 20 \leq t \leq 50\}$
(b)

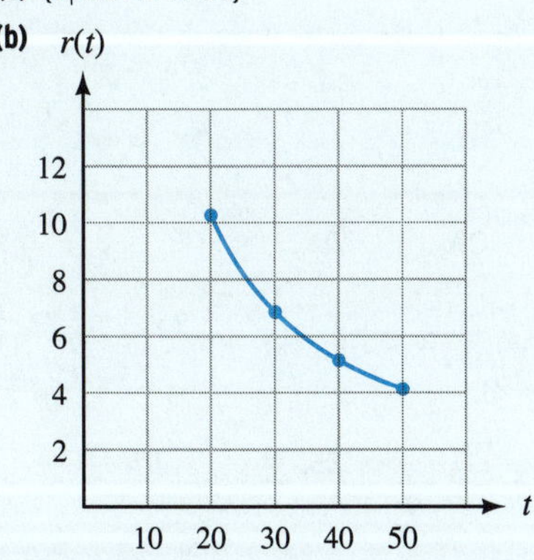

83. **(a)** $\{d \mid 1 \leq d \leq 6\}$
(b)

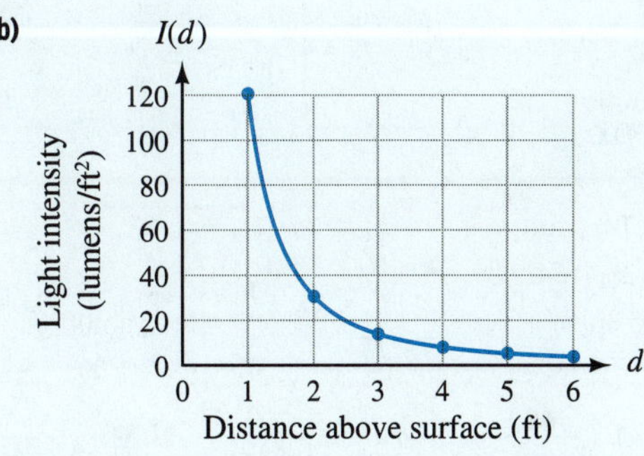

85. **(a)** $\frac{2}{5}$ **(b)** $\frac{4}{5}$ **87.** 3,080 vehicles per hour **89.** (1, 2) **91.** $(\frac{5}{2}, -1)$ **93.** (0, 3) **95.** (3, 4) **97.** $x^2 + 2xy + y^2 + 2xz + 2yz + 4z^2$
99. $a^2 - ab + ac + b^2 - 2bc + c^2$ **101.** $x - y - 1$ **103.** $\frac{(x+1)^2}{(x-1)^3}$ **105.** **(a)** 2 **(b)** −4 **(c)** Undefined **(d)** 2 **(e)** 1 **(f)** −6 **(g)** 4 **(h)** −3

Problem Set 3.2

1. $2x^2 - 4x + 3$ **3.** $-2x^2 - 3x + 4$ **5.** $2y^2 + \frac{5}{2} - \frac{3}{2y^2}$ **7.** $-\frac{5}{2}x + 4 + \frac{3}{x}$ **9.** $4ab^3 + 6a^2b$ **11.** $-xy + 2y^2 - 3xy^2$ **13.** $x + 2$
15. $a - 3$ **17.** $5x + 6y$ **19.** $x^2 + xy + y^2$ **21.** $(y^2+4)(y+2)$ **23.** $(x+2)(x+5)$ **25.** $(2x+3)(2x-3)$ **27.** $x - 7 + \frac{7}{x+2}$
29. $2x + 5 + \frac{2}{3x-4}$ **31.** $2x^2 - 5x + 1 + \frac{4}{x+1}$ **33.** $y^2 - 3y - 13$ **35.** $x - 3$ **37.** $3y^2 + 6y + 8 + \frac{37}{2y-4}$ **39.** $a^3 + 2a^2 + 4a + 6 + \frac{17}{a-2}$
41. $y^3 + 2y^2 + 4y + 8$ **43.** $x^2 - 2x + 1$ **45.** **(a)** 4 **(b)** 4 **47.** **(a)** 5 **(b)** 5 **49.** **(a)** $2x + h$ **(b)** $x + a$ **51.** **(a)** $2x + h$ **(b)** $x + a$
53. **(a)** $2x + h - 3$ **(b)** $x + a - 3$ **55.** **(a)** $4x + 2h + 3$ **(b)** $2x + 2a + 3$ **57.** $(x+3)(x+2)(x+1)$ **59.** $(x+3)(x+4)(x-2)$ **61.** yes
63. 7; it is the same.
67. **(a)**

x	1	5	10	15	20
$C(x)$	2.15	2.75	3.50	4.25	5.00

(b) $\overline{C}(x) = \frac{2}{x} + 0.15$
(c)

x	1	5	10	15	20
$\overline{C}(x)$	2.15	0.55	0.35	0.28	0.25

(d) It decreases.

(e)

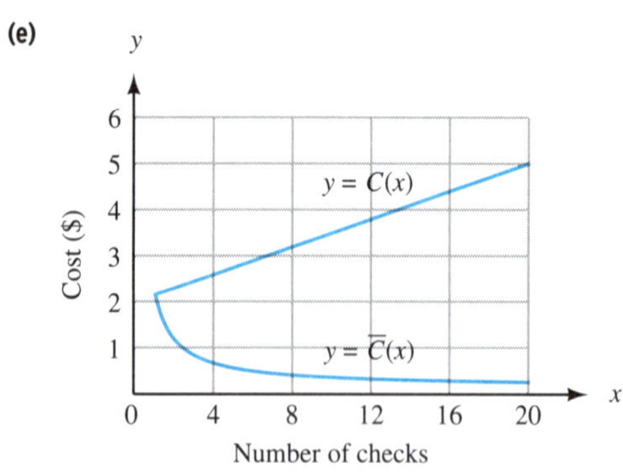

(f) $y = C(x)$: domain = $\{x \mid 1 \le x \le 20\}$; range = $\{y \mid 2.15 \le y \le 5.00\}$
$y = \overline{C}(x)$: domain = $\{x \mid 1 \le x \le 20\}$; range = $\{y \mid 0.25 \le y \le 2.15\}$

69. **(a)** $T(100) = 11.95$, $T(400) = 32.95$, $T(500) = 39.95$ **(b)** $\overline{T}(m) = \frac{4.95}{m} + 0.07$ **(c)** $\overline{T}(100) = 0.1195$, $\overline{T}(400) = 0.0824$, $\overline{T}(500) = 0.0799$
71. $\frac{21}{10}$ **73.** $\frac{11}{8}$ **75.** $\frac{1}{18}$ **77.** 32 **79.** $4x^3 - x^2 + 3$ **81.** $0.5x^2 - 0.4x + 0.3$ **83.** $\frac{3}{2}x - \frac{5}{2}$ **85.** $\frac{2}{3}x + \frac{1}{3} + \frac{2}{3x - 1}$

Problem Set 3.3

1. $\frac{1}{6}$ **3.** $\frac{9}{4}$ **5.** $\frac{1}{2}$ **7.** $\frac{15y}{x^2}$ **9.** $\frac{b}{a}$ **11.** $\frac{2y^5}{z^3}$ **13.** $\frac{x+3}{x+2}$ **15.** $y + 1$ **17.** $\frac{3(x+4)}{x-2}$ **19.** $\frac{y^2}{xy+1}$ **21.** $\frac{x^2+9}{x^2-9}$ **23.** $\frac{1}{4}$ **25.** 1
27. $\frac{(a-2)(a+2)}{a-5}$ **29.** $\frac{9t^2-6t+4}{4t^2-2t+1}$ **31.** $\frac{x+3}{x+4}$ **33.** $\frac{a-b}{5}$ **35.** $\frac{5c-1}{3c-2}$ **37.** $\frac{5a-b}{9a^2+15ab+25b^2}$ **39.** 2 **41.** $x(x^2-1)$
43. $\frac{(a+4b)(a-3b)}{(a-4b)(a+5b)}$ **45.** $\frac{2y-1}{2y-3}$ **47.** $\frac{(y-2)(y+1)}{(y+2)(y-1)}$ **49.** $\frac{x-1}{x+1}$ **51.** $\frac{x-2}{x+3}$ **53.** $\frac{w(y-1)}{w-x}$ **55.** $\frac{(m+2)(x+y)}{(2x+y)^2}$ **57.** $\frac{(1-2d)(d+c)}{d-c}$
59. $\frac{(r+s)(r-s)^2}{(r^2+s^2)^2}$ **61.** $3x$ **63.** $2(x+5)$ **65.** $x - 2$ **67.** $-(y-4)$ **69.** $(a-5)(a+1)$
71.

NUMBER OF COPIES x	PRICE PER COPY ($) $p(x)$
1	20.33
10	9.33
20	6.40
50	4.00
100	3.05

73. $305.00 **75.** **(a)** $\frac{a^2b}{a^3} \cdot \frac{ab^2}{b^3} \div \frac{a^4}{b} = \frac{b}{a^4}$ **(b)** $\frac{a^2b}{a^3} \cdot \frac{ab^2}{b^3} \cdot \frac{a^4}{b} = \frac{a^4}{b}$ **(c)** $\frac{a^2b}{a^3} \div \frac{ab^2}{b^3} \cdot \frac{a^4}{b} = a^2b$
77. 3 **79.** 2 **81.** $3a^2 - 4a + 2$ **83.** $\frac{x^4 - x^2y^2 + y^4}{x^2+y^2}$ **85.** $\frac{(a+5)(a-1)}{3a^2-2a+1}$
87. $\frac{a(c-1)}{a-b}$

Problem Set 3.4

1. $\frac{5}{4}$ **3.** $\frac{1}{3}$ **5.** $\frac{41}{24}$ **7.** $\frac{19}{144}$ **9.** $\frac{31}{24}$ **11.** 1 **13.** -1 **15.** $\frac{1}{x+y}$ **17.** 1 **19.** $\frac{a^2+2a-3}{a^3}$ **21.** 1 **23.** $\frac{4-3t}{2t^2}$ **25.** $\frac{1}{2}$ **27.** $\frac{1}{5}$
29. $\frac{x+3}{2(x+1)}$ **31.** $\frac{a-b}{a^2+ab+b^2}$ **33.** $\frac{2y-3}{4y^2+6y+9}$ **35.** $\frac{2(2x-3)}{(x-3)(x-2)}$ **37.** $\frac{1}{2t-7}$ **39.** $\frac{4}{(a-3)(a+1)}$ **41.** $\frac{-4x^2}{(2x+1)(2x-1)(4x^2+2x+1)}$
43. $\frac{2}{(2x+3)(4x+3)}$ **45.** $\frac{a}{(a+4)(a+5)}$ **47.** $\frac{x+1}{(x-2)(x+3)}$ **49.** $\frac{x-1}{(x+1)(x+2)}$ **51.** $\frac{1}{(x+2)(x+1)}$ **53.** $\frac{1}{(x+2)(x+3)}$ **55.** $\frac{4x+5}{2x+1}$ **57.** $\frac{22-5t}{4-t}$
59. $\frac{2x^2+3x-4}{2x+3}$ **61.** $\frac{2x-3}{2x}$ **63.** $\frac{1}{2}$ **65.** 5.1 **67.** $(3+4)^{-1} = \frac{1}{7}$, $3^{-1} + 4^{-1} = \frac{7}{12}$
69. **(a)** 120 months **(b)** The two objects will never meet **71.** $x + \frac{4}{x} = \frac{x^2+4}{x}$ **73.** $\frac{1}{x} + \frac{1}{x+1} = \frac{2x+1}{x(x+1)}$
75. Domain {1, 3, 4}; range {2, 4}; function **77.** Domain {1, 2, 3}; range {1, 2, 3}; function **79.** $\frac{x-1}{x+3}$ **81.** $\frac{a^2+ab+a-v}{u+v}$
83. $\frac{6x+5}{(4x-1)(3x-4)}$ **85.** $\frac{y(y^2+1)}{(y+1)^2(y-1)^2}$

Problem Set 3.5

1. $\frac{9}{8}$ **3.** $\frac{2}{15}$ **5.** $\frac{119}{20}$ **7.** $\frac{1}{x+1}$ **9.** $\frac{a+1}{a-1}$ **11.** $\frac{y-x}{y+x}$ **13.** $\frac{1}{(x+5)(x-2)}$ **15.** $\frac{1}{a^2-a+1}$ **17.** $\frac{x+3}{x+2}$ **19.** $\frac{a+3}{a-2}$ **21.** $\frac{9x^2+6x+4}{x(x+1)}$
23. $\frac{x}{x-2}$ **25.** $\frac{x-3}{x}$ **27.** $\frac{x+4}{x+2}$ **29.** $\frac{a-1}{a+1}$ **31.** $-\frac{x}{3}$ **33.** $\frac{y^2+1}{2y}$ **35.** $\frac{-x^2+x-1}{x-1}$ **37.** $\frac{2x-1}{2x+3}$ **39.** $-\frac{1}{x(x+h)}$ **41.** $\frac{3c+4a-2b}{5}$
43. $\frac{(t-4)(t+1)}{(t+6)(t-3)}$ **45.** $\frac{(5b-1)(b+5)}{2(2b-11)}$ **47.** $-\frac{3}{2x+14}$ **49.** $2m - 9$ **51.** **(a)** $-\frac{4}{ax}$ **(b)** $-\frac{1}{(x+1)(a+1)}$ **(c)** $-\frac{a+x}{a^2x^2}$
55. **(a)** As v approaches 0, the denominator approaches 1. **(b)** $v = \frac{fs}{h} - s$ **57.** **(a)** $\frac{ab}{a+b}$ **(b)** $\frac{12}{7}$ hours **59.** 1 **61.** $-3, 4$ **63.** 2, 3
65. $\frac{1}{3}$ **67.** $\frac{40}{243}$ **69.** $\frac{a-2b}{a+2b}$ **71.** $a + b$ **73.** $-\frac{qt}{q+t}$

Problem Set 3.6

1. $-\frac{35}{3}$ **3.** $-\frac{18}{5}$ **5.** $\frac{36}{11}$ **7.** 2 **9.** 5 **11.** 2 **13.** $-3, 4$ **15.** $-\frac{4}{3}, 1$ **17.** No solution (-1 does not check) **19.** 5 **21.** $-\frac{1}{2}, \frac{5}{3}$
23. $\frac{2}{3}$ **25.** 18 **27.** No solution (4 does not check) **29.** -4 (3 does not check) **31.** -6 **33.** -5 **35.** $\frac{53}{17}$
37. 2 (1 does not check) **39.** No solution (3 does not check) **41.** $\frac{22}{3}$ **43.** 2 **45.** 1, 5 **47.** $x = \frac{ab}{a-b}$ **49.** $R = \frac{R_1R_2}{R_1+R_2}$
51. $y = \frac{x-3}{x-1}$ **53.** $y = \frac{1-x}{3x-2}$

55.

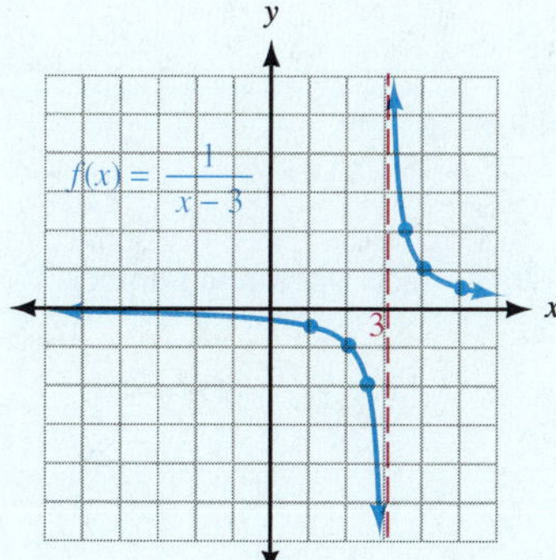

57.

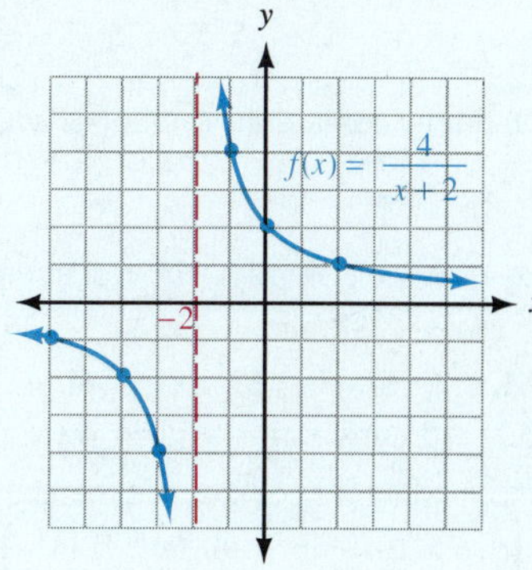

59.

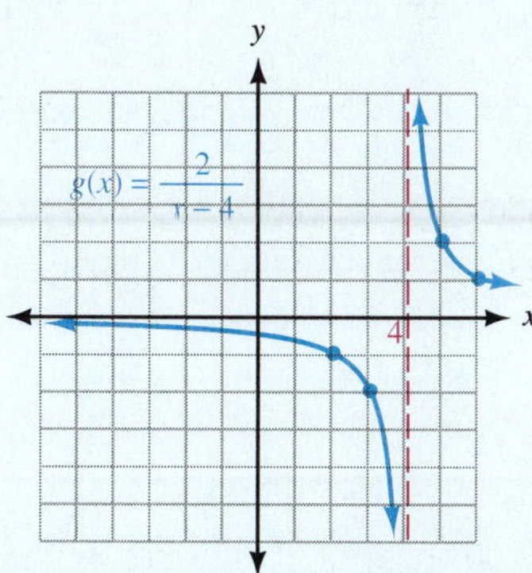

61.

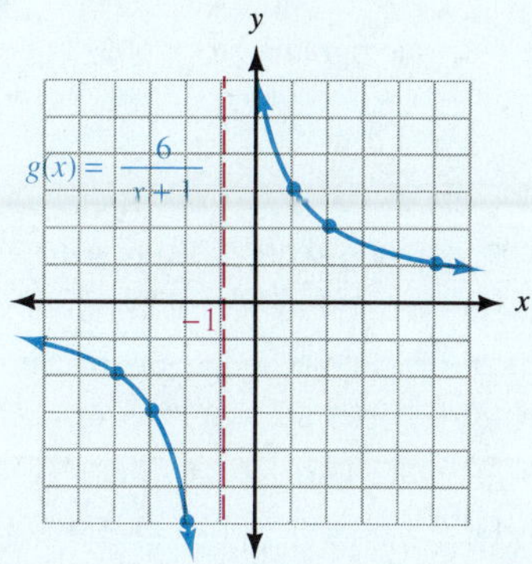

63. $f(0) = -\frac{1}{3}, f(6) = \frac{1}{3}$ **65.** $f(1) = -\frac{1}{2}, f(5) = \frac{1}{2}$ **67.** $\{x \mid x \neq 3\}$ **69.** $f(0) = 2, f(-4) = -2$ **71.** $f(2) = 1, f(-6) = -1$
73. $\{x \mid x \neq -2\}$ **75.** $\frac{24}{5}$ feet

79. (a)

TIME t (sec)	SPEED OF KAYAK RELATIVE TO THE WATER v (m/sec)	CURRENT OF THE RIVER c (m/sec)
240	4	1
300	4	2
514	4	3
338	3	1
540	3	2
impossible	3	3

(b) Equal to the time in the river because they are racing both upstream and downstream.
(c) He will not go anywhere. The denominator of $\frac{450}{v-c}$ will be 0. **81.** 60 geese, 48 ducks **83.** 150 oranges, 144 apples
85. $-\frac{5}{2}, -\frac{2}{3}, \frac{5}{2}$ **87.** $-2, 2, 3$ **89.** $x = -\frac{a}{5}$ **91.** $v = \frac{16t^2 + s}{t}$ **93.** $f = \frac{pg}{g - p}$

Problem Set 3.7

1. $\frac{1}{5}, \frac{3}{5}$ **3.** 3 or $\frac{1}{3}$ **5.** 3, 4 **7.** 3 **9.** $\frac{5}{3}$ mph **11.** 6 mph **13.** A: 75 mph; B: 60 mph **15.** 540 mph **17.** 54 mph
19. 16 hours **21.** 15 hours **23.** $5\frac{1}{4}$ minutes **25.** 135 ft² **27.** 48 fl oz **29.** 8 qt **31.** 20 pt **33.** 480 fl oz **35.** 8 gal
37. 6 qt **39.** 9 yd³ **41.** 16 cups **43.** 72 qt **45.** 51.1 acres **47.** 5.9 mph **49.** 20.7 mph **51.** 4.6 mph **53.** 2.7 mph
55. 3,241,440 minutes is a little more than 6 years **57.** $\frac{1}{3}[(x + \frac{2}{3}x) + \frac{1}{3}(x + \frac{2}{3}x)] = 10; \frac{27}{2}$ **59. (a)** 30 grams **(b)** 3.25 moles **61.** $\frac{2}{3a}$
63. $(x - 3)(x + 2)$ **65.** 1 **67.** $\frac{3 - x}{3 + x}$ **69.** No solution (3 does not check)

Chapter 3 Review

1. $\frac{25x^2}{7y^3}$ **2.** $\frac{a(a-b)}{4}$ **3.** $\frac{x-5}{x+5}$ **4.** $\frac{a+1}{a-1}$ **5.** $3x + 2 + \frac{4}{x}$ **6.** $-9b + 5a - 7a^2b^2$ **7.** $x^{3n} - x^{2n}$ **8.** $x + 2$ **9.** $5x + 6y$
10. $(y + 2)(y^2 + 4)$ **11.** $4x + 1 - \frac{2}{2x - 7}$ **12.** $y^2 - 3y - 13$ **13.** $\frac{9}{5}$ **14.** $\frac{3x}{4y^2}$ **15.** $\frac{x-1}{x^2+1}$ **16.** 1 **17.** $\frac{x+2}{x-2}$ **18.** $(2x - 3)(x + 3)$
19. $\frac{31}{30}$ **20.** -1 **21.** $\frac{x^2 + x + 1}{x^3}$ **22.** $\frac{1}{(y+4)(y+3)}$ **23.** $\frac{x-1}{2(x+1)(x+2)}$ **24.** $\frac{15x-2}{5x-2}$ **25.** 5 **26.** $\frac{1}{a^2 - a + 1}$ **27.** $\frac{x^2+x+1}{x^2+1}$ **28.** $\frac{x+3}{x+2}$
29. 6 **30.** 1 **31.** -6 **32.** Possible solution -3, which does not check; ∅ **33.** $\frac{22}{3}$ **34.** Possible solutions 4 and -5; only 4 checks
35. Car: 30 miles per hour; truck 20 miles per hour **36.** 7.5 miles per hour **37.** 742 miles per hour **38.** 15 hours **39.** 2.7 miles
40. 1,012 miles per hour

Chapter 4

Note: Some of the answers that follow contain the symbol ≈, which is read "is approximately equal to."

Problem Set 4.1

1. 6 **3.** −3 **5.** 2 **7.** −2 **9.** 2 **11.** 12 **13.** Not a real number **15.** −7 **17.** −3 **19.** 2 **21.** Not a real number
23. 0.2 **25.** 0.2 **27.** $\frac{9}{5}$ **29.** $\frac{4}{5}$ **31.** $6a^4$ **33.** $3a^4$ **35.** xy^2 **37.** $2x^2y$ **39.** $2a^3b^5$ **41.** 9 **43.** 125 **45.** 8 **47.** $\frac{1}{3}$
49. $\frac{1}{27}$ **51.** $\frac{6}{5}$ **53.** $\frac{8}{27}$ **55.** 7 **57.** $\frac{3}{4}$ **59.** $x^{4/5}$ **61.** a **63.** $\frac{1}{x^{2/5}}$ **65.** $x^{1/6}$ **67.** $x^{9/25}y^{1/2}z^{1/5}$ **69.** $\frac{b^{7/4}}{a^{1/8}}$ **71.** $y^{3/10}$
73. $\frac{1}{a^2b^4}$ **75.** $\frac{s^{1/2}}{r^{20}}$ **77.** $10b^3$ **79.** $(9^{1/2}+4^{1/2})^2 = 25;\ 9+4=13$ **83.** 25 mph **85.** 1.618
87. $\frac{13}{8}$; The numerator and denominator are consecutive members of the Fibonacci sequence.
89. **(a)** 424 pm **(b)** 600 pm **(c)** 6×10^{-10} m **91.** **(a)** $\sqrt{2}$ in. **(b)** $\sqrt{3}$ in. **93.** **(a)** B **(b)** A **(c)** C **(d)** (0, 0) and (1, 1)
95. $x^6 - x^3$ **97.** $x^2 + 2x - 15$ **99.** $x^4 - 10x^2 + 25$ **101.** $x^3 - 27$ **103.** 1.7 **105.** 5.6 **107.** (0, 0) and (1, 1)
109. **(a)** 1.62 micrograms **(b)** 0.87 micrograms **(c)** 0.00293 micrograms **(d)** 2.86×10^{-6} micrograms

Problem Set 4.2

1. $x + x^2$ **3.** $a^2 - a$ **5.** $6x^3 - 8x^2 + 10x$ **7.** $12x^2 - 36y^2$ **9.** $x^{4/3} - 2x^{2/3} - 8$ **11.** $a - 10a^{1/2} + 21$ **13.** $20y^{2/3} - 7y^{1/3} - 6$
15. $10x^{4/3} + 21x^{2/3}y^{1/2} + 9y$ **17.** $t + 10t^{1/2} + 25$ **19.** $x^3 + 8x^{3/2} + 16$ **21.** $a - 2a^{1/2}b^{1/2} + b$ **23.** $4x - 12x^{1/2}y^{1/2} + 9y$ **25.** $a - 3$
27. $x^3 - y^3$ **29.** $t - 8$ **31.** $4x^3 - 3$ **33.** $x + y$ **35.** $a - 8$ **37.** $8x + 1$ **39.** $t - 1$ **41.** $2x^{1/2} + 3$ **43.** $3x^{1/3} - 4y^{1/3}$
45. $3a - 2b$ **47.** **(a)** $3a^3(4a - 3)$ **(b)** $3(y+2)^2(4y+5)$ **(c)** $3(x-2)^{1/2}(4x-11)$
49. **(a)** $5y^7(y-3)$ **(b)** $5(n-4)^5(n-7)$ **(c)** $5(x-3)^{7/5}(x-6)$
51. **(a)** $3b^2(3ab+2)$ **(b)** $3(n+4)(3mn+12m+2)$ **(c)** $3(x+1)^{1/2}(3x^2+3x+2)$ **53.** $(x^{1/3}-2)(x^{1/3}-3)$ **55.** $(a^{1/5}-4)(a^{1/5}+2)$
57. $(2y^{1/3}+1)(y^{1/3}-3)$ **59.** $(3t^{1/5}+5)(3t^{1/5}-5)$ **61.** $(2x^{1/7}+5)^2$ **63.** $\frac{3+x}{x^{1/2}}$ **65.** $\frac{x+5}{x^{1/3}}$ **67.** $\frac{x^3+3x^2+1}{(x^3+1)^{1/2}}$ **69.** $\frac{-4}{(x^2+4)^{1/2}}$ **71.** 2
73. 27 **75.** 0.871 **77.** 15.8% **79.** 5.9% **81.** 3.54 seconds **83.** 1.87 years **85.** $\frac{1}{x^2+9}$ **87.** $3x - 4x^3y$ **89.** $5x - 4$
91. $x^2 + 5x + 25$ **93.** $x = 16, y = 9$ **95.** xy **97.** $x^{17/20}$ **99.** $m^{1/12}$

Problem Set 4.3

1. $2\sqrt{2}$ **3.** $7\sqrt{2}$ **5.** $12\sqrt{2}$ **7.** $4\sqrt{5}$ **9.** $4\sqrt{3}$ **11.** $15\sqrt{3}$ **13.** $3\sqrt[3]{2}$ **15.** $4\sqrt[3]{2}$ **17.** $6\sqrt[3]{2}$ **19.** $2\sqrt[5]{2}$ **21.** $3x\sqrt{2x}$
23. $2y\sqrt[4]{2y^3}$ **25.** $2xy^2\sqrt[3]{5xy}$ **27.** $4abc^2\sqrt{3b}$ **29.** $2bc\sqrt[3]{6a^2c}$ **31.** $2xy^2\sqrt[5]{2x^3y^2}$ **33.** $3xy^2z\sqrt[5]{x^2}$ **35.** $2\sqrt{3}$
37. $\sqrt{-20}$; not a real number **39.** $\frac{\sqrt{11}}{2}$ **41.** $\frac{2\sqrt{3}}{3}$ **43.** $\frac{5\sqrt{6}}{6}$ **45.** $\frac{\sqrt{2}}{2}$ **47.** $\frac{\sqrt{5}}{5}$ **49.** $2\sqrt[3]{4}$ **51.** $\frac{2\sqrt[3]{3}}{3}$ **53.** $\frac{\sqrt[4]{24x^2}}{2x}$
55. $\frac{\sqrt[4]{8y^3}}{y}$ **57.** $\frac{\sqrt[3]{36xy^2}}{3y}$ **59.** $\frac{\sqrt[3]{6xy^2}}{3y}$ **61.** $\frac{\sqrt[4]{2x}}{2x}$ **63.** $\frac{3x\sqrt{15xy}}{5y}$ **65.** $\frac{5xy\sqrt{6xz}}{2z}$ **67.** $\frac{2ab\sqrt{6ac^2}}{3c}$ **69.** $\frac{2xy^2\sqrt[3]{3z^2}}{3z}$ **71.** $5|x|$
73. $3|xy|\sqrt{3x}$ **75.** $|x-5|$ **77.** $|2x+3|$ **79.** $2|a(a+2)|$ **81.** $2|x|\sqrt{x-2}$ **83.** $\sqrt{9+16} = 5;\ \sqrt{9}+\sqrt{16} = 7$ **85.** $5\sqrt{13}$ feet
87. **(a)** 13 feet **(b)** $2\sqrt{14} \approx 7.5$ feet **89.** **(a)** $\sqrt{8001} \approx 89.45$ miles **(b)** $\sqrt{16{,}004} \approx 126.51$ miles **(c)** $\sqrt{24{,}009} \approx 154.95$ miles
93. $\sqrt{2}, \sqrt{3}, 2, \sqrt{5}, \sqrt{6}, \sqrt{7}$; 10th term: $\sqrt{11}$; 100th term: $\sqrt{101}$ **95.** $\frac{y^3}{x^2}$ **97.** 1 **99.** $\frac{4x^2-6x+9}{9x^2-3x+1}$ **101.** $12\sqrt[3]{5}$ **103.** $6\sqrt[3]{49}$
105. $\frac{\sqrt[10]{a^7}}{a}$ **107.** $\frac{\sqrt[20]{a^9}}{a}$ **109.**

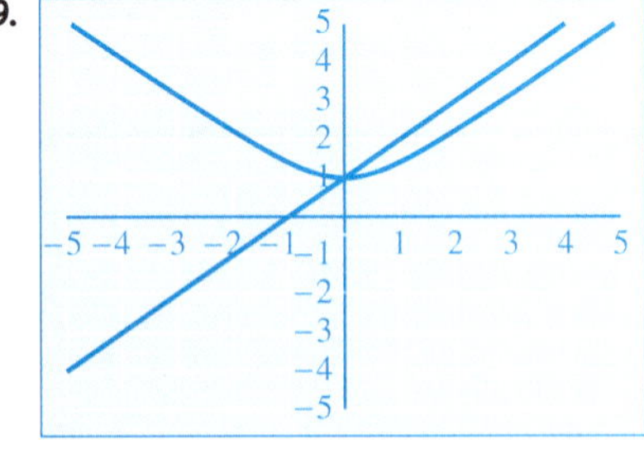

111. 0.76 **113.** 0 **115.** **(a)** $s = \frac{1}{2}(a + b + c)$ **(b)** 14.70

Problem Set 4.4

1. $7\sqrt{5}$ **3.** $-x\sqrt{7}$ **5.** $\sqrt[3]{10}$ **7.** $9\sqrt[5]{6}$ **9.** 0 **11.** $\sqrt{5}$ **13.** $-32\sqrt{2}$ **15.** $-3x\sqrt{2}$ **17.** $-2\sqrt[3]{2}$ **19.** $8x\sqrt[3]{xy^2}$ **21.** $3a^2b\sqrt{3ab}$
23. $11ab\sqrt[3]{3a^2b}$ **25.** $10xy\sqrt[4]{3y}$ **27.** $\sqrt{2}$ **29.** $\frac{8\sqrt{5}}{15}$ **31.** $\frac{(x-1)\sqrt{x}}{x}$ **33.** $\frac{3\sqrt{2}}{2}$ **35.** $\frac{5\sqrt{6}}{6}$ **37.** $\frac{8\sqrt[3]{25}}{5}$
39. $\sqrt{12} \approx 3.464;\ 2\sqrt{3} \approx 3.464$ **41.** $\sqrt{8} + \sqrt{18} \approx 7.071 \approx \sqrt{50};\ \sqrt{26} \approx 5.099$ **43.** $8\sqrt{2x}$ **45.** 5

53. $\sqrt{2}:1$ or $1.414:1$ **55. (a)** $\sqrt{2}:1$ **(b)** $5:\sqrt{2}$ **(c)** $5:4$ **57.** 1 **59.** $\frac{13-3t}{3-t}$ **61.** $\frac{6}{4x+3}$

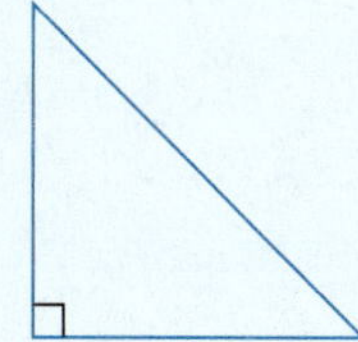

63. $\frac{x-y}{x^2+xy+y^2}$ **65.** $\frac{5ab^2\sqrt{c}}{3}$ **67.** $-\frac{b}{a}$ **69.** $\sqrt{7}$ **71.** $17ab\sqrt{3b}-2a\sqrt{2b}$ **73.** $\frac{3xy\sqrt{6}}{2}$

Problem Set 4.5

1. $3\sqrt{2}$ **3.** $10\sqrt{21}$ **5.** 720 **7.** 54 **9.** $\sqrt{6}-9$ **11.** $24+6\sqrt[3]{4}$ **13.** $7+2\sqrt{6}$ **15.** $x+2\sqrt{x}-15$ **17.** $34+20\sqrt{3}$
19. $19+8\sqrt{3}$ **21.** $x-6\sqrt{x}+9$ **23.** $4a-12\sqrt{ab}+9b$ **25.** $x+4\sqrt{x-4}$ **27.** $x-6\sqrt{x-5}+4$ **29.** 1 **31.** $a-49$
33. $25-x$ **35.** $x-8$ **37.** $10+6\sqrt{3}$ **39.** $\frac{\sqrt{3}+1}{2}$ **41.** $\frac{5-\sqrt{5}}{4}$ **43.** $\frac{x+3\sqrt{x}}{x-9}$ **45.** $\frac{10+3\sqrt{5}}{11}$ **47.** $\frac{3\sqrt{x}+3\sqrt{y}}{x-y}$ **49.** $2+\sqrt{3}$
51. $\frac{11-4\sqrt{7}}{3}$ **53.** $\frac{a+2\sqrt{ab}+b}{a-b}$ **55.** $\frac{x+4\sqrt{x}+4}{x-4}$ **57.** $\frac{5-\sqrt{21}}{4}$ **59.** $\frac{\sqrt{x}-3x+2}{1-x}$ **63.** $10\sqrt{3}$ **65.** $x+6\sqrt{x}+9$ **67.** 75
69. $\frac{5\sqrt{2}}{4}$ seconds; $\frac{5}{2}$ seconds **75.** $-\frac{1}{8}$ **77.** $\frac{y-2}{y+2}$ **79.** $\frac{2x+1}{2x-1}$ **81.** $\frac{\sqrt{xy}}{2}$ **83.** $\frac{2a^2}{b}$ **85.** $\frac{y^2\sqrt[5]{z}}{xz}$ **87.** $\frac{3\sqrt{x}+3\sqrt{xy}}{1-y}$
89. $\frac{\sqrt{b}+1}{b-1}$ **91.** $\frac{x-4-2\sqrt{x-4}}{x-8}$ **93.** $\frac{x+\sqrt{x^2-9}}{3}$ **95.** $\frac{\sqrt[3]{x^2}-2\sqrt[3]{x}+4}{x+8}$ **97.** $\frac{\sqrt[3]{9}-\sqrt[3]{6}+\sqrt[3]{4}}{5}$

Problem Set 4.6

1. 4 **3.** No solution **5.** 5 **7.** No solution **9.** $\frac{39}{2}$ **11.** No solution **13.** 5 **15.** 3 **17.** $-\frac{32}{3}$ **19.** 3, 4 **21.** −2, −1
23. −1 **25.** No solution **27.** 7 **29.** 0, 3 **31.** −4 **33.** 8 **35.** 0 **37.** 9 **39.** 0 **41.** 8
43. No solution (9 does not check) **45.** 0 (32 does not check) **47.** 6 (−2 does not check) **49.** $h=100-16t^2$ **51.** 3.24 feet
53. 5 meters **55.** 2,500 meters **57.** $0 \le y \le 100$

59.

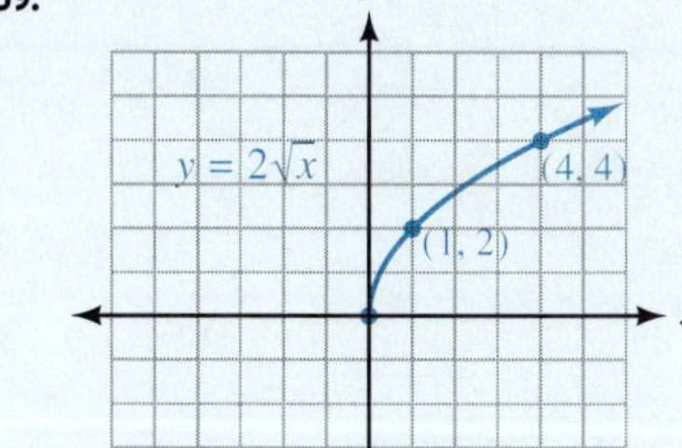

61.

y = √x − 2
4
−2
(1, −1)

63.

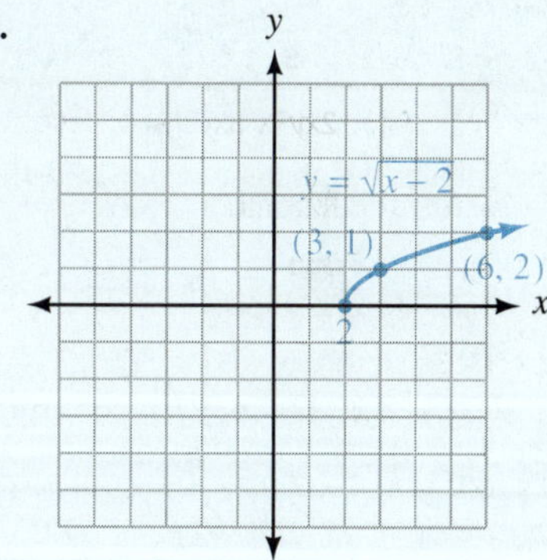

65.

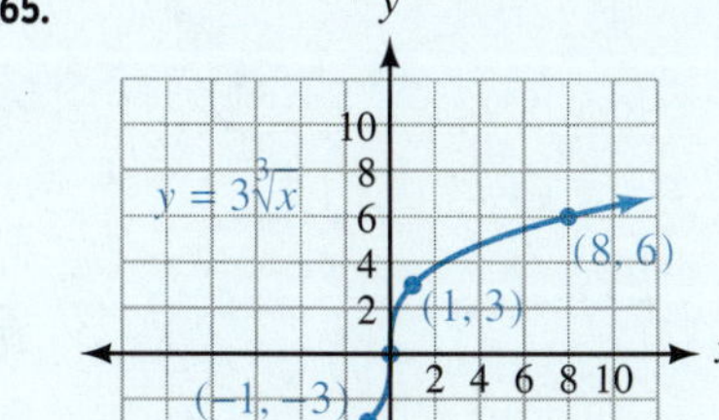

67.

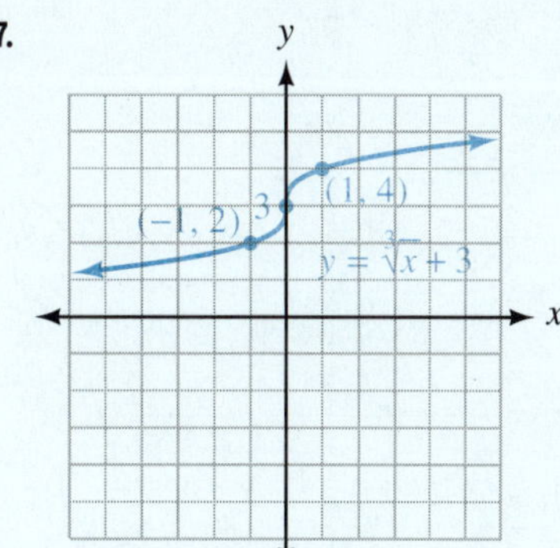

69.

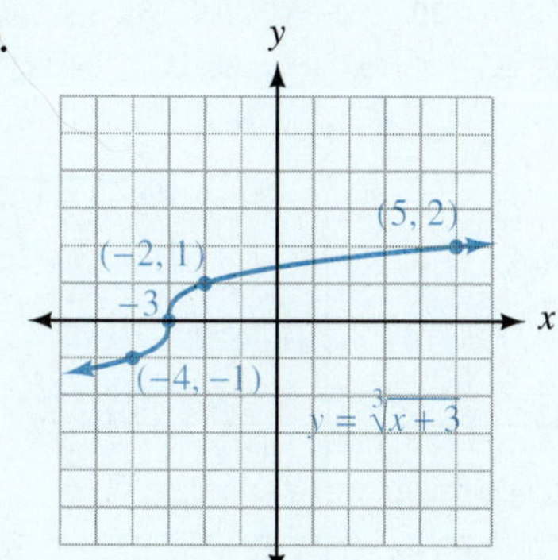

71. $\sqrt{6}-2$ **73.** $x+10\sqrt{x}+25$ **75.** $\frac{x-3\sqrt{x}}{x-9}$ **77.** 6 (2 does not check) **79.** −3, −1, 1 **81.** $y=\frac{1}{8}x^2$

83.

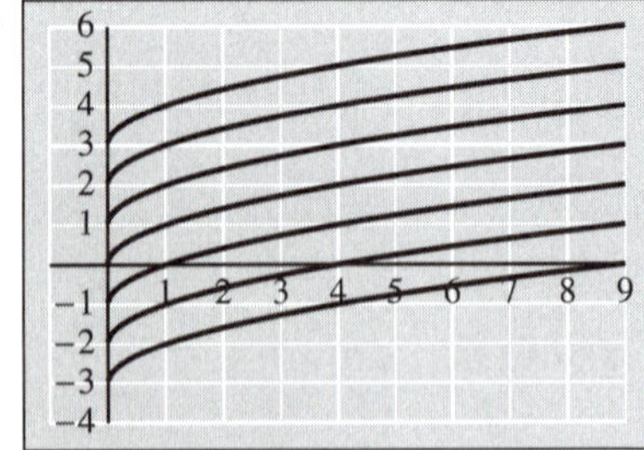

85. The value of b shifts the curve b units along the y-axis.

87.

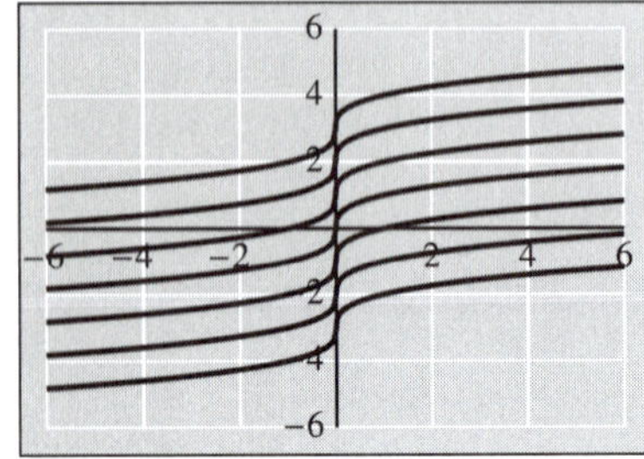

89. The value of b shifts the curve b units along the y-axis.

91.

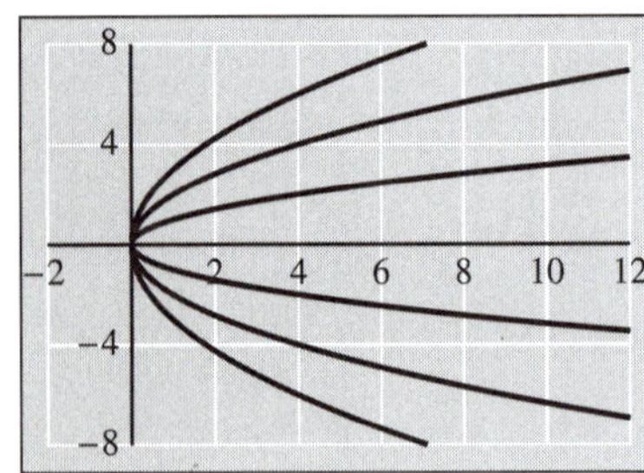

93. The smaller the absolute value of a, the more slowly the graph rises or falls. If a is negative, the graph is lies below the x-axis.

Chapter 4 Review

1. 7 **2.** -3 **3.** 2 **4.** 27 **5.** $2x^3y^2$ **6.** $\frac{1}{16}$ **7.** x^2 **8.** a^2b^4 **9.** $a^{7/20}$ **10.** $a^{5/12}b^{8/3}$ **11.** $12x + 11x^{1/2}y^{1/2} - 15y$
12. $a^{2/3} - 10a^{1/3} + 25$ **13.** $4x^{1/2} + 2x^{5/6}$ **14.** $2(x - 3)^{1/4}(4x - 13)$ **15.** $\frac{x+5}{x^{1/4}}$ **16.** $2\sqrt{3}$ **17.** $5\sqrt{2}$ **18.** $2\sqrt[3]{2}$ **19.** $3x\sqrt{2}$
20. $4ab^2c\sqrt{5a}$ **21.** $2abc\sqrt[4]{2bc^2}$ **22.** $\frac{3\sqrt{2}}{2}$ **23.** $3\sqrt[3]{4}$ **24.** $\frac{4x\sqrt{21xy}}{7y}$ **25.** $\frac{2y\sqrt[3]{45x^2z^2}}{3z}$ **26.** $-2x\sqrt{6}$ **27.** $3\sqrt{3}$ **28.** $\frac{8\sqrt{5}}{5}$
29. $7\sqrt{2}$ **30.** $11a^2b\sqrt{3ab}$ **31.** $-4xy\sqrt[3]{xz^2}$ **32.** $\sqrt{6} - 4$ **33.** $x - 5\sqrt{x} + 6$ **34.** $3\sqrt{5} + 6$ **35.** $6 + \sqrt{35}$
36. $\frac{63 + 12\sqrt{7}}{47}$ **37.** 0 **38.** 3 **39.** 5 **40.** $\varnothing$

41.

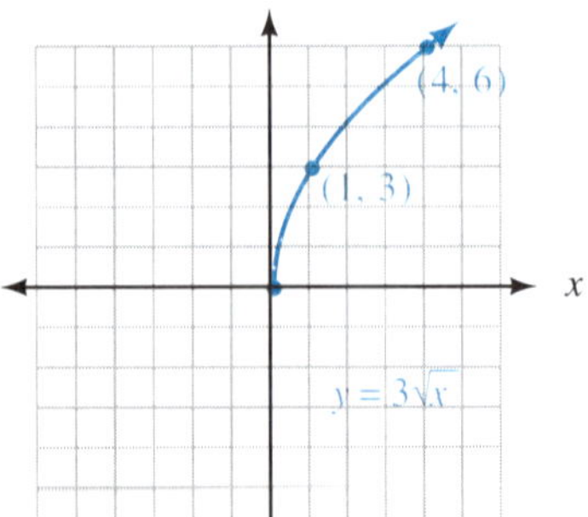

42.

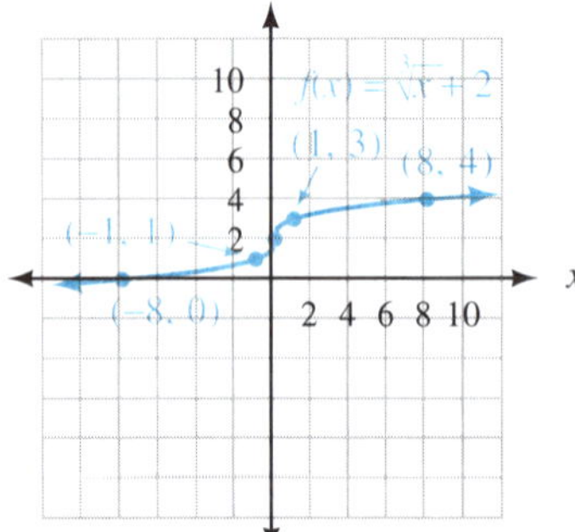

43. 22.5 feet **44.** 65 yards

Chapter 5

Problem Set 5.1

1. $6i$ **3.** $-5i$ **5.** $6i\sqrt{2}$ **7.** $-2i\sqrt{3}$ **9.** 1 **11.** -1 **13.** $-i$ **15.** $x = 3, y = -1$ **17.** $x = -2, y = -\frac{1}{2}$ **19.** $x = -8, y = -5$ **21.** $x = 7, y = \frac{1}{2}$ **23.** $x = \frac{3}{7}, y = \frac{2}{5}$ **25.** $5 + 9i$ **27.** $5 - i$ **29.** $2 - 4i$ **31.** $1 - 6i$ **33.** $2 + 2i$ **35.** $-1 - 7i$ **37.** $6 + 8i$ **39.** $2 - 24i$ **41.** $-15 + 12i$ **43.** $18 + 24i$ **45.** $10 + 11i$ **47.** $21 + 23i$ **49.** $-2 + 2i$ **51.** $2 - 11i$ **53.** $-21 + 20i$ **55.** $-2i$ **57.** $-7 - 24i$ **59.** 5 **61.** 40 **63.** 13 **65.** 164 **67.** $-3 - 2i$ **69.** $-2 + 5i$ **71.** $\frac{8}{13} + \frac{12}{13}i$ **73.** $-\frac{18}{13} - \frac{12}{13}i$ **75.** $-\frac{5}{13} + \frac{12}{13}i$ **77.** $\frac{13}{15} - \frac{2}{5}i$ **79.** $-\frac{3}{2}$ **81.** $-3, \frac{1}{2}$ **83.** $\frac{5}{4}$ or $\frac{4}{5}$

Problem Set 5.2

1. ± 5 **3.** $\pm 3i$ **5.** $\pm\frac{\sqrt{3}}{2}$ **7.** $\pm 2i\sqrt{3}$ **9.** $\pm\frac{3\sqrt{5}}{2}$ **11.** $-2, 3$ **13.** $\frac{-3 \pm 3i}{2}$ **15.** $\frac{-2 \pm 2i\sqrt{2}}{5}$ **17.** $-4 \pm 3i\sqrt{3}$ **19.** $\frac{3 \pm 2i}{2}$ **21.** $x^2 + 12x + 36 = (x + 6)^2$ **23.** $x^2 - 4x + 4 = (x - 2)^2$ **25.** $a^2 - 10a + 25 = (a - 5)^2$ **27.** $x^2 + 5x + \frac{25}{4} = \left(x + \frac{5}{2}\right)^2$ **29.** $y^2 - 7y + \frac{49}{4} = \left(y - \frac{7}{2}\right)^2$ **31.** $x^2 + \frac{1}{2}x + \frac{1}{16} = \left(x + \frac{1}{4}\right)^2$ **33.** $x^2 + \frac{2}{3}x + \frac{1}{9} = \left(x + \frac{1}{3}\right)^2$ **35.** $-6, 2$ **37.** $-9, -3$ **39.** $1 \pm 2i$ **41.** $4 \pm \sqrt{15}$ **43.** $\frac{5 \pm \sqrt{37}}{2}$ **45.** $1 \pm \sqrt{5}$ **47.** $\frac{4 \pm \sqrt{13}}{3}$ **49.** $\frac{3 \pm i\sqrt{71}}{8}$ **51.** $\frac{-2 \pm \sqrt{7}}{3}$ **53.** $\frac{5 \pm \sqrt{47}}{2}$ **55.** $\frac{5 \pm i\sqrt{19}}{4}$ **57.** $\frac{\sqrt{3}}{2}$ inch, 1 inch **59.** $x\sqrt{3}, 2x$ **61.** $\sqrt{2}$ inches **63.** $\frac{\sqrt{2}}{2}$ inch **65.** $x\sqrt{2}$ **67.** 781 feet **69.** 0.21 **71.** $20\sqrt{2} \approx 28$ feet **73.** 7.3% **75.** 19.6 inches **77.** $3\sqrt{5}$ **79.** $3y^2\sqrt{3y}$ **81.** $3x^2y\sqrt[3]{2y^2}$ **83.** 13 **85.** $\frac{3\sqrt{2}}{2}$ **87.** $\sqrt[3]{2}$ **89.** $x = \pm 2a$ **91.** $x = -a$ **93.** $x = 0, -2a$ **95.** $x = \frac{-p \pm \sqrt{p^2 - 4q}}{2}$ **97.** $x = \frac{-p \pm \sqrt{p^2 - 12q}}{6}$

Problem Set 5.3

1. (a) $\frac{-2 \pm \sqrt{10}}{3}$ **(b)** $\frac{2 \pm \sqrt{10}}{3}$ **(c)** $\frac{-2 \pm i\sqrt{2}}{3}$ **(d)** $\frac{-2 \pm \sqrt{10}}{2}$ **(e)** $\frac{2 \pm i\sqrt{2}}{2}$ **3.** $-3, -2$ **5.** $2 \pm \sqrt{3}$ **7.** 1, 2 **9.** 0, 5 **11.** $-\frac{4}{3}, 0$ **13.** $\frac{3 \pm \sqrt{5}}{4}$ **15.** $-3 \pm \sqrt{17}$ **17.** $\frac{-1 \pm i\sqrt{5}}{2}$ **19.** 1 **21.** $\frac{1 \pm i\sqrt{47}}{6}$ **23.** $4 \pm \sqrt{2}$ **25.** $\frac{1}{2}, 1$ **27.** $-\frac{1}{2}, 3$ **29.** $\frac{-1 \pm i\sqrt{7}}{2}$ **31.** $1 \pm \sqrt{2}$ **33.** $\frac{-3 \pm \sqrt{5}}{2}$ **35.** $-5, 3$ **37.** $2, -1 \pm i\sqrt{3}$ **39.** $-\frac{3}{2}, \frac{3 \pm 3i\sqrt{3}}{4}$ **41.** $\frac{1}{5}, \frac{-1 \pm i\sqrt{3}}{10}$ **43.** $0, \frac{-1 \pm i\sqrt{5}}{2}$ **45.** $0, 1 \pm i$ **47.** $0, \frac{-1 \pm i\sqrt{2}}{3}$ **49.** $\frac{-3 - 2i}{5}$ **51.** 2 seconds **53.** $\frac{1}{4}$ second, 1 second **55.** 20 items, 60 items **57.** 750 patterns, 1,000 patterns **59.** 0.49 centimeter (8.86 cm is impossible) **61. (a)** $l + w = 10, lw = 15$ **(b)** 8.16 yards, 1.84 yards **(c)** Two answers are possible because either dimension (long or short) may be considered the length. **63.** $4y + 1 - \frac{2}{2y - 7}$ **65.** $x^2 + 7x + 12$ **67.** 5 **69.** $\frac{27}{125}$ **71.** $\frac{1}{4}$ **73.** $21x^3y$ **75.** $-2\sqrt{3}, \sqrt{3}$ **77.** $\frac{-\sqrt{2} \pm \sqrt{6}}{2}$ **79.** $-2i, i$ **81.** $-i, 4i$

Chapter 5 Review

1. 1 **2.** $-i$ **3.** $x = -\frac{3}{2}, y = -\frac{1}{2}$ **4.** $x = -2, y = -4$ **5.** $9 + 3i$ **6.** $-7 + 4i$ **7.** $-6 + 12i$ **8.** $5 + 14i$ **9.** $12 + 16i$ **10.** 25 **11.** $1 - 3i$ **12.** $-\frac{6}{5} + \frac{3}{5}i$ **13.** 0, 5 **14.** $0, \frac{4}{3}$ **15.** $\frac{4 \pm 7i}{3}$ **16.** $-3 \pm \sqrt{3}$ **17.** $-5, 2$ **18.** $-3, -2$ **19.** 3 **20.** 2 **21.** $\frac{-3 \pm \sqrt{3}}{2}$ **22.** $\frac{3 \pm \sqrt{5}}{2}$ **23.** $-5, 2$ **24.** $0, \frac{9}{4}$ **25.** $\frac{2 \pm i\sqrt{15}}{2}$ **26.** $1 \pm \sqrt{2}$ **27.** $-\frac{4}{3}, \frac{1}{2}$ **28.** 3 **29.** $5 \pm \sqrt{7}$ **30.** $0, \frac{5 \pm \sqrt{21}}{2}$ **31.** $-1, \frac{3}{5}$ **32.** $3, \frac{-3 \pm 3i\sqrt{3}}{2}$ **33.** $\frac{1 \pm i\sqrt{2}}{3}$ **34.** $\frac{3 \pm \sqrt{29}}{2}$ **35.** 100 or 170 items **36.** 20 or 60 items

Index